Craftsman 2022 NATIONAL PLUMBING & HVAC ESTIMATOR

By James A. Thomson

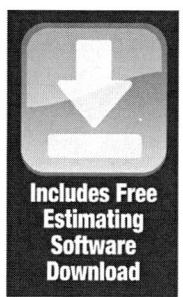

Includes Free Estimating Software Download

Includes inside the back cover:

Inside the back cover of this book you'll find a software download certificate. To access the download, follow the instructions printed there. The download includes the National Estimator, an easy to-use estimating program with all the cost estimates in this book. The software will run on PCs using Windows XP, Vista, 7, 8, or 10 operating systems.

Quarterly price updates on the Web are free and automatic all during 2022. You'll be prompted when it's time to collect the next update. A connection to the Web is required.

Download all of Craftsman's most popular costbooks for one low price with the Craftsman Site License. http://CraftsmanSiteLicense.com

- Turn your estimate into a bid.
- Turn your bid into a contract.
- ConstructionContractWriter.com

Craftsman Book Company
6058 Corte del Cedro, Carlsbad, CA 92011

Acknowledgments

The sample "Standard Form Subcontract" and "Subcontract Change Order" forms used in the final section of this book are reprinted with the permission of the publisher, the Associated General Contractors of America (National Office), 1957 E Street NW, Washington, District of Columbia 20006.

Looking for other construction reference manuals?
Craftsman has the books to fill your needs. **Call toll-free 1-800-829-8123**
Visit our Web site: http://www.craftsman-book.com

Cover design by: *Jennifer Johnson*
Photos: *iStock by Getty Images™*

© 2021 Craftsman Book Company
ISBN 978-1-57218-376-6
Published September 2021 for the year 2022.

Contents

How to Use This Book 5

Plumbing Equipment

Domestic Hot Water Heaters 19
Water Softeners 22
Kitchen Equipment 25
Kitchen Equipment Connections 26
Plumbing Fixtures 27
Plumbing Fixture Rough-In 31

Piping Systems

Copper Pipe, Type K with Brazed Joints 33
Copper Pipe, Type K with Soft-Soldered Joints.... 43
Copper Pipe, Type L with Brazed Joints 53
Copper Pipe, Type L with Soft-Soldered Joints.... 61
Copper Pipe, Type M with Brazed Joints 70
Copper Pipe, Type M with Soft-Soldered Joints ... 78
Copper, Pressfit 86
Copper Pipe, Type K & L
with Roll Grooved Joints 89
PVC, Schedule 40, with Solvent-Weld Joints 91
PVC, Schedule 80, with Solvent-Weld Joints 101
Polyethylene-Aluminum Pipe
with Crimped Joints 111
Polyethylene-Aluminum Pipe
with Compression Joints 116
Plumbing and Piping Specialties 119
Cast Iron, DWV, Service Weight, No-Hub
with Coupled Joints 135
Cast Iron, DWV, Service Weight,
Hub & Spigot with Gasketed Joints 141
Copper, DWV, with Soft-Soldered Joints 146
ABS, DWV with Solvent-Weld Joints 150
PVC, DWV with Solvent-Weld Joints 154
PVC, DWV with Gasketed Bell and
Spigot Joints 159
Polypropylene, Schedule 40,
with Heat-Fused Joints 164

Floor, Area, Roof and Planter Drains 168

Cleanouts 169

Fire Protection

Fire Protection Sprinklers 170

Fire Protection Equipment 174
Fire Protection Sprinkler Pipe and Fittings
(Roll Grooved) 176
Fire Protection Branch Pipe & Fittings 180
Fire Protection Sprinkler Pipe and
Fittings (CPVC) 183

HVAC Equipment

Commercial Boilers 185
Commercial Boiler Connections 190
Commercial Boiler Components and
Accessories 192
Centrifugal Pumps and Pump Connections 206
Heat Exchangers and Connections 207
Fan Coil Units and Connections 208
Reheat Coils and Connections 209
Unit Heaters and Connections 210
Chillers and Chiller Connections 211
Condensing Units and Cooling Towers 212
Cooling Towers and Cooling Tower
Connections 213

Steel Piping Systems

Carbon Steel, Schedule 40 with
150# Fittings & Butt-Welded Joints 214
Carbon Steel, Schedule 40 with
150# M.I. Fittings & Threaded Joints 223
Carbon Steel, Schedule 5 with
Pressfit Fittings 234
Carbon Steel, Schedule 80 with
300# Fittings & Butt-Welded Joints 237
Carbon Steel, Schedule 80 with
300# M.I. Fittings & Threaded Joints 247
Carbon Steel, Schedule 160 with
3,000-6,000# Fittings 255
Carbon Steel, Schedule 40 with
Roll-Grooved Joints 266
Carbon Steel, Schedule 10 with
Roll-Grooved Joints 273
Carbon Steel, Schedule 40 with
Cut-Grooved Joints 280

Residential HVAC Assemblies 286

Air Handling Unit Accessories 290

Heat Recovery Ventilators - Commercial..........291
Heat Recovery Ventilators - Residential..........292
Water Coil Piping..........294
Air Handling Unit Coil Connections..........297
Gas-Fired Furnaces..........299
Energy Recovery Systems, Enthalpy..........301
Unit Heaters..........302
Infrared Heaters..........304
Heat Pump Systems..........305
Water Pump Systems..........313
Geothermal/Domestic Water Wells..........316
Biomass-Fired Boilers..........319
Fans and Blowers..........324
Ventilators & Residential Exhaust Fans..........326
Apparatus Housing..........331
Air Devices, Registers & Grilles..........333
Air Devices, Diffusers & Grilles..........334
Terminal Units (VAV)..........337

Ducting Systems

Ductwork Specialties..........339
Galvanized Steel Ductwork..........344
Installed Ductwork Per Pound..........346
Galvanized Steel Spiral Ductwork..........348
Galvanized Steel Round Spiral Fittings..........349
Galvanized Steel Rectangular Ductwork..........351
Galvanized Steel Rectangular 90 Degree Elbows..........353
Galvanized Steel Spiral Duct..........356
Galvanized Steel Spiral Duct Fittings..........358
Galvanized Steel Spiral Tees..........360
Galvanized Steel Spiral Crosses..........366
Galvanized Steel Rectangular Ductwork..........369
Galvanized Steel Rectangular Elbows..........381
Galvanized Steel Drops and Tees..........391
Galvanized Steel Round Ductwork..........394
Fiberglass Ductwork..........395

Fiberglass Pipe Insulation..........397
Calcium Silicate Pipe Insulation with Aluminum Jacket..........399
Closed Cell Elastomeric Pipe Insulation..........400
Thermal Duct Insulation..........401
Balancing of HVAC Systems..........402
Temperature Controls..........405

Ductile Iron Pipe Systems

Ductile Iron, Class 153, Cement-Lined with Mechanical Joints..........407
Ductile Iron, Class 153, Double Cement-Lined with Mechanical Joints..........409
Ductile Iron, Class 110, Cement-Lined with Mechanical Joints..........411

Cast Iron, Class 150 with Mechanical Joints....412
Asbestos-Cement, Class 2400 or 3000 with Mechanical Joints..........413
Fiberglass Tanks..........415
Plastic Tanks..........416
Trenching..........418
Equipment Rental..........420
Close-Out Items..........421
HVAC & Plumbing Demolition..........422
Budget Estimating..........435

Forms and Letters

Change Estimates..........438
Subcontract Forms..........447
Purchase Orders..........451
Construction Schedules..........453
Letter of Intent..........456
Submittal Data..........458
Billing Breakdown Worksheet..........461

Index..........463

How to Use This Book

This 2022 National Plumbing & HVAC Estimator is a guide to estimating labor and material costs for plumbing, heating, ventilating and air conditioning systems in residential, commercial and industrial buildings.

 Inside the back cover of this book you'll find a software download certificate. To access the download, follow the instructions printed there. The download includes an easy to-use estimating program with all the cost estimates in this book. The software will run on PCs using Windows XP, Vista, 7, 8, or 10 operating systems.

When the *National Estimator* program has been installed, click Help on the menu bar to see a list of topics that will get you up and running. Or, go online to www.craftsman-book.com and click on Support, then Tutorials, to view an interactive tutorial for *National Estimator*.

Costs in This Manual will apply within a few percent on a wide variety of projects. Using the information given on the pages that follow will explain how to use these costs and suggest procedures to follow when compiling estimates. Reading the remainder of this section will help you produce more reliable estimates for plumbing and HVAC work.

 Manhour Estimates in This Book will be accurate for some jobs and inaccurate for others. No manhour estimate fits all jobs because every construction project is unique. Expect installation times to vary widely from job to job, from crew to crew, and even for the same crew from day to day.

There's no way to eliminate all errors when making manhour estimates. But you can minimize the risk of a major error by:

1. Understanding what's included in the manhour estimates in this book, and

2. Adjusting the manhour estimates in this book for unusual job conditions.

The Craft@Hrs Column. Manhour estimates in this book are listed in the column headed *Craft@Hrs*. For example, on page 19 you'll see an estimate for installing a 6 gallon hot water heater. In the *Craft@Hrs* column opposite 6 gallon you'll see:

P1@.500

To the left of the @ symbol you see an abbreviation for the recommended work crew.

Page 7 shows the wage rates and craft codes used in this book.

To the right of the @ symbol you see a number. The number is the estimated manhours (not crew hours) required to install each unit of material listed. In the case of a 6 gallon hot water heater, P1@.500 means that .500 manhours are required to install 1 hot water heater.

Costs in the Labor $ Column are based on manhour estimates in the *Craft@Hrs* column. Multiply the manhour estimate by the assumed hourly labor cost to find the installation cost in the *Labor $* column. For example, .500 manhours times $37.22 (the average wage for crew P1) is $18.61.

Quarterly price updates on the Web are free and automatic all during 2022. You'll be prompted by Craftsman Software Update when it's time to collect the next update. A connection to the Web is required.

Manhour Estimates include all productive labor normally associated with installing the materials described. These estimates assume normal conditions: experienced craftsmen working on reasonably well planned and managed new construction with fair to good productivity. Labor estimates also assume that materials are standard grade, appropriate tools are on hand, work done by other crafts is adequate, layout and installation are relatively uncomplicated, and working conditions don't slow progress.

All manhour estimates include tasks such as:

- Unloading and storing construction materials, tools and equipment on site.
- Working no more than two floors above or below ground level.
- Working no more than 10 feet above an uncluttered floor.
- Normal time lost due to work breaks.
- Moving tools and equipment from a storage area or truck not more than 200 feet from the work area.
- Returning tools and equipment to the storage area or truck at the end of the day.
- Planning and discussing the work to be performed.
- Normal handling, measuring, cutting and fitting.
- Regular cleanup of construction debris.
- Infrequent correction or repairs required because of faulty installation.

National Plumbing & HVAC Estimator

If the work you're estimating won't be done under these conditions, you need to apply a correction factor to adjust the manhour estimates in this book to fit your job.

Applying Correction Factors. Analyze your job carefully to determine whether a labor correction factor is needed. Failure to consider job conditions is probably the most common reason for inaccurate estimates.

Use one or more of the recommended correction factors in Table 1 to adjust for unusual job conditions. To make the adjustment, multiply the manhour estimate by the appropriate conversion factor. On some jobs, several correction factors may be needed. A correction factor less than 1.00 means that favorable working conditions will reduce the manhours required.

Supervision Expense to the installing contractor is not included in the labor cost. The cost of supervision and non-productive labor varies widely from job to job. Calculate the cost of supervision and non-productive labor and add this to the estimate.

Hourly Labor Costs also vary from job to job. This book assumes an average manhour labor cost of $43.89 for plumbers and $42.53 for sheet metal workers. If these hourly labor costs are not accurate for your jobs, adjust the labor costs up or down by an appropriate percentage. Instructions on the next page explain how to make these adjustments. If you're using the National Estimator disk, it's easy to set your own wage rates.

Hourly labor costs in this book include the basic wage, fringe benefits, the employer's contribution to welfare, pension, vacation and apprentice funds, and all tax and insurance charges based on wages. Table 2 at the top of the next page shows how hourly labor costs in this book were calculated. It's important that you understand what's included in the figures in each of the six columns in Table 2. Here's an explanation:

Column 1, the base wage per hour, is the craftsman's hourly wage. These figures are representative of what many contractors are paying plumbers, sheet metal workers and helpers in 2022.

Column 2, taxable fringe benefits, includes vacation pay, sick leave and other taxable benefits. These fringe benefits average about 5.59% of the base wage for many plumbing and HVAC contractors. This benefit is in addition to the base wage.

Condition	Correction Factor
Work in large open areas, no partitions	.85
Prefabrication under ideal conditions, bench work	.90
Large quantities of repetitive work	.90
Very capable tradesmen	.95
Work 300' from storage area	1.03
Work 400' from storage area	1.05
Work 500' from storage area	1.07
Work on 3rd through 5th floors	1.05
Work on 6th through 9th floors	1.10
Work on 10th through 13th floors	1.15
Work on 14th through 17th floors	1.20
Work on 18th through 21st floors	1.25
Work over 21 floors	1.35
Work in cramped shafts	1.30
Work in commercial kitchens	1.10
Work above a sloped floor	1.25
Work in attic space	1.50
Work in crawl space	1.20
Work in a congested equipment room	1.20
Work 15' above floor level	1.10
Work 20' above floor level	1.20
Work 25' above floor level	1.30
Work 30' above floor level	1.40
Work 35' to 40' above floor level	1.50

Table 1 Recommended Correction Factors

Column 3, insurance and employer-paid taxes in percent, shows the insurance and tax rate for the craft workers. The cost of insurance in this column includes workers' compensation and contractor's casualty and liability coverage. Insurance rates vary widely from state to state and depend on a contractor's loss experience. Note that taxes and insurance increase the hourly labor cost by approximately 30%. There is no legal way to avoid these costs.

Column 4, insurance and employer taxes in dollars, shows the hourly cost of taxes and insurance. Insurance and taxes are paid on the costs in both columns 1 and 2.

Column 5, non-taxable fringe benefits, includes employer paid non-taxable benefits such as medical coverage and tax-deferred pension and profit sharing plans. These fringe benefits average 4.94% of the base wage for many plumbing and HVAC contractors.

How to Use This Book

Column Number	1	2	3	4	5	6
Craft	Base wage per hour	Taxable fringe benefits (at 5.59% of base wage)	Insurance and employer taxes (%)	Insurance and employer taxes ($)	Non-taxable fringe benefits (at 4.94% of base wage)	Total hourly cost used in this book
Laborer	21.22	1.19	31.64%	7.09	1.05	30.55
Plumber	32.34	1.81	23.83%	8.14	1.60	43.89
Sheet Metal Worker	30.97	1.73	25.37%	8.30	1.53	42.53
Operating Engineer	31.57	1.77	24.58%	8.19	1.56	43.09
Sprinkler Fitter	31.78	1.78	24.55%	8.24	1.57	43.37
Electrician	31.28	1.75	19.53%	6.45	1.55	41.03
Cement Mason	26.67	1.49	22.72%	6.40	1.32	35.88

Craft Code	Crew Composition	Average Hourly Cost per Manhour
ER	4 building plumbers, 2 building laborers, 1 operating engineer	39.96
SN	4 building sheet metal workers, 2 building laborers, 1 operating engineer	39.19
P1	1 building plumber and 1 building laborer	37.22
ST	1 sprinkler fitter	43.37
SK	4 sprinkler fitters, 2 building laborers, 1 operating engineer	39.67
SL	1 sprinkler fitter and 1 laborer	36.96
S2	1 building sheet metal worker, 1 building laborer	36.54
BE	1 electrician	41.03
CF	1 cement mason	35.88
SW	1 sheet metal worker	42.53

Table 2 Labor Costs Used in This Book

The employer pays no taxes or insurance on these benefits.

Column 6, the total hourly cost in dollars, is the sum of columns 1, 2, 4, and 5. The labor costs in Column 6 were used to compute costs in the Labor $ column of this book.

Adjusting Costs in the Labor $ Column. The hourly labor costs used in this book may apply within a few percent on many of your jobs. But wage rates may be much higher or lower in some areas. If the hourly costs shown in Column 6 of Table 2 are not accurate for your work, adjust labor costs to fit your jobs.

For example, suppose your hourly labor costs are as follows:

Plumber	$29.00
Laborer	$19.00
Total hourly crew cost	$48.00

Your average cost per manhour would be $24.00 ($48.00 per crew hour divided by 2 because this is a crew of two).

A labor cost of $24.00 is about 65% of the $37.22 labor cost used for crew P1. Multiply costs in the Labor $ column by .645 to find your estimated cost.

For example, notice on page 19 that the labor cost for installing a 6 gallon hot water heater is $18.60 each. If installed by your plumbing crew working at an average cost of $24.00 per manhour, your estimated cost would be 64.5% of $18.60 or $12.00 per heater.

Adjusting the labor costs in this book will make your estimates much more accurate. Making adjustments to labor costs is both quick and easy if you use the National Estimator program.

Equipment Cost will vary according to need and application. It is typically $32.00 per day for a 2-ton chain hoist.

Material Costs in this manual are intended to reflect what medium- to low-volume contractors will be paying in 2022 after applying normal discounts. These costs include charges for delivery to within 25 to 30 miles of the supplier.

Overhead and Profit for the installing contractor are not included in the costs in this manual unless specifically identified in the text. Markup can vary widely with local economic conditions, competition and the installing contractor's operating expenses. Add the markup that's appropriate for your company, the job and the competitive environment.

How Accurate Are These Figures? As accurate as possible considering that the editors don't know your material suppliers, haven't seen the plans or specifications, don't know what building code applies or where the job is, had to project material costs at least six months into the future, and had no record of how much work the crew that will be assigned to the job can handle.

You wouldn't bid a job under those conditions. And I don't claim that all plumbing and HVAC work is done at these prices.

Estimating Is an Art, not a science. There is no one price that applies on all jobs. On many jobs the range between high and low bid will be 10% or more. There's room for legitimate disagreement on what the correct costs are, even when complete plans and specifications are available, the date and site are established, and labor and material costs are identical for all bidders.

No estimate fits all jobs. Good estimates are custom made for a particular project and a single contractor through judgment, analysis and experience. This book is not intended as a substitute for judgment, analysis and sound estimating practice. It's an aid in developing an informed opinion of cost, not an answer book.

Additional Costs to Consider

Here's a checklist of additional costs to consider before submitting any bid.

1. Sales taxes
2. Mobilization costs
3. Payment and performance bond costs
4. Permits and fees
5. Storage container rental costs
6. Utility costs
7. Tool costs
8. Callback costs during warranty period
9. Demobilization costs

Exclusions and Clarifications

Neither the job specifications nor the contract may identify exactly what work should be included in the plumbing and HVAC bid. Obviously, you have to identify what work is included in the job.

The most efficient way to define the scope of the work is to prepare a list of tasks not normally performed by your company and attach that list to each bid submitted. Here's a good list of work that should be excluded from your bid.

Your Bid Should Exclude

Final cleaning of plumbing fixtures
Backings for plumbing fixtures
Toilet room accessories
Electrical work, including motor starters
Electrical wiring and conduit over 100 volts
Temporary utilities
Painting, priming and surface preparation
Structural cutting, patching or repairing
Fire protection and landscape sprinklers
Equipment supports
Surveying and layout of control lines
Removal or stockpiling of excess soil
Concrete work, including forming and rebar
Setting of equipment furnished by others
Equipment, unless shown, and personnel hoisting
Wall and floor blockouts
Pitch pockets
The costs of performance or payment bonds
Site utilities
Asbestos removal or disposal
Contaminated soil removal or disposal
Major increases in copper material prices
Fire dampers not shown on the plans

Your Bid Should Include

Trash sweep-up only. Others haul it away
Site utilities from building to property line only
Piping to 5 feet outside the building only
Plumbing & HVAC permits for your work only

Beware of Price Changes

There's no way to be sure what prices will be in three to six months. All labor, equipment, material and subcontract prices in a bid should be based on costs anticipated when the project is expected to be built, not when the estimate is compiled. That presents a problem. Except for the installation of underground utilities, most plumbing and HVAC work is done six months to a year after the bid is submitted. When possible, get price protection in writing from your suppliers and subcontractors. If your suppliers and subs won't guarantee prices, include an escalation allowance in your bid to cover anticipated price increases.

Material Pricing Conditions

All equipment and material prices quoted by your vendors will be conditional. They usually don't include sales tax and are subject to specific payment and shipping terms. Every estimator should understand the meaning of common shipping terms. They define who pays the freight and who has responsibility for processing freight-damage claims. Here's a summary of important conditions you should understand.

F.O.B. Factory (Free On Board at the Factory): Title passes to the buyer when the goods are delivered by the seller to the freight carrier. The buyer pays the freight and is responsible for freight-damage claims.

F.O.B. Factory F.F.A. (Free On Board at the Factory, Full Freight Allowed): The title passes to the buyer when the goods are delivered by the seller to the freight carrier. The seller pays the freight charges, but the buyer is responsible for freight-damage claims.

F.O.B. (city of destination) (Free On Board to your city): The title passes to the buyer when the goods are delivered by the seller to the freight terminal in the city, or nearest city, of destination. The seller pays the freight and is responsible for freight-damage claims to the terminal. The buyer pays the freight charge and is responsible for freight-damage claims from the terminal to the final destination.

F.O.B. Job Site (Free On Board at job site, or contractor's shop): The title passes to the buyer when the goods are delivered to the job site (or shop). The seller pays the freight and is responsible for freight-damage claims.

F.A.S. Port [of a specific city] (Free Alongside Ship at the nearest port): The title passes to the buyer when goods are delivered to the ship dock or port terminal. The seller pays the freight and is responsible for freight-damage claims to the ship dock or port terminal only. The buyer pays the freight and is responsible for freight-damage claims from the ship dock or port terminal to the designated delivery point.

Obviously, it's to your advantage to instruct all vendors to quote costs F.O.B. the job site or your shop.

Reducing Costs

Most construction specifications allow the use of alternative equipment and materials. It's the estimator's responsibility to select the most cost-effective products. Research and compare your costs before making any decisions. Avoid selecting any material or equipment simply because that's what you've always done.

Don't recommend plastic products such as ABS, PVC, or polypropylene pipe or corrugated flexible ducts until you've checked local code requirements. Most building codes prohibit use of these materials inside public buildings such as schools, care centers and hospitals.

It's wise to select 100% factory-packaged equipment. Beware of equipment labeled "Some assembly required." Field labor costs for mounting loose coils, motors and similar equipment are very high.

Value Engineering

Let's suppose you've submitted a combined plumbing and HVAC bid for $233,000. Your cutthroat competitor put in a bid at $4,000 less, $229,000. Obviously there's no way you're going to get the job. Right?

Not so fast! Maybe value engineering can help you win that contract — while fattening your profit margin.

Suppose the proposal you submitted had two parts. Part I is the bid for $233,000, based entirely on job plans and specs, just the way they were written. But appended to your proposal is Part II, a list of suggestions for saving money without sacrificing any of the capacity or quality designed into the system. Here's an example of what might be in Part II:

1. Deduct for providing pipe hanger spacings per UPC in lieu of specified spacings: $1,750.00

2. Deduct for reducing heating hot water pipe sizes by using 40 degrees F Delta T in lieu of specified 20 degrees F Delta T: $4,600.00

3. Deduct for providing pressure/temperature taps at air handling units, pumps and chillers in lieu of specified thermometers and pressure gauges: $875.00

4. Deduct for eliminating water treatment in closed piping systems: $1,800.00

5. Deduct for piping chilled and heating hot water pumps in parallel in lieu of providing 100% standby pumps: $2,900.00

Total deductions: **$11,925.00**

Adopting these suggestions would make you low bidder by nearly $8,000. A saving like that will be tempting to most owners, especially if the owner understands that your suggestions result in a system that is every bit as good and maybe better than the system as originally designed.

You're not offering to undercut the competition. Far from it. You're using knowledge and experience to create better value for the owner. That's called value engineering and it's likely to win the respect of nearly all cost-conscious owners.

Notice that reducing costs is only part of what value engineering is all about. You don't cut costs at the expense of system quality, integrity, capacity or performance.

Don't waste your time, and your client's, by offering to substitute cheaper or lower-quality fixtures or equipment. Any cutthroat contractor with a price list can do that. Recommend the use of inferior materials and you'll be associated with the inferior goods you promote. Some owners consider even the suggestion to be insulting.

The recommendations you make (like most of those in the example) will require design changes. You can expect to be examined (or even challenged) on these points. Be ready to explain and defend each of your suggestions. Convince the client (or the design engineer) that your ideas are based on sound engineering principles and you're well on the way to winning the owner's confidence and the contract.

Now, let's go back to the list and see how we might justify the five value engineering recommendations.

1. **Pipe Hanger Spacing.** The pipe hanger spacings recommended in the Uniform Plumbing Code (UPC) are calculated by experienced, professional structural engineers. The safety factors used in these calculations are very conservative. They've been widely used for many years and have proved to be more than adequate. There's no need for more hangers than the UPC requires.

2. **Changing HHW Delta T.** In hydronic heating systems, heat measured in Btus is pumped to terminal units. The proposed change of the Delta T, from 20 degrees F to 40 degrees F, has no effect whatsoever on how many Btus the system delivers. You're not changing anything but the volume of water being pumped. At lower volume levels, the size of the pump, the pipe and the pipe insulation can all be reduced. Not one of these changes will affect the system's ability to transmit heat. Furthermore, operating costs will also drop, since less pump horsepower will be needed to run the smaller pump.

3. **Thermometers/Pressure Gauges.** Thermometers and pressure gauges installed on or near vibrating machinery have a very short life expectancy. Gauges quickly lose accuracy under harsh conditions. Readings will become less and less reliable. That's potentially dangerous. You can avoid this problem by using insertion-type pressure/temperature taps instead. Store these sensitive gauges in a desk drawer or a tool crib when not in use. Safely stored, they're protected from damage. They'll give accurate readings longer and won't need to be replaced as often. And they're simple to use. Just insert a gauge in one of the conveniently located taps. Make the reading, then remove the gauge and put it away.

4. **Water Treatment.** ITT Bell & Gossett has done studies on corrosion in closed hydronic systems that have a make-up water rate of no more than 5% per year. These studies show that corrosion virtually stops when entrained air is either removed or depleted. No water treatment is needed in this closed system.

5. **100% Standby Pumps.** Two pumps piped and operated in parallel are more economical. Even if one pump fails, the other pump can maintain delivery at 75 to 80% of the designed flow rate. That's usually adequate for emergency operation.

These cost-saving ideas are small, but could tip the balance in your favor. I hope they demonstrate the potential that value engineering has when bidding jobs. Any time you're compiling an estimate, keep an eye out for ways to save money or reduce the owner's cost. Jot a note to yourself about each potential saving you identify. Before submitting the bid, make a list of your alternate suggestions. Maybe best of all, markup on your value engineering suggestions can be higher than your normal markup. If value engineering can cut costs by $10,000, maybe as much as $4,000 of that should end up in your pocket!

How to Use This Book

Value Engineering: Surplus Materials

Value engineering doesn't begin and end with job plans and specs. Value engineering means getting the most value at the least cost, no matter whether it's value to the owner or value to the contractor. Smart mechanical contractors learn to build extra value into their jobs by controlling shrinkage of materials. Nearly every significant plumbing and HVAC job ends with at least some surplus material on hand. Material left over when the job is done tends to be discarded as waste or hauled off the job in the back of a truck that doesn't have your company name on the door. And why not? It's surplus — not needed. The owner didn't need it. So now it's up for grabs.

Not quite. Let's consider who actually owns that surplus material. When your company has been paid, every piece of material your crew installed belongs to the building owner. But what about those fittings, hangers and valves delivered to the job site but never actually used? Almost certainly, those materials were included in your bid. So aren't they the property of the owner? Not in my opinion. The owner contracted for a mechanical system and (presumably) has one. Unless it's a cost-plus job or a labor-only job, the owner didn't buy materials delivered to the job site. The owner bought a mechanical system and has one — completely separate and apart from any surplus materials. In my mind, the property owner has no more claim to left-over materials than the same owner would have claim to labor hours not expended or equipment not used on the same job.

Unless there's some provision in your contract to the contrary, surplus material belongs to the installing contractor. But your right to that material and the chance of actually getting it back to your shop are two very different propositions. I see recovery of surplus material as a training issue. As a matter of company policy, make it clear to your crews that surplus material belongs to your company. The supervisor on every job should be accountable for recovery of excess material. Every significant job will have at least some surplus. Accounting for that surplus should be part of your routine close-out procedure. Fortunately, it's not difficult. I'll explain.

Control of surplus materials begins with a good checklist, or form. I recommend the Materials, Equipment and Tool form, "MET" for short. A blank MET form appears following this section. Your MET should show both what's delivered to the job site (material, equipment and tools) and surplus "drops" returned to your shop at project close-out. A MET ensures that the estimator, the shop inventory manager and your field supervisor are on the same page. Your MET establishes accountability. Nothing falls through the cracks. Job input equals job output plus returns. Everything delivered to the job and not expended should be returned to your shop.

Here's how it works:

1. Based on the estimate that won you the job, the items needed are purchased for the job and staged for delivery to the job site.

2. As materials, equipment and tools are delivered to the job site, your supervisor completes the first three columns of the MET form: Description, Quantity and Date.

3. As work is completed, the same supervisor completes the four columns under Returned to Inventory: Quantity Returned, Date, Status Code and Value. The status code will be either "RS" (Returned and Salvaged) or "RN" (Returned New).

4. Back at your shop, both RS and RN materials should be restored to inventory.

5. If your company has an inventory manager, have that manager assign the return value to each item returned. If you're using QuickBooks Pro, the "Adjust Inventory" feature can handle this task quite easily. Add two new categories under "Inventory Stock on Hand by Vendor." The first new category is Returned Salvage. The second is Returned New. Be sure the value of RS materials includes the cost of any reconditioning done to restore salvaged materials (such as pumps and boilers) to serviceable condition.

6. Comparing MET deployed to the job site with MET returned to inventory yields MET actually used on the job. That's a very important number to every plumbing and HVAC estimator. Be sure actual usage gets entered on the Project Summary form.

7. When the take-off on your next estimate is complete, compare that materials list with a summary of RS and RN materials on hand from prior jobs.

8. Evaluate which returned materials can be redeployed on the new job.

9. It's a management decision to either (1) charge the new job for the cost of RS and RN materials already on hand, or (2) consider materials on hand as "free" and a competitive advantage in winning the new bid. Either way, RN and RS materials are an asset to your company.

Plumbing and HVAC materials are expensive. Every mechanical contractor has an interest in MET tracking. Everyone in your company should be aware of the need for good materials management. Used correctly, the MET form in this book can help engineer more value into your jobs.

Maximizing the Value of Old Estimates

There should be two profits in every job. The first is money in the bank — a return on time and expenses. The second is what you learn from the job — primarily by comparing the estimate you made with what turns out to be your actual cost. On some jobs, the value of lessons learned may outweigh net revenue.

Every plumbing and HVAC contractor has marginal jobs. That's normal. What *shouldn't* be normal is repeating mistakes. The best way to avoid trouble in your future is to keep track of your past. Keeping old estimates available for reference can help prevent errors on new estimates.

As your file of completed estimates grows, organization becomes more important. You need an easy way to find similar projects with the same components and comparable scope of work. If your estimating file is in QuickBooks Pro, searching by keyword may be enough. Otherwise, I recommend creating a short summary for each completed job, and an index that references all summaries available for comparison. You'll find a blank Project Summary form at the end of this section. To make reference easier, create an index by type of job and equipment used. You may choose to use an alphabetical index based on client name or project ID.

How to complete the Project Summary form is obvious. The many ways to use this form may not be so obvious, so here are a few pointers.

1. Use your index of Project Summary forms to find completed jobs most similar to the job you're bidding. Believe it or not, Project Summary forms with the widest margin of error will be most useful. Ask yourself: Who worked on those projects? Who was the field superintendent? Who were the vendors? Did the errors result from poor estimating or the poor performance of vendors, supervisors or crews? The most common estimating errors occur when (a) inspecting the job site, (b) examining the plans or (c) reading the specifications. What did you miss and why? Look for pitfalls to avoid in the job now being estimated. Identify the biggest two or three mistakes made when bidding that job. Make a notation about each on the Project Summary form.

2. Now look at your bid for the current job. Which mistakes made on a prior job might you expect on this job? Concentrate on the big three oversights to avoid: Inspecting the job site; examining the plans; and reading the specifications.

3. Unless there's a major error in take-off, your estimate of material costs should be within about 5 percent of the actual costs of materials. However, it's common for labor cost estimates to vary 20 percent or more from actual labor costs. This is precisely where data from old jobs comes in handy. If your Project Summary files show that some project types are consistent money-losers, either shift your company's focus to another class of work, factor more contingency into your bids, or find some way to wring inefficiencies out of the labor component. Poor staging, delivery and retrieval procedures drag down labor productivity on any job.

4. Use your file of Project Summary forms to spot any common thread that runs through either money-making jobs or money-losing jobs. For example, if the names of certain subcontractors or vendors are prominent on low-margin jobs, maybe there's a relationship between your profit margin and choice of subs and suppliers. Even the best and most reliable vendors can become complacent if not challenged occasionally.

5. Project Summary forms should note changes and extras identified after the contract was signed — both for which your company was paid and changes done without additional compensation. Projects with changes and extras that exceed about 4 percent of the contract price deserve special scrutiny. Jobs with changes beyond about 4 percent aren't good for business, at least in my opinion. Nearly all changes have a negative impact on your job schedule and require a disproportionate investment of management resources. Too many changes can antagonize the owner and design staff, even if they were responsible for the altered plans. You may know of a mechanical contractor with a reputation for capitalizing on change orders. But I've rarely seen a job plagued with changes that turned into a money-maker for anyone — except the attorneys. Your file of Project Summary forms will show job types that carry change order risk. Before finalizing and submitting any bid, consider whether the job will get mired in disputes over changes and extras. If similar jobs have ended on the courthouse steps, factor that risk into your estimate.

Utility of a Project Summary forms file is limited only by your ingenuity. The important point is to keep and organize the source of your second profit available on every job. What you learn can be more valuable than what you earn.

The Estimating Procedure

Every plumbing and HVAC estimator works under deadline pressure. You'll seldom have the luxury of spending as much time as you would like on an estimate. Estimators who aren't organized waste valuable time and tend to make careless errors. Try to be well-organized and consistent in your approach to estimating. For most projects, I recommend that you follow the procedures listed below and in the order listed:

1. Get a second set of project drawings and specifications for use by your suppliers and subcontractors. Remember that your subs and suppliers need access to the plans and specs and time to prepare their quotes.

2. Study the plans and specs carefully. Highlight important items. Make a list of specific tasks that require labor unit correction factors. The estimate is never complete until you're totally familiar with the project and the applicable construction codes.

3. Get the general contractor or owner to identify the proposed construction schedule and subcontractor lay-down (storage) area. Work schedule and site conditions always affect your costs.

4. Contact all potential suppliers and subcontractors as early as possible. Set a time when each can come to your office to make their take-offs from the spare set of contract documents.

When this important preliminary work is done, or in progress, it's time to begin your detailed take-off.

Guidelines for Good Estimating

You can compile estimates on a legal pad, a printed estimating form or on a computer. Regardless of the method, these guidelines will apply:

List Each Cost Separately on your take-off sheet. Don't combine system estimates, even if the materials are the same type. A combined system estimate may have to be completely redone if materials for one system are changed at a later date. Use the Estimate Detail Sheet on page 16 if you don't already have a good material take-off form.

Use Engineer's Identification Numbers when listing equipment. The word pump without any other description is ambiguous when there are several pumps included in the project.

Don't Forget Labor Adjustment factors if your labor costs are significantly higher or lower than the costs used in this book. See instructions on page 7 for adjusting labor costs.

Use Colored Pencils or highlighters to mark the items you've taken off and listed. Use a different color for each piping or ducting system.

Log Telephone Quotes and other important phone conversations on a telephone quote form. See the sample on page 18.

Project Estimated Costs for labor, material and equipment to the time when the work is expected to be done, not when the job is being estimated.

The only good estimate is a complete estimate. You've probably heard this saying, "He who makes the most mistakes is likely to be low bidder, and live to regret it."

Preparing the Proposal

It's both common courtesy and good business practice to deliver an unpriced copy of your bid or proposal letter to the general contractor three or four days before the bid deadline date. This gives the contractor time to study your proposal and obtain alternate pricing for items you may have excluded. To avoid misunderstandings, make sure your proposals include, as a minimum, the following elements:

1. The complete name and address of the proposed project.
2. Specification title and issue date.
3. A complete listing of drawings and their issue or revision date.
4. A complete list of addenda and their dates of issue.
5. A list of specification section numbers covered by your proposal.
6. A list of exclusions, clarifications and assumptions.

Your final bid can be phoned in or sent by fax, but it should reach the general contractor or owner no more than five or ten minutes before the bid deadline. Prices submitted too early may have to be revised because of last-minute price changes by subcontractors or suppliers.

MET Worksheet
Material, Equipment and Tool Delivery and Surplus Return Record

Project ID _____   Job Location _____
Supervisor _____   Start Date _____

Description of Material, Equipment or Tool Delivered or Returned	Delivered to Job Site		Returned to Inventory			
	Quantity Delivered	Date Delivered	Quantity Returned	Date Returned	Status Code RN or RS	Value at Return

PROJECT SUMMARY

Project ID _____ Job Location _____

Short description _____

Supervisor _____ Start Date _____

Index ID _____ Client _____

Estimator _____ Subcontractors _____

Major vendors _____

Sources of cost deviation _____

Related Projects by ID Number _____

Thumbnail Summary	Labor	Material	Equipment	Subcontract	Deployed RN/RS	Total
Actual cost						
Estimate						
Over/(Under)						
Full Summary						
Bid amount						
Estimated cost						
Projected profit						
Cost overrun						
Bid profit						
Change orders						
Cost of changes						
Total profit						
Total profit with RN/RS						
Redeployment						

Estimate Detail Sheet

Data carried forward from Take-Off Quantity Survey Sheet(s)

Company/Department _____ Estimator _____ Date _____
Project _____ Checked by _____ Date _____
Address _____ Notes: _____
Job description _____ Estimate # _____
CSI Division/Account _____ Estimate due _____

Item Description	Quantity	Unit	Crew @ MH/Unit	Manhours Ext.	Materials Unit $	Materials Ext. $	Labor Unit $	Labor Ext. $	Equipment Unit $	Equipment Ext. $	Subcontract Unit $	Subcontract Ext. $	Total $
Totals This Sheet				Manhours		Material $		Labor $		Equipment $		Subcontract $	Total $

Carry totals forward to Estimate Summary Sheet   Estimate # _____   Estimate Detail Sheet ____ of ____

Quotation Sheet

Job: _____

Supplier: _____

Salesperson: _____ **Phone No:** _____

Per Plans/Specs: _____ **Freight:** _____ **Terms:** _____

Description	Delivery Time	Price

By: _____

Record of Telephone Conversation

Date:_____ Time:_____ Project: _____

Telecon with: _____

Company: _____ **Phone No:** _____

Subject: _____

Details of Conversation: _____

_____

_____

_____

_____

_____

_____

_____

_____

_____

_____

_____

_____

_____

_____

_____

_____

_____

_____

_____

_____

_____

_____

By: _____

Domestic Hot Water Heaters

Description	Craft@Hrs	Unit	Material $	Labor $	Equipment $	Total $

Electric domestic hot water heater (residential). Set in place only (floor models). Make additional allowances for pipe and electrical connections. (See below)

Description	Craft@Hrs	Unit	Material $	Labor $	Equipment $	Total $
6 gallon 1.5 KW/110V	P1@.500	Ea	471.00	18.60	—	489.60
10 gallon 1.5 KW/110V	P1@.500	Ea	527.00	18.60	—	545.60
15 gallon 1.5 KW/110V	P1@.750	Ea	554.00	27.90	—	581.90
20 gallon 1.5 KW/110V	P1@.750	Ea	521.00	27.90	—	548.90
30 gallon 1.5 KW/110V	P1@1.00	Ea	534.00	37.20	—	571.20
40 gallon 1.5 KW/110V	P1@1.20	Ea	559.00	44.70	—	603.70
50 gallon 3 KW/110V	P1@1.30	Ea	603.00	48.40	—	651.40
12 gallon 3 KW/220V	P1@.500	Ea	464.00	18.60	—	482.60
20 gallon 3 KW/220V	P1@.750	Ea	508.00	27.90	—	535.90
30 gallon 3 KW/220V	P1@1.00	Ea	580.00	37.20	—	617.20
40 gallon 3 KW/220V	P1@1.20	Ea	631.00	44.70	—	675.70
50 gallon 3 KW/220V	P1@1.30	Ea	676.00	48.40	—	724.40

Electric domestic hot water heater (commercial), 208/240 volt. Set in place only. Make additional allowances for pipe and electrical connections. (See below)

Description	Craft@Hrs	Unit	Material $	Labor $	Equipment $	Total $
96 gallon, 12 kw	P1@1.50	Ea	2,670.00	55.80	—	2,725.80
96 gallon, 18 kw	P1@1.50	Ea	3,620.00	55.80	—	3,675.80
96 gallon, 36 kw	P1@1.50	Ea	3,750.00	55.80	—	3,805.80
120 gallon, 18 kw	P1@2.00	Ea	3,840.00	74.40	—	3,914.40
120 gallon, 36 kw	P1@2.00	Ea	3,960.00	74.40	—	4,034.40
120 gallon, 54 kw	P1@2.00	Ea	4,690.00	74.40	—	4,764.40
120 gallon, 63 kw	P1@2.00	Ea	5,060.00	74.40	—	5,134.40

Gas-fired domestic hot water heater (residential). Set in place only, Make additional allowances for pipe and combustion venting connections. (See below)

Description	Craft@Hrs	Unit	Material $	Labor $	Equipment $	Total $
30 gallon	P1@1.00	Ea	548.00	37.20	—	585.20
40 gallon	P1@1.00	Ea	886.00	37.20	—	923.20
50 gallon	P1@1.50	Ea	1,010.00	55.80	—	1,065.80

Domestic Hot Water Heaters

Description	Craft@Hrs	Unit	Material $	Labor $	Equipment $	Total $

Gas-fired domestic hot water heater (commercial), standard efficiency. Set in place only, Make additional allowances for pipe and combustion venting connections. (See below)

Description	Craft@Hrs	Unit	Material $	Labor $	Equipment $	Total $
50 gal./ 95 gph	P1@2.00	Ea	2,660.00	74.40	—	2,734.40
67 gal./106 gph	P1@2.00	Ea	3,150.00	74.40	—	3,224.40
76 gal./175 gph	P1@2.00	Ea	4,210.00	74.40	—	4,284.40
91 gal./291 gph	P1@2.00	Ea	5,090.00	74.40	—	5,164.40

Gas-fired domestic hot water heater (commercial), energy miser. Set in place only, Make additional allowances for pipe and combustion venting connections. (See below)

Description	Craft@Hrs	Unit	Material $	Labor $	Equipment $	Total $
50 gal./ 95 gph	P1@2.00	Ea	6,560.00	74.40	—	6,634.40
67 gal./106 gph	P1@2.00	Ea	6,850.00	74.40	—	6,924.40
76 gal./175 gph	P1@2.00	Ea	8,490.00	74.40	—	8,564.40
91 gal./291 gph	P1@2.00	Ea	10,100.00	74.40	—	10,174.40

Tankless natural gas water heaters. Ambient pressure. DOE and UL rated. For residential, multi-dwelling and light commercial potable water applications. Add the cost of piping, tempering valve, circulating pump, controls, and electrical connection, post-installation inspection by both the fire marshal and the mechanical inspector to validate federal, state and local energy tax credits or energy tax credit offsets. For larger arrays (laundries, institutional facilities, food processing plants), develop an estimate based on the required capacity and multiply these costs by the number of heaters required. Rated in Btus and gallons per minute capacity. (1 Mbh = 1,000 Btus)

Description	Craft@Hrs	Unit	Material $	Labor $	Equipment $	Total $
19.5-140 Mbh, .75-5.8 Gpm	P1@16.0	Ea	1,910.00	596.00	—	2,506.00
11-199 Mbh, .5-7 Gpm	P1@20.0	Ea	2,260.00	744.00	—	3,004.00
25-235 Mbh, .75-9.6 Gpm	P1@20.0	Ea	2,940.00	744.00	—	3,684.00

Tankless electric point-of-use water heaters. Ambient pressure, DOE and UL rated. For residential, multi-dwelling and light commercial potable water applications. Cost does not include piping, tempering valve, circulating pump, controls, storage tank, electrical connection. Add the cost of post-installation inspection by the mechanical inspector to validate federal, state and local energy tax credits or energy tax credit offsets. In rated gallons per minute capacity.

Description	Craft@Hrs	Unit	Material $	Labor $	Equipment $	Total $
5.5 Kw/40 Amp, .75-2 Gpm	P1@4.00	Ea	482.00	149.00	—	631.00
9.5 Kw/50 Amp, .75-2.5 Gpm	P1@4.25	Ea	571.00	158.00	—	729.00
19 Kw/100 Amp, 1-3.5 Gpm	P1@4.50	Ea	950.00	167.00	—	1,117.00
28 Kw/120 Amp, 1.5-5 Gpm	P1@4.75	Ea	1,740.00	177.00	—	1,917.00

Domestic Hot Water Heater Connections

Description	Craft@Hrs	Unit	Material $	Labor $	Equipment $	Total $

Domestic hot water heater connection assembly. Includes supply, return, recirculation and relief piping and fittings (copper), relief and isolation valves. Make additional allowances for gas and venting connections where applicable.

Description	Craft@Hrs	Unit	Material $	Labor $	Equipment $	Total $
¾" residential	P1@1.75	Ea	299.00	65.10	—	364.10
¾" commercial	P1@2.25	Ea	401.00	83.70	—	484.70
1" commercial	P1@2.75	Ea	702.00	102.00	—	804.00
1¼" commercial	P1@3.50	Ea	860.00	130.00	—	990.00
1½" commercial	P1@3.75	Ea	895.00	140.00	—	1,035.00
2" commercial	P1@4.50	Ea	954.00	167.00	—	1,121.00
2½" commercial	P1@5.75	Ea	1,980.00	214.00	—	2,194.00
3" commercial	P1@6.50	Ea	3,040.00	242.00	—	3,282.00

Domestic water heater combustion vent connection. Make additional allowances for piping distances greater than 25'.

Description	Craft@Hrs	Unit	Material $	Labor $	Equipment $	Total $
2" B-vent	P1@.090	LF	6.48	3.35	—	9.83
3" B-vent	P1@.100	LF	8.01	3.72	—	11.73
4" B-vent	P1@.110	LF	10.60	4.09	—	14.69
6" B-vent	P1@.130	LF	13.00	4.84	—	17.84
Tankless heater vent kit	P1@2.50	Ea	614.00	93.10	—	707.10
Power vent kit	P1@2.00	Ea	1,510.00	74.40	—	1,584.40

Water Softeners and Controllers

Description	Craft@Hrs	Unit	Material $	Labor $	Equipment $	Total $

Water softener, time clock controller. Including brine tank, brine well and pick-up tube. Labor includes setting in place, connecting the unit to an existing domestic water distribution system, start up and testing.

Description	Craft@Hrs	Unit	Material $	Labor $	Equipment $	Total $
20,000 grain water softener, TCC	P1@4.50	Ea	595.00	167.00	—	762.00
30,000 grain water softener, TCC	P1@4.50	Ea	634.00	167.00	—	801.00
45,000 grain water softener, TCC	P1@4.50	Ea	705.00	167.00	—	872.00
50,000 grain water softener, TCC	P1@4.75	Ea	795.00	177.00	—	972.00
60,000 grain water softener, TCC	P1@4.75	Ea	940.00	177.00	—	1,117.00
75,000 grain water softener, TCC	P1@5.00	Ea	1,010.00	186.00	—	1,196.00
90,000 grain water softener, TCC	P1@5.50	Ea	1,370.00	205.00	—	1,575.00
120,000 grain water softener, TCC	P1@5.75	Ea	1,470.00	214.00	—	1,684.00

Water softener, mechanically-metered controller. Including brine tank, brine well and pick up tube. Labor includes setting in place, connecting the unit to an existing domestic water distribution system, start up and testing.

Description	Craft@Hrs	Unit	Material $	Labor $	Equipment $	Total $
20,000 grain water softener, MMC	P1@4.50	Ea	773.00	167.00	—	940.00
30,000 grain water softener, MMC	P1@4.50	Ea	807.00	167.00	—	974.00
45,000 grain water softener, MMC	P1@4.50	Ea	878.00	167.00	—	1,045.00
50,000 grain water softener, MMC	P1@4.75	Ea	967.00	177.00	—	1,144.00
60,000 grain water softener, MMC	P1@4.75	Ea	1,130.00	177.00	—	1,307.00
75,000 grain water softener, MMC	P1@5.00	Ea	1,210.00	186.00	—	1,396.00
90,000 grain water softener, MMC	P1@5.50	Ea	1,550.00	205.00	—	1,755.00
120,000 grain water softener, MMC	P1@5.75	Ea	1,650.00	214.00	—	1,864.00

Water Softeners and Controllers

Description	Craft@Hrs	Unit	Material $	Labor $	Equipment $	Total $

Water softener, electronically-metered controller. Including brine tank, brine well and pick up tube. Labor includes setting in place, connecting the unit to an existing domestic water distribution system, start up and testing.

Description	Craft@Hrs	Unit	Material $	Labor $	Equipment $	Total $
20,000 grain water softener, EMC	P1@4.50	Ea	820.00	167.00	—	987.00
30,000 grain water softener, EMC	P1@4.50	Ea	844.00	167.00	—	1,011.00
45,000 grain water softener, EMC	P1@4.50	Ea	926.00	167.00	—	1,093.00
50,000 grain water softener, EMC	P1@4.75	Ea	1,010.00	177.00	—	1,187.00
60,000 grain water softener, EMC	P1@4.75	Ea	1,190.00	177.00	—	1,367.00
75,000 grain water softener, EMC	P1@5.00	Ea	1,250.00	186.00	—	1,436.00
90,000 grain water softener, EMC	P1@5.50	Ea	1,590.00	205.00	—	1,795.00
120,000 grain water softener, EMC	P1@5.75	Ea	1,700.00	214.00	—	1,914.00

Water softener accessories

Description	Craft@Hrs	Unit	Material $	Labor $	Equipment $	Total $
By-pass valve	P1@.400	Ea	77.90	14.90	—	92.80
Manifold adapter kit	P1@.200	Ea	21.00	7.44	—	28.44
Turbulator	P1@.400	Ea	38.40	14.90	—	53.30

Iron filter, electronically-metered controller. Manganese green sand filter. Labor includes setting in place, connecting the unit to an existing domestic water distribution system, start-up and testing.

Description	Craft@Hrs	Unit	Material $	Labor $	Equipment $	Total $
42,000 iron filter (1.5 cf media), 5 gpm	P1@4.00	Ea	775.00	149.00	—	924.00
65,000 iron filter (2.0 cf media), 6 gpm	P1@4.50	Ea	917.00	167.00	—	1,084.00
84,000 iron filter (2.5 cf media), 8 gpm	P1@4.75	Ea	979.00	177.00	—	1,156.00
Replacement green sand media	P1@1.20	CF	44.90	44.70	—	89.60

Iron filter accessories

Description	Craft@Hrs	Unit	Material $	Labor $	Equipment $	Total $
By-pass valve	P1@.400	Ea	78.90	14.90	—	93.80
Air vent	P1@.200	Ea	62.60	7.44	—	70.04
Air controller	P1@.400	Ea	70.70	14.90	—	85.60

Water Softener Accessories

Description	Craft@Hrs	Unit	Material $	Labor $	Equipment $	Total $

Combination iron filter/water softener. Zeolite resins soften water and remove iron and manganese. Controller automatically controls PH level. Labor includes setting in place, connecting the unit to an existing domestic water distribution system, start-up and testing.

Description	Craft@Hrs	Unit	Material $	Labor $	Equipment $	Total $
40,000 iron filter, 1.3 cf media	P1@4.00	Ea	1,500.00	149.00	—	1,649.00
60,000 iron filter, 1.7 cf media	P1@4.50	Ea	1,620.00	167.00	—	1,787.00
80,000 iron filter, 2.5 cf media	P1@4.75	Ea	2,350.00	177.00	—	2,527.00

Hot water softener, time clock controller. Brass valve construction. Designed for 150 F. maximum operating temperature. Includes brine tank, brine well and pick-up tube. Labor includes setting in place, connecting the unit to an existing domestic water distribution system, start-up and testing.

Description	Craft@Hrs	Unit	Material $	Labor $	Equipment $	Total $
20,000 grain hot water softener	P1@4.50	Ea	1,880.00	167.00	—	2,047.00
30,000 grain hot water softener	P1@4.50	Ea	2,000.00	167.00	—	2,167.00
40,000 grain hot water softener	P1@4.50	Ea	2,090.00	167.00	—	2,257.00
60,000 grain hot water softener	P1@4.75	Ea	2,470.00	177.00	—	2,647.00

Pressure tank, fiberglass wound. Labor includes setting in place, connecting the tank to a domestic water distribution system and testing.

Description	Craft@Hrs	Unit	Material $	Labor $	Equipment $	Total $
Fiberglass pressure tank, 20 gallon	P1@2.00	Ea	258.00	74.40	—	332.40
Fiberglass pressure tank, 30 gallon	P1@2.00	Ea	291.00	74.40	—	365.40
Fiberglass pressure tank, 80 gallon	P1@2.75	Ea	471.00	102.00	—	573.00
Fiberglass pressure tank, 120 gallon	P1@3.50	Ea	621.00	130.00	—	751.00
Brass tank tee assembly, ¾"	P1@3.50	Ea	41.60	130.00	—	171.60
Brass tank tee assembly, 1"	P1@3.50	Ea	77.50	130.00	—	207.50
Brass tank tee assembly, 1¼"	P1@3.50	Ea	132.00	130.00	—	262.00

Kitchen Equipment

Description	Craft@Hrs	Unit	Material $	Labor $	Equipment $	Total $

Ultra-violet water disinfection unit. Stainless steel reactor, audible and visible alarm, lamp end-of-life indicator and 7-day override. Gpm rating at 30,000 mj/m2 output. Labor includes setting in place, connecting to the water system and testing.

Description	Craft@Hrs	Unit	Material $	Labor $	Equipment $	Total $
UV system, 1 gpm, ¼" in/out	P1@3.00	Ea	255.00	112.00	—	367.00
UV system, 6 gpm, ½" in/out	P1@3.00	Ea	496.00	112.00	—	608.00
UV system, 8 gpm, ¾" in/out	P1@4.00	Ea	574.00	149.00	—	723.00
UV system, 12 gpm, ¾" in/out	P1@4.00	Ea	736.00	149.00	—	885.00
UV replacement lamp, 20W, 1 gpm	P1@.750	Ea	57.10	27.90	—	85.00
UV replacement lamp, 32W, 6 gpm	P1@.750	Ea	64.80	27.90	—	92.70
UV replacement lamp, 39W, 8-12 gpm	P1@.750	Ea	82.90	27.90	—	110.80
UV replacement ballast, 420 Mv/110V	P1@1.00	Ea	250.00	37.20	—	287.20

Kitchen equipment booster heater

Description	Craft@Hrs	Unit	Material $	Labor $	Equipment $	Total $
1,000 watt	P1@4.00	Ea	834.00	149.00	—	983.00

Dishwasher

Description	Craft@Hrs	Unit	Material $	Labor $	Equipment $	Total $
Built-in	P1@5.00	Ea	921.00	186.00	—	1,107.00

Garbage disposal

Description	Craft@Hrs	Unit	Material $	Labor $	Equipment $	Total $
½ HP	P1@2.00	Ea	182.00	74.40	—	256.40
¾ HP	P1@2.00	Ea	304.00	74.40	—	378.40

Grease and oil interceptor

Description	Craft@Hrs	Unit	Material $	Labor $	Equipment $	Total $
4 GPM	P1@4.00	Ea	372.00	149.00	—	521.00
10 GPM	P1@5.00	Ea	606.00	186.00	—	792.00
15 GPM	P1@7.00	Ea	904.00	261.00	—	1,165.00
20 GPM	P1@8.00	Ea	1,090.00	298.00	—	1,388.00

Hair and lint interceptor

Description	Craft@Hrs	Unit	Material $	Labor $	Equipment $	Total $
1½"	P1@.650	Ea	212.00	24.20	—	236.20
2"	P1@.750	Ea	302.00	27.90	—	329.90

All bronze ¾" to 1½" in-line NPT pump

Description	Craft@Hrs	Unit	Material $	Labor $	Equipment $	Total $
1/12 HP	P1@1.50	Ea	572.00	55.80	—	627.80
1/6 HP	P1@1.50	Ea	854.00	55.80	—	909.80
1/4 HP	P1@1.50	Ea	1,000.00	55.80	—	1,055.80

Kitchen Equipment Connections

Description	Craft@Hrs	Unit	Material $	Labor $	Equipment $	Total $
Kitchen appliance gas trim						
½"	P1@1.15	Ea	44.00	42.80	—	86.80
¾"	P1@1.30	Ea	80.50	48.40	—	128.90
1"	P1@1.60	Ea	93.30	59.60	—	152.90
1¼"	P1@2.10	Ea	154.00	78.20	—	232.20
1½"	P1@2.50	Ea	195.00	93.10	—	288.10
2"	P1@3.00	Ea	260.00	112.00	—	372.00
Hot and cold water supply						
½"	P1@1.10	Ea	49.50	40.90	—	90.40
¾"	P1@1.55	Ea	70.20	57.70	—	127.90
1"	P1@1.90	Ea	95.60	70.70	—	166.30
1¼"	P1@2.50	Ea	134.00	93.10	—	227.10
1½"	P1@3.00	Ea	168.00	112.00	—	280.00
Continuous waste						
2-part	P1@.250	Ea	63.00	9.31	—	72.31
3-part	P1@.350	Ea	107.00	13.00	—	120.00
4-part	P1@.450	Ea	137.00	16.70	—	153.70
Indirect waste						
½"	P1@1.05	Ea	16.00	39.10	—	55.10
¾"	P1@1.50	Ea	27.10	55.80	—	82.90
1"	P1@1.90	Ea	43.50	70.70	—	114.20
1¼"	P1@2.15	Ea	64.20	80.00	—	144.20
1½"	P1@2.60	Ea	84.60	96.80	—	181.40
2"	P1@3.00	Ea	129.00	112.00	—	241.00
Kitchen fixture waste tailpiece						
1½"	P1@.100	Ea	16.10	3.72	—	19.82
Kitchen fixture trap with solder bushing						
1½"	P1@.250	Ea	54.20	9.31	—	63.51
2"	P1@.300	Ea	75.10	11.20	—	86.30

Plumbing Fixtures

Description	Craft@Hrs	Unit	Material $	Labor $	Equipment $	Total $

Water closet, floor-mounted, flush tank, white vitreous china, lined tank. Complete with trim. Make additional allowances for rough-in. Based on American Standard Cadet series. ADA means American Disabilities Act compliant. (Wheelchair accessible)

Description	Craft@Hrs	Unit	Material $	Labor $	Equipment $	Total $
Round bowl	P1@2.10	Ea	277.00	78.20	—	355.20
Elongated bowl	P1@2.10	Ea	335.00	78.20	—	413.20
ADA, 18" high	P1@2.10	Ea	456.00	78.20	—	534.20

Water closet, floor-mounted, flush valve, white vitreous china. Complete with trim. Make additional allowances for rough-in. Based on American Standard. ADA means American Disabilities Act compliant. (Wheelchair accessible)

Description	Craft@Hrs	Unit	Material $	Labor $	Equipment $	Total $
Elongated bowl	P1@2.60	Ea	434.00	96.80	—	530.80
Elongated bowl, ADA 18" high	P1@2.60	Ea	516.00	96.80	—	612.80
Elongated bowl with a bedpan cleanser	P1@4.10	Ea	749.00	153.00	—	902.00
Elongated bowl, ADA 18" high with a bedpan cleanser	P1@4.10	Ea	811.00	153.00	—	964.00

Water closet, wall-hung, flush valve, white vitreous china. Complete with fixture carrier and all trim. Make additional allowances for rough-in. Based on American Standard Afwall series.

Description	Craft@Hrs	Unit	Material $	Labor $	Equipment $	Total $
Elongated bowl	P1@3.55	Ea	687.00	132.00	—	819.00
Elongated bowl with electronic flush valve	P1@3.80	Ea	1,230.00	141.00	—	1,371.00
Elongated bowl with bedpan cleanser	P1@5.05	Ea	996.00	188.00	—	1,184.00
Electronic flush valve, add	P1@.600	Ea	543.00	22.30	—	565.30

Urinal, wall-hung, flush valve, white vitreous china. Complete with trim. Make additional allowances for rough-in.

Description	Craft@Hrs	Unit	Material $	Labor $	Equipment $	Total $
Siphon-jet type	P1@3.15	Ea	687.00	117.00	—	804.00
Wash-out type	P1@3.10	Ea	561.00	115.00	—	676.00
Wash-down type	P1@3.00	Ea	396.00	112.00	—	508.00
Urinal carrier, add	P1@.600	Ea	125.00	22.30	—	147.30
Electronic flush valve, add	P1@.600	Ea	448.00	22.30	—	470.30

Urinal, stall-type, flush valve, white vitreous china. Complete with trim. Make additional allowances for rough-in.

Description	Craft@Hrs	Unit	Material $	Labor $	Equipment $	Total $
Stall urinal	P1@5.00	Ea	1,330.00	186.00	—	1,516.00

Plumbing Fixtures

Description	Craft@Hrs	Unit	Material $	Labor $	Equipment $	Total $

Lavatory, wall-hung, white vitreous china. Complete with trim and fixture carrier. Make additional allowances for rough-in. ADA means American Disabilities Act compliant. (Wheelchair accessible)

Description	Craft@Hrs	Unit	Material $	Labor $	Equipment $	Total $
Wall-hung lav	P1@2.70	Ea	575.00	100.00	—	675.00
Wall-hung, ADA	P1@2.70	Ea	837.00	100.00	—	937.00
Fixture carrier	P1@.500	Ea	119.00	18.60	—	137.60

Countertop lavatory, white. Complete with trim. Make additional allowances for rough-in.

Description	Craft@Hrs	Unit	Material $	Labor $	Equipment $	Total $
Vitreous china	P1@2.00	Ea	426.00	74.40	—	500.40
Enameled steel	P1@2.00	Ea	360.00	74.40	—	434.40
Acrylic	P1@2.00	Ea	262.00	74.40	—	336.40

Bathtub, white, 60" x 32". Complete with trim, including shower head. Make additional allowances for rough-in.

Description	Craft@Hrs	Unit	Material $	Labor $	Equipment $	Total $
Enameled steel	P1@2.50	Ea	599.00	93.10	—	692.10
Cast iron	P1@3.50	Ea	865.00	130.00	—	995.00
Fiberglass	P1@2.50	Ea	553.00	93.10	—	646.10
Acrylic	P1@2.50	Ea	591.00	93.10	—	684.10

Tub and shower combination, fiberglass, white. Complete with trim. Make additional allowances for rough-in.

Description	Craft@Hrs	Unit	Material $	Labor $	Equipment $	Total $
One-piece	P1@4.50	Ea	1,230.00	167.00	—	1,397.00
Two-piece (reno)	P1@5.50	Ea	1,580.00	205.00	—	1,785.00
Four-piece (reno)	P1@6.25	Ea	1,670.00	233.00	—	1,903.00

Shower stall, white, 36" x 36". Complete with trim. Make additional allowances for rough-in.

Description	Craft@Hrs	Unit	Material $	Labor $	Equipment $	Total $
Fiberglass one-piece	P1@3.50	Ea	765.00	130.00	—	895.00
Fiberglass three-piece	P1@4.25	Ea	987.00	158.00	—	1,145.00
Acrylic one-piece	P1@3.50	Ea	1,150.00	130.00	—	1,280.00
Acrylic three-piece	P1@4.25	Ea	1,510.00	158.00	—	1,668.00

Shower basin, 36" x 36". Complete with trim (faucet, shower head and strainer). Make additional allowances for rough-in.

Description	Craft@Hrs	Unit	Material $	Labor $	Equipment $	Total $
Fiberglass	P1@2.50	Ea	522.00	93.10	—	615.10
Acrylic	P1@2.50	Ea	561.00	93.10	—	654.10
Molded stone	P1@2.65	Ea	543.00	98.60	—	641.60

Plumbing Fixtures

Description	Craft@Hrs	Unit	Material $	Labor $	Equipment $	Total $

Kitchen sink, double compartment. Complete with trim. Make additional allowances for rough-in.

Stainless steel	P1@2.15	Ea	439.00	80.00	—	519.00
Cast iron	P1@2.50	Ea	573.00	93.10	—	666.10
Acrylic	P1@2.15	Ea	520.00	80.00	—	600.00

Kitchen sink, single compartment. Complete with trim. Make additional allowances for rough-in.

Stainless steel	P1@2.00	Ea	371.00	74.40	—	445.40
Cast iron	P1@2.10	Ea	428.00	78.20	—	506.20
Acrylic	P1@2.00	Ea	386.00	74.40	—	460.40

Bar sink. Complete with trim. Make additional allowances for rough-in.

Stainless steel	P1@2.00	Ea	315.00	74.40	—	389.40
Acrylic	P1@2.00	Ea	212.00	74.40	—	286.40

Exam room sink. Complete with trim. Make additional allowances for rough-in.

Stainless steel	P1@2.10	Ea	458.00	78.20	—	536.20
Acrylic	P1@2.10	Ea	391.00	78.20	—	469.20

Laboratory sink. Complete with trim. Make additional allowances for rough-in.

Stainless steel	P1@2.25	Ea	525.00	83.70	—	608.70
Acrylic	P1@2.25	Ea	458.00	83.70	—	541.70

Laundry sink, double compartment. Complete with trim. Make additional allowances for rough-in.

Cast iron	P1@3.50	Ea	614.00	130.00	—	744.00
Acrylic	P1@2.25	Ea	270.00	83.70	—	353.70

Laundry sink, single compartment. Complete with trim. Make additional allowances for rough-in.

Cast iron	P1@2.75	Ea	531.00	102.00	—	633.00
Acrylic	P1@2.00	Ea	186.00	74.40	—	260.40

Mop sink, floor-mounted, 36" x 24". Complete with trim. Make additional allowances for rough-in.

Molded stone	P1@2.65	Ea	747.00	98.60	—	845.60
Terrazzo	P1@2.65	Ea	824.00	98.60	—	922.60
Acrylic	P1@2.35	Ea	576.00	87.50	—	663.50

Plumbing Fixtures

Description	Craft@Hrs	Unit	Material $	Labor $	Equipment $	Total $

Slop sink, enameled cast iron with P-trap, standard. Complete with trim. Make additional allowances for rough-in.

Description	Craft@Hrs	Unit	Material $	Labor $	Equipment $	Total $
Slop sink with P-trap, std.	P1@3.50	Ea	1,280.00	130.00	—	1,410.00

Floor sink, recessed, enameled steel, white. Add 40% to material prices for acid-resisting finish. Complete with strainer. Make additional allowances for rough-in.

Description	Craft@Hrs	Unit	Material $	Labor $	Equipment $	Total $
9" x 9"	P1@1.00	Ea	110.00	37.20	—	147.20
12" x 12"	P1@1.00	Ea	128.00	37.20	—	165.20
15" x 15"	P1@1.15	Ea	87.50	42.80	—	130.30
18" x 18"	P1@1.25	Ea	115.00	46.50	—	161.50
24" x 24"	P1@1.50	Ea	156.00	55.80	—	211.80

Drinking fountain, refrigerated, stainless steel. Complete with trim. Make additional allowances for rough-in. ADA means American Disabilities Act compliant. (Wheelchair accessible)

Description	Craft@Hrs	Unit	Material $	Labor $	Equipment $	Total $
Free-standing	P1@2.00	Ea	1,310.00	74.40	—	1,384.40
Semi-recessed	P1@2.50	Ea	1,750.00	93.10	—	1,843.10
Fully-recessed	P1@2.50	Ea	3,020.00	93.10	—	3,113.10
Wall-hung	P1@2.00	Ea	1,230.00	74.40	—	1,304.40
Wall-hung, ADA	P1@2.50	Ea	3,020.00	93.10	—	3,113.10

Drinking fountain, non-refrigerated. Complete with trim. Make additional allowances for rough-in. ADA means American Disabilities Act compliant. (Wheelchair accessible) S.S. means stainless steel.

Description	Craft@Hrs	Unit	Material $	Labor $	Equipment $	Total $
Recessed, china	P1@2.50	Ea	1,020.00	93.10	—	1,113.10
Wall-hung, china	P1@2.00	Ea	579.00	74.40	—	653.40
Recessed, S.S.	P1@2.50	Ea	1,160.00	93.10	—	1,253.10
Wall-hung, S.S.	P1@2.00	Ea	617.00	74.40	—	691.40
ADA, S.S.	P1@2.50	Ea	1,050.00	93.10	—	1,143.10

Plumbing Fixtures Rough-in

Description	Craft@Hrs	Unit	Material $	Labor $	Equipment $	Total $

Commercial plumbing fixture rough-in. Includes type L copper supply pipe and DWV copper (to 2½") or cast iron (MJ) DWV (over 2½") drain and vent piping. Make additional allowances for plumbing fixtures and trim. Use these costs for preliminary estimates.

Description	Craft@Hrs	Unit	Material $	Labor $	Equipment $	Total $
Water closet, wall-hung, flush valve, with carrier	P1@2.25	Ea	926.00	83.70	—	1,009.70
Water closet, wall-hung, flush valve, no carrier	P1@1.95	Ea	849.00	72.60	—	921.60
Water closet, floor-mounted, flush valve	P1@2.75	Ea	750.00	102.00	—	852.00
Water closet, floor-mounted, tank type	P1@2.25	Ea	575.00	83.70	—	658.70
Bidet	P1@2.00	Ea	400.00	74.40	—	474.40
Urinal, wall-hung, flush valve, with carrier	P1@3.10	Ea	1,010.00	115.00	—	1,125.00
Urinal, wall-hung, flush valve, without carrier	P1@2.35	Ea	575.00	87.50	—	662.50
Lavatory, wall-hung, with carrier	P1@2.40	Ea	833.00	89.30	—	922.30
Lavatory	P1@1.90	Ea	400.00	70.70	—	470.70
Sink	P1@1.90	Ea	431.00	70.70	—	501.70
Bath tub	P1@2.35	Ea	616.00	87.50	—	703.50
Shower	P1@2.60	Ea	722.00	96.80	—	818.80
Mop sink	P1@2.40	Ea	512.00	89.30	—	601.30
Slop sink	P1@2.60	Ea	366.00	96.80	—	462.80
Laundry tub	P1@1.95	Ea	435.00	72.60	—	507.60
Wash fountain	P1@2.10	Ea	469.00	78.20	—	547.20
Lab sink, glass drainage	P1@3.80	Ea	1,850.00	141.00	—	1,991.00
Lab sink, acid resistant plastic drainage	P1@2.65	Ea	294.00	98.60	—	392.60
Drinking fountain	P1@2.20	Ea	319.00	81.90	—	400.90
Emergency eyewash and shower	P1@1.75	Ea	121.00	65.10	—	186.10
Washing machine	P1@2.25	Ea	468.00	83.70	—	551.70

Plumbing Fixtures Rough-in

Description	Craft@Hrs	Unit	Material $	Labor $	Equipment $	Total $

Commercial plumbing fixture group rough-in. Includes Type L copper supply pipe and DWV copper (to 2½") or cast iron (MJ) DWV (over 2½") drain and vent piping. Make additional allowances for plumbing fixtures and trim. Use these costs for preliminary estimates.

Description	Craft@Hrs	Unit	Material $	Labor $	Equipment $	Total $
3-piece washroom group	P1@5.50	Ea	1,100.00	205.00	—	1,305.00
3-piece washroom group back to back	P1@9.75	Ea	2,040.00	363.00	—	2,403.00
Kitchen sink, back to back	P1@2.15	Ea	594.00	80.00	—	674.00
Battery of water closets, floor-mounted, tank type, per water closet	P1@1.75	Ea	463.00	65.10	—	528.10
Battery of water closets, floor-mounted, flush valve, per water closet	P1@2.20	Ea	605.00	81.90	—	686.90
Battery of water closets, wall-hung, flush valve, with carrier, per water closet	P1@1.80	Ea	795.00	67.00	—	862.00
Battery of water closets, wall-hung, flush valve, without carrier, per water closet	P1@1.50	Ea	233.00	55.80	—	288.80
Battery of urinals, wall-hung, flush valve with carrier, per urinal	P1@2.45	Ea	956.00	91.20	—	1,047.20
Battery of urinals, wall-hung, flush valve without carrier, per urinal	P1@1.90	Ea	493.00	70.70	—	563.70
Battery of lavatory basins, wall-hung, with carrier, per lavatory	P1@2.00	Ea	744.00	74.40	—	818.40
Battery of lavatory basins, without carrier, per lavatory	P1@1.50	Ea	336.00	55.80	—	391.80

Residential plumbing fixture rough-in. Includes polyethylene (PE) supply pipe and ABS DWV drain and vent piping. Make additional allowances for plumbing fixtures and trim. Use these costs for preliminary estimates.

Description	Craft@Hrs	Unit	Material $	Labor $	Equipment $	Total $
Water closet, floor-mounted, tank type	P1@2.00	Ea	128.00	74.40	—	202.40
Bidet	P1@1.85	Ea	96.30	68.90	—	165.20
Lavatory	P1@1.75	Ea	96.30	65.10	—	161.40
Counter sink	P1@1.75	Ea	106.00	65.10	—	171.10
Bathtub	P1@2.10	Ea	96.30	78.20	—	174.50
Shower	P1@2.45	Ea	141.00	91.20	—	232.20
Laundry tub	P1@1.75	Ea	87.80	65.10	—	152.90
Washing machine	P1@2.00	Ea	109.00	74.40	—	183.40

Copper, Type K with Brazed Joints

Copper, Type K with Brazed Joints

Type K hard-drawn copper pipe with wrought copper fittings and brazed joints is used in a wide variety of plumbing and HVAC systems such as potable water, heating hot water, chilled water, compressed air, refrigerant and A.C. condensate.

Brazed joints are those made with silver or other alloy filler metals having melting points at, or above, 1,000 degrees F. Maximum working pressure/temperature relationships for brazed joints are approximately as follows:

Maximum Working Pressure (PSIG)*			
Water Temperature	Nominal Pipe Size (inches)		
(degrees F.)	Up to 1	1¼ to 2	2½ to 4
Up to 350	270	190	150

*For copper pipe and solder-type fittings using brazing alloys melting at, or above 1,000 degrees F.

This section has been arranged to save the estimator's time by including all normally-used system components such as pipe, fittings, valves, hanger assemblies, riser clamps and miscellaneous items under one heading. Additional items can be found under "Plumbing and Piping Specialties." The cost estimates in this section are based on the conditions, limitations and wage rates described in the section "How to Use This Book" beginning on page 5.

Description	Craft@Hrs	Unit	Material $	Labor $	Equipment $	Total $

Type K copper pipe with brazed joints, installed horizontally. Complete installation including six to twelve wrought copper tees and six to twelve wrought copper elbows every 100', and hangers spaced to meet plumbing code *(a tee & elbow every 8.3' for ½" dia., a tee and elbow every 16.6' for 4" dia.)*.

½"	P1@.117	LF	3.65	4.35	—	8.00
¾"	P1@.127	LF	6.38	4.73	—	11.11
1"	P1@.147	LF	10.10	5.47	—	15.57
1¼"	P1@.167	LF	13.90	6.22	—	20.12
1½"	P1@.187	LF	17.20	6.96	—	24.16
2"	P1@.217	LF	29.00	8.08	—	37.08
2½"	P1@.247	LF	47.80	9.19	—	56.99
3"	P1@.277	LF	55.20	10.30	—	65.50
4"	P1@.317	LF	100.00	11.80	—	111.80

Copper, Type K with Brazed Joints

Description	Craft@Hrs	Unit	Material $	Labor $	Equipment $	Total $

Type K copper pipe with brazed joints, installed risers. Complete installation including a wrought copper reducing tee every floor and a riser clamp every other floor.

½"	P1@.064	LF	3.29	2.38	—	5.67
¾"	P1@.074	LF	5.95	2.75	—	8.70
1"	P1@.079	LF	9.06	2.94	—	12.00
1¼"	P1@.084	LF	12.50	3.13	—	15.63
1½"	P1@.104	LF	14.90	3.87	—	18.77
2"	P1@.134	LF	25.30	4.99	—	30.29
2½"	P1@.154	LF	37.70	5.73	—	43.43
3"	P1@.174	LF	52.80	6.48	—	59.28
4"	P1@.194	LF	82.80	7.22	—	90.02

Type K copper pipe with brazed joints, pipe only

½"	P1@.032	LF	2.77	1.19	—	3.96
¾"	P1@.035	LF	5.13	1.30	—	6.43
1"	P1@.038	LF	6.46	1.41	—	7.87
1¼"	P1@.042	LF	8.63	1.56	—	10.19
1½"	P1@.046	LF	9.43	1.71	—	11.14
2"	P1@.053	LF	16.80	1.97	—	18.77
2½"	P1@.060	LF	21.90	2.23	—	24.13
3"	P1@.066	LF	28.70	2.46	—	31.16
4"	P1@.080	LF	51.30	2.98	—	54.28

Type K copper 45-degree ell C x C with brazed joints

½"	P1@.129	Ea	.82	4.80	—	5.62
¾"	P1@.181	Ea	1.56	6.74	—	8.30
1"	P1@.232	Ea	4.61	8.64	—	13.25
1¼"	P1@.284	Ea	6.24	10.60	—	16.84
1½"	P1@.335	Ea	9.29	12.50	—	21.79
2"	P1@.447	Ea	14.70	16.60	—	31.30
2½"	P1@.550	Ea	37.60	20.50	—	58.10
3"	P1@.654	Ea	49.70	24.30	—	74.00
4"	P1@.860	Ea	95.40	32.00	—	127.40

Type K copper 90-degree ell C x C with brazed joints

½"	P1@.129	Ea	.48	4.80	—	5.28
¾"	P1@.181	Ea	1.01	6.74	—	7.75
1"	P1@.232	Ea	3.52	8.64	—	12.16
1¼"	P1@.284	Ea	5.68	10.60	—	16.28
1½"	P1@.335	Ea	8.57	12.50	—	21.07
2"	P1@.447	Ea	14.90	16.60	—	31.50
2½"	P1@.550	Ea	31.10	20.50	—	51.60
3"	P1@.654	Ea	39.40	24.30	—	63.70
4"	P1@.860	Ea	83.30	32.00	—	115.30

Copper, Type K with Brazed Joints

Description	Craft@Hrs	Unit	Material $	Labor $	Equipment $	Total $

Type K copper 90-degree ell ftg. x C with brazed joints

Description	Craft@Hrs	Unit	Material $	Labor $	Equipment $	Total $
½"	P1@.129	Ea	1.01	4.80	—	5.81
¾"	P1@.181	Ea	3.32	6.74	—	10.06
1"	P1@.232	Ea	4.98	8.64	—	13.62
1¼"	P1@.284	Ea	6.94	10.60	—	17.54
1½"	P1@.335	Ea	10.30	12.50	—	22.80
2"	P1@.447	Ea	20.30	16.60	—	36.90
2½"	P1@.550	Ea	42.30	20.50	—	62.80
3"	P1@.654	Ea	47.60	24.30	—	71.90
4"	P1@.860	Ea	127.00	32.00	—	159.00

Type K copper tee C x C x C with brazed joints

Description	Craft@Hrs	Unit	Material $	Labor $	Equipment $	Total $
½"	P1@.156	Ea	.77	5.81	—	6.58
¾"	P1@.204	Ea	2.09	7.59	—	9.68
1"	P1@.263	Ea	7.85	9.79	—	17.64
1¼"	P1@.322	Ea	12.50	12.00	—	24.50
1½"	P1@.380	Ea	18.30	14.10	—	32.40
2"	P1@.507	Ea	28.20	18.90	—	47.10
2½"	P1@.624	Ea	57.20	23.20	—	80.40
3"	P1@.741	Ea	88.50	27.60	—	116.10
4"	P1@.975	Ea	148.00	36.30	—	184.30

Copper, Type K with Brazed Joints

Description	Craft@Hrs	Unit	Material $	Labor $	Equipment $	Total $

Type K copper branch reducing tee C x C x C with brazed joints

Description	Craft@Hrs	Unit	Material $	Labor $	Equipment $	Total $
½" x 3/8"	P1@.129	Ea	1.57	4.80	—	6.37
¾" x ½"	P1@.182	Ea	3.65	6.77	—	10.42
1" x ½"	P1@.209	Ea	9.32	7.78	—	17.10
1" x ¾"	P1@.234	Ea	9.32	8.71	—	18.03
1¼" x ½"	P1@.241	Ea	14.10	8.97	—	23.07
1¼" x ¾"	P1@.264	Ea	14.10	9.83	—	23.93
1¼" x 1"	P1@.287	Ea	14.10	10.70	—	24.80
1½" x ½"	P1@.291	Ea	17.30	10.80	—	28.10
1½" x ¾"	P1@.307	Ea	17.30	11.40	—	28.70
1½" x 1"	P1@.323	Ea	17.30	12.00	—	29.30
1½" x 1¼"	P1@.339	Ea	17.30	12.60	—	29.90
2" x ½"	P1@.348	Ea	22.60	13.00	—	35.60
2" x ¾"	P1@.373	Ea	22.60	13.90	—	36.50
2" x 1"	P1@.399	Ea	22.60	14.90	—	37.50
2" x 1¼"	P1@.426	Ea	22.60	15.90	—	38.50
2" x 1½"	P1@.452	Ea	22.60	16.80	—	39.40
2½" x ½"	P1@.468	Ea	67.40	17.40	—	84.80
2½" x ¾"	P1@.485	Ea	67.40	18.10	—	85.50
2½" x 1"	P1@.502	Ea	67.40	18.70	—	86.10
2½" x 1½"	P1@.537	Ea	67.40	20.00	—	87.40
2½" x 2"	P1@.555	Ea	67.40	20.70	—	88.10
3" x 1¼"	P1@.562	Ea	76.40	20.90	—	97.30
3" x 2"	P1@.626	Ea	79.30	23.30	—	102.60
3" x 2½"	P1@.660	Ea	79.30	24.60	—	103.90
4" x 1¼"	P1@.669	Ea	138.00	24.90	—	162.90
4" x 1½"	P1@.718	Ea	138.00	26.70	—	164.70
4" x 2"	P1@.768	Ea	138.00	28.60	—	166.60
4" x 2½"	P1@.819	Ea	138.00	30.50	—	168.50
4" x 3"	P1@.868	Ea	138.00	32.30	—	170.30

Copper, Type K with Brazed Joints

Description	Craft@Hrs	Unit	Material $	Labor $	Equipment $	Total $

Type K copper reducer with brazed joints

Description	Craft@Hrs	Unit	Material $	Labor $	Equipment $	Total $
½" x 3/8"	P1@.148	Ea	.56	5.51	—	6.07
¾" x ½"	P1@.155	Ea	1.17	5.77	—	6.94
1" x ¾"	P1@.206	Ea	1.91	7.67	—	9.58
1¼" x ½"	P1@.206	Ea	3.21	7.67	—	10.88
1¼" x ¾"	P1@.232	Ea	3.21	8.64	—	11.85
1¼" x 1"	P1@.258	Ea	3.21	9.60	—	12.81
1½" x ½"	P1@.232	Ea	4.90	8.64	—	13.54
1½" x ¾"	P1@.257	Ea	4.90	9.57	—	14.47
1½" x 1"	P1@.283	Ea	4.90	10.50	—	15.40
1½" x 1¼"	P1@.308	Ea	4.90	11.50	—	16.40
2" x ½"	P1@.287	Ea	7.38	10.70	—	18.08
2" x ¾"	P1@.277	Ea	7.38	10.30	—	17.68
2" x 1"	P1@.338	Ea	7.38	12.60	—	19.98
2" x 1¼"	P1@.365	Ea	7.38	13.60	—	20.98
2" x 1½"	P1@.390	Ea	7.38	14.50	—	21.88
2½" x 1"	P1@.390	Ea	20.10	14.50	—	34.60
2½" x 1¼"	P1@.416	Ea	20.10	15.50	—	35.60
2½" x 1½"	P1@.442	Ea	20.10	16.50	—	36.60
2½" x 2"	P1@.497	Ea	20.10	18.50	—	38.60
3" x 1¼"	P1@.468	Ea	23.50	17.40	—	40.90
3" x 1½"	P1@.493	Ea	23.50	18.30	—	41.80
3" x 2"	P1@.548	Ea	23.50	20.40	—	43.90
3" x 2½"	P1@.600	Ea	23.50	22.30	—	45.80
4" x 2"	P1@.652	Ea	45.30	24.30	—	69.60
4" x 2½"	P1@.704	Ea	45.30	26.20	—	71.50
4" x 3"	P1@.755	Ea	45.30	28.10	—	73.40

Type K copper adapter C x MPT

Description	Craft@Hrs	Unit	Material $	Labor $	Equipment $	Total $
½"	P1@.090	Ea	1.53	3.35	—	4.88
¾"	P1@.126	Ea	2.18	4.69	—	6.87
1"	P1@.162	Ea	5.41	6.03	—	11.44
1¼"	P1@.198	Ea	10.10	7.37	—	17.47
1½"	P1@.234	Ea	11.40	8.71	—	20.11
2"	P1@.312	Ea	14.20	11.60	—	25.80
2½"	P1@.384	Ea	46.10	14.30	—	60.40
3"	P1@.456	Ea	76.40	17.00	—	93.40

Type K copper adapter C x FPT

Description	Craft@Hrs	Unit	Material $	Labor $	Equipment $	Total $
½"	P1@.090	Ea	2.32	3.35	—	5.67
¾"	P1@.126	Ea	4.21	4.69	—	8.90
1"	P1@.162	Ea	6.35	6.03	—	12.38
1¼"	P1@.198	Ea	13.40	7.37	—	20.77
1½"	P1@.234	Ea	16.10	8.71	—	24.81
2"	P1@.312	Ea	24.00	11.60	—	35.60
2½"	P1@.384	Ea	60.50	14.30	—	74.80
3"	P1@.456	Ea	83.20	17.00	—	100.20

Copper, Type K with Brazed Joints

Description	Craft@Hrs	Unit	Material $	Labor $	Equipment $	Total $
Type K copper flush bushing with brazed joints						
½" x 3/8"	P1@.148	Ea	1.41	5.51	—	6.92
¾" x ½"	P1@.155	Ea	2.41	5.77	—	8.18
1" x ¾"	P1@.206	Ea	6.24	7.67	—	13.91
1" x ½"	P1@.206	Ea	7.12	7.67	—	14.79
1¼" x 1"	P1@.258	Ea	7.40	9.60	—	17.00
1½" x 1¼"	P1@.308	Ea	9.42	11.50	—	20.92
2" x 1½"	P1@.390	Ea	17.20	14.50	—	31.70
Type K copper union with brazed joints						
½"	P1@.146	Ea	5.67	5.43	—	11.10
¾"	P1@.205	Ea	6.76	7.63	—	14.39
1"	P1@.263	Ea	15.20	9.79	—	24.99
1¼"	P1@.322	Ea	27.60	12.00	—	39.60
1½"	P1@.380	Ea	33.90	14.10	—	48.00
2"	P1@.507	Ea	48.90	18.90	—	67.80
Type K copper dielectric union with brazed joints						
½"	P1@.146	Ea	5.31	5.43	—	10.74
¾"	P1@.205	Ea	5.31	7.63	—	12.94
1"	P1@.263	Ea	11.00	9.79	—	20.79
1¼"	P1@.322	Ea	18.70	12.00	—	30.70
1½"	P1@.380	Ea	25.80	14.10	—	39.90
2"	P1@.507	Ea	34.40	18.90	—	53.30
Type K copper pressure cap with brazed joints						
½"	P1@.083	Ea	.47	3.09	—	3.56
¾"	P1@.116	Ea	.94	4.32	—	5.26
1"	P1@.149	Ea	2.13	5.55	—	7.68
1¼"	P1@.182	Ea	3.06	6.77	—	9.83
1½"	P1@.215	Ea	4.46	8.00	—	12.46
2"	P1@.286	Ea	7.77	10.60	—	18.37
2½"	P1@.352	Ea	19.70	13.10	—	32.80
3"	P1@.418	Ea	30.80	15.60	—	46.40
4"	P1@.550	Ea	47.40	20.50	—	67.90
Type K copper coupling with brazed joints						
½"	P1@.129	Ea	.22	4.80	—	5.02
¾"	P1@.181	Ea	.55	6.74	—	7.29
1"	P1@.232	Ea	2.02	8.64	—	10.66
1¼"	P1@.284	Ea	3.53	10.60	—	14.13
1½"	P1@.335	Ea	5.10	12.50	—	17.60
2"	P1@.447	Ea	7.50	16.60	—	24.10
2½"	P1@.550	Ea	17.70	20.50	—	38.20
3"	P1@.654	Ea	27.00	24.30	—	51.30
4"	P1@.860	Ea	52.90	32.00	—	84.90

Copper, Type K with Brazed Joints

Description	Craft@Hrs	Unit	Material $	Labor $	Equipment $	Total $

Class 125 bronze body gate valve, brazed joints

½"	P1@.240	Ea	18.80	8.93	—	27.73
¾"	P1@.300	Ea	23.50	11.20	—	34.70
1"	P1@.360	Ea	33.00	13.40	—	46.40
1¼"	P1@.480	Ea	41.50	17.90	—	59.40
1½"	P1@.540	Ea	56.00	20.10	—	76.10
2"	P1@.600	Ea	93.60	22.30	—	115.90
2½"	P1@1.00	Ea	155.00	37.20	—	192.20
3"	P1@1.50	Ea	222.00	55.80	—	277.80

Class 125 iron body gate valve, flanged joints

2"	P1@.500	Ea	358.00	18.60	—	376.60
2½"	P1@.600	Ea	483.00	22.30	—	505.30
3"	P1@.750	Ea	524.00	27.90	—	551.90
4"	P1@1.35	Ea	769.00	50.20	—	819.20
5"	P1@2.00	Ea	1,470.00	74.40	—	1,544.40
6"	P1@2.50	Ea	1,470.00	93.10	—	1,563.10

Class 125 bronze body globe valve, with brazed joints

½"	P1@.240	Ea	34.80	8.93	—	43.73
¾"	P1@.300	Ea	46.30	11.20	—	57.50
1"	P1@.360	Ea	79.80	13.40	—	93.20
1¼"	P1@.480	Ea	112.00	17.90	—	129.90
1½"	P1@.540	Ea	152.00	20.10	—	172.10
2"	P1@.600	Ea	246.00	22.30	—	268.30

Class 125 iron body globe valve, flanged joints

2½"	P1@.650	Ea	406.00	24.20	—	430.20
3"	P1@.750	Ea	487.00	27.90	—	514.90
4"	P1@1.35	Ea	657.00	50.20	—	707.20

200 PSIG iron body butterfly valve, lug-type, lever operated

2"	P1@.450	Ea	161.00	16.70	—	177.70
2½"	P1@.450	Ea	167.00	16.70	—	183.70
3"	P1@.550	Ea	176.00	20.50	—	196.50
4"	P1@.550	Ea	219.00	20.50	—	239.50

200 PSIG iron body butterfly valve, wafer-type, lever operated

2"	P1@.450	Ea	147.00	16.70	—	163.70
2½"	P1@.450	Ea	150.00	16.70	—	166.70
3"	P1@.550	Ea	161.00	20.50	—	181.50
4"	P1@.550	Ea	195.00	20.50	—	215.50

Copper, Type K with Brazed Joints

Description	Craft@Hrs	Unit	Material $	Labor $	Equipment $	Total $

Class 125 bronze body 2-piece ball valve, brazed joints

½"	P1@.240	Ea	10.40	8.93	—	19.33
¾"	P1@.300	Ea	13.80	11.20	—	25.00
1"	P1@.360	Ea	24.00	13.40	—	37.40
1¼"	P1@.480	Ea	40.00	17.90	—	57.90
1½"	P1@.540	Ea	54.10	20.10	—	74.20
2"	P1@.600	Ea	68.60	22.30	—	90.90
3"	P1@1.50	Ea	470.00	55.80	—	525.80
4"	P1@1.80	Ea	614.00	67.00	—	681.00

Class 125 bronze body swing check valve, brazed joints

½"	P1@.240	Ea	20.80	8.93	—	29.73
¾"	P1@.300	Ea	29.70	11.20	—	40.90
1"	P1@.360	Ea	38.80	13.40	—	52.20
1¼"	P1@.480	Ea	55.40	17.90	—	73.30
1½"	P1@.540	Ea	78.10	20.10	—	98.20
2"	P1@.600	Ea	129.00	22.30	—	151.30
2½"	P1@1.00	Ea	261.00	37.20	—	298.20
3"	P1@1.50	Ea	375.00	55.80	—	430.80

Class 125 iron body swing check valve, flanged joints

2"	P1@.500	Ea	182.00	18.60	—	200.60
2½"	P1@.600	Ea	230.00	22.30	—	252.30
3"	P1@.750	Ea	287.00	27.90	—	314.90
4"	P1@1.35	Ea	421.00	50.20	—	471.20

Class 125 iron body silent check valve, wafer-type

2"	P1@.500	Ea	133.00	18.60	—	151.60
2½"	P1@.600	Ea	148.00	22.30	—	170.30
3"	P1@.750	Ea	170.00	27.90	—	197.90
4"	P1@1.35	Ea	227.00	50.20	—	277.20

Class 125 bronze body strainer, threaded joints

½"	P1@.230	Ea	32.00	8.56	—	40.56
¾"	P1@.260	Ea	42.00	9.68	—	51.68
1"	P1@.330	Ea	51.50	12.30	—	63.80
1¼"	P1@.440	Ea	72.00	16.40	—	88.40
1½"	P1@.495	Ea	108.00	18.40	—	126.40
2"	P1@.550	Ea	188.00	20.50	—	208.50

Class 125 iron body strainer, flanged joints

2"	P1@.500	Ea	144.00	18.60	—	162.60
2½"	P1@.600	Ea	160.00	22.30	—	182.30
3"	P1@.750	Ea	186.00	27.90	—	213.90
4"	P1@1.35	Ea	314.00	50.20	—	364.20

Copper, Type K with Brazed Joints

Description	Craft@Hrs	Unit	Material $	Labor $	Equipment $	Total $

Installation of 2-way control valve, threaded joints

½"	P1@.210	Ea	—	7.82	—	7.82
¾"	P1@.275	Ea	—	10.20	—	10.20
1"	P1@.350	Ea	—	13.00	—	13.00
1¼"	P1@.430	Ea	—	16.00	—	16.00
1½"	P1@.505	Ea	—	18.80	—	18.80
2"	P1@.675	Ea	—	25.10	—	25.10
2½"	P1@.830	Ea	—	30.90	—	30.90
3"	P1@.990	Ea	—	36.80	—	36.80

Installation of 3-way control valve, threaded joints

½"	P1@.260	Ea	—	9.68	—	9.68
¾"	P1@.365	Ea	—	13.60	—	13.60
1"	P1@.475	Ea	—	17.70	—	17.70
1¼"	P1@.575	Ea	—	21.40	—	21.40
1½"	P1@.680	Ea	—	25.30	—	25.30
2"	P1@.910	Ea	—	33.90	—	33.90
2½"	P1@1.12	Ea	—	41.70	—	41.70
3"	P1@1.33	Ea	—	49.50	—	49.50

Companion flange, 150 pound cast brass

2"	P1@.290	Ea	294.00	10.80	—	304.80
2½"	P1@.380	Ea	365.00	14.10	—	379.10
3"	P1@.460	Ea	365.00	17.10	—	382.10
4"	P1@.600	Ea	583.00	22.30	—	605.30

Bolt and gasket sets

2"	P1@.500	Ea	4.29	18.60	—	22.89
2½"	P1@.650	Ea	5.01	24.20	—	29.21
3"	P1@.750	Ea	8.50	27.90	—	36.40
4"	P1@1.00	Ea	14.20	37.20	—	51.40

Thermometer with well

7"	P1@.250	Ea	186.00	9.31	—	195.31
9"	P1@.250	Ea	192.00	9.31	—	201.31

Dial-type pressure gauge

2½"	P1@.200	Ea	38.50	7.44	—	45.94
3½"	P1@.200	Ea	51.20	7.44	—	58.64

Copper, Type K with Brazed Joints

Description	Craft@Hrs	Unit	Material $	Labor $	Equipment $	Total $
Pressure/temperature tap						
Tap	P1@.150	Ea	17.70	5.58	—	23.28
Hanger with swivel assembly						
½"	P1@.250	Ea	4.33	9.31	—	13.64
¾"	P1@.250	Ea	4.79	9.31	—	14.10
1"	P1@.250	Ea	5.00	9.31	—	14.31
1¼"	P1@.300	Ea	5.13	11.20	—	16.33
1½"	P1@.300	Ea	5.52	11.20	—	16.72
2"	P1@.300	Ea	5.76	11.20	—	16.96
2½"	P1@.350	Ea	7.84	13.00	—	20.84
3"	P1@.350	Ea	9.69	13.00	—	22.69
4"	P1@.350	Ea	10.70	13.00	—	23.70
Riser clamp						
½"	P1@.100	Ea	3.20	3.72	—	6.92
¾"	P1@.100	Ea	4.73	3.72	—	8.45
1"	P1@.100	Ea	4.77	3.72	—	8.49
1¼"	P1@.105	Ea	5.73	3.91	—	9.64
1½"	P1@.110	Ea	6.04	4.09	—	10.13
2"	P1@.115	Ea	6.43	4.28	—	10.71
2½"	P1@.120	Ea	6.73	4.47	—	11.20
3"	P1@.120	Ea	7.31	4.47	—	11.78
4"	P1@.125	Ea	9.33	4.65	—	13.98

Copper, Type K with Soft-Soldered Joints

Type K hard-drawn copper pipe with wrought copper fittings and soft-soldered joints is used in a wide variety of plumbing and HVAC systems such as potable water, heating hot water, chilled water and A.C. condensate.

Soft-soldered joints are those made with solders having melting points in the 350 degree F. to 500 degree F. range. Maximum working pressure/temperature relationships for soft-soldered joints are approximately as follows:

Maximum Working Pressures (PSIG)*				
Soft-solder Type	Water Temperature (degrees F.)	Nominal Pipe Size (inches)		
		Up to 1	1¼ to 2	2½ to 4
50-50 tin-lead**	100	200	175	150
	150	150	125	100
	200	100	90	75
	250	85	75	50
95-5 tin-antimony	100	500	400	300
	150	400	350	275
	200	300	250	200
	250	200	175	150

*For copper pipe and solder-type fittings using soft-solders melting at approximately 350 degrees F. to 500 degrees F.
**The use of any solder containing lead is not allowed in potable water systems.

This section has been arranged to save the estimator's time by including all normally-used system components such as pipe, fittings, valves, hanger assemblies, riser clamps and miscellaneous items under one heading. Additional items can be found under "Plumbing and Piping Specialties." The cost estimates in this section are based on the conditions, limitations and wage rates described in the section "How to Use This Book" beginning on page 5.

Description	Craft@Hrs	Unit	Material $	Labor $	Equipment $	Total $

Type K copper pipe with soft-soldered joints installed horizontally.
Complete installation including six to twelve wrought copper tees and six to twelve wrought copper elbows every 100', and hangers spaced to meet plumbing code (*a tee & elbow every 8.3' for ½" dia., a tee and elbow every 16.6' for 4" dia.*)

Description	Craft@Hrs	Unit	Material $	Labor $	Equipment $	Total $
½"	P1@.112	LF	3.65	4.17	—	7.82
¾"	P1@.122	LF	6.38	4.54	—	10.92
1"	P1@.142	LF	10.10	5.29	—	15.39
1¼"	P1@.162	LF	13.90	6.03	—	19.93
1½"	P1@.182	LF	17.20	6.77	—	23.97
2"	P1@.212	LF	29.00	7.89	—	36.89
2½"	P1@.242	LF	47.80	9.01	—	56.81
3"	P1@.272	LF	55.20	10.10	—	65.30
4"	P1@.312	LF	100.00	11.60	—	111.60

Copper, Type K with Soft-Soldered Joints

Description	Craft@Hrs	Unit	Material $	Labor $	Equipment $	Total $

Type K copper pipe with soft-soldered joints, installed risers.
Complete installation including a wrought copper reducing tee every floor and a riser clamp every other floor.

Size	Craft@Hrs	Unit	Material $	Labor $	Equipment $	Total $
½"	P1@.062	LF	3.29	2.31	—	5.60
¾"	P1@.072	LF	5.95	2.68	—	8.63
1"	P1@.077	LF	9.06	2.87	—	11.93
1¼"	P1@.082	LF	12.50	3.05	—	15.55
1½"	P1@.102	LF	14.90	3.80	—	18.70
2"	P1@.132	LF	25.30	4.91	—	30.21
2½"	P1@.152	LF	37.70	5.66	—	43.36
3"	P1@.172	LF	52.80	6.40	—	59.20
4"	P1@.192	LF	82.80	7.15	—	89.95

Type K copper pipe with soft-soldered joints, pipe only

Size	Craft@Hrs	Unit	Material $	Labor $	Equipment $	Total $
½"	P1@.032	LF	2.77	1.19	—	3.96
¾"	P1@.035	LF	5.13	1.30	—	6.43
1"	P1@.038	LF	6.46	1.41	—	7.87
1¼"	P1@.042	LF	8.63	1.56	—	10.19
1½"	P1@.046	LF	9.43	1.71	—	11.14
2"	P1@.053	LF	16.80	1.97	—	18.77
2½"	P1@.060	LF	21.90	2.23	—	24.13
3"	P1@.066	LF	28.70	2.46	—	31.16
4"	P1@.080	LF	51.30	2.98	—	54.28

Type K copper 45-degree ell C x C with soft-soldered joints

Size	Craft@Hrs	Unit	Material $	Labor $	Equipment $	Total $
½"	P1@.107	Ea	.82	3.98	—	4.80
¾"	P1@.150	Ea	1.55	5.58	—	7.13
1"	P1@.193	Ea	4.57	7.18	—	11.75
1¼"	P1@.236	Ea	6.21	8.78	—	14.99
1½"	P1@.278	Ea	9.23	10.30	—	19.53
2"	P1@.371	Ea	14.60	13.80	—	28.40
2½"	P1@.457	Ea	37.60	17.00	—	54.60
3"	P1@.543	Ea	49.50	20.20	—	69.70
4"	P1@.714	Ea	95.40	26.60	—	122.00

Copper, Type K with Soft-Soldered Joints

Description	Craft@Hrs	Unit	Material $	Labor $	Equipment $	Total $

Type K copper 90-degree ell C x C with soft-soldered joints

Description	Craft@Hrs	Unit	Material $	Labor $	Equipment $	Total $
½"	P1@.107	Ea	.50	3.98	—	4.48
¾"	P1@.150	Ea	1.01	5.58	—	6.59
1"	P1@.193	Ea	3.49	7.18	—	10.67
1¼"	P1@.236	Ea	5.68	8.78	—	14.46
1½"	P1@.278	Ea	8.48	10.30	—	18.78
2"	P1@.371	Ea	14.90	13.80	—	28.70
2½"	P1@.457	Ea	30.90	17.00	—	47.90
3"	P1@.543	Ea	39.10	20.20	—	59.30
4"	P1@.714	Ea	83.30	26.60	—	109.90

Type K copper 90-degree ell ftg. x C with soft-soldered joint

Description	Craft@Hrs	Unit	Material $	Labor $	Equipment $	Total $
½"	P1@.107	Ea	.77	3.98	—	4.75
¾"	P1@.150	Ea	1.66	5.58	—	7.24
1"	P1@.193	Ea	4.94	7.18	—	12.12
1¼"	P1@.236	Ea	6.86	8.78	—	15.64
1½"	P1@.278	Ea	10.20	10.30	—	20.50
2"	P1@.371	Ea	20.30	13.80	—	34.10
2½"	P1@.457	Ea	42.20	17.00	—	59.20
3"	P1@.543	Ea	47.10	20.20	—	67.30
4"	P1@.714	Ea	95.40	26.60	—	122.00

Type K copper tee C x C x C with soft-soldered joints

Description	Craft@Hrs	Unit	Material $	Labor $	Equipment $	Total $
½"	P1@.129	Ea	.77	4.80	—	5.57
¾"	P1@.181	Ea	2.09	6.74	—	8.83
1"	P1@.233	Ea	7.85	8.67	—	16.52
1¼"	P1@.285	Ea	12.50	10.60	—	23.10
1½"	P1@.337	Ea	18.00	12.50	—	30.50
2"	P1@.449	Ea	28.00	16.70	—	44.70
2½"	P1@.552	Ea	57.20	20.50	—	77.70
3"	P1@.656	Ea	87.90	24.40	—	112.30
4"	P1@.863	Ea	146.00	32.10	—	178.10

Copper, Type K with Soft-Soldered Joints

Description	Craft@Hrs	Unit	Material $	Labor $	Equipment $	Total $
Type K copper branch reducing tee C x C x C with soft-soldered joints						
½" x 3/8"	P1@.129	Ea	1.57	4.80	—	6.37
¾" x ½"	P1@.182	Ea	3.34	6.77	—	10.11
1" x ½"	P1@.195	Ea	6.53	7.26	—	13.79
1" x ¾"	P1@.219	Ea	6.53	8.15	—	14.68
1¼" x ½"	P1@.225	Ea	9.83	8.37	—	18.20
1¼" x ¾"	P1@.247	Ea	9.83	9.19	—	19.02
1¼" x 1"	P1@.268	Ea	9.83	9.97	—	19.80
1½" x ½"	P1@.272	Ea	12.10	10.10	—	22.20
1½" x ¾"	P1@.287	Ea	12.10	10.70	—	22.80
1½" x 1"	P1@.302	Ea	12.10	11.20	—	23.30
1½" x 1¼"	P1@.317	Ea	12.10	11.80	—	23.90
2" x ½"	P1@.325	Ea	15.80	12.10	—	27.90
2" x ¾"	P1@.348	Ea	15.80	13.00	—	28.80
2" x 1"	P1@.373	Ea	15.80	13.90	—	29.70
2" x 1¼"	P1@.398	Ea	15.80	14.80	—	30.60
2" x 1½"	P1@.422	Ea	15.80	15.70	—	31.50
2½" x ½"	P1@.437	Ea	47.70	16.30	—	64.00
2½" x ¾"	P1@.453	Ea	47.70	16.90	—	64.60
2½" x 1"	P1@.469	Ea	47.70	17.50	—	65.20
2½" x 1½"	P1@.502	Ea	47.70	18.70	—	66.40
2½" x 2"	P1@.519	Ea	47.70	19.30	—	67.00
3" x 1¼"	P1@.525	Ea	53.90	19.50	—	73.40
3" x 2"	P1@.585	Ea	55.80	21.80	—	77.60
3" x 2½"	P1@.617	Ea	55.80	23.00	—	78.80
4" x 1¼"	P1@.625	Ea	73.10	23.30	—	96.40
4" x 1½"	P1@.671	Ea	73.10	25.00	—	98.10
4" x 2"	P1@.718	Ea	73.10	26.70	—	99.80
4" x 2½"	P1@.765	Ea	73.10	28.50	—	101.60
4" x 3"	P1@.811	Ea	73.10	30.20	—	103.30

Copper, Type K with Soft-Soldered Joints

Description	Craft@Hrs	Unit	Material $	Labor $	Equipment $	Total $

Type K copper reducer with soft-soldered joints

Description	Craft@Hrs	Unit	Material $	Labor $	Equipment $	Total $
½" x 3/8"	P1@.124	Ea	.59	4.62	—	5.21
¾" x ½"	P1@.129	Ea	1.09	4.80	—	5.89
1" x ¾"	P1@.172	Ea	1.91	6.40	—	8.31
1¼" x ½"	P1@.172	Ea	3.21	6.40	—	9.61
1¼" x ¾"	P1@.193	Ea	3.21	7.18	—	10.39
1¼" x 1"	P1@.215	Ea	3.21	8.00	—	11.21
1½" x ½"	P1@.193	Ea	4.90	7.18	—	12.08
1½" x ¾"	P1@.214	Ea	4.90	7.97	—	12.87
1½" x 1"	P1@.236	Ea	4.90	8.78	—	13.68
1½" x 1¼"	P1@.257	Ea	4.90	9.57	—	14.47
2" x ½"	P1@.239	Ea	7.38	8.90	—	16.28
2" x ¾"	P1@.261	Ea	7.38	9.71	—	17.09
2" x 1"	P1@.282	Ea	7.38	10.50	—	17.88
2" x 1¼"	P1@.304	Ea	7.38	11.30	—	18.68
2" x 1½"	P1@.325	Ea	7.38	12.10	—	19.48
2½" x 1"	P1@.325	Ea	20.10	12.10	—	32.20
2½" x 1¼"	P1@.347	Ea	20.10	12.90	—	33.00
2½" x 1½"	P1@.368	Ea	20.10	13.70	—	33.80
2½" x 2"	P1@.414	Ea	20.10	15.40	—	35.50
3" x 2"	P1@.457	Ea	23.50	17.00	—	40.50
3" x 2½"	P1@.500	Ea	23.50	18.60	—	42.10
4" x 2"	P1@.543	Ea	45.30	20.20	—	65.50
4" x 2½"	P1@.586	Ea	45.30	21.80	—	67.10
4" x 3"	P1@.629	Ea	45.30	23.40	—	68.70

Type K copper adapter C x MPT with soft-soldered joint

Description	Craft@Hrs	Unit	Material $	Labor $	Equipment $	Total $
½"	P1@.075	Ea	1.39	2.79	—	4.18
¾"	P1@.105	Ea	2.18	3.91	—	6.09
1"	P1@.134	Ea	5.41	4.99	—	10.40
1¼"	P1@.164	Ea	10.10	6.10	—	16.20
1½"	P1@.194	Ea	11.40	7.22	—	18.62
2"	P1@.259	Ea	14.20	9.64	—	23.84
2½"	P1@.319	Ea	46.10	11.90	—	58.00
3"	P1@.378	Ea	76.40	14.10	—	90.50

Copper, Type K with Soft-Soldered Joints

Description	Craft@Hrs	Unit	Material $	Labor $	Equipment $	Total $

Type K copper adapter C x FPT with soft-soldered joint

Description	Craft@Hrs	Unit	Material $	Labor $	Equipment $	Total $
½"	P1@.075	Ea	2.32	2.79	—	5.11
¾"	P1@.105	Ea	4.21	3.91	—	8.12
1"	P1@.134	Ea	6.35	4.99	—	11.34
1¼"	P1@.164	Ea	13.40	6.10	—	19.50
1½"	P1@.194	Ea	16.10	7.22	—	23.32
2"	P1@.259	Ea	24.00	9.64	—	33.64
2½"	P1@.319	Ea	60.50	11.90	—	72.40
3"	P1@.378	Ea	83.20	14.10	—	97.30

Type K copper flush bushing with soft-soldered joints

Description	Craft@Hrs	Unit	Material $	Labor $	Equipment $	Total $
½" x 3/8"	P1@.124	Ea	1.41	4.62	—	6.03
¾" x ½"	P1@.129	Ea	2.41	4.80	—	7.21
1" x ¾"	P1@.172	Ea	6.24	6.40	—	12.64
1" x ½"	P1@.172	Ea	7.12	6.40	—	13.52
1¼" x 1"	P1@.215	Ea	7.40	8.00	—	15.40
1½" x 1¼"	P1@.257	Ea	9.42	9.57	—	18.99
2" x 1½"	P1@.325	Ea	17.20	12.10	—	29.30

Type K copper union with soft-soldered joint

Description	Craft@Hrs	Unit	Material $	Labor $	Equipment $	Total $
½"	P1@.121	Ea	5.67	4.50	—	10.17
¾"	P1@.170	Ea	6.76	6.33	—	13.09
1"	P1@.218	Ea	15.20	8.11	—	23.31
1¼"	P1@.267	Ea	27.60	9.94	—	37.54
1½"	P1@.315	Ea	33.90	11.70	—	45.60
2"	P1@.421	Ea	48.90	15.70	—	64.60

Type K copper dielectric union with soft-soldered joint

Description	Craft@Hrs	Unit	Material $	Labor $	Equipment $	Total $
½"	P1@.121	Ea	5.31	4.50	—	9.81
¾"	P1@.170	Ea	5.31	6.33	—	11.64
1"	P1@.218	Ea	11.00	8.11	—	19.11
1¼"	P1@.267	Ea	18.70	9.94	—	28.64
1½"	P1@.315	Ea	25.80	11.70	—	37.50
2"	P1@.421	Ea	34.40	15.70	—	50.10

Type K copper cap with soft-soldered joint

Description	Craft@Hrs	Unit	Material $	Labor $	Equipment $	Total $
½"	P1@.069	Ea	.47	2.57	—	3.04
¾"	P1@.096	Ea	.94	3.57	—	4.51
1"	P1@.124	Ea	2.13	4.62	—	6.75
1¼"	P1@.151	Ea	3.06	5.62	—	8.68
1½"	P1@.178	Ea	4.46	6.63	—	11.09
2"	P1@.237	Ea	7.77	8.82	—	16.59
2½"	P1@.292	Ea	19.70	10.90	—	30.60
3"	P1@.347	Ea	30.80	12.90	—	43.70
4"	P1@.457	Ea	47.40	17.00	—	64.40

Copper, Type K with Soft-Soldered Joints

Description	Craft@Hrs	Unit	Material $	Labor $	Equipment $	Total $

Type K copper coupling with soft-soldered joints

Description	Craft@Hrs	Unit	Material $	Labor $	Equipment $	Total $
½"	P1@.107	Ea	.30	3.98	—	4.28
¾"	P1@.150	Ea	.76	5.58	—	6.34
1"	P1@.193	Ea	2.03	7.18	—	9.21
1¼"	P1@.236	Ea	3.54	8.78	—	12.32
1½"	P1@.278	Ea	5.12	10.30	—	15.42
2"	P1@.371	Ea	7.52	13.80	—	21.32
2½"	P1@.457	Ea	17.80	17.00	—	34.80
3"	P1@.543	Ea	27.10	20.20	—	47.30
4"	P1@.714	Ea	53.00	26.60	—	79.60

Class 125 bronze body gate valve, soft-soldered joints

Description	Craft@Hrs	Unit	Material $	Labor $	Equipment $	Total $
½"	P1@.200	Ea	18.60	7.44	—	26.04
¾"	P1@.249	Ea	23.50	9.27	—	32.77
1"	P1@.299	Ea	33.00	11.10	—	44.10
1¼"	P1@.398	Ea	41.50	14.80	—	56.30
1½"	P1@.448	Ea	55.70	16.70	—	72.40
2"	P1@.498	Ea	93.00	18.50	—	111.50
2½"	P1@.830	Ea	154.00	30.90	—	184.90
3"	P1@1.24	Ea	221.00	46.20	—	267.20

Class 125 iron body gate valve, flanged joints

Description	Craft@Hrs	Unit	Material $	Labor $	Equipment $	Total $
2"	P1@.500	Ea	358.00	18.60	—	376.60
2½"	P1@.600	Ea	483.00	22.30	—	505.30
3"	P1@.750	Ea	524.00	27.90	—	551.90
4"	P1@1.35	Ea	769.00	50.20	—	819.20
5"	P1@2.00	Ea	1,470.00	74.40	—	1,544.40
6"	P1@2.50	Ea	1,470.00	93.10	—	1,563.10

Class 125 bronze body globe valve, with soft-soldered joints

Description	Craft@Hrs	Unit	Material $	Labor $	Equipment $	Total $
½"	P1@.200	Ea	34.80	7.44	—	42.24
¾"	P1@.249	Ea	46.30	9.27	—	55.57
1"	P1@.299	Ea	79.80	11.10	—	90.90
1¼"	P1@.398	Ea	112.00	14.80	—	126.80
1½"	P1@.448	Ea	152.00	16.70	—	168.70
2"	P1@.498	Ea	246.00	18.50	—	264.50

Class 125 iron body globe valve, flanged ends

Description	Craft@Hrs	Unit	Material $	Labor $	Equipment $	Total $
2½"	P1@.600	Ea	406.00	22.30	—	428.30
3"	P1@.750	Ea	487.00	27.90	—	514.90
4"	P1@1.35	Ea	653.00	50.20	—	703.20

Copper, Type K with Soft-Soldered Joints

Description	Craft@Hrs	Unit	Material $	Labor $	Equipment $	Total $

200 PSIG iron body butterfly valve, lug-type, lever operated

2"	P1@.450	Ea	160.00	16.70	—	176.70
2½"	P1@.450	Ea	166.00	16.70	—	182.70
3"	P1@.550	Ea	175.00	20.50	—	195.50
4"	P1@.550	Ea	218.00	20.50	—	238.50

200 PSIG iron body butterfly valve, wafer-type, lever operated

2"	P1@.450	Ea	147.00	16.70	—	163.70
2½"	P1@.450	Ea	150.00	16.70	—	166.70
3"	P1@.550	Ea	161.00	20.50	—	181.50
4"	P1@.550	Ea	195.00	20.50	—	215.50

Class 125 bronze body 2-piece ball valve, soft-soldered joints

½"	P1@.200	Ea	10.40	7.44	—	17.84
¾"	P1@.249	Ea	12.50	9.27	—	21.77
1"	P1@.299	Ea	24.00	11.10	—	35.10
1¼"	P1@.398	Ea	40.00	14.80	—	54.80
1½"	P1@.448	Ea	54.10	16.70	—	70.80
2"	P1@.498	Ea	68.60	18.50	—	87.10
3"	P1@1.24	Ea	470.00	46.20	—	516.20
4"	P1@1.45	Ea	614.00	54.00	—	668.00

Class 125 bronze body swing check valve, soft-soldered joints

½"	P1@.200	Ea	20.80	7.44	—	28.24
¾"	P1@.249	Ea	29.70	9.27	—	38.97
1"	P1@.299	Ea	38.80	11.10	—	49.90
1¼"	P1@.398	Ea	55.40	14.80	—	70.20
1½"	P1@.448	Ea	78.10	16.70	—	94.80
2"	P1@.498	Ea	129.00	18.50	—	147.50
2½"	P1@.830	Ea	261.00	30.90	—	291.90
3"	P1@1.24	Ea	375.00	46.20	—	421.20

Class 125 iron body swing check valve, flanged ends

2"	P1@.500	Ea	182.00	18.60	—	200.60
2½"	P1@.600	Ea	230.00	22.30	—	252.30
3"	P1@.750	Ea	287.00	27.90	—	314.90
4"	P1@1.35	Ea	421.00	50.20	—	471.20

Class 125 iron body silent check valve, wafer-type

2"	P1@.500	Ea	133.00	18.60	—	151.60
2½"	P1@.600	Ea	148.00	22.30	—	170.30
3"	P1@.750	Ea	170.00	27.90	—	197.90
4"	P1@1.35	Ea	227.00	50.20	—	277.20

Copper, Type K with Soft-Soldered Joints

Description	Craft@Hrs	Unit	Material $	Labor $	Equipment $	Total $

Class 125 bronze body strainer, threaded ends

½"	P1@.230	Ea	32.00	8.56	—	40.56
¾"	P1@.260	Ea	42.00	9.68	—	51.68
1"	P1@.330	Ea	51.50	12.30	—	63.80
1¼"	P1@.440	Ea	72.00	16.40	—	88.40
1½"	P1@.495	Ea	108.00	18.40	—	126.40
2"	P1@.550	Ea	188.00	20.50	—	208.50

Class 125 iron body strainer, flanged ends

2"	P1@.500	Ea	144.00	18.60	—	162.60
2½"	P1@.600	Ea	160.00	22.30	—	182.30
3"	P1@.750	Ea	186.00	27.90	—	213.90
4"	P1@1.35	Ea	314.00	50.20	—	364.20

Installation of copper 2-way control valve, threaded joints

½"	P1@.210	Ea	—	7.82	—	7.82
¾"	P1@.275	Ea	—	10.20	—	10.20
1"	P1@.350	Ea	—	13.00	—	13.00
1¼"	P1@.430	Ea	—	16.00	—	16.00
1½"	P1@.505	Ea	—	18.80	—	18.80
2"	P1@.675	Ea	—	25.10	—	25.10
2½"	P1@.830	Ea	—	30.90	—	30.90
3"	P1@.990	Ea	—	36.80	—	36.80

Installation of copper 3-way control valve, threaded joints

½"	P1@.260	Ea	—	9.68	—	9.68
¾"	P1@.365	Ea	—	13.60	—	13.60
1"	P1@.475	Ea	—	17.70	—	17.70
1¼"	P1@.575	Ea	—	21.40	—	21.40
1½"	P1@.680	Ea	—	25.30	—	25.30
2"	P1@.910	Ea	—	33.90	—	33.90
2½"	P1@1.12	Ea	—	41.70	—	41.70
3"	P1@1.33	Ea	—	49.50	—	49.50

Companion flange, 150 pound cast brass

2"	P1@.290	Ea	294.00	10.80	—	304.80
2½"	P1@.380	Ea	365.00	14.10	—	379.10
3"	P1@.460	Ea	365.00	17.10	—	382.10
4"	P1@.600	Ea	583.00	22.30	—	605.30

Copper, Type K with Soft-Soldered Joints

Description	Craft@Hrs	Unit	Material $	Labor $	Equipment $	Total $
Bolt and gasket sets						
2"	P1@.500	Ea	4.30	18.60	—	22.90
2½"	P1@.650	Ea	5.05	24.20	—	29.25
3"	P1@.750	Ea	8.60	27.90	—	36.50
4"	P1@1.00	Ea	14.40	37.20	—	51.60
Thermometer with well						
7"	P1@.250	Ea	183.00	9.31	—	192.31
9"	P1@.250	Ea	188.00	9.31	—	197.31
Dial-type pressure gauge						
2½"	P1@.200	Ea	37.80	7.44	—	45.24
3½"	P1@.200	Ea	50.20	7.44	—	57.64
Pressure/temperature tap						
Tap	P1@.150	Ea	17.40	5.58	—	22.98
Hanger with swivel assembly						
½"	P1@.250	Ea	4.25	9.31	—	13.56
¾"	P1@.250	Ea	4.71	9.31	—	14.02
1"	P1@.250	Ea	4.92	9.31	—	14.23
1¼"	P1@.300	Ea	5.05	11.20	—	16.25
1½"	P1@.300	Ea	5.43	11.20	—	16.63
2"	P1@.300	Ea	5.66	11.20	—	16.86
2½"	P1@.350	Ea	7.70	13.00	—	20.70
3"	P1@.350	Ea	9.53	13.00	—	22.53
4"	P1@.350	Ea	10.50	13.00	—	23.50
Riser clamp						
½"	P1@.100	Ea	3.15	3.72	—	6.87
¾"	P1@.100	Ea	4.65	3.72	—	8.37
1"	P1@.100	Ea	4.69	3.72	—	8.41
1¼"	P1@.105	Ea	5.66	3.91	—	9.57
1½"	P1@.110	Ea	5.98	4.09	—	10.07
2"	P1@.115	Ea	6.33	4.28	—	10.61
2½"	P1@.120	Ea	6.66	4.47	—	11.13
3"	P1@.120	Ea	7.21	4.47	—	11.68
4"	P1@.125	Ea	8.92	4.65	—	13.57

Copper, Type L with Brazed Joints

Type L hard-drawn copper pipe with wrought copper fittings and brazed joints is used in a wide variety of plumbing and HVAC systems such as potable water, heating hot water, chilled water, compressed air, refrigerant and A.C. condensate.

Brazed joints are those made with silver or other alloy filler metals having melting points at, or above, 1,000 degrees F. Maximum working pressure/temperature relationships for brazed joints are approximately as follows:

Water Temperature (degrees F.)	Maximum Working Pressure (PSIG)*		
	Nominal Pipe Size (inches)		
	Up to 1	1¼ to 2	2½ to 4
Up to 350	270	190	150

*For copper pipe and solder-type fittings using brazing alloys melting at, or above, 1,000 degrees F.

This section has been arranged to save the estimator's time by including all normally-used system components such as pipe, fittings, valves, hanger assemblies, riser clamps and miscellaneous items under one heading. Additional items can be found under "Plumbing and Piping Specialties." The cost estimates in this section are based on the conditions, limitations and wage rates described in the section "How to Use This Book" beginning on page 5.

Description	Craft@Hrs	Unit	Material $	Labor $	Equipment $	Total $

Type L copper pipe with brazed joints installed horizontally. Complete installation including six to twelve wrought copper tees and six to twelve wrought copper elbows every 100', and hangers spaced to meet plumbing code *(a tee & elbow every 8.3' for ½" dia., a tee and elbow every 16.6' for 4" dia.)*

½"	P1@.116	LF	2.51	4.32	—	6.83
¾"	P1@.126	LF	3.86	4.69	—	8.55
1"	P1@.146	LF	6.94	5.43	—	12.37
1¼"	P1@.166	LF	9.87	6.18	—	16.05
1½"	P1@.186	LF	13.10	6.92	—	20.02
2"	P1@.216	LF	23.90	8.04	—	31.94
2½"	P1@.246	LF	42.20	9.16	—	51.36
3"	P1@.276	LF	58.40	10.30	—	68.70
4"	P1@.316	LF	84.30	11.80	—	96.10

Copper, Type L with Brazed Joints

Description	Craft@Hrs	Unit	Material $	Labor $	Equipment $	Total $

Type L copper pipe with brazed joints, installed risers. Complete installation including a wrought copper reducing tee every floor and a riser clamp every other floor.

Description	Craft@Hrs	Unit	Material $	Labor $	Equipment $	Total $
½"	P1@.063	LF	2.46	2.34	—	4.80
¾"	P1@.073	LF	3.89	2.72	—	6.61
1"	P1@.073	LF	6.60	2.72	—	9.32
1¼"	P1@.083	LF	9.31	3.09	—	12.40
1½"	P1@.103	LF	11.60	3.83	—	15.43
2"	P1@.133	LF	30.50	4.95	—	35.45
2½"	P1@.153	LF	32.90	5.69	—	38.59
3"	P1@.173	LF	42.70	6.44	—	49.14
4"	P1@.193	LF	63.50	7.18	—	70.68

Type L copper pipe with brazed joints, pipe only

Description	Craft@Hrs	Unit	Material $	Labor $	Equipment $	Total $
½"	P1@.032	LF	2.00	1.19	—	3.19
¾"	P1@.035	LF	3.16	1.30	—	4.46
1"	P1@.038	LF	4.37	1.41	—	5.78
1¼"	P1@.042	LF	6.28	1.56	—	7.84
1½"	P1@.046	LF	7.99	1.71	—	9.70
2"	P1@.053	LF	13.70	1.97	—	15.67
2½"	P1@.060	LF	20.10	2.23	—	22.33
3"	P1@.066	LF	27.10	2.46	—	29.56
4"	P1@.080	LF	42.50	2.98	—	45.48

Type L copper 45-degree ell C x C with brazed joints

Description	Craft@Hrs	Unit	Material $	Labor $	Equipment $	Total $
½"	P1@.129	Ea	.82	4.80	—	5.62
¾"	P1@.181	Ea	1.56	6.74	—	8.30
1"	P1@.232	Ea	4.61	8.64	—	13.25
1¼"	P1@.284	Ea	6.24	10.60	—	16.84
1½"	P1@.335	Ea	9.29	12.50	—	21.79
2"	P1@.447	Ea	14.70	16.60	—	31.30
2½"	P1@.550	Ea	37.60	20.50	—	58.10
3"	P1@.654	Ea	49.70	24.30	—	74.00
4"	P1@.860	Ea	95.40	32.00	—	127.40

Type L copper 90-degree ell C x C with brazed joints

Description	Craft@Hrs	Unit	Material $	Labor $	Equipment $	Total $
½"	P1@.129	Ea	.48	4.80	—	5.28
¾"	P1@.181	Ea	1.01	6.74	—	7.75
1"	P1@.232	Ea	3.52	8.64	—	12.16
1¼"	P1@.284	Ea	5.68	10.60	—	16.28
1½"	P1@.335	Ea	8.57	12.50	—	21.07
2"	P1@.447	Ea	14.90	16.60	—	31.50
2½"	P1@.550	Ea	31.10	20.50	—	51.60
3"	P1@.654	Ea	39.40	24.30	—	63.70
4"	P1@.860	Ea	83.30	32.00	—	115.30

Copper, Type L with Brazed Joints

Description	Craft@Hrs	Unit	Material $	Labor $	Equipment $	Total $

Type L copper 90-degree ell ftg. x C with brazed joint

½"	P1@.129	Ea	.77	4.80	—	5.57
¾"	P1@.181	Ea	1.66	6.74	—	8.40
1"	P1@.232	Ea	4.98	8.64	—	13.62
1¼"	P1@.284	Ea	6.94	10.60	—	17.54
1½"	P1@.335	Ea	10.30	12.50	—	22.80
2"	P1@.447	Ea	20.30	16.60	—	36.90
2½"	P1@.550	Ea	42.30	20.50	—	62.80
3"	P1@.654	Ea	47.60	24.30	—	71.90
4"	P1@.860	Ea	95.40	32.00	—	127.40

Type L copper tee C x C x C with brazed joints

½"	P1@.156	Ea	.81	5.81	—	6.62
¾"	P1@.204	Ea	2.19	7.59	—	9.78
1"	P1@.263	Ea	8.24	9.79	—	18.03
1¼"	P1@.322	Ea	13.10	12.00	—	25.10
1½"	P1@.380	Ea	19.20	14.10	—	33.30
2"	P1@.507	Ea	29.60	18.90	—	48.50
2½"	P1@.624	Ea	60.10	23.20	—	83.30
3"	P1@.741	Ea	92.90	27.60	—	120.50
4"	P1@.975	Ea	155.00	36.30	—	191.30

Type L copper branch reducing tee C x C x C with brazed joints

½" x 3/8"	P1@.129	Ea	1.57	4.80	—	6.37
¾" x ½"	P1@.182	Ea	3.34	6.77	—	10.11
1" x ½"	P1@.209	Ea	6.53	7.78	—	14.31
1" x ¾"	P1@.234	Ea	6.53	8.71	—	15.24
1¼" x ½"	P1@.241	Ea	9.83	8.97	—	18.80
1¼" x ¾"	P1@.264	Ea	9.83	9.83	—	19.66
1¼" x 1"	P1@.287	Ea	9.83	10.70	—	20.53
1½" x ½"	P1@.291	Ea	12.10	10.80	—	22.90
1½" x ¾"	P1@.307	Ea	12.10	11.40	—	23.50
1½" x 1"	P1@.323	Ea	12.10	12.00	—	24.10
1½" x 1¼"	P1@.339	Ea	12.10	12.60	—	24.70
2" x ½"	P1@.348	Ea	15.80	13.00	—	28.80
2" x ¾"	P1@.373	Ea	15.80	13.90	—	29.70
2" x 1"	P1@.399	Ea	15.80	14.90	—	30.70
2" x 1¼"	P1@.426	Ea	15.80	15.90	—	31.70
2" x 1½"	P1@.452	Ea	15.80	16.80	—	32.60
2½" x ½"	P1@.468	Ea	47.70	17.40	—	65.10
2½" x ¾"	P1@.485	Ea	47.70	18.10	—	65.80
2½" x 1"	P1@.502	Ea	47.70	18.70	—	66.40
2½" x 1½"	P1@.537	Ea	47.70	20.00	—	67.70
2½" x 2"	P1@.555	Ea	47.70	20.70	—	68.40
3" x 1¼"	P1@.562	Ea	53.90	20.90	—	74.80
3" x 2"	P1@.626	Ea	55.80	23.30	—	79.10
3" x 2½"	P1@.660	Ea	55.80	24.60	—	80.40
4" x 1¼"	P1@.669	Ea	97.30	24.90	—	122.20
4" x 1½"	P1@.718	Ea	97.30	26.70	—	124.00
4" x 2"	P1@.768	Ea	97.30	28.60	—	125.90
4" x 2½"	P1@.819	Ea	97.30	30.50	—	127.80
4" x 3"	P1@.868	Ea	97.30	32.30	—	129.60

Copper, Type L with Brazed Joints

Description	Craft@Hrs	Unit	Material $	Labor $	Equipment $	Total $
Type L copper reducer with brazed joints						
½" x 3/8"	P1@.148	Ea	.56	5.51	—	6.07
¾" x ½"	P1@.155	Ea	1.06	5.77	—	6.83
1" x ¾"	P1@.206	Ea	1.91	7.67	—	9.58
1¼" x ½"	P1@.206	Ea	3.21	7.67	—	10.88
1¼" x ¾"	P1@.232	Ea	3.21	8.64	—	11.85
1¼" x 1"	P1@.258	Ea	3.21	9.60	—	12.81
1½" x ½"	P1@.232	Ea	4.90	8.64	—	13.54
1½" x ¾"	P1@.257	Ea	4.90	9.57	—	14.47
1½" x 1"	P1@.283	Ea	4.90	10.50	—	15.40
1½" x 1¼"	P1@.308	Ea	4.90	11.50	—	16.40
2" x ½"	P1@.287	Ea	7.38	10.70	—	18.08
2" x ¾"	P1@.277	Ea	7.38	10.30	—	17.68
2" x 1"	P1@.338	Ea	7.38	12.60	—	19.98
2" x 1¼"	P1@.365	Ea	7.38	13.60	—	20.98
2" x 1½"	P1@.390	Ea	7.38	14.50	—	21.88
2½" x 1"	P1@.390	Ea	20.10	14.50	—	34.60
2½" x 1¼"	P1@.416	Ea	20.10	15.50	—	35.60
2½" x 1½"	P1@.442	Ea	20.10	16.50	—	36.60
2½" x 2"	P1@.497	Ea	20.10	18.50	—	38.60
3" x 1¼"	P1@.468	Ea	23.50	17.40	—	40.90
3" x 1½"	P1@.493	Ea	23.50	18.30	—	41.80
3" x 2"	P1@.548	Ea	23.50	20.40	—	43.90
3" x 2½"	P1@.600	Ea	23.50	22.30	—	45.80
4" x 2"	P1@.652	Ea	45.30	24.30	—	69.60
4" x 2½"	P1@.704	Ea	45.30	26.20	—	71.50
4" x 3"	P1@.755	Ea	45.30	28.10	—	73.40
Type L copper adapter C x MPT with brazed joint						
½"	P1@.090	Ea	1.39	3.35	—	4.74
¾"	P1@.126	Ea	2.18	4.69	—	6.87
1"	P1@.162	Ea	5.41	6.03	—	11.44
1¼"	P1@.198	Ea	10.10	7.37	—	17.47
1½"	P1@.234	Ea	11.40	8.71	—	20.11
2"	P1@.312	Ea	14.20	11.60	—	25.80
2½"	P1@.384	Ea	46.10	14.30	—	60.40
3"	P1@.456	Ea	76.40	17.00	—	93.40
Type L copper adapter C x FPT with brazed joint						
½"	P1@.090	Ea	2.32	3.35	—	5.67
¾"	P1@.126	Ea	4.21	4.69	—	8.90
1"	P1@.162	Ea	6.35	6.03	—	12.38
1¼"	P1@.198	Ea	13.40	7.37	—	20.77
1½"	P1@.234	Ea	16.10	8.71	—	24.81
2"	P1@.312	Ea	24.00	11.60	—	35.60
2½"	P1@.384	Ea	60.50	14.30	—	74.80
3"	P1@.456	Ea	83.20	17.00	—	100.20

Copper, Type L with Brazed Joints

Description	Craft@Hrs	Unit	Material $	Labor $	Equipment $	Total $

Type L copper flush bushing with brazed joints

Description	Craft@Hrs	Unit	Material $	Labor $	Equipment $	Total $
½" x 3/8"	P1@.148	Ea	1.41	5.51	—	6.92
¾" x ½"	P1@.155	Ea	2.41	5.77	—	8.18
1" x ¾"	P1@.206	Ea	6.24	7.67	—	13.91
1" x ½"	P1@.206	Ea	7.12	7.67	—	14.79
1¼" x 1"	P1@.258	Ea	7.40	9.60	—	17.00
1½" x 1¼"	P1@.308	Ea	9.42	11.50	—	20.92
2" x 1½"	P1@.390	Ea	17.20	14.50	—	31.70

Type L copper union with brazed joints

Description	Craft@Hrs	Unit	Material $	Labor $	Equipment $	Total $
½"	P1@.146	Ea	5.67	5.43	—	11.10
¾"	P1@.205	Ea	6.76	7.63	—	14.39
1"	P1@.263	Ea	15.20	9.79	—	24.99
1¼"	P1@.322	Ea	27.60	12.00	—	39.60
1½"	P1@.380	Ea	33.90	14.10	—	48.00
2"	P1@.507	Ea	48.90	18.90	—	67.80

Type L copper dielectric union with brazed joints

Description	Craft@Hrs	Unit	Material $	Labor $	Equipment $	Total $
½"	P1@.146	Ea	5.31	5.43	—	10.74
¾"	P1@.205	Ea	5.31	7.63	—	12.94
1"	P1@.263	Ea	11.00	9.79	—	20.79
1¼"	P1@.322	Ea	18.70	12.00	—	30.70
1½"	P1@.380	Ea	25.80	14.10	—	39.90
2"	P1@.507	Ea	34.40	18.90	—	53.30

Type L copper cap with brazed joints

Description	Craft@Hrs	Unit	Material $	Labor $	Equipment $	Total $
½"	P1@.083	Ea	.47	3.09	—	3.56
¾"	P1@.116	Ea	.94	4.32	—	5.26
1"	P1@.149	Ea	2.13	5.55	—	7.68
1¼"	P1@.182	Ea	3.06	6.77	—	9.83
1½"	P1@.215	Ea	4.46	8.00	—	12.46
2"	P1@.286	Ea	7.77	10.60	—	18.37
2½"	P1@.352	Ea	19.70	13.10	—	32.80
3"	P1@.418	Ea	30.80	15.60	—	46.40
4"	P1@.550	Ea	47.40	20.50	—	67.90

Type L copper coupling with brazed joints

Description	Craft@Hrs	Unit	Material $	Labor $	Equipment $	Total $
½"	P1@.129	Ea	.32	4.80	—	5.12
¾"	P1@.181	Ea	.75	6.74	—	7.49
1"	P1@.232	Ea	2.02	8.64	—	10.66
1¼"	P1@.284	Ea	3.53	10.60	—	14.13
1½"	P1@.335	Ea	5.10	12.50	—	17.60
2"	P1@.447	Ea	7.50	16.60	—	24.10
2½"	P1@.550	Ea	17.70	20.50	—	38.20
3"	P1@.654	Ea	27.00	24.30	—	51.30
4"	P1@.860	Ea	52.90	32.00	—	84.90

Copper, Type L with Brazed Joints

Description	Craft@Hrs	Unit	Material $	Labor $	Equipment $	Total $
Class 125 bronze body gate valve, brazed joints						
½"	P1@.240	Ea	18.80	8.93	—	27.73
¾"	P1@.300	Ea	23.50	11.20	—	34.70
1"	P1@.360	Ea	33.00	13.40	—	46.40
1¼"	P1@.480	Ea	41.50	17.90	—	59.40
1½"	P1@.540	Ea	56.00	20.10	—	76.10
2"	P1@.600	Ea	93.60	22.30	—	115.90
2½"	P1@1.00	Ea	155.00	37.20	—	192.20
3"	P1@1.50	Ea	222.00	55.80	—	277.80
Class 125 iron body gate valve, flanged joints						
2"	P1@.500	Ea	358.00	18.60	—	376.60
2½"	P1@.600	Ea	483.00	22.30	—	505.30
3"	P1@.750	Ea	524.00	27.90	—	551.90
4"	P1@1.35	Ea	769.00	50.20	—	819.20
5"	P1@2.00	Ea	1,470.00	74.40	—	1,544.40
6"	P1@2.50	Ea	1,470.00	93.10	—	1,563.10
Class 125 bronze body globe valve, with brazed joints						
½"	P1@.240	Ea	34.80	8.93	—	43.73
¾"	P1@.300	Ea	46.30	11.20	—	57.50
1"	P1@.360	Ea	79.80	13.40	—	93.20
1¼"	P1@.480	Ea	112.00	17.90	—	129.90
1½"	P1@.540	Ea	152.00	20.10	—	172.10
2"	P1@.600	Ea	246.00	22.30	—	268.30
Class 125 iron body globe valve, flanged joints						
2½"	P1@.650	Ea	406.00	24.20	—	430.20
3"	P1@.750	Ea	487.00	27.90	—	514.90
4"	P1@1.35	Ea	657.00	50.20	—	707.20
200 PSIG iron body butterfly valve, lug-type, lever operated						
2"	P1@.450	Ea	161.00	16.70	—	177.70
2½"	P1@.450	Ea	167.00	16.70	—	183.70
3"	P1@.550	Ea	176.00	20.50	—	196.50
4"	P1@.550	Ea	219.00	20.50	—	239.50
200 PSIG iron body butterfly valve, wafer-type, lever operated						
2"	P1@.450	Ea	147.00	16.70	—	163.70
2½"	P1@.450	Ea	150.00	16.70	—	166.70
3"	P1@.550	Ea	161.00	20.50	—	181.50
4"	P1@.550	Ea	195.00	20.50	—	215.50

Copper, Type L with Brazed Joints

Description	Craft@Hrs	Unit	Material $	Labor $	Equipment $	Total $

Class 125 bronze body 2-piece ball valve, brazed joints

Description	Craft@Hrs	Unit	Material $	Labor $	Equipment $	Total $
½"	P1@.240	Ea	10.40	8.93	—	19.33
¾"	P1@.300	Ea	13.80	11.20	—	25.00
1"	P1@.360	Ea	24.00	13.40	—	37.40
1¼"	P1@.480	Ea	40.00	17.90	—	57.90
1½"	P1@.540	Ea	54.10	20.10	—	74.20
2"	P1@.600	Ea	68.60	22.30	—	90.90
3"	P1@1.50	Ea	470.00	55.80	—	525.80
4"	P1@1.80	Ea	614.00	67.00	—	681.00

Class 125 bronze body swing check valve, brazed joints

Description	Craft@Hrs	Unit	Material $	Labor $	Equipment $	Total $
½"	P1@.240	Ea	20.80	8.93	—	29.73
¾"	P1@.300	Ea	29.70	11.20	—	40.90
1"	P1@.360	Ea	38.80	13.40	—	52.20
1¼"	P1@.480	Ea	55.40	17.90	—	73.30
1½"	P1@.540	Ea	78.10	20.10	—	98.20
2"	P1@.600	Ea	129.00	22.30	—	151.30
2½"	P1@1.00	Ea	261.00	37.20	—	298.20
3"	P1@1.50	Ea	375.00	55.80	—	430.80

Class 125 iron body swing check valve, flanged joints

Description	Craft@Hrs	Unit	Material $	Labor $	Equipment $	Total $
2"	P1@.500	Ea	182.00	18.60	—	200.60
2½"	P1@.600	Ea	230.00	22.30	—	252.30
3"	P1@.750	Ea	287.00	27.90	—	314.90
4"	P1@1.35	Ea	421.00	50.20	—	471.20

Class 125 iron body silent check valve, wafer-type

Description	Craft@Hrs	Unit	Material $	Labor $	Equipment $	Total $
2"	P1@.500	Ea	133.00	18.60	—	151.60
2½"	P1@.600	Ea	148.00	22.30	—	170.30
3"	P1@.750	Ea	170.00	27.90	—	197.90
4"	P1@1.35	Ea	227.00	50.20	—	277.20

Class 125 bronze body strainer, threaded joints

Description	Craft@Hrs	Unit	Material $	Labor $	Equipment $	Total $
½"	P1@.230	Ea	32.00	8.56	—	40.56
¾"	P1@.260	Ea	42.00	9.68	—	51.68
1"	P1@.330	Ea	51.50	12.30	—	63.80
1¼"	P1@.440	Ea	72.00	16.40	—	88.40
1½"	P1@.495	Ea	108.00	18.40	—	126.40
2"	P1@.550	Ea	188.00	20.50	—	208.50

Class 125 iron body strainer, flanged joints

Description	Craft@Hrs	Unit	Material $	Labor $	Equipment $	Total $
2"	P1@.500	Ea	144.00	18.60	—	162.60
2½"	P1@.600	Ea	160.00	22.30	—	182.30
3"	P1@.750	Ea	186.00	27.90	—	213.90
4"	P1@1.35	Ea	314.00	50.20	—	364.20

Copper, Type L with Brazed Joints

Description	Craft@Hrs	Unit	Material $	Labor $	Equipment $	Total $
Installation of 2-way control valve, threaded joints						
½"	P1@.210	Ea	—	7.82	—	7.82
¾"	P1@.275	Ea	—	10.20	—	10.20
1"	P1@.350	Ea	—	13.00	—	13.00
1¼"	P1@.430	Ea	—	16.00	—	16.00
1½"	P1@.505	Ea	—	18.80	—	18.80
2"	P1@.675	Ea	—	25.10	—	25.10
2½"	P1@.830	Ea	—	30.90	—	30.90
3"	P1@.990	Ea	—	36.80	—	36.80
Installation of 3-way control valve, threaded joints						
½"	P1@.260	Ea	—	9.68	—	9.68
¾"	P1@.365	Ea	—	13.60	—	13.60
1"	P1@.475	Ea	—	17.70	—	17.70
1¼"	P1@.575	Ea	—	21.40	—	21.40
1½"	P1@.680	Ea	—	25.30	—	25.30
2"	P1@.910	Ea	—	33.90	—	33.90
2½"	P1@1.12	Ea	—	41.70	—	41.70
3"	P1@1.33	Ea	—	49.50	—	49.50
Companion flange, 150 pound cast brass						
2"	P1@.290	Ea	291.00	10.80	—	301.80
2½"	P1@.380	Ea	359.00	14.10	—	373.10
3"	P1@.460	Ea	359.00	17.10	—	376.10
4"	P1@.600	Ea	571.00	22.30	—	593.30
Bolt and gasket sets						
2"	P1@.500	Ea	4.25	18.60	—	22.85
2½"	P1@.650	Ea	4.97	24.20	—	29.17
3"	P1@.750	Ea	8.43	27.90	—	36.33
4"	P1@1.00	Ea	14.10	37.20	—	51.30
Thermometer with well						
7"	P1@.250	Ea	183.00	9.31	—	192.31
9"	P1@.250	Ea	188.00	9.31	—	197.31
Dial-type pressure gauge						
2½"	P1@.200	Ea	38.30	7.44	—	45.74
3½"	P1@.200	Ea	50.50	7.44	—	57.94

Copper, Type L with Soft-Soldered Joints

Type L hard-drawn copper pipe with wrought copper fittings and soft-soldered joints is used in a wide variety of plumbing and HVAC systems such as potable water, heating hot water, chilled water, and A.C. condensate.

Soft-soldered joints are those made with solders having melting points in the 350 degree F. to 500 degree F. range. Maximum working temperature/pressure relationships for soft-soldered joints are approximately as follows:

Maximum Working Pressures (PSIG)*				
Soft-solder Type	Water Temperature (degrees F.)	Nominal Pipe Size (inches)		
		Up to 1	1¼ to 2	2½ x 4
50-50 tin-lead**	100	200	175	150
	150	150	125	100
	200	100	90	75
	250	85	75	50
95-5 tin-antimony	100	500	400	300
	150	400	350	275
	200	300	250	200
	250	200	175	150

*For copper pipe and solder-type fittings using soft-solders melting at approximately 350 degrees F. to 500 degrees F.

**The use of any solder containing lead is not allowed in potable water systems.

This section has been arranged to save the estimator's time by including all normally-used system components such as pipe, fittings, valves, hanger assemblies, riser clamps and miscellaneous items under one heading. Additional items can be found under "Plumbing and Piping Specialties." The cost estimates in this section are based on the conditions, limitations and wage rates described in the section "How to Use This Book" beginning on page 5.

Description	Craft@Hrs	Unit	Material $	Labor $	Equipment $	Total $

Type L copper pipe with soft-soldered joints installed horizontally.
Complete installation including eight wrought copper tees and eight wrought copper elbows every 100', and hangers spaced to meet plumbing code. Use these figures for preliminary estimates.

½"	P1@.111	LF	2.31	4.13	—	6.44
¾"	P1@.121	LF	3.58	4.50	—	8.08
1"	P1@.141	LF	6.47	5.25	—	11.72
1¼"	P1@.161	LF	9.23	5.99	—	15.22
1½"	P1@.181	LF	12.50	6.74	—	19.24
2"	P1@.211	LF	22.50	7.85	—	30.35
2½"	P1@.241	LF	38.90	8.97	—	47.87
3"	P1@.271	LF	54.40	10.10	—	64.50
4"	P1@.311	LF	78.50	11.60	—	90.10

Copper, Type L with Soft-Soldered Joints

Description	Craft@Hrs	Unit	Material $	Labor $	Equipment $	Total $

Type L copper pipe with soft-soldered joints, installed risers.
Complete installation including a wrought copper reducing tee every floor and a riser clamp every other floor.

½"	P1@.061	LF	2.51	2.27	—	4.78
¾"	P1@.071	LF	3.89	2.64	—	6.53
1"	P1@.076	LF	6.60	2.83	—	9.43
1¼"	P1@.081	LF	9.31	3.01	—	12.32
1½"	P1@.101	LF	11.60	3.76	—	15.36
2"	P1@.131	LF	30.50	4.88	—	35.38
2½"	P1@.151	LF	32.90	5.62	—	38.52
3"	P1@.171	LF	42.70	6.36	—	49.06
4"	P1@.191	LF	63.50	7.11	—	70.61

Type L copper pipe with soft-soldered joints, pipe only

½"	P1@.032	LF	2.00	1.19	—	3.19
¾"	P1@.035	LF	3.16	1.30	—	4.46
1"	P1@.038	LF	4.37	1.41	—	5.78
1¼"	P1@.042	LF	6.28	1.56	—	7.84
1½"	P1@.046	LF	7.99	1.71	—	9.70
2"	P1@.053	LF	13.70	1.97	—	15.67
2½"	P1@.060	LF	20.10	2.23	—	22.33
3"	P1@.066	LF	27.10	2.46	—	29.56
4"	P1@.080	LF	42.50	2.98	—	45.48

Type L copper 45-degree ell C x C with soft-soldered joints

½"	P1@.107	Ea	.82	3.98	—	4.80
¾"	P1@.150	Ea	1.56	5.58	—	7.14
1"	P1@.193	Ea	5.63	7.18	—	12.81
1¼"	P1@.236	Ea	7.63	8.78	—	16.41
1½"	P1@.278	Ea	9.29	10.30	—	19.59
2"	P1@.371	Ea	14.70	13.80	—	28.50
2½"	P1@.457	Ea	37.60	17.00	—	54.60
3"	P1@.543	Ea	49.70	20.20	—	69.90
4"	P1@.714	Ea	95.40	26.60	—	122.00

Type L copper 90-degree ell C x C with soft-soldered joints

½"	P1@.107	Ea	.48	3.98	—	4.46
¾"	P1@.150	Ea	1.01	5.58	—	6.59
1"	P1@.193	Ea	3.52	7.18	—	10.70
1¼"	P1@.236	Ea	5.68	8.78	—	14.46
1½"	P1@.278	Ea	8.57	10.30	—	18.87
2"	P1@.371	Ea	14.90	13.80	—	28.70
2½"	P1@.457	Ea	31.10	17.00	—	48.10
3"	P1@.543	Ea	39.40	20.20	—	59.60
4"	P1@.714	Ea	83.30	26.60	—	109.90

Copper, Type L with Soft-Soldered Joints

Description	Craft@Hrs	Unit	Material $	Labor $	Equipment $	Total $

Type L copper 90-degree ell ftg. x C with soft-soldered joint

Description	Craft@Hrs	Unit	Material $	Labor $	Equipment $	Total $
½"	P1@.107	Ea	1.01	3.98	—	4.99
¾"	P1@.150	Ea	3.32	5.58	—	8.90
1"	P1@.193	Ea	4.98	7.18	—	12.16
1¼"	P1@.236	Ea	6.94	8.78	—	15.72
1½"	P1@.278	Ea	10.30	10.30	—	20.60
2"	P1@.371	Ea	20.30	13.80	—	34.10
2½"	P1@.457	Ea	42.30	17.00	—	59.30
3"	P1@.543	Ea	47.60	20.20	—	67.80
4"	P1@.714	Ea	127.00	26.60	—	153.60

Type L copper tee C x C x C with soft-soldered joints

Description	Craft@Hrs	Unit	Material $	Labor $	Equipment $	Total $
½"	P1@.129	Ea	.77	4.80	—	5.57
¾"	P1@.181	Ea	2.09	6.74	—	8.83
1"	P1@.233	Ea	7.85	8.67	—	16.52
1¼"	P1@.285	Ea	12.50	10.60	—	23.10
1½"	P1@.337	Ea	18.30	12.50	—	30.80
2"	P1@.449	Ea	28.20	16.70	—	44.90
2½"	P1@.552	Ea	57.20	20.50	—	77.70
3"	P1@.656	Ea	88.50	24.40	—	112.90
4"	P1@.863	Ea	148.00	32.10	—	180.10

Type L copper branch reducing tee C x C x C with soft-soldered joints

Description	Craft@Hrs	Unit	Material $	Labor $	Equipment $	Total $
½" x 3/8"	P1@.129	Ea	1.57	4.80	—	6.37
¾" x ½"	P1@.182	Ea	3.34	6.77	—	10.11
1" x ½"	P1@.195	Ea	6.53	7.26	—	13.79
1" x ¾"	P1@.219	Ea	6.53	8.15	—	14.68
1¼" x ½"	P1@.225	Ea	9.83	8.37	—	18.20
1¼" x ¾"	P1@.247	Ea	9.83	9.19	—	19.02
1¼" x 1"	P1@.268	Ea	9.83	9.97	—	19.80
1½" x ½"	P1@.272	Ea	12.10	10.10	—	22.20
1½" x ¾"	P1@.287	Ea	12.10	10.70	—	22.80
1½" x 1"	P1@.302	Ea	12.10	11.20	—	23.30
1½" x 1¼"	P1@.317	Ea	12.10	11.80	—	23.90
2" x ½"	P1@.325	Ea	15.80	12.10	—	27.90
2" x ¾"	P1@.348	Ea	15.80	13.00	—	28.80
2" x 1"	P1@.373	Ea	15.80	13.90	—	29.70
2" x 1¼"	P1@.398	Ea	15.80	14.80	—	30.60
2" x 1½"	P1@.422	Ea	15.80	15.70	—	31.50
2½" x ½"	P1@.437	Ea	47.70	16.30	—	64.00
2½" x ¾"	P1@.453	Ea	47.70	16.90	—	64.60
2½" x 1"	P1@.469	Ea	47.70	17.50	—	65.20
2½" x 1½"	P1@.502	Ea	47.70	18.70	—	66.40
2½" x 2"	P1@.519	Ea	47.70	19.30	—	67.00
3" x 1¼"	P1@.525	Ea	53.90	19.50	—	73.40
3" x 2"	P1@.585	Ea	55.80	21.80	—	77.60
3" x 2½"	P1@.617	Ea	55.80	23.00	—	78.80
4" x 1¼"	P1@.625	Ea	97.30	23.30	—	120.60
4" x 1½"	P1@.671	Ea	97.30	25.00	—	122.30
4" x 2"	P1@.718	Ea	97.30	26.70	—	124.00
4" x 2½"	P1@.765	Ea	97.30	28.50	—	125.80
4" x 3"	P1@.811	Ea	97.30	30.20	—	127.50

Copper, Type L with Soft-Soldered Joints

Description	Craft@Hrs	Unit	Material $	Labor $	Equipment $	Total $

Type L copper reducer with soft-soldered joints

Description	Craft@Hrs	Unit	Material $	Labor $	Equipment $	Total $
½" x 3/8"	P1@.124	Ea	.56	4.62	—	5.18
¾" x ½"	P1@.129	Ea	1.06	4.80	—	5.86
1" x ¾"	P1@.172	Ea	1.91	6.40	—	8.31
1¼" x ½"	P1@.172	Ea	3.21	6.40	—	9.61
1¼" x ¾"	P1@.193	Ea	3.21	7.18	—	10.39
1¼" x 1"	P1@.215	Ea	3.21	8.00	—	11.21
1½" x ½"	P1@.193	Ea	4.90	7.18	—	12.08
1½" x ¾"	P1@.214	Ea	4.90	7.97	—	12.87
1½" x 1"	P1@.236	Ea	4.90	8.78	—	13.68
1½" x 1¼"	P1@.257	Ea	4.90	9.57	—	14.47
2" x ½"	P1@.239	Ea	7.38	8.90	—	16.28
2" x ¾"	P1@.261	Ea	7.38	9.71	—	17.09
2" x 1"	P1@.282	Ea	7.38	10.50	—	17.88
2" x 1¼"	P1@.304	Ea	7.38	11.30	—	18.68
2" x 1½"	P1@.325	Ea	7.38	12.10	—	19.48
2½" x 1"	P1@.325	Ea	20.10	12.10	—	32.20
2½" x 1¼"	P1@.347	Ea	20.10	12.90	—	33.00
2½" x 1½"	P1@.368	Ea	20.10	13.70	—	33.80
2½" x 2"	P1@.414	Ea	20.10	15.40	—	35.50
3" x 2"	P1@.457	Ea	23.50	17.00	—	40.50
3" x 2½"	P1@.500	Ea	23.50	18.60	—	42.10
4" x 2"	P1@.543	Ea	45.30	20.20	—	65.50
4" x 2½"	P1@.586	Ea	45.30	21.80	—	67.10
4" x 3"	P1@.629	Ea	45.30	23.40	—	68.70

Type L copper adapter C x MPT with soft-soldered joint

Description	Craft@Hrs	Unit	Material $	Labor $	Equipment $	Total $
½"	P1@.075	Ea	1.39	2.79	—	4.18
¾"	P1@.105	Ea	2.18	3.91	—	6.09
1"	P1@.134	Ea	5.41	4.99	—	10.40
1¼"	P1@.164	Ea	10.10	6.10	—	16.20
1½"	P1@.194	Ea	11.40	7.22	—	18.62
2"	P1@.259	Ea	14.20	9.64	—	23.84
2½"	P1@.319	Ea	46.10	11.90	—	58.00
3"	P1@.378	Ea	76.40	14.10	—	90.50

Type L copper adapter C x FPT with soft-soldered joint

Description	Craft@Hrs	Unit	Material $	Labor $	Equipment $	Total $
½"	P1@.075	Ea	2.32	2.79	—	5.11
¾"	P1@.105	Ea	4.21	3.91	—	8.12
1"	P1@.134	Ea	6.35	4.99	—	11.34
1¼"	P1@.164	Ea	13.40	6.10	—	19.50
1½"	P1@.194	Ea	16.10	7.22	—	23.32
2"	P1@.259	Ea	24.00	9.64	—	33.64
2½"	P1@.319	Ea	60.50	11.90	—	72.40
3"	P1@.378	Ea	83.20	14.10	—	97.30

Copper, Type L with Soft-Soldered Joints

Description	Craft@Hrs	Unit	Material $	Labor $	Equipment $	Total $

Type L copper flush bushing with soft-soldered joints

Description	Craft@Hrs	Unit	Material $	Labor $	Equipment $	Total $
½" x 3/8"	P1@.124	Ea	1.41	4.62	—	6.03
¾" x ½"	P1@.129	Ea	2.41	4.80	—	7.21
1" x ¾"	P1@.172	Ea	6.24	6.40	—	12.64
1" x ½"	P1@.172	Ea	7.12	6.40	—	13.52
1¼" x 1"	P1@.215	Ea	7.40	8.00	—	15.40
1½" x 1¼"	P1@.257	Ea	9.42	9.57	—	18.99
2" x 1½"	P1@.325	Ea	17.20	12.10	—	29.30

Type L copper union with soft-soldered joint

Description	Craft@Hrs	Unit	Material $	Labor $	Equipment $	Total $
½"	P1@.121	Ea	5.67	4.50	—	10.17
¾"	P1@.170	Ea	6.76	6.33	—	13.09
1"	P1@.218	Ea	15.20	8.11	—	23.31
1¼"	P1@.267	Ea	27.60	9.94	—	37.54
1½"	P1@.315	Ea	33.90	11.70	—	45.60
2"	P1@.421	Ea	48.90	15.70	—	64.60

Type L copper dielectric union with soft-soldered joint

Description	Craft@Hrs	Unit	Material $	Labor $	Equipment $	Total $
½"	P1@.121	Ea	5.31	4.50	—	9.81
¾"	P1@.170	Ea	5.31	6.33	—	11.64
1"	P1@.218	Ea	11.00	8.11	—	19.11
1¼"	P1@.267	Ea	18.70	9.94	—	28.64
1½"	P1@.315	Ea	25.80	11.70	—	37.50
2"	P1@.421	Ea	34.40	15.70	—	50.10

Type L copper cap with soft-soldered joint

Description	Craft@Hrs	Unit	Material $	Labor $	Equipment $	Total $
½"	P1@.069	Ea	.47	2.57	—	3.04
¾"	P1@.096	Ea	.94	3.57	—	4.51
1"	P1@.124	Ea	2.13	4.62	—	6.75
1¼"	P1@.151	Ea	3.06	5.62	—	8.68
1½"	P1@.178	Ea	4.46	6.63	—	11.09
2"	P1@.237	Ea	7.77	8.82	—	16.59
2½"	P1@.292	Ea	19.70	10.90	—	30.60
3"	P1@.347	Ea	30.80	12.90	—	43.70
4"	P1@.457	Ea	47.40	17.00	—	64.40

Type L copper coupling with soft-soldered joints

Description	Craft@Hrs	Unit	Material $	Labor $	Equipment $	Total $
½"	P1@.107	Ea	.32	3.98	—	4.30
¾"	P1@.150	Ea	.75	5.58	—	6.33
1"	P1@.193	Ea	2.02	7.18	—	9.20
1¼"	P1@.236	Ea	3.53	8.78	—	12.31
1½"	P1@.278	Ea	5.10	10.30	—	15.40
2"	P1@.371	Ea	7.50	13.80	—	21.30
2½"	P1@.457	Ea	17.70	17.00	—	34.70
3"	P1@.543	Ea	27.00	20.20	—	47.20
4"	P1@.714	Ea	52.90	26.60	—	79.50

Copper, Type L with Soft-Soldered Joints

Description	Craft@Hrs	Unit	Material $	Labor $	Equipment $	Total $

Class 125 bronze body gate valve, soft-soldered joints

½"	P1@.200	Ea	17.90	7.44	—	25.34
¾"	P1@.249	Ea	22.20	9.27	—	31.47
1"	P1@.299	Ea	31.30	11.10	—	42.40
1¼"	P1@.398	Ea	39.70	14.80	—	54.50
1½"	P1@.448	Ea	53.30	16.70	—	70.00
2"	P1@.498	Ea	88.90	18.50	—	107.40
2½"	P1@.830	Ea	149.00	30.90	—	179.90
3"	P1@1.24	Ea	213.00	46.20	—	259.20

Class 125 iron body gate valve, flanged ends

2"	P1@.500	Ea	358.00	18.60	—	376.60
2½"	P1@.600	Ea	483.00	22.30	—	505.30
3"	P1@.750	Ea	524.00	27.90	—	551.90
4"	P1@1.35	Ea	769.00	50.20	—	819.20
5"	P1@2.00	Ea	1,470.00	74.40	—	1,544.40
6"	P1@2.50	Ea	1,470.00	93.10	—	1,563.10

Class 125 bronze body globe valve, with soft-soldered joints

½"	P1@.200	Ea	34.80	7.44	—	42.24
¾"	P1@.249	Ea	46.30	9.27	—	55.57
1"	P1@.299	Ea	79.80	11.10	—	90.90
1¼"	P1@.398	Ea	112.00	14.80	—	126.80
1½"	P1@.448	Ea	152.00	16.70	—	168.70
2"	P1@.498	Ea	246.00	18.50	—	264.50

Class 125 iron body globe valve, flanged ends

2½"	P1@.600	Ea	406.00	22.30	—	428.30
3"	P1@.750	Ea	487.00	27.90	—	514.90
4"	P1@1.35	Ea	653.00	50.20	—	703.20

200 PSIG iron body butterfly valve, lug-type, lever operated

2"	P1@.450	Ea	161.00	16.70	—	177.70
2½"	P1@.450	Ea	167.00	16.70	—	183.70
3"	P1@.550	Ea	176.00	20.50	—	196.50
4"	P1@.550	Ea	219.00	20.50	—	239.50

200 PSIG iron body butterfly valve, wafer-type, lever operated

2"	P1@.450	Ea	147.00	16.70	—	163.70
2½"	P1@.450	Ea	150.00	16.70	—	166.70
3"	P1@.550	Ea	161.00	20.50	—	181.50
4"	P1@.550	Ea	195.00	20.50	—	215.50

Copper, Type L with Soft-Soldered Joints

Description	Craft@Hrs	Unit	Material $	Labor $	Equipment $	Total $

Class 125 bronze body 2-piece ball valve, soft-soldered joints

Description	Craft@Hrs	Unit	Material $	Labor $	Equipment $	Total $
½"	P1@.200	Ea	10.40	7.44	—	17.84
¾"	P1@.249	Ea	13.80	9.27	—	23.07
1"	P1@.299	Ea	24.00	11.10	—	35.10
1¼"	P1@.398	Ea	40.00	14.80	—	54.80
1½"	P1@.448	Ea	54.10	16.70	—	70.80
2"	P1@.498	Ea	68.60	18.50	—	87.10
3"	P1@1.24	Ea	470.00	46.20	—	516.20
4"	P1@1.45	Ea	614.00	54.00	—	668.00

Class 125 bronze body swing check valve, soft-soldered joints

Description	Craft@Hrs	Unit	Material $	Labor $	Equipment $	Total $
½"	P1@.200	Ea	20.80	7.44	—	28.24
¾"	P1@.249	Ea	29.70	9.27	—	38.97
1"	P1@.299	Ea	38.80	11.10	—	49.90
1¼"	P1@.398	Ea	55.40	14.80	—	70.20
1½"	P1@.448	Ea	78.10	16.70	—	94.80
2"	P1@.498	Ea	129.00	18.50	—	147.50
2½"	P1@.830	Ea	261.00	30.90	—	291.90
3"	P1@1.24	Ea	375.00	46.20	—	421.20

Class 125 iron body swing check valve, flanged ends

Description	Craft@Hrs	Unit	Material $	Labor $	Equipment $	Total $
2"	P1@.500	Ea	182.00	18.60	—	200.60
2½"	P1@.600	Ea	230.00	22.30	—	252.30
3"	P1@.750	Ea	287.00	27.90	—	314.90
4"	P1@1.35	Ea	421.00	50.20	—	471.20

Class 125 iron body silent check valve, wafer-type

Description	Craft@Hrs	Unit	Material $	Labor $	Equipment $	Total $
2"	P1@.500	Ea	133.00	18.60	—	151.60
2½"	P1@.600	Ea	148.00	22.30	—	170.30
3"	P1@.750	Ea	170.00	27.90	—	197.90
4"	P1@1.35	Ea	227.00	50.20	—	277.20

Class 125 bronze body strainer, threaded ends

Description	Craft@Hrs	Unit	Material $	Labor $	Equipment $	Total $
½"	P1@.230	Ea	32.00	8.56	—	40.56
¾"	P1@.260	Ea	42.00	9.68	—	51.68
1"	P1@.330	Ea	51.50	12.30	—	63.80
1¼"	P1@.440	Ea	72.00	16.40	—	88.40
1½"	P1@.495	Ea	108.00	18.40	—	126.40
2"	P1@.550	Ea	188.00	20.50	—	208.50

Class 125 iron body strainer, flanged ends

Description	Craft@Hrs	Unit	Material $	Labor $	Equipment $	Total $
2"	P1@.500	Ea	143.00	18.60	—	161.60
2½"	P1@.600	Ea	159.00	22.30	—	181.30
3"	P1@.750	Ea	185.00	27.90	—	212.90
4"	P1@1.35	Ea	312.00	50.20	—	362.20

Copper, Type L with Soft-Soldered Joints

Description	Craft@Hrs	Unit	Material $	Labor $	Equipment $	Total $
Installation of 2-way control valve, threaded joints						
½"	P1@.210	Ea	—	7.82	—	7.82
¾"	P1@.275	Ea	—	10.20	—	10.20
1"	P1@.350	Ea	—	13.00	—	13.00
1¼"	P1@.430	Ea	—	16.00	—	16.00
1½"	P1@.505	Ea	—	18.80	—	18.80
2"	P1@.675	Ea	—	25.10	—	25.10
2½"	P1@.830	Ea	—	30.90	—	30.90
3"	P1@.990	Ea	—	36.80	—	36.80
Installation of 3-way control valve, threaded joints						
½"	P1@.260	Ea	—	9.68	—	9.68
¾"	P1@.365	Ea	—	13.60	—	13.60
1"	P1@.475	Ea	—	17.70	—	17.70
1¼"	P1@.575	Ea	—	21.40	—	21.40
1½"	P1@.680	Ea	—	25.30	—	25.30
2"	P1@.910	Ea	—	33.90	—	33.90
2½"	P1@1.12	Ea	—	41.70	—	41.70
3"	P1@1.33	Ea	—	49.50	—	49.50
Companion flange, 150 pound cast brass						
2"	P1@.290	Ea	288.00	10.80	—	298.80
2½"	P1@.380	Ea	355.00	14.10	—	369.10
3"	P1@.460	Ea	355.00	17.10	—	372.10
4"	P1@.600	Ea	565.00	22.30	—	587.30
Bolt and gasket sets						
2"	P1@.500	Ea	4.25	18.60	—	22.85
2½"	P1@.650	Ea	4.97	24.20	—	29.17
3"	P1@.750	Ea	8.43	27.90	—	36.33
4"	P1@1.00	Ea	14.10	37.20	—	51.30
Thermometer with well						
7"	P1@.250	Ea	183.00	9.31	—	192.31
9"	P1@.250	Ea	188.00	9.31	—	197.31
Dial-type pressure gauge						
2½"	P1@.200	Ea	38.30	7.44	—	45.74
3½"	P1@.200	Ea	50.40	7.44	—	57.84
Pressure/temperature tap						
Tap	P1@.150	Ea	17.60	5.58	—	23.18

Copper, Type L with Soft-Soldered Joints

Description	Craft@Hrs	Unit	Material $	Labor $	Equipment $	Total $
Hanger with swivel assembly						
½"	P1@.250	Ea	4.25	9.31	—	13.56
¾"	P1@.250	Ea	4.71	9.31	—	14.02
1"	P1@.250	Ea	4.92	9.31	—	14.23
1¼"	P1@.300	Ea	5.05	11.20	—	16.25
1½"	P1@.300	Ea	5.43	11.20	—	16.63
2"	P1@.300	Ea	5.66	11.20	—	16.86
2½"	P1@.350	Ea	7.70	13.00	—	20.70
3"	P1@.350	Ea	9.53	13.00	—	22.53
4"	P1@.350	Ea	10.50	13.00	—	23.50
Riser clamp						
½"	P1@.100	Ea	3.15	3.72	—	6.87
¾"	P1@.100	Ea	4.65	3.72	—	8.37
1"	P1@.100	Ea	4.69	3.72	—	8.41
1¼"	P1@.105	Ea	5.66	3.91	—	9.57
1½"	P1@.110	Ea	5.98	4.09	—	10.07
2"	P1@.115	Ea	6.33	4.28	—	10.61
2½"	P1@.120	Ea	6.66	4.47	—	11.13
3"	P1@.120	Ea	7.21	4.47	—	11.68
4"	P1@.125	Ea	9.19	4.65	—	13.84

Copper, Type M with Brazed Joints

Type M hard-drawn copper pipe with wrought copper fittings and brazed joints is used in a wide variety of plumbing and HVAC systems such as potable water, heating hot water and chilled water.

Brazed joints are those made with silver or other alloy filler metals having melting points at, or above, 1,000 degrees F. Maximum working pressure/temperature relationships for brazed joints are approximately as follows:

Maximum Working Pressure (PSIG)*			
Water Temperature (degrees F.)	Nominal Pipe Size (inches)		
	Up to 1	1¼ to 2	2½ to 4
Up to 350	270	190	150

*For copper pipe and solder-type fittings using brazing alloys melting at, or above, 1,000 degrees F.

This section has been arranged to save the estimator's time by including all normally-used system components such as pipe, fittings, valves, hanger assemblies, riser clamps and miscellaneous items under one heading. Additional items can be found under "Plumbing and Piping Specialties." The cost estimates in this section are based on the conditions, limitations and wage rates described in the section "How to Use This Book" beginning on page 5.

Description	Craft@Hrs	Unit	Material $	Labor $	Equipment $	Total $

Type M copper pipe with brazed joints installed horizontally. Complete installation including six to twelve wrought copper tees and six to twelve wrought copper elbows every 100', and hangers spaced to meet plumbing code *(a tee & elbow every 8.3' for ½" dia., a tee and elbow every 16.6' for 4" dia.)*.

Description	Craft@Hrs	Unit	Material $	Labor $	Equipment $	Total $
½"	P1@.113	LF	2.03	4.21	—	6.24
¾"	P1@.123	LF	3.14	4.58	—	7.72
1"	P1@.143	LF	6.50	5.32	—	11.82
1¼"	P1@.163	LF	11.00	6.07	—	17.07
1½"	P1@.183	LF	12.90	6.81	—	19.71
2"	P1@.213	LF	27.20	7.93	—	35.13
2½"	P1@.243	LF	39.90	9.04	—	48.94
3"	P1@.273	LF	50.00	10.20	—	60.20
4"	P1@.313	LF	73.60	11.60	—	85.20

Type M copper pipe with brazed joints, installed risers. Complete installation including a wrought copper reducing tee every floor and a riser clamp every other floor.

Description	Craft@Hrs	Unit	Material $	Labor $	Equipment $	Total $
½"	P1@.062	LF	1.72	2.31	—	4.03
¾"	P1@.072	LF	2.78	2.68	—	5.46
1"	P1@.077	LF	5.44	2.87	—	8.31
1¼"	P1@.082	LF	8.02	3.05	—	11.07
1½"	P1@.102	LF	10.60	3.80	—	14.40
2"	P1@.132	LF	14.00	4.91	—	18.91
2½"	P1@.152	LF	30.50	5.66	—	36.16
3"	P1@.172	LF	37.40	6.40	—	43.80
4"	P1@.192	LF	55.40	7.15	—	62.55

Copper, Type M with Brazed Joints

Description	Craft@Hrs	Unit	Material $	Labor $	Equipment $	Total $

Type M copper pipe with brazed joints, pipe only

½"	P1@.032	LF	1.29	1.19	—	2.48
¾"	P1@.035	LF	2.10	1.30	—	3.40
1"	P1@.038	LF	3.28	1.41	—	4.69
1¼"	P1@.042	LF	4.93	1.56	—	6.49
1½"	P1@.046	LF	6.90	1.71	—	8.61
2"	P1@.053	LF	12.80	1.97	—	14.77
2½"	P1@.060	LF	17.60	2.23	—	19.83
3"	P1@.066	LF	22.00	2.46	—	24.46
4"	P1@.080	LF	35.40	2.98	—	38.38

Type M copper 45-degree ell C x C with brazed joints

½"	P1@.129	Ea	.82	4.80	—	5.62
¾"	P1@.181	Ea	1.56	6.74	—	8.30
1"	P1@.232	Ea	4.61	8.64	—	13.25
1¼"	P1@.284	Ea	6.24	10.60	—	16.84
1½"	P1@.335	Ea	9.29	12.50	—	21.79
2"	P1@.447	Ea	14.70	16.60	—	31.30
2½"	P1@.550	Ea	37.60	20.50	—	58.10
3"	P1@.654	Ea	49.70	24.30	—	74.00
4"	P1@.860	Ea	95.40	32.00	—	127.40

Type M copper 90-degree ell C x C with brazed joints

½"	P1@.129	Ea	.47	4.80	—	5.27
¾"	P1@.181	Ea	.98	6.74	—	7.72
1"	P1@.232	Ea	3.52	8.64	—	12.16
1¼"	P1@.284	Ea	5.68	10.60	—	16.28
1½"	P1@.335	Ea	8.57	12.50	—	21.07
2"	P1@.447	Ea	14.90	16.60	—	31.50
2½"	P1@.550	Ea	31.10	20.50	—	51.60
3"	P1@.654	Ea	39.40	24.30	—	63.70
4"	P1@.860	Ea	83.30	32.00	—	115.30

Type M copper 90-degree ell ftg. x C with brazed joints

½"	P1@.129	Ea	.77	4.80	—	5.57
¾"	P1@.181	Ea	1.66	6.74	—	8.40
1"	P1@.232	Ea	4.98	8.64	—	13.62
1¼"	P1@.284	Ea	6.94	10.60	—	17.54
1½"	P1@.335	Ea	10.30	12.50	—	22.80
2"	P1@.447	Ea	20.30	16.60	—	36.90
2½"	P1@.550	Ea	42.30	20.50	—	62.80
3"	P1@.654	Ea	47.60	24.30	—	71.90
4"	P1@.860	Ea	95.40	32.00	—	127.40

Copper, Type M with Brazed Joints

Description	Craft@Hrs	Unit	Material $	Labor $	Equipment $	Total $
Type M copper tee C x C x C with brazed joints						
½"	P1@.156	Ea	.77	5.81	—	6.58
¾"	P1@.204	Ea	2.09	7.59	—	9.68
1"	P1@.263	Ea	7.85	9.79	—	17.64
1¼"	P1@.322	Ea	12.50	12.00	—	24.50
1½"	P1@.380	Ea	18.30	14.10	—	32.40
2"	P1@.507	Ea	28.20	18.90	—	47.10
2½"	P1@.624	Ea	57.20	23.20	—	80.40
3"	P1@.741	Ea	88.50	27.60	—	116.10
4"	P1@.975	Ea	148.00	36.30	—	184.30

Description	Craft@Hrs	Unit	Material $	Labor $	Equipment $	Total $
Type M copper reducing branch tee C x C x C with brazed joints						
½" x 3/8"	P1@.129	Ea	1.57	4.80	—	6.37
¾" x ½"	P1@.182	Ea	3.34	6.77	—	10.11
1" x ½"	P1@.209	Ea	6.53	7.78	—	14.31
1" x ¾"	P1@.234	Ea	6.53	8.71	—	15.24
1¼" x ½"	P1@.241	Ea	9.83	8.97	—	18.80
1¼" x ¾"	P1@.264	Ea	9.83	9.83	—	19.66
1¼" x 1"	P1@.287	Ea	9.83	10.70	—	20.53
1½" x ½"	P1@.291	Ea	12.10	10.80	—	22.90
1½" x ¾"	P1@.307	Ea	12.10	11.40	—	23.50
1½" x 1"	P1@.323	Ea	12.10	12.00	—	24.10
1½" x 1¼"	P1@.339	Ea	12.10	12.60	—	24.70
2" x ½"	P1@.348	Ea	15.80	13.00	—	28.80
2" x ¾"	P1@.373	Ea	15.80	13.90	—	29.70
2" x 1"	P1@.399	Ea	15.80	14.90	—	30.70
2" x 1¼"	P1@.426	Ea	15.80	15.90	—	31.70
2" x 1½"	P1@.452	Ea	15.80	16.80	—	32.60
2½" x ½"	P1@.468	Ea	47.70	17.40	—	65.10
2½" x ¾"	P1@.485	Ea	47.70	18.10	—	65.80
2½" x 1"	P1@.502	Ea	47.70	18.70	—	66.40
2½" x 1½"	P1@.537	Ea	47.70	20.00	—	67.70
2½" x 2"	P1@.555	Ea	47.70	20.70	—	68.40
3" x 1¼"	P1@.562	Ea	53.90	20.90	—	74.80
3" x 2"	P1@.626	Ea	55.80	23.30	—	79.10
3" x 2½"	P1@.660	Ea	55.80	24.60	—	80.40
4" x 1¼"	P1@.669	Ea	97.30	24.90	—	122.20
4" x 1½"	P1@.718	Ea	97.30	26.70	—	124.00
4" x 2"	P1@.768	Ea	97.30	28.60	—	125.90
4" x 2½"	P1@.819	Ea	97.30	30.50	—	127.80
4" x 3"	P1@.868	Ea	97.30	32.30	—	129.60

Copper, Type M with Brazed Joints

Description	Craft@Hrs	Unit	Material $	Labor $	Equipment $	Total $

Type M copper reducer with brazed joints

Description	Craft@Hrs	Unit	Material $	Labor $	Equipment $	Total $
½" x 3/8"	P1@.148	Ea	.56	5.51	—	6.07
¾" x ½"	P1@.155	Ea	1.06	5.77	—	6.83
1" x ¾"	P1@.206	Ea	1.91	7.67	—	9.58
1¼" x ½"	P1@.206	Ea	3.21	7.67	—	10.88
1¼" x ¾"	P1@.232	Ea	3.21	8.64	—	11.85
1¼" x 1"	P1@.258	Ea	3.21	9.60	—	12.81
1½" x ½"	P1@.232	Ea	4.90	8.64	—	13.54
1½" x ¾"	P1@.257	Ea	4.90	9.57	—	14.47
1½" x 1"	P1@.283	Ea	4.90	10.50	—	15.40
1½" x 1¼"	P1@.308	Ea	4.90	11.50	—	16.40
2" x ½"	P1@.287	Ea	7.38	10.70	—	18.08
2" x ¾"	P1@.277	Ea	7.38	10.30	—	17.68
2" x 1"	P1@.338	Ea	7.38	12.60	—	19.98
2" x 1¼"	P1@.365	Ea	7.38	13.60	—	20.98
2" x 1½"	P1@.390	Ea	7.38	14.50	—	21.88
2½" x 1"	P1@.390	Ea	20.10	14.50	—	34.60
2½" x 1¼"	P1@.416	Ea	20.10	15.50	—	35.60
2½" x 1½"	P1@.442	Ea	20.10	16.50	—	36.60
2½" x 2"	P1@.497	Ea	20.10	18.50	—	38.60
3" x 1¼"	P1@.468	Ea	23.50	17.40	—	40.90
3" x 1½"	P1@.493	Ea	23.50	18.30	—	41.80
3" x 2"	P1@.548	Ea	23.50	20.40	—	43.90
3" x 2½"	P1@.600	Ea	23.50	22.30	—	45.80
4" x 2"	P1@.652	Ea	45.30	24.30	—	69.60
4" x 2½"	P1@.704	Ea	45.30	26.20	—	71.50
4" x 3"	P1@.755	Ea	45.30	28.10	—	73.40

Type M copper adapter C x MPT with brazed joint

Description	Craft@Hrs	Unit	Material $	Labor $	Equipment $	Total $
½"	P1@.090	Ea	1.39	3.35	—	4.74
¾"	P1@.126	Ea	2.18	4.69	—	6.87
1"	P1@.162	Ea	5.41	6.03	—	11.44
1¼"	P1@.198	Ea	10.10	7.37	—	17.47
1½"	P1@.234	Ea	11.40	8.71	—	20.11
2"	P1@.312	Ea	14.20	11.60	—	25.80
2½"	P1@.384	Ea	46.10	14.30	—	60.40
3"	P1@.456	Ea	76.40	17.00	—	93.40

Type M copper adapter C x FPT with brazed joint

Description	Craft@Hrs	Unit	Material $	Labor $	Equipment $	Total $
½"	P1@.090	Ea	2.32	3.35	—	5.67
¾"	P1@.126	Ea	4.21	4.69	—	8.90
1"	P1@.162	Ea	6.35	6.03	—	12.38
1¼"	P1@.198	Ea	13.40	7.37	—	20.77
1½"	P1@.234	Ea	16.10	8.71	—	24.81
2"	P1@.312	Ea	24.00	11.60	—	35.60
2½"	P1@.384	Ea	60.50	14.30	—	74.80
3"	P1@.456	Ea	83.20	17.00	—	100.20

Copper, Type M with Brazed Joints

Description	Craft@Hrs	Unit	Material $	Labor $	Equipment $	Total $
Type M copper flush bushing with brazed joints						
½" x 3/8"	P1@.148	Ea	1.41	5.51	—	6.92
¾" x ½"	P1@.155	Ea	2.41	5.77	—	8.18
1" x ¾"	P1@.206	Ea	6.24	7.67	—	13.91
1" x ½"	P1@.206	Ea	7.12	7.67	—	14.79
1¼" x 1"	P1@.258	Ea	7.40	9.60	—	17.00
1½" x 1¼"	P1@.308	Ea	9.42	11.50	—	20.92
2" x 1½"	P1@.390	Ea	17.20	14.50	—	31.70
Type M copper union with brazed joints						
½"	P1@.146	Ea	5.67	5.43	—	11.10
¾"	P1@.205	Ea	6.76	7.63	—	14.39
1"	P1@.263	Ea	15.20	9.79	—	24.99
1¼"	P1@.322	Ea	27.60	12.00	—	39.60
1½"	P1@.380	Ea	33.90	14.10	—	48.00
2"	P1@.507	Ea	48.90	18.90	—	67.80
Type M copper dielectric union with brazed joints						
½"	P1@.146	Ea	5.31	5.43	—	10.74
¾"	P1@.205	Ea	5.31	7.63	—	12.94
1"	P1@.263	Ea	11.00	9.79	—	20.79
1¼"	P1@.322	Ea	18.70	12.00	—	30.70
1½"	P1@.380	Ea	25.80	14.10	—	39.90
2"	P1@.507	Ea	34.40	18.90	—	53.30
Type M copper cap with brazed joints						
½"	P1@.083	Ea	.47	3.09	—	3.56
¾"	P1@.116	Ea	.94	4.32	—	5.26
1"	P1@.149	Ea	2.13	5.55	—	7.68
1¼"	P1@.182	Ea	3.06	6.77	—	9.83
1½"	P1@.215	Ea	4.46	8.00	—	12.46
2"	P1@.286	Ea	7.77	10.60	—	18.37
2½"	P1@.352	Ea	19.70	13.10	—	32.80
3"	P1@.418	Ea	30.80	15.60	—	46.40
4"	P1@.550	Ea	47.40	20.50	—	67.90
Type M copper coupling with brazed joints						
½"	P1@.129	Ea	.30	4.80	—	5.10
¾"	P1@.181	Ea	.75	6.74	—	7.49
1"	P1@.232	Ea	2.02	8.64	—	10.66
1¼"	P1@.284	Ea	3.53	10.60	—	14.13
1½"	P1@.335	Ea	5.10	12.50	—	17.60
2"	P1@.447	Ea	7.50	16.60	—	24.10
2½"	P1@.550	Ea	17.70	20.50	—	38.20
3"	P1@.654	Ea	27.00	24.30	—	51.30
4"	P1@.860	Ea	52.90	32.00	—	84.90

Copper, Type M with Brazed Joints

Description	Craft@Hrs	Unit	Material $	Labor $	Equipment $	Total $

Class 125 bronze body gate valve, brazed joints

Size	Craft@Hrs	Unit	Material $	Labor $	Equipment $	Total $
½"	P1@.240	Ea	18.80	8.93	—	27.73
¾"	P1@.300	Ea	23.50	11.20	—	34.70
1"	P1@.360	Ea	33.00	13.40	—	46.40
1¼"	P1@.480	Ea	41.50	17.90	—	59.40
1½"	P1@.540	Ea	56.00	20.10	—	76.10
2"	P1@.600	Ea	93.60	22.30	—	115.90
2½"	P1@1.00	Ea	155.00	37.20	—	192.20
3"	P1@1.50	Ea	222.00	55.80	—	277.80

Class 125 iron body gate valve, flanged ends

Size	Craft@Hrs	Unit	Material $	Labor $	Equipment $	Total $
2"	P1@.500	Ea	358.00	18.60	—	376.60
2½"	P1@.600	Ea	483.00	22.30	—	505.30
3"	P1@.750	Ea	524.00	27.90	—	551.90
4"	P1@1.35	Ea	769.00	50.20	—	819.20
5"	P1@2.00	Ea	1,470.00	74.40	—	1,544.40
6"	P1@2.50	Ea	1,470.00	93.10	—	1,563.10

Class 125 bronze body globe valve, brazed joints

Size	Craft@Hrs	Unit	Material $	Labor $	Equipment $	Total $
½"	P1@.240	Ea	34.80	8.93	—	43.73
¾"	P1@.300	Ea	46.30	11.20	—	57.50
1"	P1@.360	Ea	79.80	13.40	—	93.20
1¼"	P1@.480	Ea	112.00	17.90	—	129.90
1½"	P1@.540	Ea	152.00	20.10	—	172.10
2"	P1@.600	Ea	246.00	22.30	—	268.30

Class 125 iron body globe valve, flanged ends

Size	Craft@Hrs	Unit	Material $	Labor $	Equipment $	Total $
2½"	P1@.650	Ea	406.00	24.20	—	430.20
3"	P1@.750	Ea	487.00	27.90	—	514.90
4"	P1@1.35	Ea	657.00	50.20	—	707.20

200 PSIG iron body butterfly valve, lug-type, lever operated

Size	Craft@Hrs	Unit	Material $	Labor $	Equipment $	Total $
2"	P1@.450	Ea	161.00	16.70	—	177.70
2½"	P1@.450	Ea	167.00	16.70	—	183.70
3"	P1@.550	Ea	176.00	20.50	—	196.50
4"	P1@.550	Ea	219.00	20.50	—	239.50

200 PSIG iron body butterfly valve, wafer-type, lever operated

Size	Craft@Hrs	Unit	Material $	Labor $	Equipment $	Total $
2"	P1@.450	Ea	147.00	16.70	—	163.70
2½"	P1@.450	Ea	150.00	16.70	—	166.70
3"	P1@.550	Ea	161.00	20.50	—	181.50
4"	P1@.550	Ea	195.00	20.50	—	215.50

Copper, Type M with Brazed Joints

Description	Craft@Hrs	Unit	Material $	Labor $	Equipment $	Total $

Class 125 bronze body 2-piece ball valve, brazed joints

½"	P1@.240	Ea	10.40	8.93	—	19.33
¾"	P1@.300	Ea	13.80	11.20	—	25.00
1"	P1@.360	Ea	24.00	13.40	—	37.40
1¼"	P1@.480	Ea	40.00	17.90	—	57.90
1½"	P1@.540	Ea	54.10	20.10	—	74.20
2"	P1@.600	Ea	68.60	22.30	—	90.90
3"	P1@1.50	Ea	470.00	55.80	—	525.80
4"	P1@1.80	Ea	614.00	67.00	—	681.00

Class 125 bronze body swing check valve, brazed joints

½"	P1@.240	Ea	20.80	8.93	—	29.73
¾"	P1@.300	Ea	29.70	11.20	—	40.90
1"	P1@.360	Ea	38.80	13.40	—	52.20
1¼"	P1@.480	Ea	55.40	17.90	—	73.30
1½"	P1@.540	Ea	78.10	20.10	—	98.20
2"	P1@.600	Ea	129.00	22.30	—	151.30
2½"	P1@1.00	Ea	261.00	37.20	—	298.20
3"	P1@1.50	Ea	375.00	55.80	—	430.80

Class 125 iron body swing check valve, flanged ends

2"	P1@.500	Ea	173.00	18.60	—	191.60
2½"	P1@.600	Ea	230.00	22.30	—	252.30
3"	P1@.750	Ea	287.00	27.90	—	314.90
4"	P1@1.35	Ea	421.00	50.20	—	471.20

Class 125 iron body silent check valve, wafer-type

2"	P1@.500	Ea	133.00	18.60	—	151.60
2½"	P1@.600	Ea	148.00	22.30	—	170.30
3"	P1@.750	Ea	170.00	27.90	—	197.90
4"	P1@1.35	Ea	227.00	50.20	—	277.20

Class 125 bronze body strainer, threaded ends

½"	P1@.230	Ea	32.00	8.56	—	40.56
¾"	P1@.260	Ea	42.00	9.68	—	51.68
1"	P1@.330	Ea	51.50	12.30	—	63.80
1¼"	P1@.440	Ea	72.00	16.40	—	88.40
1½"	P1@.495	Ea	108.00	18.40	—	126.40
2"	P1@.550	Ea	188.00	20.50	—	208.50

Class 125 iron body strainer, flanged ends

2"	P1@.500	Ea	144.00	18.60	—	162.60
2½"	P1@.600	Ea	160.00	22.30	—	182.30
3"	P1@.750	Ea	186.00	27.90	—	213.90
4"	P1@1.35	Ea	314.00	50.20	—	364.20

Copper, Type M with Brazed Joints

Description	Craft@Hrs	Unit	Material $	Labor $	Equipment $	Total $

Installation of 2-way control valve, threaded joints

Description	Craft@Hrs	Unit	Material $	Labor $	Equipment $	Total $
½"	P1@.210	Ea	—	7.82	—	7.82
¾"	P1@.275	Ea	—	10.20	—	10.20
1"	P1@.350	Ea	—	13.00	—	13.00
1¼"	P1@.430	Ea	—	16.00	—	16.00
1½"	P1@.505	Ea	—	18.80	—	18.80
2"	P1@.675	Ea	—	25.10	—	25.10
2½"	P1@.830	Ea	—	30.90	—	30.90
3"	P1@.990	Ea	—	36.80	—	36.80

Installation of 3-way control valve, threaded joints

Description	Craft@Hrs	Unit	Material $	Labor $	Equipment $	Total $
½"	P1@.260	Ea	—	9.68	—	9.68
¾"	P1@.365	Ea	—	13.60	—	13.60
1"	P1@.475	Ea	—	17.70	—	17.70
1¼"	P1@.575	Ea	—	21.40	—	21.40
1½"	P1@.680	Ea	—	25.30	—	25.30
2"	P1@.910	Ea	—	33.90	—	33.90
2½"	P1@1.12	Ea	—	41.70	—	41.70
3"	P1@1.33	Ea	—	49.50	—	49.50

Companion flange, 150 pound cast brass

Description	Craft@Hrs	Unit	Material $	Labor $	Equipment $	Total $
2"	P1@.290	Ea	291.00	10.80	—	301.80
2½"	P1@.380	Ea	359.00	14.10	—	373.10
3"	P1@.460	Ea	359.00	17.10	—	376.10
4"	P1@.600	Ea	571.00	22.30	—	593.30

Bolt and gasket sets

Description	Craft@Hrs	Unit	Material $	Labor $	Equipment $	Total $
2"	P1@.500	Ea	3.60	18.60	—	22.20
2½"	P1@.650	Ea	4.22	24.20	—	28.42
3"	P1@.750	Ea	7.14	27.90	—	35.04
4"	P1@1.00	Ea	12.00	37.20	—	49.20

Thermometer with well

Description	Craft@Hrs	Unit	Material $	Labor $	Equipment $	Total $
7"	P1@.250	Ea	153.00	9.31	—	162.31
9"	P1@.250	Ea	188.00	9.31	—	197.31

Dial-type pressure gauge

Description	Craft@Hrs	Unit	Material $	Labor $	Equipment $	Total $
2½"	P1@.200	Ea	38.30	7.44	—	45.74
3½"	P1@.200	Ea	50.40	7.44	—	57.84

Pressure/temperature tap

Description	Craft@Hrs	Unit	Material $	Labor $	Equipment $	Total $
Tap	P1@.150	Ea	17.60	5.58	—	23.18

Copper, Type M with Soft-Soldered Joints

Type M hard-drawn copper pipe with wrought copper fittings and soft-soldered joints is used in a wide variety of plumbing and HVAC systems such as potable water, heating hot water, chilled water, and A.C. condensate.

Soft-soldered joints are those made with solders having melting points in the 350 degree F. to 500 degree F. range. Maximum working pressure/temperature relationships for soft-soldered joints are approximately as follows:

		Maximum Working Pressures (PSIG)*		
Soft-solder Type	Water Temperature (degrees F.)	Nominal Pipe Size (inches)		
		Up to 1	1¼ to 2	2½ x 4
50-50 tin-lead**	100	200	175	150
	150	150	125	100
	200	100	90	75
	250	85	75	50
95-5 tin-antimony	100	500	400	300
	150	400	350	275
	200	300	250	200
	250	200	175	150

*For copper pipe and solder-type fittings using soft-solders melting at approximately 350 degrees F. to 500 degrees F.

**The use of any solder containing lead is not allowed in potable water systems.

This section has been arranged to save the estimator's time by including all normally-used system components such as pipe, fittings, valves, hanger assemblies, riser clamps and miscellaneous items under one heading. Additional items can be found under "Plumbing and Piping Specialties." The cost estimates in this section are based on the conditions, limitations and wage rates described in the section "How to Use This Book" beginning on page 5.

Description	Craft@Hrs	Unit	Material $	Labor $	Equipment $	Total $

Type M copper pipe with soft-soldered joints installed horizontally.
Complete installation including six to twelve wrought copper tees and six to twelve wrought copper elbows every 100', and hangers spaced to meet plumbing code *(a tee & elbow every 8.3' for ½" dia., a tee and elbow every 16.6' for 4" dia.).*

Description	Craft@Hrs	Unit	Material $	Labor $	Equipment $	Total $
½"	P1@.110	LF	2.03	4.09	—	6.12
¾"	P1@.120	LF	3.14	4.47	—	7.61
1"	P1@.140	LF	6.50	5.21	—	11.71
1¼"	P1@.160	LF	11.00	5.96	—	16.96
1½"	P1@.180	LF	12.90	6.70	—	19.60
2"	P1@.210	LF	27.20	7.82	—	35.02
2½"	P1@.240	LF	39.90	8.93	—	48.83
3"	P1@.270	LF	50.00	10.00	—	60.00
4"	P1@.312	LF	73.60	11.60	—	85.20

Copper, Type M with Soft-Soldered Joints

Description	Craft@Hrs	Unit	Material $	Labor $	Equipment $	Total $

Type M copper pipe with soft-soldered joints, installed risers.
Complete installation including a wrought copper reducing tee every floor and a riser clamp every other floor.

Size	Craft@Hrs	Unit	Material $	Labor $	Equipment $	Total $
½"	P1@.060	LF	1.54	2.23	—	3.77
¾"	P1@.070	LF	2.78	2.61	—	5.39
1"	P1@.075	LF	5.44	2.79	—	8.23
1¼"	P1@.080	LF	8.02	2.98	—	11.00
1½"	P1@.100	LF	10.60	3.72	—	14.32
2"	P1@.130	LF	14.00	4.84	—	18.84
2½"	P1@.150	LF	30.50	5.58	—	36.08
3"	P1@.170	LF	37.40	6.33	—	43.73
4"	P1@.190	LF	55.40	7.07	—	62.47

Type M copper pipe with soft-soldered joints, pipe only

Size	Craft@Hrs	Unit	Material $	Labor $	Equipment $	Total $
½"	P1@.032	LF	1.17	1.19	—	2.36
¾"	P1@.035	LF	1.81	1.30	—	3.11
1"	P1@.038	LF	2.83	1.41	—	4.24
1¼"	P1@.042	LF	4.25	1.56	—	5.81
1½"	P1@.046	LF	5.95	1.71	—	7.66
2"	P1@.053	LF	11.00	1.97	—	12.97
2½"	P1@.060	LF	15.20	2.23	—	17.43
3"	P1@.066	LF	19.00	2.46	—	21.46
4"	P1@.080	LF	30.50	2.98	—	33.48

Type M copper 45-degree ell C x C with soft-soldered joints

Size	Craft@Hrs	Unit	Material $	Labor $	Equipment $	Total $
½"	P1@.107	Ea	.82	3.98	—	4.80
¾"	P1@.150	Ea	1.42	5.58	—	7.00
1"	P1@.193	Ea	4.61	7.18	—	11.79
1¼"	P1@.236	Ea	5.19	8.78	—	13.97
1½"	P1@.278	Ea	9.29	10.30	—	19.59
2"	P1@.371	Ea	14.70	13.80	—	28.50
2½"	P1@.457	Ea	37.60	17.00	—	54.60
3"	P1@.543	Ea	49.70	20.20	—	69.90
4"	P1@.714	Ea	95.40	26.60	—	122.00

Type M copper 90-degree ell C x C with soft-soldered joints

Size	Craft@Hrs	Unit	Material $	Labor $	Equipment $	Total $
½"	P1@.107	Ea	.44	3.98	—	4.42
¾"	P1@.150	Ea	.92	5.58	—	6.50
1"	P1@.193	Ea	3.52	7.18	—	10.70
1¼"	P1@.236	Ea	5.68	8.78	—	14.46
1½"	P1@.278	Ea	8.57	10.30	—	18.87
2"	P1@.371	Ea	14.90	13.80	—	28.70
2½"	P1@.457	Ea	31.10	17.00	—	48.10
3"	P1@.543	Ea	39.40	20.20	—	59.60
4"	P1@.714	Ea	83.30	26.60	—	109.90

Copper, Type M with Soft-Soldered Joints

Description	Craft@Hrs	Unit	Material $	Labor $	Equipment $	Total $

Type M copper 90-degree ell ftg. x C with soft-soldered joint

Description	Craft@Hrs	Unit	Material $	Labor $	Equipment $	Total $
½"	P1@.107	Ea	.77	3.98	—	4.75
¾"	P1@.150	Ea	1.66	5.58	—	7.24
1"	P1@.193	Ea	4.98	7.18	—	12.16
1¼"	P1@.236	Ea	6.94	8.78	—	15.72
1½"	P1@.278	Ea	10.30	10.30	—	20.60
2"	P1@.371	Ea	20.30	13.80	—	34.10
2½"	P1@.457	Ea	42.30	17.00	—	59.30
3"	P1@.543	Ea	47.60	20.20	—	67.80
4"	P1@.714	Ea	95.40	26.60	—	122.00

Type M copper tee C x C x C with soft-soldered joints

Description	Craft@Hrs	Unit	Material $	Labor $	Equipment $	Total $
½"	P1@.129	Ea	.71	4.80	—	5.51
¾"	P1@.181	Ea	1.90	6.74	—	8.64
1"	P1@.233	Ea	7.85	8.67	—	16.52
1¼"	P1@.285	Ea	12.50	10.60	—	23.10
1½"	P1@.337	Ea	18.30	12.50	—	30.80
2"	P1@.449	Ea	28.20	16.70	—	44.90
2½"	P1@.552	Ea	57.20	20.50	—	77.70
3"	P1@.656	Ea	88.50	24.40	—	112.90
4"	P1@.863	Ea	111.00	32.10	—	143.10

Type M copper branch reducing tee C x C x C with soft-soldered joints

Description	Craft@Hrs	Unit	Material $	Labor $	Equipment $	Total $
½" x 3/8"	P1@.129	Ea	1.57	4.80	—	6.37
¾" x ½"	P1@.182	Ea	3.34	6.77	—	10.11
1" x ½"	P1@.195	Ea	6.53	7.26	—	13.79
1" x ¾"	P1@.219	Ea	6.53	8.15	—	14.68
1¼" x ½"	P1@.225	Ea	9.83	8.37	—	18.20
1¼" x ¾"	P1@.247	Ea	9.83	9.19	—	19.02
1¼" x 1"	P1@.268	Ea	9.83	9.97	—	19.80
1½" x ½"	P1@.272	Ea	12.10	10.10	—	22.20
1½" x ¾"	P1@.287	Ea	12.10	10.70	—	22.80
1½" x 1"	P1@.302	Ea	12.10	11.20	—	23.30
1½" x 1¼"	P1@.317	Ea	12.10	11.80	—	23.90
2" x ½"	P1@.325	Ea	15.80	12.10	—	27.90
2" x ¾"	P1@.348	Ea	15.80	13.00	—	28.80
2" x 1"	P1@.373	Ea	15.80	13.90	—	29.70
2" x 1¼"	P1@.398	Ea	15.80	14.80	—	30.60
2" x 1½"	P1@.422	Ea	15.80	15.70	—	31.50
2½" x ½"	P1@.437	Ea	47.70	16.30	—	64.00
2½" x ¾"	P1@.453	Ea	47.70	16.90	—	64.60
2½" x 1"	P1@.469	Ea	47.70	17.50	—	65.20
2½" x 1½"	P1@.502	Ea	47.70	18.70	—	66.40
2½" x 2"	P1@.519	Ea	47.70	19.30	—	67.00
3" x 1¼"	P1@.525	Ea	53.90	19.50	—	73.40
3" x 2"	P1@.585	Ea	55.80	21.80	—	77.60
3" x 2½"	P1@.617	Ea	55.80	23.00	—	78.80
4" x 1¼"	P1@.625	Ea	97.30	23.30	—	120.60
4" x 1½"	P1@.671	Ea	97.30	25.00	—	122.30
4" x 2"	P1@.718	Ea	97.30	26.70	—	124.00
4" x 2½"	P1@.765	Ea	97.30	28.50	—	125.80
4" x 3"	P1@.811	Ea	97.30	30.20	—	127.50

Copper, Type M with Soft-Soldered Joints

Description	Craft@Hrs	Unit	Material $	Labor $	Equipment $	Total $
Type M copper reducer with soft-soldered joints						
½" x 3/8"	P1@.124	Ea	.56	4.62	—	5.18
¾" x ½"	P1@.129	Ea	1.06	4.80	—	5.86
1" x ¾"	P1@.172	Ea	1.91	6.40	—	8.31
1¼" x ½"	P1@.172	Ea	3.21	6.40	—	9.61
1¼" x ¾"	P1@.193	Ea	3.21	7.18	—	10.39
1¼" x 1"	P1@.215	Ea	3.21	8.00	—	11.21
1½" x ½"	P1@.193	Ea	4.90	7.18	—	12.08
1½" x ¾"	P1@.214	Ea	4.90	7.97	—	12.87
1½" x 1"	P1@.236	Ea	4.90	8.78	—	13.68
1½" x 1¼"	P1@.257	Ea	4.90	9.57	—	14.47
2" x ½"	P1@.239	Ea	7.38	8.90	—	16.28
2" x ¾"	P1@.261	Ea	7.38	9.71	—	17.09
2" x 1"	P1@.282	Ea	7.38	10.50	—	17.88
2" x 1¼"	P1@.304	Ea	7.38	11.30	—	18.68
2" x 1½"	P1@.325	Ea	7.38	12.10	—	19.48
2½" x 1"	P1@.325	Ea	20.10	12.10	—	32.20
2½" x 1¼"	P1@.347	Ea	20.10	12.90	—	33.00
2½" x 1½"	P1@.368	Ea	20.10	13.70	—	33.80
2½" x 2"	P1@.414	Ea	20.10	15.40	—	35.50
3" x 2"	P1@.457	Ea	23.50	17.00	—	40.50
3" x 2½"	P1@.500	Ea	23.50	18.60	—	42.10
4" x 2"	P1@.543	Ea	45.30	20.20	—	65.50
4" x 2½"	P1@.586	Ea	45.30	21.80	—	67.10
4" x 3"	P1@.629	Ea	45.30	23.40	—	68.70
Type M copper adapter C x MPT with soft-soldered joint						
½"	P1@.075	Ea	1.39	2.79	—	4.18
¾"	P1@.105	Ea	2.18	3.91	—	6.09
1"	P1@.134	Ea	5.41	4.99	—	10.40
1¼"	P1@.164	Ea	10.10	6.10	—	16.20
1½"	P1@.194	Ea	11.40	7.22	—	18.62
2"	P1@.259	Ea	14.20	9.64	—	23.84
2½"	P1@.319	Ea	46.10	11.90	—	58.00
3"	P1@.378	Ea	76.40	14.10	—	90.50
Type M copper adapter C x FPT with soft-soldered joint						
½"	P1@.075	Ea	2.32	2.79	—	5.11
¾"	P1@.105	Ea	4.21	3.91	—	8.12
1"	P1@.134	Ea	6.35	4.99	—	11.34
1¼"	P1@.164	Ea	13.40	6.10	—	19.50
1½"	P1@.194	Ea	16.10	7.22	—	23.32
2"	P1@.259	Ea	24.00	9.64	—	33.64
2½"	P1@.319	Ea	60.50	11.90	—	72.40
3"	P1@.378	Ea	83.20	14.10	—	97.30

Copper, Type M with Soft-Soldered Joints

Description	Craft@Hrs	Unit	Material $	Labor $	Equipment $	Total $
Type M copper flush bushing with soft-soldered joints						
½" x 3/8"	P1@.124	Ea	1.41	4.62	—	6.03
¾" x ½"	P1@.129	Ea	2.41	4.80	—	7.21
1" x ¾"	P1@.172	Ea	6.24	6.40	—	12.64
1" x ½"	P1@.172	Ea	7.12	6.40	—	13.52
1¼" x 1"	P1@.215	Ea	7.40	8.00	—	15.40
1½" x 1¼"	P1@.257	Ea	9.42	9.57	—	18.99
2" x 1½"	P1@.325	Ea	17.20	12.10	—	29.30
Type M copper union with soft-soldered joint						
½"	P1@.121	Ea	5.67	4.50	—	10.17
¾"	P1@.170	Ea	6.76	6.33	—	13.09
1"	P1@.218	Ea	15.20	8.11	—	23.31
1¼"	P1@.267	Ea	27.60	9.94	—	37.54
1½"	P1@.315	Ea	33.90	11.70	—	45.60
2"	P1@.421	Ea	48.90	15.70	—	64.60
Type M copper dielectric union with soft-soldered joint						
½"	P1@.121	Ea	5.31	4.50	—	9.81
¾"	P1@.170	Ea	5.31	6.33	—	11.64
1"	P1@.218	Ea	11.00	8.11	—	19.11
1¼"	P1@.267	Ea	18.70	9.94	—	28.64
1½"	P1@.315	Ea	25.80	11.70	—	37.50
2"	P1@.421	Ea	34.40	15.70	—	50.10
Type M copper cap with soft-soldered joint						
½"	P1@.069	Ea	.47	2.57	—	3.04
¾"	P1@.096	Ea	.94	3.57	—	4.51
1"	P1@.124	Ea	2.13	4.62	—	6.75
1¼"	P1@.151	Ea	3.06	5.62	—	8.68
1½"	P1@.178	Ea	4.46	6.63	—	11.09
2"	P1@.237	Ea	7.77	8.82	—	16.59
2½"	P1@.292	Ea	19.70	10.90	—	30.60
3"	P1@.347	Ea	30.80	12.90	—	43.70
4"	P1@.457	Ea	47.40	17.00	—	64.40
Type M copper coupling with soft-soldered joints						
½"	P1@.107	Ea	.30	3.98	—	4.28
¾"	P1@.150	Ea	.75	5.58	—	6.33
1"	P1@.193	Ea	2.02	7.18	—	9.20
1¼"	P1@.236	Ea	3.53	8.78	—	12.31
1½"	P1@.278	Ea	5.10	10.30	—	15.40
2"	P1@.371	Ea	7.50	13.80	—	21.30
2½"	P1@.457	Ea	17.70	17.00	—	34.70
3"	P1@.543	Ea	27.00	20.20	—	47.20
4"	P1@.714	Ea	52.90	26.60	—	79.50

Copper, Type M with Soft-Soldered Joints

Description	Craft@Hrs	Unit	Material $	Labor $	Equipment $	Total $

Class 125 bronze body gate valve with soft-soldered joints

½"	P1@.200	Ea	18.80	7.44	—	26.24
¾"	P1@.249	Ea	23.50	9.27	—	32.77
1"	P1@.299	Ea	33.00	11.10	—	44.10
1¼"	P1@.398	Ea	41.50	14.80	—	56.30
1½"	P1@.448	Ea	56.00	16.70	—	72.70
2"	P1@.498	Ea	93.60	18.50	—	112.10
2½"	P1@.830	Ea	155.00	30.90	—	185.90
3"	P1@1.24	Ea	222.00	46.20	—	268.20

Class 125 iron body gate valve with flanged ends

2"	P1@.500	Ea	358.00	18.60	—	376.60
2½"	P1@.600	Ea	483.00	22.30	—	505.30
3"	P1@.750	Ea	524.00	27.90	—	551.90
4"	P1@1.35	Ea	769.00	50.20	—	819.20
5"	P1@2.00	Ea	1,470.00	74.40	—	1,544.40
6"	P1@2.50	Ea	1,470.00	93.10	—	1,563.10

Class 125 bronze body globe valve with soft-soldered joints

½"	P1@.200	Ea	34.80	7.44	—	42.24
¾"	P1@.249	Ea	46.30	9.27	—	55.57
1"	P1@.299	Ea	79.80	11.10	—	90.90
1¼"	P1@.398	Ea	112.00	14.80	—	126.80
1½"	P1@.448	Ea	152.00	16.70	—	168.70
2"	P1@.498	Ea	246.00	18.50	—	264.50

Class 125 iron body globe valve with flanged ends

2½"	P1@.600	Ea	406.00	22.30	—	428.30
3"	P1@.750	Ea	487.00	27.90	—	514.90
4"	P1@1.35	Ea	653.00	50.20	—	703.20

200 PSIG iron body butterfly valve, lug-type, lever operated

2"	P1@.450	Ea	161.00	16.70	—	177.70
2½"	P1@.450	Ea	167.00	16.70	—	183.70
3"	P1@.550	Ea	176.00	20.50	—	196.50
4"	P1@.550	Ea	219.00	20.50	—	239.50

200 PSIG iron body butterfly valve, wafer-type, lever operated

2"	P1@.450	Ea	147.00	16.70	—	163.70
2½"	P1@.450	Ea	150.00	16.70	—	166.70
3"	P1@.550	Ea	161.00	20.50	—	181.50
4"	P1@.550	Ea	195.00	20.50	—	215.50

Copper, Type M with Soft-Soldered Joints

Description	Craft@Hrs	Unit	Material $	Labor $	Equipment $	Total $

Class 125 bronze body 2-piece ball valve with soft-soldered joints

Description	Craft@Hrs	Unit	Material $	Labor $	Equipment $	Total $
½"	P1@.200	Ea	10.40	7.44	—	17.84
¾"	P1@.249	Ea	13.80	9.27	—	23.07
1"	P1@.299	Ea	24.00	11.10	—	35.10
1¼"	P1@.398	Ea	40.00	14.80	—	54.80
1½"	P1@.448	Ea	54.10	16.70	—	70.80
2"	P1@.498	Ea	68.60	18.50	—	87.10
3"	P1@1.24	Ea	470.00	46.20	—	516.20
4"	P1@1.45	Ea	614.00	54.00	—	668.00

Class 125 bronze body swing check valve with soft-soldered joints

Description	Craft@Hrs	Unit	Material $	Labor $	Equipment $	Total $
½"	P1@.200	Ea	20.80	7.44	—	28.24
¾"	P1@.249	Ea	29.70	9.27	—	38.97
1"	P1@.299	Ea	38.80	11.10	—	49.90
1¼"	P1@.398	Ea	55.40	14.80	—	70.20
1½"	P1@.448	Ea	78.10	16.70	—	94.80
2"	P1@.498	Ea	129.00	18.50	—	147.50
2½"	P1@.830	Ea	261.00	30.90	—	291.90
3"	P1@1.24	Ea	375.00	46.20	—	421.20

Class 125 iron body swing check valve, flanged ends

Description	Craft@Hrs	Unit	Material $	Labor $	Equipment $	Total $
2"	P1@.500	Ea	182.00	18.60	—	200.60
2½"	P1@.600	Ea	230.00	22.30	—	252.30
3"	P1@.750	Ea	287.00	27.90	—	314.90
4"	P1@1.35	Ea	421.00	50.20	—	471.20

Class 125 iron body silent check valve, wafer-type

Description	Craft@Hrs	Unit	Material $	Labor $	Equipment $	Total $
2"	P1@.500	Ea	133.00	18.60	—	151.60
2½"	P1@.600	Ea	148.00	22.30	—	170.30
3"	P1@.750	Ea	170.00	27.90	—	197.90
4"	P1@1.35	Ea	227.00	50.20	—	277.20

Class 125 bronze body strainer, threaded ends

Description	Craft@Hrs	Unit	Material $	Labor $	Equipment $	Total $
½"	P1@.230	Ea	32.00	8.56	—	40.56
¾"	P1@.260	Ea	42.00	9.68	—	51.68
1"	P1@.330	Ea	51.50	12.30	—	63.80
1¼"	P1@.440	Ea	72.00	16.40	—	88.40
1½"	P1@.495	Ea	108.00	18.40	—	126.40
2"	P1@.550	Ea	188.00	20.50	—	208.50

Class 125 iron body strainer, flanged ends

Description	Craft@Hrs	Unit	Material $	Labor $	Equipment $	Total $
2"	P1@.500	Ea	144.00	18.60	—	162.60
2½"	P1@.600	Ea	160.00	22.30	—	182.30
3"	P1@.750	Ea	186.00	27.90	—	213.90
4"	P1@1.35	Ea	314.00	50.20	—	364.20

Copper, Type M with Soft-Soldered Joints

Description	Craft@Hrs	Unit	Material $	Labor $	Equipment $	Total $

Installation of 2-way control valve, threaded joints

Description	Craft@Hrs	Unit	Material $	Labor $	Equipment $	Total $
½"	P1@.210	Ea	—	7.82	—	7.82
¾"	P1@.275	Ea	—	10.20	—	10.20
1"	P1@.350	Ea	—	13.00	—	13.00
1¼"	P1@.430	Ea	—	16.00	—	16.00
1½"	P1@.505	Ea	—	18.80	—	18.80
2"	P1@.675	Ea	—	25.10	—	25.10
2½"	P1@.830	Ea	—	30.90	—	30.90
3"	P1@.990	Ea	—	36.80	—	36.80

Installation of 3-way control valve, threaded joints

Description	Craft@Hrs	Unit	Material $	Labor $	Equipment $	Total $
½"	P1@.260	Ea	—	9.68	—	9.68
¾"	P1@.365	Ea	—	13.60	—	13.60
1"	P1@.475	Ea	—	17.70	—	17.70
1¼"	P1@.575	Ea	—	21.40	—	21.40
1½"	P1@.680	Ea	—	25.30	—	25.30
2"	P1@.910	Ea	—	33.90	—	33.90
2½"	P1@1.12	Ea	—	41.70	—	41.70
3"	P1@1.33	Ea	—	49.50	—	49.50

Companion flange, 150 pound cast brass

Description	Craft@Hrs	Unit	Material $	Labor $	Equipment $	Total $
2"	P1@.290	Ea	301.00	10.80	—	311.80
2½"	P1@.380	Ea	368.00	14.10	—	382.10
3"	P1@.460	Ea	368.00	17.10	—	385.10
4"	P1@.600	Ea	588.00	22.30	—	610.30

Bolt and gasket sets

Description	Craft@Hrs	Unit	Material $	Labor $	Equipment $	Total $
2"	P1@.500	Ea	4.50	18.60	—	23.10
2½"	P1@.650	Ea	5.00	24.20	—	29.20
3"	P1@.750	Ea	8.49	27.90	—	36.39
4"	P1@1.00	Ea	14.10	37.20	—	51.30

Thermometer with well

Description	Craft@Hrs	Unit	Material $	Labor $	Equipment $	Total $
7"	P1@.250	Ea	183.00	9.31	—	192.31
9"	P1@.250	Ea	188.00	9.31	—	197.31

Dial-type pressure gauge

Description	Craft@Hrs	Unit	Material $	Labor $	Equipment $	Total $
2½"	P1@.200	Ea	38.30	7.44	—	45.74
3½"	P1@.200	Ea	50.40	7.44	—	57.84

Pressure/temperature tap

Description	Craft@Hrs	Unit	Material $	Labor $	Equipment $	Total $
Tap	P1@.150	Ea	17.60	5.58	—	23.18

Copper, Press Fit Fittings

Press Fit (Pressfast) Copper Pipe Fittings. For joining ½" to 4" type K, L or M hard drawn or annealed copper pipe and tube (ASTM 888). No brazing required. Includes a square cut and de-burr of the pipe ends and joining with a hand-held electric press fit tool. Equipment cost, when shown, is $1.95 per hour for pressfit tool. Fittings include an O-ring for each joint.

Description	Craft@Hrs	Unit	Material $	Labor $	Equipment $	Total $

45-degree Copper press fit ell (P x P)

Description	Craft@Hrs	Unit	Material $	Labor $	Equipment $	Total $
½"	P1@.170	Ea	3.05	6.33	.16	9.54
¾"	P1@.170	Ea	3.96	6.33	.16	10.45
1"	P1@.180	Ea	12.60	6.70	.17	19.47
1¼"	P1@.190	Ea	18.10	7.07	.18	25.35
1½"	P1@.200	Ea	29.30	7.44	.19	36.93
2"	P1@.200	Ea	41.00	7.44	.19	48.63
2½"	P1@.225	Ea	90.20	8.37	.22	98.79
3"	P1@.240	Ea	126.00	8.93	.23	135.16
4"	P1@.260	Ea	180.00	9.68	.25	189.93

90-degree Copper press fit ell (P x P)

Description	Craft@Hrs	Unit	Material $	Labor $	Equipment $	Total $
½"	P1@.170	Ea	2.52	6.33	.16	9.01
¾"	P1@.170	Ea	4.23	6.33	.16	10.72
1"	P1@.180	Ea	9.35	6.70	.17	16.22
1¼"	P1@.190	Ea	18.90	7.07	.18	26.15
1½"	P1@.200	Ea	35.50	7.44	.19	43.13
2"	P1@.200	Ea	49.60	7.44	.19	57.23
2½"	P1@.225	Ea	105.00	8.37	.22	113.59
3"	P1@.240	Ea	196.00	8.93	.23	205.16
4"	P1@.260	Ea	239.00	9.68	.25	248.93

Coupling, Copper, press fit (P x P)

Description	Craft@Hrs	Unit	Material $	Labor $	Equipment $	Total $
½"	P1@.170	Ea	2.26	6.33	.16	8.75
¾"	P1@.170	Ea	3.44	6.33	.16	9.93
1"	P1@.180	Ea	6.90	6.70	.17	13.77
1¼"	P1@.190	Ea	8.56	7.07	.18	15.81
1½"	P1@.200	Ea	16.00	7.44	.19	23.63
2"	P1@.200	Ea	20.30	7.44	.19	27.93
2½"	P1@.225	Ea	67.60	8.37	.22	76.19
3"	P1@.240	Ea	87.10	8.93	.23	96.26
4"	P1@.260	Ea	120.00	9.68	.25	129.93

Copper, Press Fit Fittings

Description	Craft@Hrs	Unit	Material $	Labor $	Equipment $	Total $

Tee, Copper, press fit (P x P)

Description	Craft@Hrs	Unit	Material $	Labor $	Equipment $	Total $
½"	P1@.170	Ea	3.05	6.33	.16	9.54
¾"	P1@.170	Ea	3.96	6.33	.16	10.45
1"	P1@.180	Ea	12.60	6.70	.17	19.47
1¼"	P1@.190	Ea	18.10	7.07	.18	25.35
1½"	P1@.200	Ea	29.30	7.44	.19	36.93
2"	P1@.200	Ea	41.00	7.44	.19	48.63
2½"	P1@.225	Ea	90.20	8.37	.22	98.79
3"	P1@.240	Ea	126.00	8.93	.23	135.16
4"	P1@.260	Ea	180.00	9.68	.25	189.93

Reducing tee, Copper press fit (P x P x P)

Description	Craft@Hrs	Unit	Material $	Labor $	Equipment $	Total $
¾ x ½"	P1@.200	Ea	9.24	7.44	.19	16.87
1 x ¾"	P1@.200	Ea	11.70	7.44	.19	19.33
1¼" x ¾"	P1@.240	Ea	23.80	8.93	.23	32.96
1¼" x 1"	P1@.240	Ea	17.00	8.93	.23	26.16
1½" x ¾"	P1@.255	Ea	36.30	9.49	.25	46.04
1½ x 1"	P1@.255	Ea	37.70	9.49	.25	47.44
2" x ¾"	P1@.270	Ea	58.70	10.00	.26	68.96
2" x 1"	P1@.270	Ea	60.80	10.00	.26	71.06
2" x 1½"	P1@.280	Ea	46.00	10.40	.27	56.67
3" x 1½"	P1@.380	Ea	229.00	14.10	.37	243.47
3" x 2"	P1@.400	Ea	203.00	14.90	.39	218.29
4" x 1½"	P1@.400	Ea	273.00	14.90	.39	288.29
4" x 2"	P1@.420	Ea	258.00	15.60	.41	274.01
4" x 3"	P1@.450	Ea	239.00	16.70	.44	256.14

Union, Copper press fit (P x M)

Description	Craft@Hrs	Unit	Material $	Labor $	Equipment $	Total $
½"	P1@.220	Ea	20.60	8.19	.21	29.00
¾"	P1@.220	Ea	27.00	8.19	.21	35.40
1"	P1@.260	Ea	43.20	9.68	.25	53.13
1¼"	P1@.280	Ea	61.60	10.40	.27	72.27
1½"	P1@.280	Ea	84.10	10.40	.27	94.77
2"	P1@.300	Ea	137.00	11.20	.29	148.49
2½"	P1@.320	Ea	209.00	11.90	.31	221.21
3"	P1@.380	Ea	226.00	14.10	.37	240.47
4"	P1@.420	Ea	261.00	15.60	.41	277.01

Copper, Press Fit Fittings

Description	Craft@Hrs	Unit	Material $	Labor $	Equipment $	Total $
Male threaded Copper press fit adapter (P x M)						
½"	P1@.220	Ea	3.25	8.19	.21	11.65
¾" x ½"	P1@.220	Ea	6.63	8.19	.21	15.03
¾"	P1@.220	Ea	5.86	8.19	.21	14.26
¾" x 1"	P1@.240	Ea	32.00	8.93	.23	41.16
1 x ¾"	P1@.250	Ea	17.00	9.31	.24	26.55
1"	P1@.260	Ea	10.70	9.68	.25	20.63
1¼"	P1@.280	Ea	23.30	10.40	.27	33.97
1½"	P1@.280	Ea	33.70	10.40	.27	44.37
2"	P1@.300	Ea	63.20	11.20	.29	74.69
2½"	P1@.320	Ea	140.00	11.90	.31	152.21
3"	P1@.380	Ea	175.00	14.10	.37	189.47
4"	P1@.420	Ea	212.00	15.60	.41	228.01
Female threaded Copper press fit adapter (P x F)						
½"	P1@.200	Ea	3.81	7.44	.19	11.44
¾"	P1@.220	Ea	6.21	8.19	.21	14.61
¾" x ½"	P1@.220	Ea	6.92	8.19	.21	15.32
¾"	P1@.220	Ea	5.86	8.19	.21	14.26
1" x ½"	P1@.240	Ea	20.40	8.93	.23	29.56
1 x ¾"	P1@.240	Ea	19.30	8.93	.23	28.46
1"	P1@.260	Ea	11.30	9.68	.25	21.23
1¼"	P1@.280	Ea	25.80	10.40	.27	36.47
1½"	P1@.280	Ea	37.60	10.40	.27	48.27
2"	P1@.300	Ea	65.60	11.20	.29	77.09
3"	P1@.380	Ea	175.00	14.10	.37	189.47
4"	P1@.420	Ea	213.00	15.60	.41	229.01
Ball Valve, Brass press fit (P x P)						
½"	P1@.220	Ea	27.40	8.19	.21	35.80
¾"	P1@.225	Ea	36.40	8.37	.22	44.99
1"	P1@.275	Ea	47.60	10.20	.27	58.07
1¼"	P1@.295	Ea	93.10	11.00	.29	104.39
1½"	P1@.295	Ea	116.00	11.00	.29	127.29
2"	P1@.330	Ea	190.00	12.30	.32	202.62
3"	P1@.395	Ea	406.00	14.70	.38	421.08
4"	P1@.435	Ea	515.00	16.20	.42	531.62

Copper, Type K & L with Roll-Grooved Joints

Hard drawn rigid copper roll-grooved pipe with factory grooved copper fittings is commonly used for domestic or potable water distribution systems.

Maximum operating pressure for fittings is 300 psi (2065kPa). Operating temperature for fittings ranges from −30 F. to 250 F. (−34 C. – 121 C.).

Consult manufacturer for pipe pressure and temperature ratings and for the various combinations of pipe, fittings and system applications.

The cost estimates in this section are based on the conditions, limitations and wage rates described in the section "How to Use This Book" beginning on page 5.

Description	Craft@Hrs	Unit	Material $	Labor $	Equipment $	Total $

Type K Copper pipe only, roll-grooved, no hangers or fittings

Size	Craft@Hrs	Unit	Material $	Labor $	Equipment $	Total $
2"	P1@.080	LF	16.70	2.98	—	19.68
2½"	P1@.090	LF	24.40	3.35	—	27.75
3"	P1@.105	LF	33.10	3.91	—	37.01
4"	P1@.145	LF	51.70	5.40	—	57.10
6"	P1@.195	LF	79.00	7.26	—	86.26

Type L Copper pipe only, roll-grooved, no hangers or fittings

Size	Craft@Hrs	Unit	Material $	Labor $	Equipment $	Total $
2"	P1@.070	LF	15.20	2.61	—	17.81
2½"	P1@.070	LF	21.90	2.61	—	24.51
3"	P1@.090	LF	29.70	3.35	—	33.05
4"	P1@.130	LF	46.40	4.84	—	51.24
6"	P1@.180	LF	71.00	6.70	—	77.70

90-degree copper elbow, roll-grooved, style #610, (Victaulic)

Size	Craft@Hrs	Unit	Material $	Labor $	Equipment $	Total $
2"	P1@.410	Ea	101.00	15.30	—	116.30
2½"	P1@.430	Ea	112.00	16.00	—	128.00
3"	P1@.500	Ea	157.00	18.60	—	175.60
4"	P1@.690	Ea	336.00	25.70	—	361.70
6"	P1@1.19	Ea	1,980.00	44.30	—	2,024.30

45-degree copper elbow, roll-grooved, style #611, (Victaulic)

Size	Craft@Hrs	Unit	Material $	Labor $	Equipment $	Total $
2"	P1@.410	Ea	89.30	15.30	—	104.60
2½"	P1@.430	Ea	97.30	16.00	—	113.30
3"	P1@.500	Ea	134.00	18.60	—	152.60
4"	P1@.690	Ea	308.00	25.70	—	333.70
6"	P1@1.19	Ea	1,690.00	44.30	—	1,734.30

Tee, copper, roll-grooved, style #620, (Victaulic)

Size	Craft@Hrs	Unit	Material $	Labor $	Equipment $	Total $
2"	P1@.510	Ea	173.00	19.00	—	192.00
2½"	P1@.540	Ea	183.00	20.10	—	203.10
3"	P1@.630	Ea	276.00	23.40	—	299.40
4"	P1@.860	Ea	609.00	32.00	—	641.00
6"	P1@1.49	Ea	2,260.00	55.50	—	2,315.50

Copper, Type K & L with Roll-Grooved Joints

Description	Craft@Hrs	Unit	Material $	Labor $	Equipment $	Total $
Reducing tee, copper, roll-grooved, style #625, (Victaulic)						
2½" x 2"	P1@.540	Ea	281.00	20.10	—	301.10
3" x 2½"	P1@.630	Ea	302.00	23.40	—	325.40
4" x 3"	P1@.860	Ea	709.00	32.00	—	741.00
Reducing tee, copper, roll-grooved, style #626, (Victaulic)						
2½" x 1"	P1@.500	Ea	358.00	18.60	—	376.60
2½" x 1½"	P1@.520	Ea	358.00	19.40	—	377.40
3" x 1½"	P1@.600	Ea	378.00	22.30	—	400.30
Reducer, copper, roll-grooved, style #650, (Victaulic)						
2½" x 2"	P1@.430	Ea	151.00	16.00	—	167.00
3" x 2"	P1@.500	Ea	151.00	18.60	—	169.60
3" x 2½"	P1@.520	Ea	151.00	19.40	—	170.40
4" x 2"	P1@.690	Ea	297.00	25.70	—	322.70
4" x 3"	P1@.720	Ea	297.00	26.80	—	323.80
6" x 3"	P1@1.20	Ea	1,030.00	44.70	—	1,074.70
Roll-grooved copper coupling with gasket, style #650, (Victaulic)						
2"	P1@.300	Ea	49.90	11.20	—	61.10
2½"	P1@.350	Ea	56.40	13.00	—	69.40
3"	P1@.400	Ea	60.90	14.90	—	75.80
4"	P1@.500	Ea	78.30	18.60	—	96.90
5"	P1@.600	Ea	158.00	22.30	—	180.30
6"	P1@.700	Ea	181.00	26.10	—	207.10
8"	P1@.900	Ea	280.00	33.50	—	313.50
10"	P1@1.10	Ea	414.00	40.90	—	454.90
12"	P1@1.35	Ea	620.00	50.20	—	670.20
Flange adapter, copper, roll-grooved, style #641, (Victaulic)						
2½"	P1@.300	Ea	300.00	11.20	—	311.20
3"	P1@.400	Ea	320.00	14.90	—	334.90
4"	P1@.550	Ea	439.00	20.50	—	459.50
Butterfly valve, brass, roll-grooved, lever handle, style #608, (Victaulic)						
2½"	P1@.450	Ea	964.00	16.70	—	980.70
3"	P1@.450	Ea	1,240.00	16.70	—	1,256.70
4"	P1@.550	Ea	1,660.00	20.50	—	1,680.50

PVC, Schedule 40, with Solvent-Weld Joints

PVC (Polyvinyl Chloride) Schedule 40 pipe with Schedule 40 socket-type fittings and solvent-welded joints is widely used in process, chemical, A.C. condensate, potable water and irrigation piping systems. PVC is available in Type I (normal impact) and Type II (high impact). Type I has a maximum temperature rating of 150 degrees F., and Type II is rated at 140 degrees F.

Consult the manufacturers for maximum pressure ratings and recommended joint solvents for specific applications.

Because of the current unreliability of many plastic valves, most engineers specify standard bronze-body and iron-body valves for use in PVC water piping systems. Plastic valves, however, have to be used in systems conveying liquids that may be injurious to metallic valves.

This section has been arranged to save the estimator's time by including all normally-used system components such as pipe, fittings, valves, hanger assemblies, riser clamps and miscellaneous items under one heading. Additional items can be found under "Plumbing and Piping Specialties." The cost estimates in this section are based on the conditions, limitations and wage rates described in the section "How to Use This Book" beginning on page 5.

Equipment cost, where shown, is $4.45 per hour for a 2-ton chain hoist.

Description	Craft@Hrs	Unit	Material $	Labor $	Equipment $	Total $

Schedule 40 PVC pipe assembly with solvent-weld joints installed horizontally. Complete installation including three to eight tees and three to eight elbows every 100' and hangers spaced to meet plumbing code *(a tee and elbow every 12' for 1/2" dia., a tee and elbow every 40' for 8" dia.)*.

Description	Craft@Hrs	Unit	Material $	Labor $	Equipment $	Total $
½"	P1@.084	LF	1.87	3.13	—	5.00
¾"	P1@.091	LF	2.23	3.39	—	5.62
1"	P1@.099	LF	2.34	3.68	—	6.02
1¼"	P1@.114	LF	2.45	4.24	—	6.69
1½"	P1@.121	LF	2.78	4.50	—	7.28
2"	P1@.130	LF	2.86	4.84	—	7.70
2½"	P1@.134	LF	4.62	4.99	—	9.61
3"	P1@.146	LF	7.24	5.43	—	12.67
4"	P1@.177	LF	9.19	6.59	—	15.78
6"	P1@.231	LF	16.90	8.60	—	25.50
8"	P1@.236	LF	30.00	8.78	—	38.78

Schedule 40 PVC pipe assembly with solvent-weld joints, installed risers. Complete installation including a reducing tee every floor and a riser clamp every other.

Description	Craft@Hrs	Unit	Material $	Labor $	Equipment $	Total $
½"	P1@.040	LF	.94	1.49	—	2.43
¾"	P1@.047	LF	1.24	1.75	—	2.99
1"	P1@.052	LF	1.58	1.94	—	3.52
1¼"	P1@.061	LF	1.99	2.27	—	4.26
1½"	P1@.069	LF	2.44	2.57	—	5.01
2"	P1@.077	LF	2.56	2.87	—	5.43
2½"	P1@.088	LF	4.07	3.28	—	7.35
3"	P1@.099	LF	6.18	3.68	—	9.86
4"	P1@.125	LF	7.95	4.65	—	12.60
6"	P1@.202	LF	18.40	7.52	—	25.92
8"	P1@.266	LF	33.90	9.90	—	43.80

PVC, Schedule 40, with Solvent-Weld Joints

Description	Craft@Hrs	Unit	Material $	Labor $	Equipment $	Total $
Schedule 40 PVC pipe with solvent-weld joints						
½"	P1@.020	LF	.55	.74	—	1.29
¾"	P1@.025	LF	.82	.93	—	1.75
1"	P1@.030	LF	1.11	1.12	—	2.23
1¼"	P1@.035	LF	1.35	1.30	—	2.65
1½"	P1@.040	LF	1.60	1.49	—	3.09
2"	P1@.045	LF	1.66	1.67	—	3.33
2½"	P1@.050	LF	2.61	1.86	—	4.47
3"	P1@.055	LF	4.19	2.05	—	6.24
4"	P1@.070	LF	4.88	2.61	—	7.49
5"	P1@.095	LF	9.19	3.54	—	12.73
6"	P1@.120	LF	9.19	4.47	—	13.66
8"	P1@.160	LF	13.00	5.96	—	18.96
Schedule 40 PVC 45-degree ell S x S with solvent-weld joints						
½"	P1@.100	Ea	.61	3.72	—	4.33
¾"	P1@.115	Ea	.93	4.28	—	5.21
1"	P1@.120	Ea	1.12	4.47	—	5.59
1¼"	P1@.160	Ea	1.58	5.96	—	7.54
1½"	P1@.180	Ea	1.98	6.70	—	8.68
2"	P1@.200	Ea	2.69	7.44	—	10.13
2½"	P1@.250	Ea	6.70	9.31	—	16.01
3"	P1@.300	Ea	10.40	11.20	—	21.60
4"	P1@.500	Ea	18.60	18.60	—	37.20
5"	P1@.550	Ea	36.90	20.50	—	57.40
6"	P1@.600	Ea	46.10	22.30	—	68.40
8"	P1@.800	Ea	110.00	29.80	—	139.80
Schedule 40 PVC 90-degree ell S x S with solvent-weld joints						
½"	P1@.100	Ea	.38	3.72	—	4.10
¾"	P1@.115	Ea	.42	4.28	—	4.70
1"	P1@.120	Ea	.72	4.47	—	5.19
1¼"	P1@.160	Ea	1.32	5.96	—	7.28
1½"	P1@.180	Ea	1.39	6.70	—	8.09
2"	P1@.200	Ea	2.19	7.44	—	9.63
2½"	P1@.250	Ea	6.68	9.31	—	15.99
3"	P1@.300	Ea	7.98	11.20	—	19.18
4"	P1@.500	Ea	14.40	18.60	—	33.00
5"	P1@.550	Ea	36.90	20.50	—	57.40
6"	P1@.600	Ea	45.60	22.30	—	67.90
8"	P1@.800	Ea	117.00	29.80	—	146.80

PVC, Schedule 40, with Solvent-Weld Joints

Description	Craft@Hrs	Unit	Material $	Labor $	Equipment $	Total $

Schedule 40 PVC tee S x S x S with solvent-weld joints

Description	Craft@Hrs	Unit	Material $	Labor $	Equipment $	Total $
½"	P1@.130	Ea	.49	4.84	—	5.33
¾"	P1@.140	Ea	.53	5.21	—	5.74
1"	P1@.170	Ea	.97	6.33	—	7.30
1¼"	P1@.220	Ea	1.53	8.19	—	9.72
1½"	P1@.250	Ea	1.70	9.31	—	11.01
2"	P1@.280	Ea	2.19	10.40	—	12.59
2½"	P1@.350	Ea	9.02	13.00	—	22.02
3"	P1@.420	Ea	11.50	15.60	—	27.10
4"	P1@.560	Ea	21.40	20.80	—	42.20
5"	P1@.710	Ea	51.50	26.40	—	77.90
6"	P1@.840	Ea	74.70	31.30	—	106.00
8"	P1@1.12	Ea	165.00	41.70	—	206.70

Schedule 40 PVC reducing tee S x S x S with solvent-weld joints

Description	Craft@Hrs	Unit	Material $	Labor $	Equipment $	Total $
¾" x ½"	P1@.140	Ea	.61	5.21	—	5.82
1" x ½"	P1@.170	Ea	1.12	6.33	—	7.45
1" x ¾"	P1@.170	Ea	1.68	6.33	—	8.01
1¼" x ½"	P1@.220	Ea	1.68	8.19	—	9.87
1¼" x ¾"	P1@.220	Ea	1.68	8.19	—	9.87
1¼" x 1"	P1@.220	Ea	1.68	8.19	—	9.87
1½" x ½"	P1@.250	Ea	2.90	9.31	—	12.21
1½" x ¾"	P1@.250	Ea	2.90	9.31	—	12.21
1½" x 1"	P1@.250	Ea	2.90	9.31	—	12.21
1½" x 1¼"	P1@.250	Ea	2.90	9.31	—	12.21
2" x ½"	P1@.280	Ea	4.13	10.40	—	14.53
2" x ¾"	P1@.280	Ea	4.13	10.40	—	14.53
2" x 1"	P1@.280	Ea	4.13	10.40	—	14.53
2" x 1¼"	P1@.280	Ea	4.13	10.40	—	14.53
2" x 1½"	P1@.280	Ea	4.13	10.40	—	14.53
2½" x 1"	P1@.350	Ea	8.59	13.00	—	21.59
2½" x 1¼"	P1@.350	Ea	8.59	13.00	—	21.59
2½" x 1½"	P1@.350	Ea	8.59	13.00	—	21.59
2½" x 2"	P1@.350	Ea	8.59	13.00	—	21.59
3" x 2"	P1@.420	Ea	12.70	15.60	—	28.30
3" x 2½"	P1@.420	Ea	12.70	15.60	—	28.30
4" x 2"	P1@.560	Ea	21.40	20.80	—	42.20
4" x 2½"	P1@.560	Ea	21.40	20.80	—	42.20
4" x 3"	P1@.560	Ea	21.40	20.80	—	42.20
5" x 3"	P1@.710	Ea	38.00	26.40	—	64.40
5" x 4"	P1@.710	Ea	38.00	26.40	—	64.40
6" x 3"	P1@.840	Ea	62.30	31.30	—	93.60
6" x 4"	P1@.840	Ea	62.30	31.30	—	93.60
6" x 5"	P1@.840	Ea	62.30	31.30	—	93.60
8" x 4"	P1@1.12	Ea	138.00	41.70	—	179.70
8" x 5"	P1@1.12	Ea	138.00	41.70	—	179.70
8" x 6"	P1@1.12	Ea	138.00	41.70	—	179.70

PVC, Schedule 40, with Solvent-Weld Joints

Description	Craft@Hrs	Unit	Material $	Labor $	Equipment $	Total $

Schedule 40 PVC reducing bushing SPIG x S with solvent-weld joints

Description	Craft@Hrs	Unit	Material $	Labor $	Equipment $	Total $
¾" x ½"	P1@.050	Ea	.40	1.86	—	2.26
1" x ½"	P1@.060	Ea	.70	2.23	—	2.93
1" x ¾"	P1@.060	Ea	.70	2.23	—	2.93
1¼" x ½"	P1@.080	Ea	.93	2.98	—	3.91
1¼" x ¾"	P1@.080	Ea	.93	2.98	—	3.91
1¼" x 1"	P1@.080	Ea	.93	2.98	—	3.91
1½" x ½"	P1@.100	Ea	.97	3.72	—	4.69
1½" x ¾"	P1@.100	Ea	.97	3.72	—	4.69
1½" x 1"	P1@.100	Ea	.97	3.72	—	4.69
1½" x 1¼"	P1@.100	Ea	.97	3.72	—	4.69
2" x 1"	P1@.110	Ea	1.63	4.09	—	5.72
2" x 1¼"	P1@.110	Ea	1.63	4.09	—	5.72
2" x 1½"	P1@.110	Ea	1.63	4.09	—	5.72
2½" x 1¼"	P1@.130	Ea	2.60	4.84	—	7.44
2½" x 1½"	P1@.130	Ea	2.60	4.84	—	7.44
2½" x 2"	P1@.130	Ea	2.60	4.84	—	7.44
3" x 2"	P1@.150	Ea	3.88	5.58	—	9.46
3" x 2½"	P1@.150	Ea	3.88	5.58	—	9.46
4" x 3"	P1@.200	Ea	8.67	7.44	—	16.11
5" x 3"	P1@.250	Ea	17.10	9.31	—	26.41
5" x 4"	P1@.250	Ea	17.10	9.31	—	26.41
6" x 4"	P1@.300	Ea	25.00	11.20	—	36.20
6" x 5"	P1@.300	Ea	25.00	11.20	—	36.20
8" x 4"	P1@.400	Ea	75.00	14.90	—	89.90
8" x 5"	P1@.400	Ea	75.00	14.90	—	89.90

Schedule 40 PVC adapter MPT x S with solvent-weld joints

Description	Craft@Hrs	Unit	Material $	Labor $	Equipment $	Total $
½"	P1@.060	Ea	.34	2.23	—	2.57
¾"	P1@.065	Ea	.38	2.42	—	2.80
1"	P1@.070	Ea	.66	2.61	—	3.27
1¼"	P1@.100	Ea	.81	3.72	—	4.53
1½"	P1@.110	Ea	1.09	4.09	—	5.18
2"	P1@.120	Ea	3.57	4.47	—	8.04
2½"	P1@.150	Ea	4.18	5.58	—	9.76
3"	P1@.180	Ea	6.08	6.70	—	12.78
4"	P1@.240	Ea	7.75	8.93	—	16.68
5"	P1@.370	Ea	13.20	13.80	—	27.00
6"	P1@.420	Ea	20.20	15.60	—	35.80

Schedule 40 PVC adapter FPT x S with solvent-weld joints

Description	Craft@Hrs	Unit	Material $	Labor $	Equipment $	Total $
½"	P1@.110	Ea	.42	4.09	—	4.51
¾"	P1@.120	Ea	.53	4.47	—	5.00
1"	P1@.130	Ea	.61	4.84	—	5.45
1¼"	P1@.170	Ea	.94	6.33	—	7.27
1½"	P1@.210	Ea	1.09	7.82	—	8.91
2"	P1@.290	Ea	1.47	10.80	—	12.27
2½"	P1@.330	Ea	3.67	12.30	—	15.97
3"	P1@.470	Ea	7.31	17.50	—	24.81
4"	P1@.630	Ea	13.50	23.40	—	36.90
5"	P1@.890	Ea	21.40	33.10	—	54.50
6"	P1@.970	Ea	30.10	36.10	—	66.20

PVC, Schedule 40, with Solvent-Weld Joints

Description	Craft@Hrs	Unit	Material $	Labor $	Equipment $	Total $

Schedule 40 PVC cap S with solvent-weld joints

½"	P1@.060	Ea	.34	2.23	—	2.57
¾"	P1@.065	Ea	.40	2.42	—	2.82
1"	P1@.070	Ea	.61	2.61	—	3.22
1¼"	P1@.100	Ea	.93	3.72	—	4.65
1½"	P1@.110	Ea	.94	4.09	—	5.03
2"	P1@.120	Ea	1.13	4.47	—	5.60
2½"	P1@.150	Ea	3.57	5.58	—	9.15
3"	P1@.180	Ea	3.94	6.70	—	10.64
4"	P1@.240	Ea	9.02	8.93	—	17.95
5"	P1@.370	Ea	15.00	13.80	—	28.80
6"	P1@.420	Ea	21.50	15.60	—	37.10

Schedule 40 PVC plug MPT with solvent-weld joints

½"	P1@.150	Ea	.99	5.58	—	6.57
¾"	P1@.160	Ea	1.08	5.96	—	7.04
1"	P1@.180	Ea	1.76	6.70	—	8.46
1¼"	P1@.210	Ea	1.82	7.82	—	9.64
1½"	P1@.230	Ea	1.96	8.56	—	10.52
2"	P1@.240	Ea	2.56	8.93	—	11.49
2½"	P1@.260	Ea	3.52	9.68	—	13.20
3"	P1@.300	Ea	5.33	11.20	—	16.53
4"	P1@.400	Ea	12.20	14.90	—	27.10

Schedule 40 PVC coupling S x S with solvent-weld joints

½"	P1@.100	Ea	.23	3.72	—	3.95
¾"	P1@.115	Ea	.34	4.28	—	4.62
1"	P1@.120	Ea	.59	4.47	—	5.06
1¼"	P1@.160	Ea	.81	5.96	—	6.77
1½"	P1@.180	Ea	.85	6.70	—	7.55
2"	P1@.200	Ea	1.33	7.44	—	8.77
2½"	P1@.250	Ea	2.90	9.31	—	12.21
3"	P1@.300	Ea	4.57	11.20	—	15.77
4"	P1@.500	Ea	6.56	18.60	—	25.16
5"	P1@.550	Ea	12.20	20.50	—	32.70
6"	P1@.600	Ea	20.70	22.30	—	43.00
8"	P1@.800	Ea	38.80	29.80	—	68.60

Schedule 40 PVC union S x S with solvent-weld joints

1/2"	P1@.125	Ea	3.05	4.65	—	7.70
3/4"	P1@.140	Ea	3.79	5.21	—	9.00
1"	P1@.160	Ea	3.94	5.96	—	9.90
1¼"	P1@.200	Ea	8.12	7.44	—	15.56
1½"	P1@.225	Ea	9.25	8.37	—	17.62
2"	P1@.250	Ea	14.40	9.31	—	23.71
2½"	P1@.275	Ea	21.40	10.20	—	31.60
3"	P1@.300	Ea	26.10	11.20	—	37.30

PVC, Schedule 40, with Solvent-Weld Joints

Description	Craft@Hrs	Unit	Material $	Labor $	Equipment $	Total $

Class 125 bronze body gate valve, threaded ends

½"	P1@.210	Ea	18.80	7.82	—	26.62
¾"	P1@.250	Ea	23.50	9.31	—	32.81
1"	P1@.300	Ea	33.00	11.20	—	44.20
1¼"	P1@.400	Ea	41.50	14.90	—	56.40
1½"	P1@.450	Ea	56.00	16.70	—	72.70
2"	P1@.500	Ea	93.60	18.60	—	112.20
2½"	P1@.750	Ea	155.00	27.90	—	182.90
3"	P1@.950	Ea	222.00	35.40	—	257.40

Class 125 iron body gate valve, flanged ends

2"	P1@.500	Ea	358.00	18.60	—	376.60
2½"	P1@.600	Ea	483.00	22.30	—	505.30
3"	P1@.750	Ea	524.00	27.90	—	551.90
4"	P1@1.35	Ea	769.00	50.20	—	819.20
5"	ER@2.00	Ea	1,470.00	79.90	1.30	1,551.20
6"	ER@2.50	Ea	1,470.00	99.90	1.63	1,571.53
8"	ER@3.00	Ea	2,420.00	120.00	1.96	2,541.96

Class 125 bronze body globe valve, threaded ends

½"	P1@.210	Ea	34.80	7.82	—	42.62
¾"	P1@.250	Ea	46.30	9.31	—	55.61
1"	P1@.300	Ea	66.40	11.20	—	77.60
1¼"	P1@.400	Ea	93.40	14.90	—	108.30
1½"	P1@.450	Ea	124.00	16.70	—	140.70
2"	P1@.500	Ea	204.00	18.60	—	222.60

200 PSIG iron body butterfly valve, lug-type, lever operated

2"	P1@.450	Ea	161.00	16.70	—	177.70
2½"	P1@.450	Ea	167.00	16.70	—	183.70
3"	P1@.550	Ea	176.00	20.50	—	196.50
4"	P1@.550	Ea	219.00	20.50	—	239.50
5"	ER@.800	Ea	288.00	32.00	.52	320.52
6"	ER@.800	Ea	352.00	32.00	.52	384.52
8"	ER@.900	Ea	481.00	36.00	.59	517.59

200 PSIG iron body butterfly valve, wafer-type, lever operated

2"	P1@.450	Ea	147.00	16.70	—	163.70
2½"	P1@.450	Ea	150.00	16.70	—	166.70
3"	P1@.550	Ea	161.00	20.50	—	181.50
4"	P1@.550	Ea	195.00	20.50	—	215.50
5"	ER@.800	Ea	260.00	32.00	.52	292.52
6"	ER@.800	Ea	326.00	32.00	.52	358.52
8"	ER@.900	Ea	452.00	36.00	.59	488.59

PVC, Schedule 40, with Solvent-Weld Joints

Description	Craft@Hrs	Unit	Material $	Labor $	Equipment $	Total $

Class 125 bronze body 2-piece ball valve, threaded ends

½"	P1@.210	Ea	10.40	7.82	—	18.22
¾"	P1@.250	Ea	13.80	9.31	—	23.11
1"	P1@.300	Ea	24.00	11.20	—	35.20
1¼"	P1@.400	Ea	40.00	14.90	—	54.90
1½"	P1@.450	Ea	54.00	16.70	—	70.70
2"	P1@.500	Ea	68.60	18.60	—	87.20
3"	P1@.625	Ea	469.00	23.30	—	492.30
4"	P1@.690	Ea	613.00	25.70	—	638.70

PVC ball valve, solid body, solvent weld joints, EDPM (female socket)

½"	P1@.170	Ea	5.86	6.33	—	12.19
¾"	P1@.200	Ea	6.63	7.44	—	14.07
1"	P1@.220	Ea	9.64	8.19	—	17.83
1¼"	P1@.280	Ea	11.10	10.40	—	21.50
1½"	P1@.315	Ea	18.70	11.70	—	30.40
2"	P1@.385	Ea	23.30	14.30	—	37.60

PVC ball valve, solid body, threaded joints, EDPM (female pipe thread)

½"	P1@.200	Ea	5.86	7.44	—	13.30
¾"	P1@.240	Ea	6.63	8.93	—	15.56
1"	P1@.290	Ea	9.64	10.80	—	20.44
1¼"	P1@.390	Ea	11.10	14.50	—	25.60
1½"	P1@.420	Ea	18.70	15.60	—	34.30
2"	P1@.490	Ea	23.30	18.20	—	41.50

PVC ball valve, tru-union, threaded joints, EDPM (female pipe thread)

2"	P1@.490	Ea	83.10	18.20	—	101.30
2½"	P1@.520	Ea	96.00	19.40	—	115.40
3"	P1@.590	Ea	222.00	22.00	—	244.00
4"	P1@.630	Ea	409.00	23.40	—	432.40

PVC, Schedule 40, with Solvent-Weld Joints

Description	Craft@Hrs	Unit	Material $	Labor $	Equipment $	Total $

PVC ball valve, union type body, solvent-weld joints (female socket)

½"	P1@.185	Ea	5.86	6.89	—	12.75
¾"	P1@.225	Ea	6.63	8.37	—	15.00
1"	P1@.270	Ea	9.64	10.00	—	19.64
1¼"	P1@.360	Ea	11.10	13.40	—	24.50
1½"	P1@.400	Ea	18.70	14.90	—	33.60
2"	P1@.465	Ea	23.30	17.30	—	40.60
2½"	P1@.490	Ea	96.00	18.20	—	114.20
3"	P1@.540	Ea	222.00	20.10	—	242.10
4"	P1@.600	Ea	409.00	22.30	—	431.30

Class 125 bronze body swing check valve, threaded

½"	P1@.210	Ea	20.80	7.82	—	28.62
¾"	P1@.250	Ea	29.70	9.31	—	39.01
1"	P1@.300	Ea	38.80	11.20	—	50.00
1¼"	P1@.400	Ea	55.40	14.90	—	70.30
1½"	P1@.450	Ea	78.10	16.70	—	94.80
2"	P1@.500	Ea	129.00	18.60	—	147.60

Class 125 iron body swing check valve, flanged ends

2"	P1@.500	Ea	182.00	18.60	—	200.60
2½"	P1@.600	Ea	230.00	22.30	—	252.30
3"	P1@.750	Ea	287.00	27.90	—	314.90
4"	P1@1.35	Ea	421.00	50.20	—	471.20
5"	ER@2.00	Ea	811.00	79.90	1.30	892.20
6"	ER@2.50	Ea	811.00	99.90	1.63	912.53
8"	ER@3.00	Ea	1,470.00	120.00	1.96	1,591.96

Class 125 iron body silent check valve, wafer-type

2"	P1@.500	Ea	133.00	18.60	—	151.60
2½"	P1@.600	Ea	148.00	22.30	—	170.30
3"	P1@.750	Ea	170.00	27.90	—	197.90
4"	P1@1.35	Ea	227.00	50.20	—	277.20
5"	ER@2.00	Ea	348.00	79.90	1.30	429.20
6"	ER@2.50	Ea	474.00	99.90	1.63	575.53
8"	ER@3.00	Ea	811.00	120.00	1.96	932.96

Class 125 bronze body strainer, threaded ends

½"	P1@.210	Ea	32.00	7.82	—	39.82
¾"	P1@.250	Ea	42.00	9.31	—	51.31
1"	P1@.300	Ea	51.50	11.20	—	62.70
1¼"	P1@.400	Ea	72.00	14.90	—	86.90
1½"	P1@.450	Ea	108.00	16.70	—	124.70
2"	P1@.500	Ea	188.00	18.60	—	206.60

PVC, Schedule 40, with Solvent-Weld Joints

Description	Craft@Hrs	Unit	Material $	Labor $	Equipment $	Total $

Class 125 iron body strainer, flanged ends

2"	P1@.500	Ea	144.00	18.60	—	162.60
2½"	P1@.600	Ea	160.00	22.30	—	182.30
3"	P1@.750	Ea	186.00	27.90	—	213.90
4"	P1@1.35	Ea	314.00	50.20	—	364.20
5"	ER@2.00	Ea	640.00	79.90	1.30	721.20
6"	ER@2.50	Ea	640.00	99.90	1.63	741.53
8"	ER@3.00	Ea	1,080.00	120.00	1.96	1,201.96

Installation of 2-way control valve, threaded joints

½"	P1@.210	Ea	—	7.82	—	7.82
¾"	P1@.275	Ea	—	10.20	—	10.20
1"	P1@.350	Ea	—	13.00	—	13.00
1¼"	P1@.430	Ea	—	16.00	—	16.00
1½"	P1@.505	Ea	—	18.80	—	18.80
2"	P1@.675	Ea	—	25.10	—	25.10
2½"	P1@.830	Ea	—	30.90	—	30.90
3"	P1@.990	Ea	—	36.80	—	36.80

Installation of 3-way control valve, threaded joints

½"	P1@.260	Ea	—	9.68	—	9.68
¾"	P1@.365	Ea	—	13.60	—	13.60
1"	P1@.475	Ea	—	17.70	—	17.70
1¼"	P1@.575	Ea	—	21.40	—	21.40
1½"	P1@.680	Ea	—	25.30	—	25.30
2"	P1@.910	Ea	—	33.90	—	33.90
2½"	P1@1.12	Ea	—	41.70	—	41.70
3"	P1@1.33	Ea	—	49.50	—	49.50

PVC companion flange

2"	P1@.140	Ea	13.80	5.21	—	19.01
2½"	P1@.180	Ea	41.60	6.70	—	48.30
3"	P1@.210	Ea	42.90	7.82	—	50.72
4"	P1@.280	Ea	44.00	10.40	—	54.40
5"	P1@.310	Ea	99.20	11.50	—	110.70
6"	P1@.420	Ea	99.20	15.60	—	114.80
8"	P1@.560	Ea	148.00	20.80	—	168.80

Bolt and gasket sets

2"	P1@.500	Ea	4.96	18.60	—	23.56
2½"	P1@.650	Ea	5.82	24.20	—	30.02
3"	P1@.750	Ea	9.91	27.90	—	37.81
4"	P1@1.00	Ea	16.60	37.20	—	53.80
5"	P1@1.10	Ea	27.40	40.90	—	68.30
6"	P1@1.20	Ea	27.40	44.70	—	72.10
8"	P1@1.25	Ea	30.50	46.50	—	77.00

PVC, Schedule 40, with Solvent-Weld Joints

Description	Craft@Hrs	Unit	Material $	Labor $	Equipment $	Total $
Thermometer with well						
7"	P1@.250	Ea	183.00	9.31	—	192.31
9"	P1@.250	Ea	188.00	9.31	—	197.31
Dial-type pressure gauge						
2½"	P1@.200	Ea	38.30	7.44	—	45.74
3½"	P1@.200	Ea	50.40	7.44	—	57.84
Pressure/temperature tap						
Tap	P1@.150	Ea	17.60	5.58	—	23.18
Hanger with swivel assembly						
½"	P1@.250	Ea	4.25	9.31	—	13.56
¾"	P1@.250	Ea	4.71	9.31	—	14.02
1"	P1@.250	Ea	4.92	9.31	—	14.23
1¼"	P1@.300	Ea	5.05	11.20	—	16.25
1½"	P1@.300	Ea	5.43	11.20	—	16.63
2"	P1@.300	Ea	5.66	11.20	—	16.86
2½"	P1@.350	Ea	7.70	13.00	—	20.70
3"	P1@.350	Ea	9.53	13.00	—	22.53
4"	P1@.350	Ea	10.50	13.00	—	23.50
5"	P1@.450	Ea	15.20	16.70	—	31.90
6"	P1@.450	Ea	17.20	16.70	—	33.90
8"	P1@.450	Ea	21.50	16.70	—	38.20
Riser clamp						
½"	P1@.100	Ea	3.15	3.72	—	6.87
¾"	P1@.100	Ea	4.65	3.72	—	8.37
1"	P1@.100	Ea	4.69	3.72	—	8.41
1¼"	P1@.105	Ea	5.66	3.91	—	9.57
1½"	P1@.110	Ea	5.98	4.09	—	10.07
2"	P1@.115	Ea	6.33	4.28	—	10.61
2½"	P1@.120	Ea	6.66	4.47	—	11.13
3"	P1@.120	Ea	7.21	4.47	—	11.68
4"	P1@.125	Ea	9.19	4.65	—	13.84
5"	P1@.180	Ea	13.20	6.70	—	19.90
6"	P1@.200	Ea	15.90	7.44	—	23.34
8"	P1@.200	Ea	25.90	7.44	—	33.34

PVC, Schedule 80, with Solvent-Weld Joints

PVC (Polyvinyl Chloride) Schedule 80 pipe with Schedule 80 socket-type fittings and solvent-welded joints is widely used in process, chemical, A.C. condensate, potable water and irrigation piping systems. PVC is available in Type I (normal impact) and Type II (high impact). Type I has a maximum temperature rating of 150 degrees F., and Type II is rated at 140 degrees F.

Schedule 80 PVC can be threaded, but it is more economical to use solvent-welded joints.

Manufacturers should be consulted for maximum pressure ratings and recommended joints solvents for specific applications.

Because of the current unreliability of many plastic valves, most engineers specify standard bronze-body and iron-body valves for use in PVC water piping systems. Plastic valves, however, have to be used in systems conveying liquids that may be injurious to metallic valves.

This section has been arranged to save the estimator's time by including all normally-used system components such as pipe, fittings, valves, hanger assemblies, riser clamps and miscellaneous items under one heading. Additional items can be found under "Plumbing and Piping Specialties." The cost estimates in this section are based on the conditions, limitations and wage rates described in the section "How to Use This Book" beginning on page 5.

Equipment cost, where shown, is $4.45 per hour for a 2-ton chain hoist.

Description	Craft@Hrs	Unit	Material $	Labor $	Equipment $	Total $

Schedule 80 PVC pipe assembly with solvent-weld joints installed horizontally. Complete installation including three to eight tees and three to eight elbows every 100' and hangers spaced to meet plumbing code *(a tee and elbow every 12' for 1/2" dia., a tee and elbow every 40' for 8"dia.)*.

Description	Craft@Hrs	Unit	Material $	Labor $	Equipment $	Total $
½"	P1@.085	LF	2.53	3.16	—	5.69
¾"	P1@.093	LF	3.17	3.46	—	6.63
1"	P1@.099	LF	3.49	3.68	—	7.17
1¼"	P1@.115	LF	4.08	4.28	—	8.36
1½"	P1@.117	LF	4.37	4.35	—	8.72
2"	P1@.120	LF	4.61	4.47	—	9.08
2½"	P1@.125	LF	7.60	4.65	—	12.25
3"	P1@.129	LF	10.30	4.80	—	15.10
4"	P1@.161	LF	14.80	5.99	—	20.79
6"	P1@.253	LF	24.60	9.42	—	34.02
8"	P1@.299	LF	41.30	11.10	—	52.40

PVC, Schedule 80, with Solvent-Weld Joints

Description	Craft@Hrs	Unit	Material $	Labor $	Equipment $	Total $

Schedule 80 PVC pipe assembly with solvent-weld joints, installed risers. Complete installation including a reducing tee every floor and a riser clamp every other.

Size	Craft@Hrs	Unit	Material $	Labor $	Equipment $	Total $
½"	P1@.041	LF	1.28	1.53	—	2.81
¾"	P1@.048	LF	1.85	1.79	—	3.64
1"	P1@.054	LF	2.24	2.01	—	4.25
1¼"	P1@.063	LF	2.63	2.34	—	4.97
1½"	P1@.071	LF	3.44	2.64	—	6.08
2"	P1@.079	LF	3.68	2.94	—	6.62
2½"	P1@.090	LF	6.02	3.35	—	9.37
3"	P1@.101	LF	7.80	3.76	—	11.56
4"	P1@.129	LF	11.10	4.80	—	15.90
6"	P1@.242	LF	21.90	9.01	—	30.91
8"	P1@.296	LF	37.00	11.00	—	48.00

Schedule 80 PVC pipe with solvent-weld joints

Size	Craft@Hrs	Unit	Material $	Labor $	Equipment $	Total $
½"	P1@.021	LF	.91	.78	—	1.69
¾"	P1@.026	LF	1.22	.97	—	2.19
1"	P1@.032	LF	1.60	1.19	—	2.79
1¼"	P1@.037	LF	1.89	1.38	—	3.27
1½"	P1@.042	LF	2.54	1.56	—	4.10
2"	P1@.047	LF	2.63	1.75	—	4.38
2½"	P1@.052	LF	4.19	1.94	—	6.13
3"	P1@.057	LF	5.60	2.12	—	7.72
4"	P1@.074	LF	8.26	2.75	—	11.01
6"	P1@.160	LF	17.30	5.96	—	23.26
8"	P1@.190	LF	26.30	7.07	—	33.37

Schedule 80 PVC 45-degree ell S x S with solvent-weld joints

Size	Craft@Hrs	Unit	Material $	Labor $	Equipment $	Total $
½"	P1@.105	Ea	2.95	3.91	—	6.86
¾"	P1@.120	Ea	4.45	4.47	—	8.92
1"	P1@.125	Ea	6.71	4.65	—	11.36
1¼"	P1@.170	Ea	8.59	6.33	—	14.92
1½"	P1@.190	Ea	10.10	7.07	—	17.17
2"	P1@.210	Ea	13.10	7.82	—	20.92
2½"	P1@.262	Ea	27.50	9.75	—	37.25
3"	P1@.315	Ea	33.40	11.70	—	45.10
4"	P1@.520	Ea	60.20	19.40	—	79.60
6"	P1@.630	Ea	75.70	23.40	—	99.10
8"	P1@.840	Ea	164.00	31.30	—	195.30

PVC, Schedule 80, with Solvent-Weld Joints

Description	Craft@Hrs	Unit	Material $	Labor $	Equipment $	Total $

Schedule 80 PVC 90-degree ell S x S with solvent-weld joints

Description	Craft@Hrs	Unit	Material $	Labor $	Equipment $	Total $
½"	P1@.105	Ea	1.60	3.91	—	5.51
¾"	P1@.120	Ea	2.02	4.47	—	6.49
1"	P1@.125	Ea	3.18	4.65	—	7.83
1¼"	P1@.170	Ea	4.34	6.33	—	10.67
1½"	P1@.190	Ea	4.62	7.07	—	11.69
2"	P1@.210	Ea	5.54	7.82	—	13.36
2½"	P1@.262	Ea	12.90	9.75	—	22.65
3"	P1@.315	Ea	14.60	11.70	—	26.30
4"	P1@.520	Ea	22.20	19.40	—	41.60
6"	P1@.630	Ea	63.30	23.40	—	86.70
8"	P1@.840	Ea	173.00	31.30	—	204.30

Schedule 80 PVC tee S x S x S with solvent-weld joints

Description	Craft@Hrs	Unit	Material $	Labor $	Equipment $	Total $
½"	P1@.135	Ea	4.41	5.02	—	9.43
¾"	P1@.145	Ea	4.62	5.40	—	10.02
1"	P1@.180	Ea	5.72	6.70	—	12.42
1¼"	P1@.230	Ea	15.80	8.56	—	24.36
1½"	P1@.265	Ea	15.80	9.86	—	25.66
2"	P1@.295	Ea	19.70	11.00	—	30.70
2½"	P1@.368	Ea	21.50	13.70	—	35.20
3"	P1@.440	Ea	27.00	16.40	—	43.40
4"	P1@.585	Ea	31.20	21.80	—	53.00
6"	P1@.880	Ea	105.00	32.80	—	137.80
8"	P1@1.18	Ea	246.00	43.90	—	289.90

Schedule 80 PVC reducing tee S x S x S with solvent-weld joints

Description	Craft@Hrs	Unit	Material $	Labor $	Equipment $	Total $
¾" x ½"	P1@.145	Ea	4.99	5.40	—	10.39
1" x ½"	P1@.175	Ea	5.38	6.51	—	11.89
1" x ¾"	P1@.175	Ea	5.38	6.51	—	11.89
1¼" x ½"	P1@.230	Ea	17.70	8.56	—	26.26
1¼" x ¾"	P1@.230	Ea	17.70	8.56	—	26.26
1¼" x 1"	P1@.230	Ea	17.70	8.56	—	26.26
1½" x ¾"	P1@.260	Ea	18.90	9.68	—	28.58
1½" x 1"	P1@.260	Ea	18.90	9.68	—	28.58
1½" x 1¼"	P1@.260	Ea	18.90	9.68	—	28.58
2" x ½"	P1@.290	Ea	23.20	10.80	—	34.00
2" x ¾"	P1@.290	Ea	23.20	10.80	—	34.00
2" x 1"	P1@.290	Ea	23.20	10.80	—	34.00
2" x 1½"	P1@.290	Ea	23.20	10.80	—	34.00
3" x 2"	P1@.430	Ea	32.90	16.00	—	48.90
4" x 2"	P1@.570	Ea	37.40	21.20	—	58.60
4" x 3"	P1@.570	Ea	37.40	21.20	—	58.60
6" x 4"	P1@.850	Ea	126.00	31.60	—	157.60
8" x 4"	P1@1.16	Ea	283.00	43.20	—	326.20
8" x 5"	P1@1.16	Ea	283.00	43.20	—	326.20
8" x 6"	P1@1.16	Ea	283.00	43.20	—	326.20

PVC, Schedule 80, with Solvent-Weld Joints

Description	Craft@Hrs	Unit	Material $	Labor $	Equipment $	Total $

Schedule 80 PVC reducing bushing SPIG x S with solvent-weld joints

Description	Craft@Hrs	Unit	Material $	Labor $	Equipment $	Total $
¾" x ½"	P1@.053	Ea	.91	1.97	—	2.88
1" x ½"	P1@.063	Ea	2.60	2.34	—	4.94
1" x ¾"	P1@.063	Ea	2.60	2.34	—	4.94
1¼" x ½"	P1@.084	Ea	4.17	3.13	—	7.30
1¼" x ¾"	P1@.084	Ea	4.17	3.13	—	7.30
1¼" x 1"	P1@.084	Ea	4.17	3.13	—	7.30
1½" x ½"	P1@.105	Ea	5.54	3.91	—	9.45
1½" x ¾"	P1@.105	Ea	5.54	3.91	—	9.45
1½" x 1"	P1@.105	Ea	5.54	3.91	—	9.45
1½" x 1¼"	P1@.105	Ea	5.54	3.91	—	9.45
2" x 1"	P1@.116	Ea	7.93	4.32	—	12.25
2" x 1¼"	P1@.116	Ea	7.93	4.32	—	12.25
2" x 1½"	P1@.116	Ea	7.93	4.32	—	12.25
3" x 2"	P1@.158	Ea	21.90	5.88	—	27.78
4" x 2½"	P1@.210	Ea	30.20	7.82	—	38.02
4" x 3"	P1@.210	Ea	30.20	7.82	—	38.02
6" x 3"	P1@.315	Ea	42.00	11.70	—	53.70
6" x 4"	P1@.315	Ea	42.00	11.70	—	53.70
8" x 4"	P1@.420	Ea	101.00	15.60	—	116.60
8" x 6"	P1@.420	Ea	101.00	15.60	—	116.60

Schedule 80 PVC adapter MPT x S with solvent-weld joints

Description	Craft@Hrs	Unit	Material $	Labor $	Equipment $	Total $
½"	P1@.063	Ea	3.31	2.34	—	5.65
¾"	P1@.068	Ea	3.62	2.53	—	6.15
1"	P1@.074	Ea	6.22	2.75	—	8.97
1¼"	P1@.105	Ea	7.36	3.91	—	11.27
1½"	P1@.116	Ea	10.50	4.32	—	14.82
2"	P1@.126	Ea	15.20	4.69	—	19.89
2½"	P1@.158	Ea	17.60	5.88	—	23.48
3"	P1@.189	Ea	19.20	7.03	—	26.23
4"	P1@.252	Ea	34.00	9.38	—	43.38
5"	P1@.385	Ea	47.50	14.30	—	61.80
6"	P1@.435	Ea	109.00	16.20	—	125.20

Schedule 80 PVC adapter FPT x S with solvent-weld joints

Description	Craft@Hrs	Unit	Material $	Labor $	Equipment $	Total $
½"	P1@.116	Ea	2.66	4.32	—	6.98
¾"	P1@.126	Ea	3.94	4.69	—	8.63
1"	P1@.137	Ea	5.75	5.10	—	10.85
1¼"	P1@.179	Ea	9.33	6.66	—	15.99
1½"	P1@.221	Ea	11.30	8.23	—	19.53
2"	P1@.305	Ea	19.90	11.40	—	31.30
2½"	P1@.345	Ea	31.40	12.80	—	44.20
3"	P1@.490	Ea	35.30	18.20	—	53.50
4"	P1@.655	Ea	60.80	24.40	—	85.20
5"	P1@.920	Ea	98.90	34.20	—	133.10
6"	P1@1.10	Ea	121.00	40.90	—	161.90

PVC, Schedule 80, with Solvent-Weld Joints

Description	Craft@Hrs	Unit	Material $	Labor $	Equipment $	Total $

Schedule 80 PVC cap S with solvent-weld joints

Description	Craft@Hrs	Unit	Material $	Labor $	Equipment $	Total $
½"	P1@.063	Ea	2.78	2.34	—	5.12
¾"	P1@.070	Ea	2.88	2.61	—	5.49
1"	P1@.075	Ea	5.13	2.79	—	7.92
1¼"	P1@.105	Ea	6.22	3.91	—	10.13
1½"	P1@.115	Ea	6.22	4.28	—	10.50
2"	P1@.125	Ea	12.20	4.65	—	16.85
2½"	P1@.158	Ea	24.80	5.88	—	30.68
3"	P1@.189	Ea	29.50	7.03	—	36.53
4"	P1@.250	Ea	49.70	9.31	—	59.01
5"	P1@.390	Ea	102.00	14.50	—	116.50
6"	P1@.450	Ea	123.00	16.70	—	139.70

Schedule 80 PVC plug MPT with solvent-weld joints

Description	Craft@Hrs	Unit	Material $	Labor $	Equipment $	Total $
½"	P1@.155	Ea	2.69	5.77	—	8.46
¾"	P1@.165	Ea	2.78	6.14	—	8.92
1"	P1@.190	Ea	3.39	7.07	—	10.46
1¼"	P1@.220	Ea	4.99	8.19	—	13.18
1½"	P1@.240	Ea	6.03	8.93	—	14.96
2"	P1@.255	Ea	6.21	9.49	—	15.70
2½"	P1@.275	Ea	15.60	10.20	—	25.80
3"	P1@.320	Ea	20.20	11.90	—	32.10
4"	P1@.425	Ea	39.40	15.80	—	55.20

Schedule 80 PVC coupling S x S with solvent-weld joints

Description	Craft@Hrs	Unit	Material $	Labor $	Equipment $	Total $
½"	P1@.105	Ea	2.83	3.91	—	6.74
¾"	P1@.120	Ea	3.79	4.47	—	8.26
1"	P1@.125	Ea	3.93	4.65	—	8.58
1¼"	P1@.170	Ea	6.00	6.33	—	12.33
1½"	P1@.190	Ea	6.43	7.07	—	13.50
2"	P1@.210	Ea	6.88	7.82	—	14.70
2½"	P1@.262	Ea	16.90	9.75	—	26.65
3"	P1@.315	Ea	19.50	11.70	—	31.20
4"	P1@.520	Ea	24.60	19.40	—	44.00
6"	P1@.630	Ea	52.50	23.40	—	75.90
8"	P1@.840	Ea	72.20	31.30	—	103.50

Schedule 80 PVC union S x S with solvent-weld joints

Description	Craft@Hrs	Unit	Material $	Labor $	Equipment $	Total $
½"	P1@.125	Ea	5.67	4.65	—	10.32
¾"	P1@.140	Ea	7.81	5.21	—	13.02
1"	P1@.160	Ea	8.23	5.96	—	14.19
1¼"	P1@.200	Ea	16.40	7.44	—	23.84
1½"	P1@.225	Ea	18.60	8.37	—	26.97
2"	P1@.250	Ea	25.10	9.31	—	34.41
2½"	P1@.275	Ea	39.70	10.20	—	49.90
3"	P1@.300	Ea	46.80	11.20	—	58.00

PVC, Schedule 80, with Solvent-Weld Joints

Description	Craft@Hrs	Unit	Material $	Labor $	Equipment $	Total $

Class 125 bronze body gate valve, threaded ends

½"	P1@.210	Ea	18.80	7.82	—	26.62
¾"	P1@.250	Ea	23.50	9.31	—	32.81
1"	P1@.300	Ea	33.00	11.20	—	44.20
1¼"	P1@.400	Ea	41.50	14.90	—	56.40
1½"	P1@.450	Ea	56.00	16.70	—	72.70
2"	P1@.500	Ea	93.60	18.60	—	112.20
2½"	P1@.750	Ea	155.00	27.90	—	182.90
3"	P1@.950	Ea	222.00	35.40	—	257.40

Class 125 iron body gate valve, flanged ends

2"	P1@.500	Ea	358.00	18.60	—	376.60
2½"	P1@.600	Ea	483.00	22.30	—	505.30
3"	P1@.750	Ea	524.00	27.90	—	551.90
4"	P1@1.35	Ea	769.00	50.20	—	819.20
5"	ER@2.00	Ea	1,470.00	79.90	1.30	1,551.20
6"	ER@2.50	Ea	1,470.00	99.90	1.63	1,571.53
8"	ER@3.00	Ea	2,420.00	120.00	1.96	2,541.96

Class 125 bronze body globe valve, threaded ends

½"	P1@.210	Ea	34.80	7.82	—	42.62
¾"	P1@.250	Ea	46.30	9.31	—	55.61
1"	P1@.300	Ea	66.40	11.20	—	77.60
1¼"	P1@.400	Ea	93.40	14.90	—	108.30
1½"	P1@.450	Ea	124.00	16.70	—	140.70
2"	P1@.500	Ea	204.00	18.60	—	222.60

200 PSIG iron body butterfly valve, lug-type, lever operated

2"	P1@.450	Ea	161.00	16.70	—	177.70
2½"	P1@.450	Ea	167.00	16.70	—	183.70
3"	P1@.550	Ea	176.00	20.50	—	196.50
4"	P1@.550	Ea	219.00	20.50	—	239.50
5"	ER@.800	Ea	288.00	32.00	.52	320.52
6"	ER@.800	Ea	352.00	32.00	.52	384.52
8"	ER@.900	Ea	481.00	36.00	.59	517.59

200 PSIG iron body butterfly valve, wafer-type, lever operated

2"	P1@.450	Ea	147.00	16.70	—	163.70
2½"	P1@.450	Ea	150.00	16.70	—	166.70
3"	P1@.550	Ea	161.00	20.50	—	181.50
4"	P1@.550	Ea	195.00	20.50	—	215.50
5"	ER@.800	Ea	260.00	32.00	.52	292.52
6"	ER@.800	Ea	326.00	32.00	.52	358.52
8"	ER@.900	Ea	452.00	36.00	.59	488.59

PVC, Schedule 80, with Solvent-Weld Joints

Description	Craft@Hrs	Unit	Material $	Labor $	Equipment $	Total $

Class 125 bronze body 2-piece ball valve, threaded ends

½"	P1@.210	Ea	10.40	7.82	—	18.22
¾"	P1@.250	Ea	13.80	9.31	—	23.11
1"	P1@.300	Ea	24.10	11.20	—	35.30
1¼"	P1@.400	Ea	40.20	14.90	—	55.10
1½"	P1@.450	Ea	54.30	16.70	—	71.00
2"	P1@.500	Ea	68.80	18.60	—	87.40
3"	P1@.625	Ea	471.00	23.30	—	494.30
4"	P1@.690	Ea	616.00	25.70	—	641.70

PVC ball valve, solid body, solvent-weld joints, EDPM (female socket)

½"	P1@.170	Ea	6.24	6.33	—	12.57
¾"	P1@.200	Ea	7.05	7.44	—	14.49
1"	P1@.220	Ea	10.30	8.19	—	18.49
1¼"	P1@.280	Ea	11.80	10.40	—	22.20
1½"	P1@.315	Ea	19.80	11.70	—	31.50
2"	P1@.385	Ea	24.60	14.30	—	38.90

PVC ball valve, solid body, threaded joints, EDPM (female pipe thread)

½"	P1@.200	Ea	6.24	7.44	—	13.68
¾"	P1@.240	Ea	7.05	8.93	—	15.98
1"	P1@.290	Ea	10.30	10.80	—	21.10
1¼"	P1@.390	Ea	11.80	14.50	—	26.30
1½"	P1@.420	Ea	19.80	15.60	—	35.40
2"	P1@.490	Ea	24.60	18.20	—	42.80

PVC ball valve, tru-union, threaded joints, EDPM (female pipe thread)

2"	P1@.490	Ea	83.10	18.20	—	101.30
2½"	P1@.520	Ea	96.00	19.40	—	115.40
3"	P1@.590	Ea	222.00	22.00	—	244.00
4"	P1@.630	Ea	409.00	23.40	—	432.40

PVC ball valve, union-type body, solvent-weld joints (female socket)

½"	P1@.185	Ea	6.24	6.89	—	13.13
¾"	P1@.225	Ea	7.05	8.37	—	15.42
1"	P1@.270	Ea	10.30	10.00	—	20.30
1¼"	P1@.360	Ea	11.80	13.40	—	25.20
1½"	P1@.400	Ea	19.80	14.90	—	34.70
2"	P1@.465	Ea	24.60	17.30	—	41.90
2½"	P1@.490	Ea	96.00	18.20	—	114.20
3"	P1@.540	Ea	222.00	20.10	—	242.10
4"	P1@.600	Ea	409.00	22.30	—	431.30

PVC, Schedule 80, with Solvent-Weld Joints

Description	Craft@Hrs	Unit	Material $	Labor $	Equipment $	Total $

Class 125 bronze body swing check valve, threaded ends

½"	P1@.210	Ea	20.80	7.82	—	28.62
¾"	P1@.250	Ea	29.70	9.31	—	39.01
1"	P1@.300	Ea	38.80	11.20	—	50.00
1¼"	P1@.400	Ea	55.40	14.90	—	70.30
1½"	P1@.450	Ea	78.10	16.70	—	94.80
2"	P1@.500	Ea	129.00	18.60	—	147.60

Class 125 iron body swing check valve, flanged ends

2"	P1@.500	Ea	182.00	18.60	—	200.60
2½"	P1@.600	Ea	230.00	22.30	—	252.30
3"	P1@.750	Ea	287.00	27.90	—	314.90
4"	P1@1.35	Ea	421.00	50.20	—	471.20
5"	ER@2.00	Ea	811.00	79.90	1.30	892.20
6"	ER@2.50	Ea	811.00	99.90	1.63	912.53
8"	ER@3.00	Ea	1,470.00	120.00	1.96	1,591.96

Class 125 iron body silent check valve, wafer-type

2"	P1@.500	Ea	133.00	18.60	—	151.60
2½"	P1@.600	Ea	148.00	22.30	—	170.30
3"	P1@.750	Ea	170.00	27.90	—	197.90
4"	P1@1.35	Ea	227.00	50.20	—	277.20
5"	ER@2.00	Ea	348.00	79.90	1.30	429.20
6"	ER@2.50	Ea	474.00	99.90	1.63	575.53
8"	ER@3.00	Ea	811.00	120.00	1.96	932.96

Class 125 bronze body strainer, threaded ends

½"	P1@.210	Ea	32.00	7.82	—	39.82
¾"	P1@.250	Ea	42.00	9.31	—	51.31
1"	P1@.300	Ea	51.50	11.20	—	62.70
1¼"	P1@.400	Ea	72.00	14.90	—	86.90
1½"	P1@.450	Ea	108.00	16.70	—	124.70
2"	P1@.500	Ea	188.00	18.60	—	206.60

Class 125 iron body strainer, flanged ends

2"	P1@.500	Ea	191.00	18.60	—	209.60
2½"	P1@.600	Ea	213.00	22.30	—	235.30
3"	P1@.750	Ea	246.00	27.90	—	273.90
4"	P1@1.35	Ea	420.00	50.20	—	470.20
5"	ER@2.00	Ea	853.00	79.90	1.30	934.20
6"	ER@2.50	Ea	853.00	99.90	1.63	954.53
8"	ER@3.00	Ea	1,450.00	120.00	1.96	1,571.96

PVC, Schedule 80, with Solvent-Weld Joints

Description	Craft@Hrs	Unit	Material $	Labor $	Equipment $	Total $

Installation of 2-way control valve, threaded joints

Description	Craft@Hrs	Unit	Material $	Labor $	Equipment $	Total $
½"	P1@.210	Ea	—	7.82	—	7.82
¾"	P1@.275	Ea	—	10.20	—	10.20
1"	P1@.350	Ea	—	13.00	—	13.00
1¼"	P1@.430	Ea	—	16.00	—	16.00
1½"	P1@.505	Ea	—	18.80	—	18.80
2"	P1@.675	Ea	—	25.10	—	25.10
2½"	P1@.830	Ea	—	30.90	—	30.90
3"	P1@.990	Ea	—	36.80	—	36.80

Installation of 3-way control valve, threaded joints

Description	Craft@Hrs	Unit	Material $	Labor $	Equipment $	Total $
½"	P1@.260	Ea	—	9.68	—	9.68
¾"	P1@.365	Ea	—	13.60	—	13.60
1"	P1@.475	Ea	—	17.70	—	17.70
1¼"	P1@.575	Ea	—	21.40	—	21.40
1½"	P1@.680	Ea	—	25.30	—	25.30
2"	P1@.910	Ea	—	33.90	—	33.90
2½"	P1@1.12	Ea	—	41.70	—	41.70
3"	P1@1.33	Ea	—	49.50	—	49.50

PVC companion flange

Description	Craft@Hrs	Unit	Material $	Labor $	Equipment $	Total $
2"	P1@.140	Ea	14.70	5.21	—	19.91
2½"	P1@.180	Ea	43.60	6.70	—	50.30
3"	P1@.210	Ea	45.50	7.82	—	53.32
4"	P1@.280	Ea	47.10	10.40	—	57.50
5"	P1@.310	Ea	105.00	11.50	—	116.50
6"	P1@.420	Ea	105.00	15.60	—	120.60
8"	P1@.560	Ea	156.00	20.80	—	176.80

Bolt and gasket set

Description	Craft@Hrs	Unit	Material $	Labor $	Equipment $	Total $
2"	P1@.500	Ea	4.89	18.60	—	23.49
2½"	P1@.650	Ea	5.72	24.20	—	29.92
3"	P1@.750	Ea	9.66	27.90	—	37.56
4"	P1@1.00	Ea	16.10	37.20	—	53.30
5"	P1@1.10	Ea	26.90	40.90	—	67.80
6"	P1@1.20	Ea	26.90	44.70	—	71.60
8"	P1@1.25	Ea	30.10	46.50	—	76.60

Thermometer with well

Description	Craft@Hrs	Unit	Material $	Labor $	Equipment $	Total $
7"	P1@.250	Ea	183.00	9.31	—	192.31
9"	P1@.250	Ea	188.00	9.31	—	197.31

Dial-type pressure gauge

Description	Craft@Hrs	Unit	Material $	Labor $	Equipment $	Total $
2½"	P1@.200	Ea	38.30	7.44	—	45.74
3½"	P1@.200	Ea	50.40	7.44	—	57.84

PVC, Schedule 80, with Solvent-Weld Joints

Description	Craft@Hrs	Unit	Material $	Labor $	Equipment $	Total $
Pressure/temperature tap						
Tap	P1@.150	Ea	17.60	5.58	—	23.18
Hanger with swivel assembly						
½"	P1@.250	Ea	4.25	9.31	—	13.56
¾"	P1@.250	Ea	4.71	9.31	—	14.02
1"	P1@.250	Ea	4.92	9.31	—	14.23
1¼"	P1@.300	Ea	5.05	11.20	—	16.25
1½"	P1@.300	Ea	5.43	11.20	—	16.63
2"	P1@.300	Ea	5.66	11.20	—	16.86
2½"	P1@.350	Ea	7.70	13.00	—	20.70
3"	P1@.350	Ea	9.53	13.00	—	22.53
4"	P1@.350	Ea	10.50	13.00	—	23.50
5"	P1@.450	Ea	15.20	16.70	—	31.90
6"	P1@.450	Ea	17.20	16.70	—	33.90
8"	P1@.450	Ea	21.50	16.70	—	38.20
Riser clamp						
½"	P1@.100	Ea	3.15	3.72	—	6.87
¾"	P1@.100	Ea	4.65	3.72	—	8.37
1"	P1@.100	Ea	4.69	3.72	—	8.41
1¼"	P1@.105	Ea	5.66	3.91	—	9.57
1½"	P1@.110	Ea	5.98	4.09	—	10.07
2"	P1@.115	Ea	6.33	4.28	—	10.61
2½"	P1@.120	Ea	6.66	4.47	—	11.13
3"	P1@.120	Ea	7.21	4.47	—	11.68
4"	P1@.125	Ea	9.19	4.65	—	13.84
5"	P1@.180	Ea	13.20	6.70	—	19.90
6"	P1@.200	Ea	15.90	7.44	—	23.34
8"	P1@.200	Ea	25.90	7.44	—	33.34

Polyethylene-Aluminum Pipe with Crimped Joints

HDPE – High Density Polyethylene pipe, or PEX pipe, is cross-linked at the molecular level. Composite pressure pipe is aluminum tube laminated between 2 layers of plastic pipe. An additional external layer of an EVOH polymer creates an oxygen barrier and restricts oxygen permeability. It is manufactured and sold in various coiled lengths (200' to 1,000' rolls). The jointing method can be either crimped or compression connections. It will not rust or corrode and, because of its flexibility, installations require approximately 40% less fittings. Expansion rates are similar to copper, but its thermal coefficient (heat loss) is over 800 times less than copper. It is also chemically resistant to most acids, salt solutions, alkalis, fats and oils.

Common applications include ice or snow melting systems, radiant floor heating systems, water service tubing, hot & cold domestic water service, chilled water systems, compressed air systems, solar and process piping applications. Ice or snow melting systems require an oxygen barrier type PEX pipe selection.

Maximum temperature	Maximum pressure	Pipe selection
73 degrees F	160 psi	PEX
140 degrees F	120 psi	PEX
180 degrees F	100 psi	PEX
210 degrees F	115 psi	PEX-AL

PEX is the designation for HDPE pipe
PEX-AL is the designation for cross linked polyethylene-aluminum composite pipe

Description	Craft@Hrs	Unit	Material $	Labor $	Equipment $	Total $

Cross linked HD Polyethylene-PEX pipe with crimped joints (PEX), pipe only

3/8"	P1@.028	LF	.42	1.04	—	1.46
1/2"	P1@.028	LF	.61	1.04	—	1.65
5/8"	P1@.030	LF	.85	1.12	—	1.97
3/4"	P1@.030	LF	1.21	1.12	—	2.33
1"	P1@.034	LF	1.94	1.27	—	3.21
Add for oxygen barrier type		%	25.0	—	—	—
Add for PEX-AL		%	50.0	—	—	—

90-degree brass ell with crimped joints (PEX)

1/2"	P1@.080	Ea	1.32	2.98	—	4.30
5/8"	P1@.085	Ea	1.59	3.16	—	4.75
3/4"	P1@.090	Ea	3.06	3.35	—	6.41
1"	P1@.105	Ea	3.33	3.91	—	7.24

90-degree brass ell with crimped joints x male pipe thread (PEX x MPT)

1/2" x 1/2"	P1@.085	Ea	1.44	3.16	—	4.60
1/2" x 3/8"	P1@.095	Ea	1.82	3.54	—	5.36

Polyethylene-Aluminum Pipe with Crimped Joints

Description	Craft@Hrs	Unit	Material $	Labor $	Equipment $	Total $

90-degree brass ell with crimped joints x copper socket (PEX x C)

1/2"	P1@.085	Ea	1.44	3.16	—	4.60
1/2" x 3/4"	P1@.090	Ea	1.82	3.35	—	5.17
5/8" x 3/4"	P1@.095	Ea	1.96	3.54	—	5.50
3/4"	P1@.095	Ea	1.60	3.54	—	5.14

90-degree brass wingback ell with crimped joints x female pipe thread (PEX x FPT)

1/2"	P1@.085	Ea	2.81	3.16	—	5.97
3/4"	P1@.095	Ea	3.33	3.54	—	6.87

Tee (brass) with crimped joints (PEX)

1/2"	P1@.095	Ea	1.41	3.54	—	4.95
5/8"	P1@.105	Ea	1.65	3.91	—	5.56
3/4"	P1@.110	Ea	3.33	4.09	—	7.42
1"	P1@.120	Ea	4.08	4.47	—	8.55

Reducing tee (brass) with crimped joints (PEX)

1/2" x 1/2" x 5/8"	P1@.110	Ea	1.65	4.09	—	5.74
1/2" x 1/2" x 3/4"	P1@.115	Ea	1.92	4.28	—	6.20
5/8" x 1/2" x 1/2"	P1@.115	Ea	1.65	4.28	—	5.93
5/8" x 1/2" x 5/8"	P1@.115	Ea	1.65	4.28	—	5.93
5/8" x 5/8" x 1/2"	P1@.115	Ea	1.65	4.28	—	5.93
3/4" x 1/2" x 1/2"	P1@.115	Ea	1.92	4.28	—	6.20
3/4" x 1/2" x 3/4"	P1@.115	Ea	2.05	4.28	—	6.33
3/4" x 3/4" x 1/2"	P1@.115	Ea	2.05	4.28	—	6.33
3/4" x 3/4" x 5/8"	P1@.115	Ea	2.05	4.28	—	6.33
3/4" x 3/4" x 1"	P1@.120	Ea	3.95	4.47	—	8.42
1" x 1/2" x 1/2"	P1@.120	Ea	3.84	4.47	—	8.31
1" x 1/2" x 3/4"	P1@.120	Ea	3.58	4.47	—	8.05
1" x 1/2" x 1"	P1@.120	Ea	3.95	4.47	—	8.42
1" x 3/4" x 1/2"	P1@.120	Ea	3.95	4.47	—	8.42
1" x 3/4" x 3/4"	P1@.120	Ea	3.95	4.47	—	8.42
1" x 3/4" x 1"	P1@.120	Ea	3.95	4.47	—	8.42
1" x 1" x 1/2"	P1@.120	Ea	3.56	4.47	—	8.03
1" x 1" x 3/4"	P1@.120	Ea	4.08	4.47	—	8.55

Coupling (brass) with crimped joints (PEX)

1/2"	P1@.080	Ea	.64	2.98	—	3.62
5/8"	P1@.085	Ea	.81	3.16	—	3.97
3/4"	P1@.090	Ea	1.80	3.35	—	5.15
1"	P1@.105	Ea	1.84	3.91	—	5.75

Polyethylene-Aluminum Pipe with Crimped Joints

Description	Craft@Hrs	Unit	Material $	Labor $	Equipment $	Total $

Reducing coupling (brass) with crimped joints (PEX)

1/2"	P1@.080	Ea	.81	2.98	—	3.79
5/8"	P1@.085	Ea	1.15	3.16	—	4.31
3/4"	P1@.090	Ea	1.23	3.35	—	4.58
1"	P1@.105	Ea	1.50	3.91	—	5.41

Adapter (brass) with crimped joints x male pipe thread (PEX x MPT)

1/2" x 1/2"	P1@.080	Ea	.96	2.98	—	3.94
1/2" x 3/4"	P1@.085	Ea	1.23	3.16	—	4.39
5/8" x 3/4"	P1@.085	Ea	1.22	3.16	—	4.38
3/4" x 1/2"	P1@.090	Ea	1.65	3.35	—	5.00
3/4" x 3/4"	P1@.090	Ea	1.75	3.35	—	5.10
3/4" x 1"	P1@.105	Ea	2.81	3.91	—	6.72
1" x 1"	P1@.105	Ea	3.06	3.91	—	6.97

Adapter (brass) with crimped joints x female pipe thread (PEX x FPT)

1/2" x 1/2"	P1@.080	Ea	1.00	2.98	—	3.98
1/2" x 3/4"	P1@.085	Ea	1.41	3.16	—	4.57
5/8" x 3/4"	P1@.085	Ea	1.41	3.16	—	4.57
3/4" x 1/2"	P1@.090	Ea	1.65	3.35	—	5.00
3/4" x 3/4"	P1@.090	Ea	1.82	3.35	—	5.17
3/4" x 1"	P1@.105	Ea	3.04	3.91	—	6.95
1" x 1"	P1@.105	Ea	3.33	3.91	—	7.24

Adapter (brass) with crimped joints x copper fitting (spigot) (PEX x C ftg)

1/2" x 1/2"	P1@.080	Ea	1.15	2.98	—	4.13
1/2" x 3/4"	P1@.085	Ea	1.42	3.16	—	4.58
5/8" x 3/4"	P1@.085	Ea	1.41	3.16	—	4.57
3/4" x 3/4"	P1@.090	Ea	2.81	3.35	—	6.16
1" x 1"	P1@.105	Ea	3.33	3.91	—	7.24

Adapter (brass) with crimped joints x copper socket (PEX x C)

1/2" x 1/2"	P1@.080	Ea	.96	2.98	—	3.94
1/2" x 3/4"	P1@.085	Ea	1.43	3.16	—	4.59
5/8" x 3/4"	P1@.085	Ea	1.49	3.16	—	4.65
3/4" x 1/2"	P1@.090	Ea	1.59	3.35	—	4.94
3/4" x 3/4"	P1@.090	Ea	2.30	3.35	—	5.65
3/4" x 1"	P1@.105	Ea	3.33	3.91	—	7.24
1" x 1"	P1@.105	Ea	3.84	3.91	—	7.75

Polyethylene-Aluminum Pipe with Crimped Joints

Description	Craft@Hrs	Unit	Material $	Labor $	Equipment $	Total $

Cap (brass) with crimped joints (PEX)

1/2"	P1@.060	Ea	.44	2.23	—	2.67
5/8"	P1@.065	Ea	.62	2.42	—	3.04
3/4"	P1@.070	Ea	.81	2.61	—	3.42
1"	P1@.085	Ea	1.00	3.16	—	4.16

Mini ball valve (brass) with crimped joints (PEX)

1/2" x 3/8" straight	P1@.150	Ea	7.30	5.58	—	12.88
1/2" x 3/8" angle	P1@.150	Ea	6.72	5.58	—	12.30
1/2" x copper	P1@.150	Ea	7.70	5.58	—	13.28
1/2" x MPT	P1@.150	Ea	8.11	5.58	—	13.69
1/2"	P1@.150	Ea	8.57	5.58	—	14.15
1/2" x comp	P1@.150	Ea	8.69	5.58	—	14.27
5/8"	P1@.165	Ea	12.20	6.14	—	18.34
5/8" x 3/4" copper	P1@.170	Ea	9.73	6.33	—	16.06
5/8" x 3/4" MPT	P1@.170	Ea	10.60	6.33	—	16.93
3/4"	P1@.170	Ea	13.10	6.33	—	19.43
3/4" x copper	P1@.170	Ea	12.30	6.33	—	18.63

Balancing valve (brass) with crimped joints (PEX)

1/2" x copper	P1@.150	Ea	44.40	5.58	—	49.98
1/2" x MPT	P1@.150	Ea	51.30	5.58	—	56.88

Manifolds (headers) copper with crimped joints (PEX)

3 - 1/2" outlets	P1@.750	Ea	36.10	27.90	—	64.00
3 - 5/8" outlets	P1@.750	Ea	40.80	27.90	—	68.70
4 - 1/2" outlets	P1@.850	Ea	46.40	31.60	—	78.00
4 - 5/8" outlets	P1@.850	Ea	53.60	31.60	—	85.20
5 - 1/2" outlets	P1@1.00	Ea	59.30	37.20	—	96.50
6 - 1/2" outlets	P1@1.20	Ea	67.40	44.70	—	112.10
8 - 1/2" outlets	P1@1.45	Ea	96.40	54.00	—	150.40
10 - 1/2" outlets	P1@1.65	Ea	124.00	61.40	—	185.40
12 - 1/2" outlets	P1@1.85	Ea	150.00	68.90	—	218.90

Manifolds (headers) copper with mini ball valves and crimped joints (PEX)

3 - 1/2" outlets	P1@.750	Ea	135.00	27.90	—	162.90
3 - 5/8" outlets	P1@.750	Ea	163.00	27.90	—	190.90
4 - 1/2" outlets	P1@.850	Ea	176.00	31.60	—	207.60
4 - 5/8" outlets	P1@.850	Ea	214.00	31.60	—	245.60
5 - 1/2" outlets	P1@1.00	Ea	224.00	37.20	—	261.20
6 - 1/2" outlets	P1@1.20	Ea	256.00	44.70	—	300.70

Polyethylene-Aluminum Pipe with Crimped Joints

Description	Craft@Hrs	Unit	Material $	Labor $	Equipment $	Total $
Crimp rings, nickel-plated soft copper						
1/2"	—	Ea	.18	—	—	.18
5/8"	—	Ea	.19	—	—	.19
3/4"	—	Ea	.22	—	—	.22
1"	—	Ea	.33	—	—	.33
Replacement O-rings						
1/2"	—	Ea	.12	—	—	.12
5/8"	—	Ea	.12	—	—	.12
3/4"	—	Ea	.18	—	—	.18
1"	—	Ea	.18	—	—	.18
Pipe hangers (polypropylene)						
1/2"	P1@.050	Ea	.06	1.86	—	1.92
5/8"	P1@.050	Ea	.06	1.86	—	1.92
3/4"	P1@.050	Ea	.06	1.86	—	1.92
1"	P1@.050	Ea	.08	1.86	—	1.94
Pipe clips (nail clips)						
1/2"	P1@.030	Ea	.13	1.12	—	1.25
5/8"	P1@.030	Ea	.22	1.12	—	1.34
1"	P1@.030	Ea	.49	1.12	—	1.61
Miscellaneous tools						
Crimp tool	—	Ea	279.00	—	—	279.00
Beveling tool	—	Ea	5.65	—	—	5.65
Reaming tool	—	Ea	33.80	—	—	33.80
Pipe bender kit	—	Ea	499.00	—	—	499.00
Bending spring	—	Ea	41.80	—	—	41.80
Pipe cutter	—	Ea	20.30	—	—	20.30

Polyethylene-Aluminum Pipe with Compression Joints

Description   Craft@Hrs   Unit   Material $ Labor $ Equipment $ Total $

HDPE – High Density Polyethylene pipe, or PEX pipe, is cross-linked at the molecular level. Composite pressure pipe is aluminum tube laminated between 2 layers of plastic pipe. An additional external layer of an EVOH polymer creates an oxygen barrier and restricts oxygen permeability. It is manufactured and sold in various coiled lengths (200' to 1,000' rolls). The jointing method can be either crimped or compression connections. It will not rust or corrode and, because of its flexibility, installations require approximately 40% less fittings. Expansion rates are similar to copper, but its thermal coefficient (heat loss) is over 800 times less than copper. It is also chemically resistant to most acids, salt solutions, alkalis, fats and oils.

Common applications include ice or snow melting systems, radiant floor heating systems, water service tubing, hot & cold domestic water service, chilled water systems, compressed air systems, solar and process piping applications. Ice or snow melting systems require an oxygen barrier type PEX pipe selection.

Maximum temperature	Maximum pressure	Pipe selection
73 degrees F	160 psi	PEX
140 degrees F	120 psi	PEX
180 degrees F	100 psi	PEX
210 degrees F	115 psi	PEX-AL

PEX is the designation for HDPE pipe
PEX-AL is the designation for cross linked polyethylene-aluminum composite pipe

Description   Craft@Hrs   Unit   Material $ Labor $ Equipment $ Total $

Cross linked HD Polyurethane PEX pipe with compression joints (PEX), pipe only

Size	Craft@Hrs	Unit	Material $	Labor $	Equipment $	Total $
3/8"	P1@.028	LF	.38	1.04	—	1.42
1/2"	P1@.028	LF	.57	1.04	—	1.61
5/8"	P1@.030	LF	.77	1.12	—	1.89
3/4"	P1@.030	LF	1.07	1.12	—	2.19
1"	P1@.034	LF	1.73	1.27	—	3.00

90-degree brass ell with compression joints (PEX)

Size	Craft@Hrs	Unit	Material $	Labor $	Equipment $	Total $
1/2"	P1@.092	Ea	3.00	3.42	—	6.42
5/8"	P1@.098	Ea	4.16	3.65	—	7.81
3/4"	P1@.104	Ea	6.07	3.87	—	9.94
1"	P1@.121	Ea	10.60	4.50	—	15.10

90-degree brass ell with compression joints x male pipe thread (PEX x MPT)

Size	Craft@Hrs	Unit	Material $	Labor $	Equipment $	Total $
½" x ½"	P1@.098	Ea	4.68	3.65	—	8.33
½" x ¾"	P1@.105	Ea	5.67	3.91	—	9.58
¾" x ¾"	P1@.109	Ea	5.23	4.06	—	9.29

Polyethylene-Aluminum Pipe with Compression Joints

Description	Craft@Hrs	Unit	Material $	Labor $	Equipment $	Total $

90-degree brass ell with compression joints x copper socket (PEX x C)

1/2"	P1@.098	Ea	2.56	3.65	—	6.21
1/2" x 3/4"	P1@.109	Ea	2.72	4.06	—	6.78
5/8" x 3/4"	P1@.112	Ea	3.24	4.17	—	7.41
3/4"	P1@.115	Ea	4.91	4.28	—	9.19

90-degree brass wingback ell with compression joints x female pipe thread (PEX x FPT)

1/2"	P1@.098	Ea	2.22	3.65	—	5.87
5/8"	P1@.105	Ea	3.81	3.91	—	7.72
3/4"	P1@.115	Ea	5.84	4.28	—	10.12

Tee (brass) with compression joints (PEX)

1/2"	P1@.109	Ea	3.66	4.06	—	7.72
5/8"	P1@.121	Ea	5.88	4.50	—	10.38
3/4"	P1@.127	Ea	8.69	4.73	—	13.42
1"	P1@.141	Ea	14.40	5.25	—	19.65

Reducing tee (brass) with compression joints (PEX)

5/8" x 5/8" x 1/2"	P1@.138	Ea	5.39	5.14	—	10.53
3/4" x 1/2" x 3/4"	P1@.138	Ea	7.65	5.14	—	12.79
3/4" x 3/4" x 1/2"	P1@.138	Ea	7.82	5.14	—	12.96
3/4" x 1/2" x 3/4"	P1@.138	Ea	6.92	5.14	—	12.06
3/4" x 3/4" x 5/8"	P1@.138	Ea	7.92	5.14	—	13.06
1" x 1/2" x 1/2"	P1@.142	Ea	11.80	5.29	—	17.09
1" x 1" x 1/2"	P1@.142	Ea	15.90	5.29	—	21.19
1" x 3/4" x 3/4"	P1@.142	Ea	8.91	5.29	—	14.20
1" x 1" x 3/4"	P1@.142	Ea	13.80	5.29	—	19.09

Coupling (brass) with compression joints (PEX)

3/8"	P1@.090	Ea	2.53	3.35	—	5.88
1/2"	P1@.092	Ea	2.76	3.42	—	6.18
5/8"	P1@.098	Ea	3.76	3.65	—	7.41
3/4"	P1@.104	Ea	5.57	3.87	—	9.44
1"	P1@.121	Ea	9.92	4.50	—	14.42

Reducing coupling (brass) with compression joints (PEX)

5/8" x 1/2"	P1@.092	Ea	3.30	3.42	—	6.72
3/4" x 1/2"	P1@.098	Ea	4.99	3.65	—	8.64
3/4" x 5/8"	P1@.104	Ea	4.83	3.87	—	8.70
1" x 3/4"	P1@.121	Ea	6.72	4.50	—	11.22

Cap (brass) with compression joints (PEX)

1/2"	P1@.069	Ea	1.86	2.57	—	4.43
5/8"	P1@.075	Ea	2.56	2.79	—	5.35
3/4"	P1@.081	Ea	3.35	3.01	—	6.36
1"	P1@.098	Ea	4.99	3.65	—	8.64

Polyethylene-Aluminum Pipe with Compression Joints

Description	Craft@Hrs	Unit	Material $	Labor $	Equipment $	Total $
Mini ball valve (brass) with compression joints (PEX)						
1/2" x 3/8" straight	P1@.173	Ea	8.82	6.44	—	15.26
1/2" x 3/8" angle	P1@.173	Ea	8.32	6.44	—	14.76
1/2" x copper	P1@.173	Ea	8.28	6.44	—	14.72
1/2" x MPT	P1@.173	Ea	8.71	6.44	—	15.15
1/2"	P1@.173	Ea	9.73	6.44	—	16.17
5/8"	P1@.173	Ea	12.80	6.44	—	19.24
5/8" x 3/4" copper	P1@.196	Ea	10.00	7.30	—	17.30
5/8" x 3/4" MPT	P1@.196	Ea	10.70	7.30	—	18.00
3/4"	P1@.196	Ea	12.00	7.30	—	19.30
3/4" x copper	P1@.196	Ea	12.30	7.30	—	19.60
Balancing valve (brass) with compression joints (PEX)						
1/2" x copper	P1@.173	Ea	49.10	6.44	—	55.54
1/2" x MPT	P1@.173	Ea	54.60	6.44	—	61.04
Crimp rings, nickel plated soft copper						
1/2"	—	Ea	.18	—	—	.18
5/8"	—	Ea	.20	—	—	.20
3/4"	—	Ea	.23	—	—	.23
1"	—	Ea	.34	—	—	.34
Replacement O-rings						
1/2"	—	Ea	.12	—	—	.12
5/8"	—	Ea	.12	—	—	.12
3/4"	—	Ea	.18	—	—	.18
1"	—	Ea	.18	—	—	.18
Pipe hangers (polypropylene)						
1/2"	P1@.050	Ea	.06	1.86	—	1.92
5/8"	P1@.050	Ea	.06	1.86	—	1.92
3/4"	P1@.050	Ea	.06	1.86	—	1.92
1"	P1@.050	Ea	.08	1.86	—	1.94
Pipe clips (nail clips)						
1/2"	P1@.030	Ea	.13	1.12	—	1.25
5/8"	P1@.030	Ea	.22	1.12	—	1.34
1"	P1@.030	Ea	.49	1.12	—	1.61
Miscellaneous tools						
Crimp tool	—	Ea	279.00	—	—	279.00
Beveling tool	—	Ea	5.65	—	—	5.65
Reaming tool	—	Ea	33.80	—	—	33.80
Pipe bender kit	—	Ea	499.00	—	—	499.00
Bending spring	—	Ea	41.80	—	—	41.80
Pipe cutter	—	Ea	20.30	—	—	20.30

Plumbing and Piping Specialties

Description	Craft@Hrs	Unit	Material $	Labor $	Equipment $	Total $

Water meters, turbine-type. Including brass connection unions

½"	P1@.500	Ea	175.00	18.60	—	193.60
¾"	P1@.500	Ea	284.00	18.60	—	302.60
1"	P1@.650	Ea	380.00	24.20	—	404.20
1½"	P1@.700	Ea	1,020.00	26.10	—	1,046.10
2"	P1@.800	Ea	1,530.00	29.80	—	1,559.80
3"	P1@.900	Ea	2,350.00	33.50	—	2,383.50
4"	P1@.975	Ea	4,370.00	36.30	—	4,406.30

Water meters, compound-type. Including brass connection unions

2"	P1@.800	Ea	3,710.00	29.80	—	3,739.80
3"	P1@.900	Ea	4,830.00	33.50	—	4,863.50
4"	P1@.975	Ea	7,620.00	36.30	—	7,656.30

Water meter by-pass and connection assembly. Includes three isolation ball valves, two tees, two 90-degree elbows, 10' of Type L copper pipe and standard support devices. Make additional allowances for the cost of the water meter.

½"	P1@1.00	Ea	83.30	37.20	—	120.50
¾"	P1@1.50	Ea	126.00	55.80	—	181.80
1"	P1@1.95	Ea	218.00	72.60	—	290.60
1½"	P1@2.45	Ea	401.00	91.20	—	492.20
2"	P1@3.25	Ea	590.00	121.00	—	711.00
3"	P1@5.50	Ea	1,760.00	205.00	—	1,965.00
4"	P1@6.75	Ea	2,990.00	251.00	—	3,241.00

Backflow preventers – reduced pressure. Including valves and test ports

¾"	P1@1.00	Ea	347.00	37.20	—	384.20
1"	P1@1.25	Ea	429.00	46.50	—	475.50
1½"	P1@2.00	Ea	690.00	74.40	—	764.40
2"	P1@2.15	Ea	857.00	80.00	—	937.00
2½"	P1@4.00	Ea	3,110.00	149.00	—	3,259.00
3"	P1@4.75	Ea	4,100.00	177.00	—	4,277.00
4"	P1@6.00	Ea	4,670.00	223.00	—	4,893.00
6"	P1@7.95	Ea	5,100.00	296.00	—	5,396.00

Backflow preventers – reduced pressure. Including integral ball valves and test ports

½"	P1@.900	Ea	211.00	33.50	—	244.50
¾"	P1@1.00	Ea	249.00	37.20	—	286.20
1"	P1@1.25	Ea	312.00	46.50	—	358.50
1¼"	P1@1.50	Ea	458.00	55.80	—	513.80
1½"	P1@2.00	Ea	498.00	74.40	—	572.40
2"	P1@2.15	Ea	560.00	80.00	—	640.00

Plumbing and Piping Specialties

Description	Craft@Hrs	Unit	Material $	Labor $	Equipment $	Total $

Backflow preventers, double-check valve assembly. Including integral ball valves and test ports

Description	Craft@Hrs	Unit	Material $	Labor $	Equipment $	Total $
½"	P1@1.00	Ea	158.00	37.20	—	195.20
¾"	P1@1.00	Ea	175.00	37.20	—	212.20
1"	P1@1.25	Ea	211.00	46.50	—	257.50
1¼"	P1@1.95	Ea	354.00	72.60	—	426.60
1½"	P1@2.00	Ea	367.00	74.40	—	441.40
2"	P1@2.15	Ea	445.00	80.00	—	525.00
2½"	P1@2.95	Ea	1,830.00	110.00	—	1,940.00
3"	P1@3.45	Ea	2,340.00	128.00	—	2,468.00
4"	P1@3.95	Ea	3,170.00	147.00	—	3,317.00
6"	P1@5.75	Ea	5,350.00	214.00	—	5,564.00

Vacuum breakers, atmospheric. Female pipe thread

Description	Craft@Hrs	Unit	Material $	Labor $	Equipment $	Total $
¼"	P1@.350	Ea	52.90	13.00	—	65.90
½"	P1@.350	Ea	73.80	13.00	—	86.80
¾"	P1@.350	Ea	90.20	13.00	—	103.20

Vacuum breakers, hose connection. Male/female hose thread, polished brass

Description	Craft@Hrs	Unit	Material $	Labor $	Equipment $	Total $
½"	P1@.150	Ea	22.50	5.58	—	28.08
¾"	P1@.150	Ea	27.50	5.58	—	33.08

Suction diffusers

Description	Craft@Hrs	Unit	Material $	Labor $	Equipment $	Total $
2"	P1@1.25	Ea	364.00	46.50	—	410.50
3"	P1@2.50	Ea	611.00	93.10	—	704.10
4"	P1@3.25	Ea	777.00	121.00	—	898.00
6"	P1@4.80	Ea	1,110.00	179.00	11.00	1,300.00
8"	P1@5.95	Ea	2,070.00	221.00	13.60	2,304.60
10"	P1@6.75	Ea	2,800.00	251.00	15.40	3,066.40

Triple-duty valves

Description	Craft@Hrs	Unit	Material $	Labor $	Equipment $	Total $
2"	P1@1.25	Ea	481.00	46.50	—	527.50
3"	P1@2.50	Ea	653.00	93.10	—	746.10
4"	P1@3.25	Ea	1,180.00	121.00	—	1,301.00
6"	P1@4.80	Ea	1,950.00	179.00	11.00	2,140.00
8"	P1@5.95	Ea	2,850.00	221.00	13.60	3,084.60
10"	P1@6.75	Ea	4,190.00	251.00	15.40	4,456.40

In-line circulating pump, all bronze. 115 volt, including flange kit

Description	Craft@Hrs	Unit	Material $	Labor $	Equipment $	Total $
1/25 HP	P1@1.50	Ea	432.00	55.80	—	487.80
1/16 HP	P1@1.55	Ea	545.00	57.70	—	602.70
1/12 HP	P1@1.60	Ea	750.00	59.60	—	809.60
1/6 HP	P1@1.75	Ea	1,270.00	65.10	—	1,335.10
1/4 HP	P1@1.95	Ea	1,910.00	72.60	—	1,982.60
1 HP	P1@2.25	Ea	2,540.00	83.70	—	2,623.70
1½ HP	P1@2.85	Ea	3,070.00	106.00	—	3,176.00

Plumbing and Piping Specialties

Description	Craft@Hrs	Unit	Material $	Labor $	Equipment $	Total $

In-line circulating pump, iron body. 115 volt, including flange kit

1/25 HP	P1@1.50	Ea	203.00	55.80	—	258.80
1/16 HP	P1@1.55	Ea	227.00	57.70	—	284.70
1/12 HP	P1@1.60	Ea	427.00	59.60	—	486.60
1/6 HP	P1@1.75	Ea	726.00	65.10	—	791.10
1/4 HP	P1@1.95	Ea	841.00	72.60	—	913.60
1 HP	P1@2.25	Ea	1,430.00	83.70	—	1,513.70
1½ HP	P1@2.85	Ea	1,870.00	106.00	—	1,976.00

Primed steel cam-lock non-fire-rated access doors

8" x 8"	P1@.400	Ea	68.30	14.90	—	83.20
12" x 12"	P1@.500	Ea	78.70	18.60	—	97.30
16" x 16"	P1@.600	Ea	96.80	22.30	—	119.10
18" x 18"	P1@.800	Ea	111.00	29.80	—	140.80
24" x 24"	P1@1.25	Ea	165.00	46.50	—	211.50
36" x 36"	P1@1.60	Ea	304.00	59.60	—	363.60
Add for Allen key lock		Ea	4.87	—	—	4.87
Add for cylinder lock		Ea	11.90	—	—	11.90

Primed steel cam-lock 1½ hour fire-rated access doors

8" x 8"	P1@.400	Ea	252.00	14.90	—	266.90
12" x 12"	P1@.500	Ea	265.00	18.60	—	283.60
16" x 16"	P1@.600	Ea	313.00	22.30	—	335.30
18" x 18"	P1@.800	Ea	318.00	29.80	—	347.80
24" x 24"	P1@1.25	Ea	446.00	46.50	—	492.50
24" x 36"	P1@1.50	Ea	575.00	55.80	—	630.80
48" x 48"	P1@2.00	Ea	890.00	74.40	—	964.40
Add for cylinder lock		Ea	12.00	—	—	12.00

Threaded automatic air vent

Float type	P1@.150	Ea	13.10	5.58	—	18.68
35 PSIG	P1@.150	Ea	83.00	5.58	—	88.58
150 PSIG	P1@.150	Ea	135.00	5.58	—	140.58

Threaded manual air vent

175 PSIG	P1@.150	Ea	7.87	5.58	—	13.45

150 pound forged steel slip-on companion flange, welding-type

2½"	P1@.610	Ea	41.10	22.70	—	63.80
3"	P1@.730	Ea	41.10	27.20	—	68.30
4"	P1@.980	Ea	56.50	36.50	—	93.00
6"	ER@1.47	Ea	74.70	58.70	.96	134.36
8"	ER@1.77	Ea	138.00	70.70	1.15	209.85
10"	ER@2.20	Ea	231.00	87.90	1.43	320.33
12"	ER@2.64	Ea	345.00	105.00	1.72	451.72

Plumbing and Piping Specialties

Description	Craft@Hrs	Unit	Material $	Labor $	Equipment $	Total $
150 pound forged steel companion flange, threaded						
2"	P1@.290	Ea	44.00	10.80	—	54.80
2½"	P1@.380	Ea	66.30	14.10	—	80.40
3"	P1@.460	Ea	73.60	17.10	—	90.70
4"	P1@.600	Ea	98.70	22.30	—	121.00
6"	ER@.680	Ea	146.00	27.20	.44	173.64
8"	ER@.760	Ea	200.00	30.40	.50	230.90
300 pound forged steel slip-on companion flange, welding-type						
2½"	P1@.810	Ea	47.20	30.10	—	77.30
3"	P1@.980	Ea	47.20	36.50	—	83.70
4"	P1@1.30	Ea	68.90	48.40	—	117.30
6"	ER@1.95	Ea	126.00	77.90	1.27	205.17
8"	ER@2.35	Ea	178.00	93.90	1.53	273.43
10"	ER@2.90	Ea	293.00	116.00	1.89	410.89
12"	ER@3.50	Ea	467.00	140.00	2.28	609.28
PVC companion flange						
2"	P1@.140	Ea	9.94	5.21	—	15.15
2½"	P1@.180	Ea	29.90	6.70	—	36.60
3"	P1@.210	Ea	31.20	7.82	—	39.02
4"	P1@.280	Ea	32.10	10.40	—	42.50
5"	P1@.310	Ea	71.90	11.50	—	83.40
6"	P1@.420	Ea	71.90	15.60	—	87.50
8"	P1@.560	Ea	107.00	20.80	—	127.80
Bolt and gasket set						
2"	P1@.500	Ea	5.18	18.60	—	23.78
2½"	P1@.650	Ea	5.99	24.20	—	30.19
3"	P1@.750	Ea	10.20	27.90	—	38.10
4"	P1@1.00	Ea	17.10	37.20	—	54.30
5"	P1@1.10	Ea	19.70	40.90	—	60.60
6"	P1@1.20	Ea	19.70	44.70	—	64.40
8"	P1@1.25	Ea	25.50	46.50	—	72.00
10"	P1@1.70	Ea	30.60	63.30	—	93.90
12"	P1@2.20	Ea	41.10	81.90	—	123.00
Dielectric union, brazed joints						
1/2"	P1@.146	Ea	16.60	5.43	—	22.03
3/4"	P1@.205	Ea	16.60	7.63	—	24.23
1"	P1@.263	Ea	34.50	9.79	—	44.29
1¼"	P1@.322	Ea	59.30	12.00	—	71.30
1½"	P1@.380	Ea	80.70	14.10	—	94.80
2"	P1@.507	Ea	111.00	18.90	—	129.90

Plumbing and Piping Specialties

Description	Craft@Hrs	Unit	Material $	Labor $	Equipment $	Total $

125/150# galvanized ASME expansion tank

15 gallon	P1@1.75	Ea	333.00	65.10	—	398.10
30 gallon	P1@2.00	Ea	600.00	74.40	—	674.40
40 gallon	P1@2.35	Ea	665.00	87.50	—	752.50
60 gallon	P1@2.65	Ea	903.00	98.60	—	1,001.60
80 gallon	P1@2.95	Ea	1,660.00	110.00	—	1,770.00
100 gallon	P1@3.25	Ea	2,000.00	121.00	—	2,121.00
120 gallon	P1@3.50	Ea	2,150.00	130.00	—	2,280.00

Expansion tank fitting

¾"	P1@.300	Ea	75.30	11.20	—	86.50

Lead roof pipe flashing

1½"	P1@.250	Ea	33.50	9.31	—	42.81
2"	P1@.250	Ea	44.10	9.31	—	53.41
3"	P1@.250	Ea	51.10	9.31	—	60.41
4"	P1@.250	Ea	56.30	9.31	—	65.61
6"	P1@.350	Ea	72.10	13.00	—	85.10
8"	P1@.400	Ea	91.20	14.90	—	106.10

Roof flashing, 20" x 20"

Pitched roof, neoprene						
all pipe	P1@.300	Ea	20.90	11.20	—	32.10
Pitched roof, aluminum/neoprene						
all pipe	P1@.300	Ea	22.10	11.20	—	33.30
Flat roof, aluminum/neoprene						
all pipe	P1@.300	Ea	42.10	11.20	—	53.30

Braided stainless steel pipe connector, solder ends, brazed joints

½"	P1@.240	Ea	47.30	8.93	—	56.23
¾"	P1@.300	Ea	62.70	11.20	—	73.90
1"	P1@.360	Ea	93.90	13.40	—	107.30
1¼"	P1@.480	Ea	109.00	17.90	—	126.90
1½"	P1@.540	Ea	125.00	20.10	—	145.10
2"	P1@.600	Ea	182.00	22.30	—	204.30

Braided stainless steel pipe connector, threaded joints

½"	P1@.210	Ea	53.40	7.82	—	61.22
¾"	P1@.250	Ea	57.70	9.31	—	67.01
1"	P1@.300	Ea	71.90	11.20	—	83.10
1¼"	P1@.400	Ea	73.30	14.90	—	88.20
1½"	P1@.450	Ea	99.50	16.70	—	116.20
2"	P1@.500	Ea	142.00	18.60	—	160.60

Plumbing and Piping Specialties

Description	Craft@Hrs	Unit	Material $	Labor $	Equipment $	Total $

Braided stainless steel pipe connector, flanged joints

2"	P1@.500	Ea	126.00	18.60	—	144.60
2½"	P1@.630	Ea	181.00	23.40	—	204.40
3"	P1@.750	Ea	235.00	27.90	—	262.90
4"	P1@1.50	Ea	307.00	55.80	—	362.80
6"	P1@2.50	Ea	610.00	93.10	—	703.10
8"	P1@3.00	Ea	1,370.00	112.00	—	1,482.00

Galvanized steel band pipe hanger

½"	P1@.250	Ea	2.14	9.31	—	11.45
¾"	P1@.250	Ea	2.14	9.31	—	11.45
1"	P1@.250	Ea	2.14	9.31	—	11.45
1¼"	P1@.300	Ea	2.27	11.20	—	13.47
1½"	P1@.300	Ea	2.41	11.20	—	13.61
2"	P1@.300	Ea	2.53	11.20	—	13.73
2½"	P1@.350	Ea	3.10	13.00	—	16.10
3"	P1@.350	Ea	3.56	13.00	—	16.56
4"	P1@.350	Ea	4.52	13.00	—	17.52
6"	P1@.450	Ea	7.11	16.70	—	23.81
8"	P1@.450	Ea	10.60	16.70	—	27.30

Trapeze pipe hanger

24"	P1@1.00	Ea	48.00	37.20	—	85.20
36"	P1@1.25	Ea	61.10	46.50	—	107.60
48"	P1@1.50	Ea	93.00	55.80	—	148.80

Wall bracket support with anchors and bolts

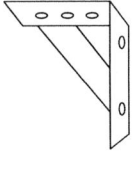

6" x 6"	P1@.600	Ea	22.30	22.30	—	44.60
8" x 8"	P1@.650	Ea	25.80	24.20	—	50.00
10" x 10"	P1@.700	Ea	29.10	26.10	—	55.20
12" x 12"	P1@.800	Ea	35.10	29.80	—	64.90
15" x 15"	P1@.950	Ea	40.40	35.40	—	75.80
18" x 18"	P1@1.10	Ea	57.10	40.90	—	98.00
24" x 24"	P1@1.60	Ea	82.60	59.60	—	142.20

Galvanized U-bolts with nuts

½"	P1@.160	Ea	1.94	5.96	—	7.90
¾"	P1@.165	Ea	1.94	6.14	—	8.08
1"	P1@.170	Ea	2.11	6.33	—	8.44
1¼"	P1@.175	Ea	2.41	6.51	—	8.92
1½"	P1@.180	Ea	2.92	6.70	—	9.62
2"	P1@.190	Ea	3.12	7.07	—	10.19
2½"	P1@.210	Ea	4.82	7.82	—	12.64
3"	P1@.250	Ea	5.57	9.31	—	14.88
4"	P1@.300	Ea	6.30	11.20	—	17.50
6"	P1@.400	Ea	9.86	14.90	—	24.76

Plumbing and Piping Specialties

Description	Craft@Hrs	Unit	Material $	Labor $	Equipment $	Total $

Nail-on wire pipe hooks

Description	Craft@Hrs	Unit	Material $	Labor $	Equipment $	Total $
½"	P1@.100	Ea	.09	3.72	—	3.81
¾"	P1@.100	Ea	.11	3.72	—	3.83
1"	P1@.100	Ea	.12	3.72	—	3.84

Galvanized steel pipe sleeves

Description	Craft@Hrs	Unit	Material $	Labor $	Equipment $	Total $
1"	P1@.120	Ea	4.52	4.47	—	8.99
1¼"	P1@.125	Ea	5.00	4.65	—	9.65
1½"	P1@.125	Ea	5.66	4.65	—	10.31
2"	P1@.130	Ea	6.11	4.84	—	10.95
2½"	P1@.150	Ea	6.20	5.58	—	11.78
3"	P1@.180	Ea	6.34	6.70	—	13.04
4"	P1@.220	Ea	7.21	8.19	—	15.40
5"	P1@.250	Ea	8.99	9.31	—	18.30
6"	P1@.270	Ea	9.72	10.00	—	19.72
8"	P1@.270	Ea	11.20	10.00	—	21.20
10"	P1@.290	Ea	13.00	10.80	—	23.80
12"	P1@.310	Ea	17.70	11.50	—	29.20
14"	P1@.330	Ea	32.90	12.30	—	45.20

Dial-type pressure gauge

Description	Craft@Hrs	Unit	Material $	Labor $	Equipment $	Total $
2½"	P1@.200	Ea	37.80	7.44	—	45.24
3½"	P1@.200	Ea	50.00	7.44	—	57.44

Riser clamp

Description	Craft@Hrs	Unit	Material $	Labor $	Equipment $	Total $
½"	P1@.100	Ea	3.07	3.72	—	6.79
¾"	P1@.100	Ea	4.49	3.72	—	8.21
1"	P1@.100	Ea	4.53	3.72	—	8.25
1¼"	P1@.105	Ea	5.50	3.91	—	9.41
1½"	P1@.110	Ea	5.80	4.09	—	9.89
2"	P1@.115	Ea	6.13	4.28	—	10.41
2½"	P1@.120	Ea	6.45	4.47	—	10.92
3"	P1@.120	Ea	7.00	4.47	—	11.47
4"	P1@.125	Ea	8.89	4.65	—	13.54
5"	P1@.180	Ea	12.90	6.70	—	19.60
6"	P1@.200	Ea	15.40	7.44	—	22.84
8"	P1@.200	Ea	25.10	7.44	—	32.54
10"	P1@.250	Ea	37.30	9.31	—	46.61
12"	P1@.250	Ea	44.20	9.31	—	53.51

15 PSIG float and thermostatic steam trap, threaded

Description	Craft@Hrs	Unit	Material $	Labor $	Equipment $	Total $
¾"	P1@.250	Ea	335.00	9.31	—	344.31
1"	P1@.300	Ea	335.00	11.20	—	346.20
1¼"	P1@.400	Ea	404.00	14.90	—	418.90
1½"	P1@.450	Ea	758.00	16.70	—	774.70
2"	P1@.500	Ea	972.00	18.60	—	990.60

Plumbing and Piping Specialties

Description	Craft@Hrs	Unit	Material $	Labor $	Equipment $	Total $
15 PSIG inverted bucket steam trap, threaded joints						
½"	P1@.210	Ea	194.00	7.82	—	201.82
¾"	P1@.250	Ea	477.00	9.31	—	486.31
1"	P1@.300	Ea	683.00	11.20	—	694.20
1¼"	P1@.400	Ea	1,140.00	14.90	—	1,154.90
1½"	P1@.450	Ea	1,140.00	16.70	—	1,156.70
2"	P1@.500	Ea	1,710.00	18.60	—	1,728.60
Thermometer with well						
7"	P1@.250	Ea	183.00	9.31	—	192.31
9"	P1@.250	Ea	188.00	9.31	—	197.31
Class 125 bronze body 2-piece ball valve, brazed joints						
½"	P1@.240	Ea	10.40	8.93	—	19.33
¾"	P1@.300	Ea	13.80	11.20	—	25.00
1"	P1@.360	Ea	24.00	13.40	—	37.40
1¼"	P1@.480	Ea	40.00	17.90	—	57.90
1½"	P1@.540	Ea	54.10	20.10	—	74.20
2"	P1@.600	Ea	68.60	22.30	—	90.90
3"	P1@1.50	Ea	470.00	55.80	—	525.80
4"	P1@1.80	Ea	614.00	67.00	—	681.00
Class 125 bronze body 2-piece ball valve, soft-soldered joints						
½"	P1@.200	Ea	10.40	7.44	—	17.84
¾"	P1@.249	Ea	13.80	9.27	—	23.07
1"	P1@.299	Ea	24.00	11.10	—	35.10
1¼"	P1@.308	Ea	40.00	11.50	—	51.50
1½"	P1@.498	Ea	54.10	18.50	—	72.60
2"	P1@.498	Ea	68.60	18.50	—	87.10
3"	P1@1.24	Ea	470.00	46.20	—	516.20
4"	P1@1.45	Ea	614.00	54.00	—	668.00
Class 125 bronze body 2-piece ball valve, threaded joints						
½"	P1@.210	Ea	10.40	7.82	—	18.22
¾"	P1@.250	Ea	13.80	9.31	—	23.11
1"	P1@.300	Ea	24.00	11.20	—	35.20
1¼"	P1@.400	Ea	40.00	14.90	—	54.90
1½"	P1@.450	Ea	54.10	16.70	—	70.80
2"	P1@.500	Ea	68.60	18.60	—	87.20
3"	P1@.625	Ea	470.00	23.30	—	493.30
4"	P1@.690	Ea	614.00	25.70	—	639.70

Plumbing and Piping Specialties

Description	Craft@Hrs	Unit	Material $	Labor $	Equipment $	Total $

Class 150, 600 pound W.O.G. bronze body 2-piece ball valve, threaded joints

½"	P1@.210	Ea	17.00	7.82	—	24.82
¾"	P1@.250	Ea	27.90	9.31	—	37.21
1"	P1@.300	Ea	35.40	11.20	—	46.60
1¼"	P1@.400	Ea	44.10	14.90	—	59.00
1½"	P1@.450	Ea	59.30	16.70	—	76.00
2"	P1@.500	Ea	75.10	18.60	—	93.70

Bronze body circuit balance valve

½"	P1@.210	Ea	74.30	7.82	—	82.12
¾"	P1@.250	Ea	79.00	9.31	—	88.31
1"	P1@.300	Ea	91.80	11.20	—	103.00
1¼"	P1@.400	Ea	114.00	14.90	—	128.90
1½"	P1@.400	Ea	145.00	14.90	—	159.90
2"	P1@.500	Ea	202.00	18.60	—	220.60
2½"	P1@1.75	Ea	766.00	65.10	—	831.10
3"	P1@2.50	Ea	1,180.00	93.10	—	1,273.10
4"	P1@3.15	Ea	2,010.00	117.00	—	2,127.00

200 PSIG iron body butterfly valve, lug-type, lever operated

2"	P1@.450	Ea	161.00	16.70	—	177.70
2½"	P1@.450	Ea	167.00	16.70	—	183.70
3"	P1@.550	Ea	176.00	20.50	—	196.50
4"	P1@.505	Ea	219.00	18.80	—	237.80
6"	ER@.800	Ea	352.00	32.00	.52	384.52
8"	ER@.800	Ea	481.00	32.00	.52	513.52
10"	ER@.900	Ea	669.00	36.00	.59	705.59
12"	ER@1.00	Ea	875.00	40.00	.65	915.65

200 PSIG iron body butterfly valve, wafer-type, lever operated

2"	P1@.450	Ea	147.00	16.70	—	163.70
2½"	P1@.450	Ea	150.00	16.70	—	166.70
3"	P1@.550	Ea	161.00	20.50	—	181.50
4"	P1@.550	Ea	195.00	20.50	—	215.50
6"	ER@.800	Ea	326.00	32.00	.52	358.52
8"	ER@.800	Ea	452.00	32.00	.52	484.52
10"	ER@.900	Ea	632.00	36.00	.59	668.59
12"	ER@1.00	Ea	831.00	40.00	.65	871.65

Class 125 bronze body swing check valve, with brazed joints

½"	P1@.240	Ea	20.80	8.93	—	29.73
¾"	P1@.300	Ea	29.70	11.20	—	40.90
1"	P1@.360	Ea	38.80	13.40	—	52.20
1¼"	P1@.480	Ea	55.40	17.90	—	73.30
1½"	P1@.540	Ea	78.10	20.10	—	98.20
2"	P1@.600	Ea	129.00	22.30	—	151.30
2½"	P1@1.00	Ea	261.00	37.20	—	298.20
3"	P1@1.50	Ea	375.00	55.80	—	430.80

Plumbing and Piping Specialties

Description	Craft@Hrs	Unit	Material $	Labor $	Equipment $	Total $

Class 125 bronze body swing check valve with soft-soldered joints

½"	P1@.200	Ea	20.80	7.44	—	28.24
¾"	P1@.249	Ea	29.70	9.27	—	38.97
1"	P1@.299	Ea	38.80	11.10	—	49.90
1¼"	P1@.398	Ea	55.40	14.80	—	70.20
1½"	P1@.448	Ea	78.10	16.70	—	94.80
2"	P1@.498	Ea	129.00	18.50	—	147.50
2½"	P1@.830	Ea	261.00	30.90	—	291.90
3"	P1@1.24	Ea	375.00	46.20	—	421.20

Class 125 bronze body swing check valve, threaded joints

½"	P1@.210	Ea	20.80	7.82	—	28.62
¾"	P1@.250	Ea	29.70	9.31	—	39.01
1"	P1@.300	Ea	38.80	11.20	—	50.00
1¼"	P1@.400	Ea	55.40	14.90	—	70.30
1½"	P1@.450	Ea	78.10	16.70	—	94.80
2"	P1@.500	Ea	129.00	18.60	—	147.60

Class 300 bronze body swing check valve, threaded joints

½"	P1@.210	Ea	70.10	7.82	—	77.92
¾"	P1@.250	Ea	82.90	9.31	—	92.21
1"	P1@.300	Ea	101.00	11.20	—	112.20
1¼"	P1@.400	Ea	147.00	14.90	—	161.90
1½"	P1@.450	Ea	211.00	16.70	—	227.70
2"	P1@.500	Ea	293.00	18.60	—	311.60

Class 125 iron body swing check valve, flanged joints

2"	P1@.500	Ea	182.00	18.60	—	200.60
2½"	P1@.600	Ea	230.00	22.30	—	252.30
3"	P1@.750	Ea	287.00	27.90	—	314.90
4"	P1@1.35	Ea	421.00	50.20	—	471.20
6"	ER@2.50	Ea	817.00	99.90	1.63	918.53
8"	ER@3.00	Ea	1,470.00	120.00	1.96	1,591.96
10"	ER@4.00	Ea	2,390.00	160.00	2.61	2,552.61
12"	ER@4.50	Ea	3,280.00	180.00	2.93	3,462.93

Class 250 iron body swing check valve, flanged joints

2"	P1@.500	Ea	469.00	18.60	—	487.60
2½"	P1@.600	Ea	547.00	22.30	—	569.30
3"	P1@.750	Ea	679.00	27.90	—	706.90
4"	P1@1.35	Ea	833.00	50.20	—	883.20
6"	ER@2.50	Ea	1,600.00	99.90	1.63	1,701.53
8"	ER@3.00	Ea	2,970.00	120.00	1.96	3,091.96

Plumbing and Piping Specialties

Description	Craft@Hrs	Unit	Material $	Labor $	Equipment $	Total $

Class 125 iron body silent check valve, wafer-type

2"	P1@.500	Ea	133.00	18.60	—	151.60
2½"	P1@.600	Ea	148.00	22.30	—	170.30
3"	P1@.750	Ea	170.00	27.90	—	197.90
4"	P1@1.35	Ea	227.00	50.20	—	277.20
5"	P1@2.00	Ea	348.00	74.40	—	422.40
6"	P1@2.50	Ea	474.00	93.10	—	567.10
8"	P1@3.00	Ea	811.00	112.00	—	923.00

Class 125 iron body silent check valve, flanged joints

2"	P1@.500	Ea	133.00	18.60	—	151.60
2½"	P1@.600	Ea	153.00	22.30	—	175.30
3"	P1@.750	Ea	173.00	27.90	—	200.90
4"	P1@1.35	Ea	226.00	50.20	—	276.20
6"	ER@2.50	Ea	406.00	99.90	1.63	507.53
8"	ER@3.00	Ea	732.00	120.00	1.96	853.96
10"	ER@4.00	Ea	1,130.00	160.00	2.61	1,292.61

Class 250 iron body silent check valve, flanged joints

2"	P1@.500	Ea	196.00	18.60	—	214.60
2½"	P1@.600	Ea	226.00	22.30	—	248.30
3"	P1@.750	Ea	251.00	27.90	—	278.90
4"	P1@1.35	Ea	341.00	50.20	—	391.20
6"	ER@2.50	Ea	599.00	99.90	1.63	700.53
8"	ER@3.00	Ea	1,090.00	120.00	1.96	1,211.96
10"	ER@4.00	Ea	1,730.00	160.00	2.61	1,892.61

Class 125 bronze body gate valve, solder ends

½"	P1@.240	Ea	19.30	8.93	—	28.23
¾"	P1@.300	Ea	24.10	11.20	—	35.30
1"	P1@.360	Ea	33.90	13.40	—	47.30
1¼"	P1@.480	Ea	42.60	17.90	—	60.50
1½"	P1@.540	Ea	57.40	20.10	—	77.50
2"	P1@.600	Ea	96.10	22.30	—	118.40
2½"	P1@1.00	Ea	159.00	37.20	—	196.20
3"	P1@1.50	Ea	228.00	55.80	—	283.80

Class 125 bronze body gate valve, threaded joints

½"	P1@.210	Ea	19.30	7.82	—	27.12
¾"	P1@.250	Ea	24.10	9.31	—	33.41
1"	P1@.300	Ea	33.90	11.20	—	45.10
1¼"	P1@.400	Ea	42.60	14.90	—	57.50
1½"	P1@.450	Ea	57.40	16.70	—	74.10
2"	P1@.500	Ea	96.10	18.60	—	114.70
2½"	P1@.750	Ea	159.00	27.90	—	186.90
3"	P1@.950	Ea	228.00	35.40	—	263.40

Plumbing and Piping Specialties

Description	Craft@Hrs	Unit	Material $	Labor $	Equipment $	Total $

Class 300 bronze body gate valve, threaded joints

½"	P1@.210	Ea	39.20	7.82	—	47.02
¾"	P1@.250	Ea	51.20	9.31	—	60.51
1"	P1@.300	Ea	62.00	11.20	—	73.20
1¼"	P1@.400	Ea	84.50	14.90	—	99.40
1½"	P1@.450	Ea	107.00	16.70	—	123.70
2"	P1@.500	Ea	157.00	18.60	—	175.60
2½"	P1@.750	Ea	320.00	27.90	—	347.90
3"	P1@.950	Ea	455.00	35.40	—	490.40

Class 125 iron body gate valve, flanged joints

2"	P1@.500	Ea	306.00	18.60	—	324.60
2½"	P1@.600	Ea	414.00	22.30	—	436.30
3"	P1@.750	Ea	449.00	27.90	—	476.90
4"	P1@1.35	Ea	658.00	50.20	—	708.20
6"	ER@2.50	Ea	1,240.00	99.90	1.63	1,341.53
8"	ER@3.00	Ea	2,060.00	120.00	1.96	2,181.96
10"	ER@4.00	Ea	3,390.00	160.00	2.61	3,552.61
12"	ER@4.50	Ea	4,680.00	180.00	2.93	4,862.93

Class 250 iron body gate valve, flanged joints

2"	P1@.500	Ea	536.00	18.60	—	554.60
2½"	P1@.600	Ea	387.00	22.30	—	409.30
3"	P1@.750	Ea	489.00	27.90	—	516.90
4"	P1@1.35	Ea	829.00	50.20	—	879.20
6"	ER@2.50	Ea	1,550.00	99.90	1.63	1,651.53
8"	ER@3.00	Ea	2,390.00	120.00	1.96	2,511.96
10"	ER@4.00	Ea	4,250.00	160.00	2.61	4,412.61
12"	ER@4.50	Ea	6,930.00	180.00	2.93	7,112.93

Class 125 bronze body globe valve, brazed joints

½"	P1@.240	Ea	35.70	8.93	—	44.63
¾"	P1@.300	Ea	47.50	11.20	—	58.70
1"	P1@.360	Ea	68.10	13.40	—	81.50
1¼"	P1@.480	Ea	95.90	17.90	—	113.80
1½"	P1@.540	Ea	128.00	20.10	—	148.10
2"	P1@.600	Ea	209.00	22.30	—	231.30

Class 125 bronze body globe valve, soft-soldered joints

½"	P1@.200	Ea	35.70	7.44	—	43.14
¾"	P1@.249	Ea	47.50	9.27	—	56.77
1"	P1@.299	Ea	68.10	11.10	—	79.20
1¼"	P1@.398	Ea	95.90	14.80	—	110.70
1½"	P1@.448	Ea	128.00	16.70	—	144.70
2"	P1@.498	Ea	209.00	18.50	—	227.50

Plumbing and Piping Specialties

Description	Craft@Hrs	Unit	Material $	Labor $	Equipment $	Total $
Class 125 bronze body globe valve, threaded ends						
½"	P1@.210	Ea	35.70	7.82	—	43.52
¾"	P1@.250	Ea	47.50	9.31	—	56.81
1"	P1@.300	Ea	68.10	11.20	—	79.30
1¼"	P1@.400	Ea	95.90	14.90	—	110.80
1½"	P1@.450	Ea	128.00	16.70	—	144.70
2"	P1@.500	Ea	209.00	18.60	—	227.60

Description	Craft@Hrs	Unit	Material $	Labor $	Equipment $	Total $
Class 300 bronze body globe valve, threaded joints						
½"	P1@.210	Ea	67.00	7.82	—	74.82
¾"	P1@.250	Ea	85.10	9.31	—	94.41
1"	P1@.300	Ea	120.00	11.20	—	131.20
1¼"	P1@.400	Ea	175.00	14.90	—	189.90
1½"	P1@.450	Ea	235.00	16.70	—	251.70
2"	P1@.500	Ea	323.00	18.60	—	341.60

Description	Craft@Hrs	Unit	Material $	Labor $	Equipment $	Total $
Class 125 iron body globe valve, flanged joints						
2"	P1@.500	Ea	246.00	18.60	—	264.60
2½"	P1@.600	Ea	379.00	22.30	—	401.30
3"	P1@.750	Ea	457.00	27.90	—	484.90
4"	P1@1.35	Ea	614.00	50.20	—	664.20
6"	ER@2.50	Ea	1,080.00	99.90	1.63	1,181.53
8"	ER@3.00	Ea	1,650.00	120.00	1.96	1,771.96
10"	ER@4.00	Ea	2,520.00	160.00	2.61	2,682.61

Description	Craft@Hrs	Unit	Material $	Labor $	Equipment $	Total $
Class 250 iron body globe valve, flanged joints						
2"	P1@.500	Ea	539.00	18.60	—	557.60
2½"	P1@.600	Ea	671.00	22.30	—	693.30
3"	P1@.750	Ea	798.00	27.90	—	825.90
4"	P1@1.35	Ea	1,100.00	50.20	—	1,150.20
6"	ER@2.50	Ea	2,170.00	99.90	1.63	2,271.53
8"	ER@3.00	Ea	4,820.00	120.00	1.96	4,941.96

Description	Craft@Hrs	Unit	Material $	Labor $	Equipment $	Total $
Class 125 bronze body strainer, threaded joints						
½"	P1@.230	Ea	32.00	8.56	—	40.56
¾"	P1@.260	Ea	42.00	9.68	—	51.68
1"	P1@.330	Ea	51.50	12.30	—	63.80
1¼"	P1@.440	Ea	72.00	16.40	—	88.40
1½"	P1@.495	Ea	108.00	18.40	—	126.40
2"	P1@.550	Ea	188.00	20.50	—	208.50

Description	Craft@Hrs	Unit	Material $	Labor $	Equipment $	Total $
Class 125 iron body strainer, flanged joints						
2"	P1@.500	Ea	132.00	18.60	—	150.60
2½"	P1@.600	Ea	147.00	22.30	—	169.30
3"	P1@.750	Ea	169.00	27.90	—	196.90
4"	P1@1.35	Ea	290.00	50.20	—	340.20
6"	ER@2.50	Ea	587.00	99.90	1.63	688.53
8"	ER@3.00	Ea	992.00	120.00	1.96	1,113.96

Plumbing and Piping Specialties

Description	Craft@Hrs	Unit	Material $	Labor $	Equipment $	Total $

Class 250 iron body strainer, flanged joints

2"	P1@.500	Ea	224.00	18.60	—	242.60
2½"	P1@.600	Ea	245.00	22.30	—	267.30
3"	P1@.750	Ea	288.00	27.90	—	315.90
4"	P1@1.35	Ea	509.00	50.20	—	559.20
6"	ER@2.50	Ea	815.00	99.90	1.63	916.53
8"	ER@3.00	Ea	1,490.00	120.00	1.96	1,611.96

Installation of 2-way control valve, threaded joints

½"	P1@.210	Ea	—	7.82	—	7.82
¾"	P1@.250	Ea	—	9.31	—	9.31
1"	P1@.300	Ea	—	11.20	—	11.20
1¼"	P1@.400	Ea	—	14.90	—	14.90
1½"	P1@.450	Ea	—	16.70	—	16.70
2"	P1@.500	Ea	—	18.60	—	18.60
2½"	P1@.830	Ea	—	30.90	—	30.90
3"	P1@.990	Ea	—	36.80	—	36.80

Installation of 2-way control valve, flanged joints

2"	P1@.500	Ea	—	18.60	—	18.60
2½"	P1@.600	Ea	—	22.30	—	22.30
3"	P1@.750	Ea	—	27.90	—	27.90
4"	P1@1.35	Ea	—	50.20	—	50.20
6"	P1@2.50	Ea	—	93.10	5.70	98.80
8"	P1@3.00	Ea	—	112.00	6.84	118.84
10"	P1@4.00	Ea	—	149.00	9.12	158.12
12"	P1@4.50	Ea	—	167.00	10.30	177.30

Installation of 3-way control valve, threaded joints

½"	P1@.260	Ea	—	9.68	—	9.68
¾"	P1@.365	Ea	—	13.60	—	13.60
1"	P1@.475	Ea	—	17.70	—	17.70
1¼"	P1@.575	Ea	—	21.40	—	21.40
1½"	P1@.680	Ea	—	25.30	—	25.30
2"	P1@.910	Ea	—	33.90	—	33.90
2½"	P1@1.12	Ea	—	41.70	—	41.70
3"	P1@1.33	Ea	—	49.50	—	49.50

Plumbing and Piping Specialties

Description	Craft@Hrs	Unit	Material $	Labor $	Equipment $	Total $

Installation of 3-way control valve, flanged joints

2"	P1@.910	Ea	—	33.90	—	33.90
2½"	P1@1.12	Ea	—	41.70	—	41.70
3"	P1@1.33	Ea	—	49.50	—	49.50
4"	P1@2.00	Ea	—	74.40	—	74.40
6"	P1@3.70	Ea	—	138.00	8.44	146.44
8"	P1@4.40	Ea	—	164.00	10.00	174.00
10"	P1@5.90	Ea	—	220.00	13.50	233.50
12"	P1@6.50	Ea	—	242.00	14.80	256.80

Lever-handled gas valve, threaded joints

½"	P1@.210	Ea	31.60	7.82	—	39.42
¾"	P1@.250	Ea	35.00	9.31	—	44.31
1"	P1@.300	Ea	43.40	11.20	—	54.60
1¼"	P1@.400	Ea	58.20	14.90	—	73.10
1½"	P1@.450	Ea	78.40	16.70	—	95.10
2"	P1@.500	Ea	114.00	18.60	—	132.60

Plug-type gas valve, flanged joints, lubricated

2"	P1@.500	Ea	237.00	18.60	—	255.60
2½"	P1@.600	Ea	384.00	22.30	—	406.30
3"	P1@.750	Ea	540.00	27.90	—	567.90
4"	P1@1.35	Ea	739.00	50.20	—	789.20
6"	P1@2.50	Ea	1,320.00	93.10	—	1,413.10

¾" hose bibb

Low quality	P1@.350	Ea	12.40	13.00	—	25.40
Med. quality	P1@.350	Ea	17.90	13.00	—	30.90
High quality	P1@.350	Ea	23.60	13.00	—	36.60

Water pressure reducing valve, threaded joints, 25-75 PSI

½"	P1@.210	Ea	110.00	7.82	—	117.82
¾"	P1@.250	Ea	142.00	9.31	—	151.31
1"	P1@.300	Ea	231.00	11.20	—	242.20
1¼"	P1@.400	Ea	282.00	14.90	—	296.90
1½"	P1@.450	Ea	402.00	16.70	—	418.70
2"	P1@.500	Ea	636.00	18.60	—	654.60

Water hammer arrester, threaded joints

½"	P1@.160	Ea	49.00	5.96	—	54.96
¾"	P1@.190	Ea	56.90	7.07	—	63.97
1"	P1@.230	Ea	144.00	8.56	—	152.56
1¼"	P1@.300	Ea	168.00	11.20	—	179.20
1½"	P1@.340	Ea	241.00	12.70	—	253.70
2"	P1@.380	Ea	367.00	14.10	—	381.10

Plumbing and Piping Specialties

Description	Craft@Hrs	Unit	Material $	Labor $	Equipment $	Total $
Carbon steel weldolet, Schedule 40						
½"	P1@.330	Ea	16.40	12.30	1.04	29.74
¾"	P1@.500	Ea	17.00	18.60	1.58	37.18
1"	P1@.670	Ea	17.80	24.90	2.12	44.82
1¼"	P1@.830	Ea	21.40	30.90	2.63	54.93
1½"	P1@1.00	Ea	21.40	37.20	3.16	61.76
2"	P1@1.33	Ea	23.70	49.50	4.21	77.41
2½"	P1@1.67	Ea	52.10	62.20	5.28	119.58
3"	P1@2.00	Ea	59.20	74.40	6.33	139.93
4"	P1@2.67	Ea	76.40	99.40	8.45	184.25
6"	P1@4.00	Ea	315.00	149.00	12.70	476.70
8"	P1@4.80	Ea	463.00	179.00	15.20	657.20
10"	P1@6.00	Ea	641.00	223.00	19.00	883.00
12"	P1@7.20	Ea	788.00	268.00	22.80	1,078.80
Carbon steel weldolet, Schedule 80						
½"	P1@.440	Ea	18.20	16.40	1.39	35.99
¾"	P1@.670	Ea	18.60	24.90	2.12	45.62
1"	P1@.890	Ea	20.40	33.10	2.82	56.32
1¼"	P1@1.11	Ea	25.70	41.30	3.51	70.51
1½"	P1@1.33	Ea	25.70	49.50	4.21	79.41
2"	P1@1.77	Ea	27.70	65.90	5.60	99.20
2½"	P1@2.22	Ea	61.80	82.60	7.02	151.42
3"	P1@2.66	Ea	63.60	99.00	8.42	171.02
4"	P1@3.55	Ea	141.00	132.00	11.23	284.23
6"	P1@5.32	Ea	443.00	198.00	16.40	657.40
8"	P1@6.38	Ea	592.00	237.00	20.00	849.00
Carbon steel threadolet, Schedule 40						
¾"	P1@.330	Ea	10.20	12.30	—	22.50
1"	P1@.440	Ea	11.30	16.40	—	27.70
1¼"	P1@.560	Ea	15.80	20.80	—	36.60
1½"	P1@.670	Ea	17.80	24.90	—	42.70
2"	P1@.890	Ea	20.70	33.10	—	53.80
2½"	P1@1.11	Ea	63.80	41.30	—	105.10
Carbon steel threadolet, Schedule 80						
½"	P1@.300	Ea	15.30	11.20	—	26.50
¾"	P1@.440	Ea	16.90	16.40	—	33.30
1"	P1@.590	Ea	20.80	22.00	—	42.80
1¼"	P1@.740	Ea	93.60	27.50	—	121.10
1½"	P1@.890	Ea	93.60	33.10	—	126.70
2"	P1@1.18	Ea	122.00	43.90	—	165.90

Cast Iron, DWV, Service Weight, No-Hub with Coupled Joints

Service weight cast iron soil pipe with hub-less coupled joints is widely used for non-pressurized underground and in-building drain, waste and vent (DWV) systems.

This section has been arranged to save the estimator's time by including all normally-used system components such as pipe, fittings, hanger assemblies and riser clamps under one heading. Additional items can be found under "Plumbing and Piping Specialties." The cost estimates in this section are based on the conditions, limitations and wage rates described in the section "How to Use This Book" beginning on page 5.

Equipment cost, where shown, is $4.45 per hour for a 2-ton chain hoist.

Description	Craft@Hrs	Unit	Material $	Labor $	Equipment $	Total $

DWV cast iron no-hub (MJ) horizontal pipe assembly. Horizontally hung in a building. Assembly includes fittings, couplings, hangers and rod. Based on a reducing wye and a 45-degree elbow every 15 feet and a hanger every 6 feet for 2" pipe. A reducing wye and a 45-degree elbow every 40 feet and a hanger every 8 feet for 10" pipe. *(Distance between fittings in the pipe assembly increases as pipe diameter increases.)* Use these figures for preliminary estimates.

Description	Craft@Hrs	Unit	Material $	Labor $	Equipment $	Total $
2"	P1@.130	LF	11.80	4.84	—	16.64
3"	P1@.160	LF	13.00	5.96	—	18.96
4"	P1@.190	LF	15.20	7.07	—	22.27
6"	ER@.220	LF	29.50	8.79	.14	38.43
8"	ER@.260	LF	56.70	10.40	.17	67.27
10"	ER@.310	LF	87.80	12.40	.20	100.40

DWV cast iron no-hub (MJ), pipe only

Description	Craft@Hrs	Unit	Material $	Labor $	Equipment $	Total $
2"	P1@.050	LF	6.76	1.86	—	8.62
3"	P1@.060	LF	8.05	2.23	—	10.28
4"	P1@.080	LF	10.50	2.98	—	13.48
6"	ER@.110	LF	19.70	4.40	.14	24.24
8"	ER@.130	LF	36.30	5.19	.17	41.66
10"	ER@.140	LF	63.30	5.59	.20	69.09

DWV cast iron no-hub 1/8 bend

Description	Craft@Hrs	Unit	Material $	Labor $	Equipment $	Total $
2"	P1@.250	Ea	6.82	9.31	—	16.13
3"	P1@.350	Ea	8.88	13.00	—	21.88
4"	P1@.500	Ea	11.20	18.60	—	29.80
6"	ER@.650	Ea	36.00	26.00	.42	62.42
8"	ER@.710	Ea	96.60	28.40	.46	125.46
10"	ER@.800	Ea	144.00	32.00	.52	176.52

Cast Iron, DWV, Service Weight, No-Hub with Coupled Joints

Description	Craft@Hrs	Unit	Material $	Labor $	Equipment $	Total $
DWV cast iron no-hub ¼ bend						
2"	P1@.250	Ea	8.32	9.31	—	17.63
3"	P1@.350	Ea	10.30	13.00	—	23.30
4"	P1@.500	Ea	15.70	18.60	—	34.30
6"	ER@.650	Ea	38.20	26.00	.42	64.62
8"	ER@.710	Ea	148.00	28.40	.46	176.86
10"	ER@.710	Ea	162.00	28.40	.46	190.86
DWV cast iron no-hub long sweep ¼ bend						
2"	P1@.250	Ea	14.50	9.31	—	23.81
3"	P1@.350	Ea	17.50	13.00	—	30.50
4"	P1@.500	Ea	27.70	18.60	—	46.30
5"	ER@.580	Ea	50.80	23.20	.38	74.38
6"	ER@.650	Ea	61.80	26.00	.42	88.22
DWV cast iron no-hub low-heel outlet ¼ bend						
3"	P1@.470	Ea	14.70	17.50	—	32.20
4"	P1@.650	Ea	18.00	24.20	—	42.20
DWV cast iron no-hub closet bend						
3"	P1@.350	Ea	33.90	13.00	—	46.90
4"	P1@.500	Ea	34.80	18.60	—	53.40
DWV cast iron no-hub closet flange						
4" x 3"	P1@.500	Ea	49.60	18.60	—	68.20
DWV cast iron no-hub P-trap						
2"	P1@.250	Ea	13.50	9.31	—	22.81
3"	P1@.350	Ea	41.80	13.00	—	54.80
4"	P1@.500	Ea	51.70	18.60	—	70.30
6"	ER@.650	Ea	184.00	26.00	.42	210.42
DWV cast iron no-hub wye						
2"	P1@.380	Ea	10.60	14.10	—	24.70
3"	P1@.520	Ea	14.80	19.40	—	34.20
4"	P1@.700	Ea	23.30	26.10	—	49.40
6"	ER@.900	Ea	62.70	36.00	.59	99.29
8"	ER@1.15	Ea	159.00	46.00	.75	205.75
10"	ER@1.50	Ea	382.00	59.90	.98	442.88

Cast Iron, DWV, Service Weight, No-Hub with Coupled Joints

Description	Craft@Hrs	Unit	Material $	Labor $	Equipment $	Total $

DWV cast iron no-hub reducing wye

3" x 2"	P1@.470	Ea	14.20	17.50	—	31.70
4" x 2"	P1@.600	Ea	19.00	22.30	—	41.30
4" x 3"	P1@.650	Ea	20.00	24.20	—	44.20
6" x 2"	ER@.770	Ea	50.90	30.80	.50	82.20
6" x 3"	ER@.770	Ea	55.30	30.80	.50	86.60
6" x 4"	ER@.800	Ea	54.30	32.00	.52	86.82
8" x 2"	ER@1.00	Ea	130.00	40.00	.65	170.65
8" x 3"	ER@1.00	Ea	132.00	40.00	.65	172.65
8" x 4"	ER@1.00	Ea	134.00	40.00	.65	174.65
8" x 6"	ER@1.10	Ea	138.00	44.00	.72	182.72

DWV cast iron no-hub combination upright wye and 1/8 bend

2"	P1@.380	Ea	26.30	14.10	—	40.40
3"	P1@.520	Ea	42.40	19.40	—	61.80
4"	P1@.700	Ea	56.00	26.10	—	82.10

DWV cast iron no-hub combination upright reducing wye and 1/8 bend

3" x 2"	P1@.470	Ea	46.00	17.50	—	63.50
4" x 2"	P1@.600	Ea	54.80	22.30	—	77.10
4" x 3"	P1@.650	Ea	57.30	24.20	—	81.50

DWV cast iron no-hub double wye

2"	P1@.500	Ea	32.70	18.60	—	51.30
3"	P1@.700	Ea	33.10	26.10	—	59.20
4"	P1@.950	Ea	51.40	35.40	—	86.80
6"	ER@1.10	Ea	179.00	44.00	.72	223.72
8"	ER@1.45	Ea	326.00	57.90	.94	384.84

DWV cast iron no-hub double reducing wye

3" x 2"	P1@.520	Ea	32.70	19.40	—	52.10
4" x 2"	P1@.740	Ea	45.50	27.50	—	73.00
4" x 3"	P1@1.00	Ea	46.80	37.20	—	84.00
6" x 4"	ER@1.75	Ea	176.00	69.90	1.14	247.04

DWV cast iron no-hub combination wye and 1/8 bend

2"	P1@.380	Ea	20.40	14.10	—	34.50
3"	P1@.520	Ea	29.10	19.40	—	48.50
4"	P1@.700	Ea	46.40	26.10	—	72.50
6"	ER@.900	Ea	158.00	36.00	.59	194.59
8"	ER@1.15	Ea	318.00	46.00	.75	364.75

Cast Iron, DWV, Service Weight, No-Hub with Coupled Joints

Description	Craft@Hrs	Unit	Material $	Labor $	Equipment $	Total $
DWV cast iron no-hub combination reducing wye and 1/8 bend						
3" x 2"	P1@.470	Ea	24.70	17.50	—	42.20
4" x 2"	P1@.600	Ea	37.60	22.30	—	59.90
4" x 3"	P1@.650	Ea	40.30	24.20	—	64.50
6" x 2"	ER@.770	Ea	180.00	30.80	.50	211.30
6" x 3"	ER@.770	Ea	189.00	30.80	.50	220.30
6" x 4"	ER@.800	Ea	194.00	32.00	.52	226.52
8" x 4"	ER@1.20	Ea	353.00	48.00	.78	401.78
DWV cast iron no-hub double combination wye and 1/8 bend						
2"	P1@.500	Ea	35.40	18.60	—	54.00
3"	P1@.700	Ea	56.80	26.10	—	82.90
4"	P1@.950	Ea	69.90	35.40	—	105.30
DWV cast iron no-hub double combination reducing wye and 1/8 bend						
3" x 2"	P1@.600	Ea	40.10	22.30	—	62.40
4" x 2"	P1@1.25	Ea	63.20	46.50	—	109.70
4" x 3"	P1@1.35	Ea	66.10	50.20	—	116.30
DWV cast iron no-hub sanitary tee (TY)						
2"	P1@.380	Ea	12.40	14.10	—	26.50
3"	P1@.520	Ea	14.80	19.40	—	34.20
4"	P1@.700	Ea	24.50	26.10	—	50.60
6"	ER@.900	Ea	74.30	36.00	.59	110.89
8"	ER@1.15	Ea	196.00	46.00	.75	242.75
DWV cast iron no-hub reducing sanitary tee (TY)						
3" x 2"	P1@.470	Ea	14.00	17.50	—	31.50
4" x 2"	P1@.600	Ea	18.80	22.30	—	41.10
4" x 3"	P1@.650	Ea	21.60	24.20	—	45.80
6" x 2"	ER@.770	Ea	58.90	30.80	.50	90.20
6" x 3"	ER@.770	Ea	55.70	30.80	.50	87.00
6" x 4"	ER@.800	Ea	60.40	32.00	.52	92.92
8" x 3"	ER@1.00	Ea	155.00	40.00	.65	195.65
8" x 4"	ER@1.00	Ea	151.00	40.00	.65	191.65
8" x 6"	ER@1.10	Ea	159.00	44.00	.72	203.72
DWV cast iron no-hub tapped sanitary tee (tapped TY)						
2" x 1½"	P1@.200	Ea	34.60	7.44	—	42.04
2" x 2"	P1@.250	Ea	34.60	9.31	—	43.91
3" x 2"	P1@.350	Ea	44.10	13.00	—	57.10
4" x 2"	P1@.425	Ea	52.20	15.80	—	68.00
6" x 2"	ER@.700	Ea	135.00	28.00	.46	163.46

Cast Iron, DWV, Service Weight, No-Hub with Coupled Joints

Description	Craft@Hrs	Unit	Material $	Labor $	Equipment $	Total $

DWV cast iron no-hub test tee

2"	P1@.370	Ea	43.80	13.80	—	57.60
3"	P1@.520	Ea	61.80	19.40	—	81.20
4"	P1@.750	Ea	78.80	27.90	—	106.70
6"	ER@1.12	Ea	282.00	44.80	.73	327.53
8"	ER@1.25	Ea	512.00	50.00	.81	562.81

DWV cast iron no-hub cross (double TY)

2"	P1@.500	Ea	34.80	18.60	—	53.40
3"	P1@.700	Ea	37.00	26.10	—	63.10
4"	P1@.950	Ea	53.40	35.40	—	88.80
6"	ER@1.30	Ea	198.00	51.90	.85	250.75

DWV cast iron no-hub reducing cross (double TY)

3" x 2"	P1@.600	Ea	32.80	22.30	—	55.10
4" x 2"	P1@.900	Ea	45.40	33.50	—	78.90
4" x 3"	P1@.900	Ea	46.40	33.50	—	79.90
6" x 4"	ER@1.20	Ea	182.00	48.00	.78	230.78
8" x 4"	ER@1.50	Ea	457.00	59.90	.98	517.88

DWV cast iron no-hub reducer

2"	P1@.120	Ea	6.84	4.47	—	11.31
3"	P1@.130	Ea	7.37	4.84	—	12.21
4"	P1@.150	Ea	10.90	5.58	—	16.48
6"	ER@.200	Ea	29.20	7.99	.13	37.32
8"	ER@.400	Ea	75.00	16.00	.26	91.26
10"	ER@.650	Ea	116.00	26.00	.42	142.42

DWV cast iron no-hub cap

2"	P1@.125	Ea	4.17	4.65	—	8.82
3"	P1@.175	Ea	7.16	6.51	—	13.67
4"	P1@.250	Ea	9.32	9.31	—	18.63
6"	ER@.375	Ea	17.70	15.00	.24	32.94
8"	ER@.500	Ea	32.70	20.00	.33	53.03
10"	ER@.750	Ea	50.50	30.00	.49	80.99

MJ (mechanical joint) coupling with gasket (labor included with fittings)

2"	—	Ea	8.56	—	—	8.56
3"	—	Ea	9.36	—	—	9.36
4"	—	Ea	10.00	—	—	10.00
6"	—	Ea	27.30	—	—	27.30
8"	—	Ea	50.90	—	—	50.90
10"	—	Ea	70.60	—	—	70.60

Cast Iron, DWV, Service Weight, No-Hub with Coupled Joints

Description	Craft@Hrs	Unit	Material $	Labor $	Equipment $	Total $
Hanger with swivel assembly						
2"	P1@.400	Ea	7.22	14.90	—	22.12
3"	P1@.450	Ea	9.31	16.70	—	26.01
4"	P1@.450	Ea	11.60	16.70	—	28.30
5"	P1@.500	Ea	17.70	18.60	—	36.30
6"	P1@.500	Ea	20.00	18.60	—	38.60
8"	P1@.550	Ea	26.10	20.50	—	46.60
10"	P1@.650	Ea	34.40	24.20	—	58.60
Riser clamp						
2"	P1@.115	Ea	6.65	4.28	—	10.93
3"	P1@.120	Ea	7.59	4.47	—	12.06
4"	P1@.125	Ea	9.66	4.65	—	14.31
5"	P1@.180	Ea	13.80	6.70	—	20.50
6"	P1@.200	Ea	16.70	7.44	—	24.14
8"	P1@.200	Ea	27.30	7.44	—	34.74
10"	P1@.250	Ea	40.40	9.31	—	49.71

Cast Iron, DWV, Service Weight, Hub & Spigot with Gasketed Joints

Service weight hub and spigot cast iron soil pipe with gasketed joints is widely used for non-pressurized underground and in-building drain, waste and vent (DWV) systems.

This section has been arranged to save the estimator's time by including all normally-used system components such as pipe, fittings, valves, hanger assemblies and riser clamps under one heading. Additional items can be found under "Plumbing and Piping Specialties." The cost estimates in this section are based on the conditions, limitations and wage rates described in the section "How to Use This Book" beginning on page 5.

Equipment cost, where shown, is $4.45 per hour for a 2-ton chain hoist.

Description	Craft@Hrs	Unit	Material $	Labor $	Equipment $	Total $

DWV cast iron pipe, hub and spigot, gasketed joints

2"	P1@.060	LF	12.10	2.23	—	14.33
3"	P1@.070	LF	13.30	2.61	—	15.91
4"	P1@.090	LF	17.50	3.35	—	20.85
6"	ER@.110	LF	33.40	4.40	.07	37.87
8"	ER@.140	LF	56.00	5.59	.09	61.68
10"	ER@.180	LF	89.50	7.19	.12	96.81
12"	ER@.230	LF	116.00	9.19	.15	125.34
15"	ER@.280	LF	174.00	11.20	.18	185.38

DWV cast iron 1/8 bend, hub and spigot, gasketed joints

2"	P1@.300	Ea	14.20	11.20	—	25.40
3"	P1@.400	Ea	19.20	14.90	—	34.10
4"	P1@.550	Ea	27.00	20.50	—	47.50
6"	ER@.700	Ea	67.90	28.00	.46	96.36
8"	ER@1.15	Ea	161.00	46.00	.75	207.75
10"	ER@1.25	Ea	218.00	50.00	.81	268.81
12"	ER@2.25	Ea	502.00	89.90	1.47	593.37
15"	ER@2.50	Ea	841.00	99.90	1.63	942.53

DWV cast iron ¼ bend, hub and spigot, gasketed joints

2"	P1@.300	Ea	15.70	11.20	—	26.90
3"	P1@.400	Ea	24.10	14.90	—	39.00
4"	P1@.550	Ea	42.50	20.50	—	63.00
6"	ER@.900	Ea	80.60	36.00	.59	117.19
8"	ER@1.15	Ea	216.00	46.00	.75	262.75
10"	ER@1.25	Ea	299.00	50.00	.81	349.81
12"	ER@2.25	Ea	626.00	89.90	1.47	717.37
15"	ER@2.50	Ea	1,260.00	99.90	1.63	1,361.53

DWV cast iron heel inlet ¼ bend, hub and spigot, gasketed joints

3" x 2"	P1@.550	Ea	29.50	20.50	—	50.00
4" x 2"	P1@.700	Ea	35.90	26.10	—	62.00

Cast Iron, DWV, Service Weight, Hub & Spigot with Gasketed Joints

Description	Craft@Hrs	Unit	Material $	Labor $	Equipment $	Total $

DWV cast iron 1/16 bend, hub and spigot, gasketed joints

2"	P1@.300	Ea	8.89	11.20	—	20.09
3"	P1@.400	Ea	15.00	14.90	—	29.90
4"	P1@.550	Ea	20.80	20.50	—	41.30
6"	ER@.700	Ea	34.60	28.00	.46	63.06
8"	ER@1.15	Ea	122.00	46.00	.75	168.75

DWV cast iron closet bend, hub and spigot, gasketed joints

4" x 4"	P1@.400	Ea	51.60	14.90	—	66.50

DWV cast iron closet flange, hub and spigot, gasketed joints

4" x 2"	P1@.075	Ea	16.70	2.79	—	19.49
4" x 3"	P1@.100	Ea	23.20	3.72	—	26.92
4" x 4"	P1@.125	Ea	29.90	4.65	—	34.55

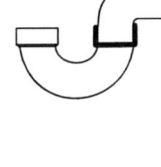

DWV cast iron regular P-trap, hub and spigot, gasketed joints

2"	P1@.300	Ea	59.80	11.20	—	71.00
3"	P1@.400	Ea	83.00	14.90	—	97.90
4"	P1@.550	Ea	128.00	20.50	—	148.50
6"	ER@.700	Ea	282.00	28.00	.46	310.46
8"	ER@1.15	Ea	830.00	46.00	.75	876.75
10"	ER@2.00	Ea	1,140.00	79.90	1.30	1,221.20

DWV cast iron sanitary tee (TY), hub and spigot, gasketed joints

2"	P1@.400	Ea	28.70	14.90	—	43.60
3"	P1@.600	Ea	50.20	22.30	—	72.50
4"	P1@.750	Ea	101.00	27.90	—	128.90
6"	ER@.900	Ea	232.00	36.00	.59	268.59
8"	ER@1.50	Ea	335.00	59.90	.98	395.88

DWV cast iron sanitary reducing tee (TY), hub and spigot, gasketed joints

3" x 2"	P1@.400	Ea	40.70	14.90	—	55.60
4" x 2"	P1@.650	Ea	106.00	24.20	—	130.20
4" x 3"	P1@.700	Ea	106.00	26.10	—	132.10
6" x 2"	ER@.750	Ea	144.00	30.00	.49	174.49
6" x 3"	ER@.800	Ea	144.00	32.00	.52	176.52
6" x 4"	ER@.850	Ea	144.00	34.00	.55	178.55

DWV cast iron sanitary tapped tee (tapped TY), hub and spigot, gasketed joints

2"	P1@.300	Ea	24.60	11.20	—	35.80
3"	P1@.400	Ea	36.20	14.90	—	51.10
4"	P1@.550	Ea	35.90	20.50	—	56.40

Cast Iron, DWV, Service Weight, Hub & Spigot with Gasketed Joints

Description	Craft@Hrs	Unit	Material $	Labor $	Equipment $	Total $

DWV cast iron combination wye and 1/8 bend, hub and spigot, gasketed joints

2"	P1@.400	Ea	25.40	14.90	—	40.30
3"	P1@.600	Ea	38.40	22.30	—	60.70
4"	P1@.750	Ea	53.30	27.90	—	81.20
5"	ER@.850	Ea	100.00	34.00	.55	134.55
6"	ER@.900	Ea	127.00	36.00	.59	163.59

DWV cast iron combination reducing wye and 1/8 bend, hub and spigot, gasketed joints

3" x 2"	P1@.550	Ea	28.70	20.50	—	49.20
4" x 2"	P1@.650	Ea	39.40	24.20	—	63.60
4" x 3"	P1@.700	Ea	45.70	26.10	—	71.80
5" x 2"	ER@.700	Ea	81.00	28.00	.46	109.46
5" x 4"	ER@.800	Ea	95.90	32.00	.52	128.42
6" x 2"	ER@.750	Ea	82.10	30.00	.49	112.59
6" x 3"	ER@.750	Ea	87.20	30.00	.49	117.69
6" x 4"	ER@.800	Ea	91.60	32.00	.52	124.12

DWV cast iron combination double wye and 1/8 bend, hub and spigot, gasketed joints

2"	P1@.500	Ea	48.70	18.60	—	67.30
3"	P1@.750	Ea	66.60	27.90	—	94.50
4"	P1@.950	Ea	94.70	35.40	—	130.10
6"	ER@1.10	Ea	325.00	44.00	.72	369.72

DWV cast iron combination double reducing wye and 1/8 bend, hub and spigot, gasketed joints

3" x 2"	P1@.550	Ea	53.50	20.50	—	74.00
4" x 2"	P1@.650	Ea	67.30	24.20	—	91.50
4" x 3"	P1@.700	Ea	76.30	26.10	—	102.40
6" x 4"	ER@.800	Ea	215.00	32.00	.52	247.52

DWV cast iron wye, hub and spigot, gasketed joints

2"	P1@.400	Ea	23.50	14.90	—	38.40
3"	P1@.600	Ea	40.40	22.30	—	62.70
4"	P1@.750	Ea	54.10	27.90	—	82.00
6"	ER@.900	Ea	130.00	36.00	.59	166.59
8"	ER@1.50	Ea	290.00	59.90	.98	350.88
10"	ER@2.00	Ea	487.00	79.90	1.30	568.20
12"	ER@2.50	Ea	1,100.00	99.90	1.63	1,201.53
15"	ER@3.00	Ea	2,180.00	120.00	1.96	2,301.96

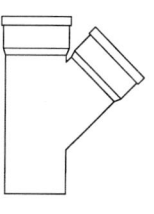

Cast Iron, DWV, Service Weight, Hub & Spigot with Gasketed Joints

Description	Craft@Hrs	Unit	Material $	Labor $	Equipment $	Total $

DWV cast iron reducing wye, hub and spigot, gasketed joints

Description	Craft@Hrs	Unit	Material $	Labor $	Equipment $	Total $
4" x 2"	P1@.650	Ea	40.70	24.20	—	64.90
4" x 3"	P1@.700	Ea	49.90	26.10	—	76.00
6" x 2"	ER@.750	Ea	104.00	30.00	.49	134.49
6" x 3"	ER@.750	Ea	104.00	30.00	.49	134.49
6" x 4"	ER@.800	Ea	105.00	32.00	.52	137.52
8" x 4"	ER@1.35	Ea	241.00	53.90	.88	295.78
8" x 6"	ER@1.45	Ea	257.00	57.90	.94	315.84
10" x 4"	ER@1.85	Ea	416.00	73.90	1.21	491.11
10" x 6"	ER@1.90	Ea	428.00	75.90	1.24	505.14
10" x 8"	ER@1.95	Ea	457.00	77.90	1.27	536.17
12" x 6"	ER@2.35	Ea	774.00	93.90	1.53	869.43
12" x 8"	ER@2.40	Ea	850.00	95.90	1.56	947.46
12" x 10"	ER@2.45	Ea	916.00	97.90	1.60	1,015.50
15" x 6"	ER@2.85	Ea	1,250.00	114.00	1.86	1,365.86
15" x 8"	ER@2.85	Ea	1,260.00	114.00	1.86	1,375.86
15" x 10"	ER@2.90	Ea	1,430.00	116.00	1.86	1,547.86
15" x 12"	ER@2.95	Ea	1,440.00	118.00	1.92	1,559.92

DWV cast iron double wye, hub and spigot, gasketed joints

Description	Craft@Hrs	Unit	Material $	Labor $	Equipment $	Total $
2"	P1@.500	Ea	30.50	18.60	—	49.10
3"	P1@.750	Ea	52.90	27.90	—	80.80
4"	P1@.950	Ea	73.40	35.40	—	108.80
6"	ER@1.10	Ea	170.00	44.00	.72	214.72
8"	ER@1.85	Ea	421.00	73.90	1.21	496.11
10"	ER@2.50	Ea	588.00	99.90	1.63	689.53
12"	ER@2.75	Ea	851.00	110.00	1.79	962.79
15"	ER@3.50	Ea	2,570.00	140.00	2.28	2,712.28

DWV cast iron reducing double wye, hub and spigot, gasketed joints

Description	Craft@Hrs	Unit	Material $	Labor $	Equipment $	Total $
3" x 2"	P1@.550	Ea	44.60	20.50	—	65.10
4" x 2"	P1@.650	Ea	54.60	24.20	—	78.80
4" x 3"	P1@.700	Ea	60.60	26.10	—	86.70
6" x 4"	ER@.800	Ea	128.00	32.00	.52	160.52

DWV cast iron reducer, hub and spigot, gasketed joints

Description	Craft@Hrs	Unit	Material $	Labor $	Equipment $	Total $
3"	P1@.150	Ea	15.70	5.58	—	21.28
4"	P1@.200	Ea	40.70	7.44	—	48.14
6"	ER@.320	Ea	63.60	12.80	.21	76.61
8"	ER@.350	Ea	134.00	14.00	.23	148.23
10"	ER@.570	Ea	284.00	22.80	.37	307.17
12"	ER@.620	Ea	377.00	24.80	.40	402.20
15"	ER@1.12	Ea	697.00	44.80	.73	742.53

Cast Iron, DWV, Service Weight, Hub & Spigot with Gasketed Joints

Description	Craft@Hrs	Unit	Material $	Labor $	Equipment $	Total $
Gaskets for hub and spigot joints						
2"	—	Ea	3.83	—	—	3.83
3"	—	Ea	5.07	—	—	5.07
4"	—	Ea	6.45	—	—	6.45
5"	—	Ea	9.89	—	—	9.89
6"	—	Ea	10.40	—	—	10.40
8"	—	Ea	22.50	—	—	22.50
10"	—	Ea	35.10	—	—	35.10
12"	—	Ea	44.70	—	—	44.70
15"	—	Ea	53.40	—	—	53.40
Hanger with swivel assembly						
2"	P1@.400	Ea	7.22	14.90	—	22.12
3"	P1@.450	Ea	9.31	16.70	—	26.01
4"	P1@.450	Ea	11.60	16.70	—	28.30
5"	P1@.500	Ea	17.70	18.60	—	36.30
6"	P1@.500	Ea	20.00	18.60	—	38.60
8"	P1@.550	Ea	26.10	20.50	—	46.60
10"	P1@.650	Ea	34.40	24.20	—	58.60
12"	P1@.750	Ea	47.40	27.90	—	75.30
15"	P1@.850	Ea	62.80	31.60	—	94.40
Riser clamp						
2"	P1@.115	Ea	6.65	4.28	—	10.93
3"	P1@.120	Ea	7.59	4.47	—	12.06
4"	P1@.125	Ea	9.66	4.65	—	14.31
5"	P1@.180	Ea	13.80	6.70	—	20.50
6"	P1@.200	Ea	16.70	7.44	—	24.14
8"	P1@.200	Ea	27.30	7.44	—	34.74
10"	P1@.250	Ea	40.40	9.31	—	49.71
12"	P1@.250	Ea	50.30	9.31	—	59.61

Copper, DWV with Soft-Soldered Joints

Copper drainage tubing is used for non-pressurized above-grade, interior building drain, waste and vent (DWV) systems.

Because of its thin walls, it must be handled carefully and protected from damage during building construction.

This section has been arranged to save the estimator's time by including all normally-used system components such as pipe, fittings, hanger assemblies and riser clamps under one heading. Additional items can be found under "Plumbing and Piping Specialties." The cost estimates in this section are based on the conditions, limitations and wage rates described in the section "How to Use This Book" beginning on page 5.

Description	Craft@Hrs	Unit	Material $	Labor $	Equipment $	Total $

DWV copper pipe assembly, installed horizontally.
Assembly includes fittings, couplings, hangers and rod. Based on a reducing wye and a 45-degree elbow every 12 feet for 1¼" pipe, a reducing wye and a 45-degree elbow every 25 feet for 4" pipe, and hangers spaced to meet plumbing code.

Description	Craft@Hrs	Unit	Material $	Labor $	Equipment $	Total $
1¼"	P1@.130	LF	10.90	4.84	—	15.74
1½"	P1@.132	LF	11.10	4.91	—	16.01
2"	P1@.137	LF	14.30	5.10	—	19.40
3"	P1@.179	LF	25.60	6.66	—	32.26
4"	P1@.202	LF	48.00	7.52	—	55.52

DWV copper pipe assembly, installed risers.
Complete installation, including a reducing tee every floor and a riser clamp every other.

Description	Craft@Hrs	Unit	Material $	Labor $	Equipment $	Total $
1¼"	P1@.069	LF	8.51	2.57	—	11.08
1½"	P1@.074	LF	9.59	2.75	—	12.34
2"	P1@.086	LF	13.10	3.20	—	16.30
3"	P1@.110	LF	25.50	4.09	—	29.59
4"	P1@.134	LF	42.60	4.99	—	47.59

DWV copper pipe, soft-soldered joints

Description	Craft@Hrs	Unit	Material $	Labor $	Equipment $	Total $
1¼"	P1@.040	LF	4.49	1.49	—	5.98
1½"	P1@.045	LF	5.70	1.67	—	7.37
2"	P1@.050	LF	7.50	1.86	—	9.36
3"	P1@.060	LF	11.80	2.23	—	14.03
4"	P1@.070	LF	15.10	2.61	—	17.71

DWV copper 1/8 bend, soft-soldered joints

Description	Craft@Hrs	Unit	Material $	Labor $	Equipment $	Total $
1¼"	P1@.170	Ea	5.04	6.33	—	11.37
1½"	P1@.200	Ea	4.63	7.44	—	12.07
2"	P1@.270	Ea	8.38	10.00	—	18.38
3"	P1@.400	Ea	20.70	14.90	—	35.60
4"	P1@.540	Ea	85.20	20.10	—	105.30

Copper, DWV with Soft-Soldered Joints

Description	Craft@Hrs	Unit	Material $	Labor $	Equipment $	Total $

DWV copper ¼ bend, soft-soldered joints

Description	Craft@Hrs	Unit	Material $	Labor $	Equipment $	Total $
1¼"	P1@.170	Ea	5.66	6.33	—	11.99
1½"	P1@.200	Ea	6.46	7.44	—	13.90
2"	P1@.270	Ea	10.40	10.00	—	20.40
3"	P1@.400	Ea	27.40	14.90	—	42.30
4"	P1@.540	Ea	132.00	20.10	—	152.10

DWV copper closet flange, soft-soldered joints

Description	Craft@Hrs	Unit	Material $	Labor $	Equipment $	Total $
4" x 3"	P1@.680	Ea	27.00	25.30	—	52.30
4" x 4"	P1@.980	Ea	33.20	36.50	—	69.70

DWV copper P-trap, soft-soldered joints

Description	Craft@Hrs	Unit	Material $	Labor $	Equipment $	Total $
1¼"	P1@.240	Ea	38.20	8.93	—	47.13
1½"	P1@.240	Ea	33.30	8.93	—	42.23
2"	P1@.240	Ea	50.70	8.93	—	59.63
3"	P1@.320	Ea	146.00	11.90	—	157.90

DWV copper sanitary tee (TY), soft-soldered joints

Description	Craft@Hrs	Unit	Material $	Labor $	Equipment $	Total $
1½"	P1@.240	Ea	11.80	8.93	—	20.73
2"	P1@.320	Ea	16.10	11.90	—	28.00
3"	P1@.490	Ea	55.40	18.20	—	73.60
4"	P1@.650	Ea	140.00	24.20	—	164.20

DWV copper sanitary cross, soft-soldered joints

Description	Craft@Hrs	Unit	Material $	Labor $	Equipment $	Total $
1½"	P1@.640	Ea	77.90	23.80	—	101.70
2"	P1@.900	Ea	112.00	33.50	—	145.50
3"	P1@1.36	Ea	318.00	50.60	—	368.60

DWV copper combination wye and 1/8 bend, soft-soldered joints

Description	Craft@Hrs	Unit	Material $	Labor $	Equipment $	Total $
1½"	P1@.240	Ea	24.80	8.93	—	33.73
2"	P1@.320	Ea	42.30	11.90	—	54.20
3"	P1@.490	Ea	140.00	18.20	—	158.20
4"	P1@.650	Ea	315.00	24.20	—	339.20

DWV copper wye, soft-soldered joints

Description	Craft@Hrs	Unit	Material $	Labor $	Equipment $	Total $
1½"	P1@.240	Ea	19.30	8.93	—	28.23
2"	P1@.320	Ea	27.60	11.90	—	39.50
3"	P1@.490	Ea	69.20	18.20	—	87.40
4"	P1@.650	Ea	140.00	24.20	—	164.20

DWV copper cleanout with threaded plug

Description	Craft@Hrs	Unit	Material $	Labor $	Equipment $	Total $
1¼"	P1@.100	Ea	9.80	3.72	—	13.52
1½"	P1@.160	Ea	14.60	5.96	—	20.56
2"	P1@.230	Ea	19.00	8.56	—	27.56

Copper, DWV with Soft-Soldered Joints

Description	Craft@Hrs	Unit	Material $	Labor $	Equipment $	Total $
DWV copper female adapter						
1½"	P1@.130	Ea	7.53	4.84	—	12.37
2"	P1@.170	Ea	14.40	6.33	—	20.73
3"	P1@.230	Ea	57.20	8.56	—	65.76
4"	P1@.300	Ea	98.00	11.20	—	109.20
DWV copper male adapter						
1½"	P1@.130	Ea	8.52	4.84	—	13.36
2"	P1@.170	Ea	10.20	6.33	—	16.53
3"	P1@.230	Ea	50.20	8.56	—	58.76
4"	P1@.300	Ea	83.80	11.20	—	95.00
DWV copper soil adapter (copper x MJ)						
2"	P1@.170	Ea	12.20	6.33	—	18.53
3"	P1@.250	Ea	25.10	9.31	—	34.41
4"	P1@.320	Ea	46.80	11.90	—	58.70
DWV copper test cap, soft-soldered joint						
1½"	P1@.100	Ea	.29	3.72	—	4.01
2"	P1@.130	Ea	.41	4.84	—	5.25
3"	P1@.200	Ea	.53	7.44	—	7.97
4"	P1@.250	Ea	1.09	9.31	—	10.40
DWV copper test tee, soft-soldered joints						
1½"	P1@.200	Ea	28.10	7.44	—	35.54
2"	P1@.270	Ea	36.60	10.00	—	46.60
3"	P1@.400	Ea	98.80	14.90	—	113.70
DWV copper fitting reducer, soft-soldered joints						
1½"	P1@.100	Ea	4.63	3.72	—	8.35
2"	P1@.140	Ea	5.94	5.21	—	11.15
3"	P1@.200	Ea	20.70	7.44	—	28.14
4"	P1@.270	Ea	33.00	10.00	—	43.00
DWV copper coupling, soft-soldered joints						
1¼"	P1@.170	Ea	2.48	6.33	—	8.81
1½"	P1@.200	Ea	2.85	7.44	—	10.29
2"	P1@.270	Ea	4.06	10.00	—	14.06
3"	P1@.400	Ea	7.53	14.90	—	22.43
4"	P1@.540	Ea	29.60	20.10	—	49.70

Copper, DWV with Soft-Soldered Joints

Description	Craft@Hrs	Unit	Material $	Labor $	Equipment $	Total $
Hanger with swivel assembly						
1¼"	P1@.300	Ea	5.05	11.20	—	16.25
1½"	P1@.300	Ea	5.43	11.20	—	16.63
2"	P1@.300	Ea	5.66	11.20	—	16.86
3"	P1@.350	Ea	9.53	13.00	—	22.53
4"	P1@.350	Ea	10.50	13.00	—	23.50
Riser clamp						
1¼"	P1@.105	Ea	5.66	3.91	—	9.57
1½"	P1@.110	Ea	5.98	4.09	—	10.07
2"	P1@.115	Ea	6.33	4.28	—	10.61
3"	P1@.120	Ea	7.21	4.47	—	11.68
4"	P1@.125	Ea	8.92	4.65	—	13.57

ABS, DWV with Solvent-Weld Joints

ABS (Acrylonitrile Butadiene Styrene) ASTM D 2661-73 is made from various grades of rubber-based resins and is primarily used for non-pressurized drain, waste and vent (DWV) systems.

ABS can also be used for low-temperature (100 degree F. maximum) water distribution systems, but consult the manufacturers for the proper ASTM type, maximum pressure limitations and recommended joint solvent.

The current Uniform Plumbing Code (UPC) states in Sections 401 and 503 that "ABS piping systems shall be limited to those structures where combustible construction is allowed." This does not preclude its use for underground DWV systems.

This section has been arranged to save the estimator's time by including all normally-used system components such as pipe, fittings, hanger assemblies and riser clamps under one heading. The cost estimates in this section are based on the conditions, limitations and wage rates described in the section "How to Use This Book" beginning on page 5.

Description	Craft@Hrs	Unit	Material $	Labor $	Equipment $	Total $
ABS DWV pipe with solvent-weld joints						
1½"	P1@.020	LF	.94	.74	—	1.68
2"	P1@.025	LF	1.47	.93	—	2.40
3"	P1@.030	LF	2.34	1.12	—	3.46
4"	P1@.040	LF	3.81	1.49	—	5.30
6"	P1@.055	LF	7.07	2.05	—	9.12
ABS DWV 1/8 bend with solvent-weld joints						
1½"	P1@.120	Ea	.91	4.47	—	5.38
2"	P1@.125	Ea	1.61	4.65	—	6.26
3"	P1@.175	Ea	3.89	6.51	—	10.40
4"	P1@.250	Ea	7.35	9.31	—	16.66
6"	P1@.325	Ea	58.30	12.10	—	70.40
ABS DWV 1/8 street bend with solvent-weld joints						
1½"	P1@.060	Ea	.98	2.23	—	3.21
2"	P1@.065	Ea	1.88	2.42	—	4.30
3"	P1@.088	Ea	4.13	3.28	—	7.41
4"	P1@.125	Ea	11.40	4.65	—	16.05
6"	P1@.163	Ea	57.10	6.07	—	63.17
ABS DWV ¼ bend with solvent-weld joints						
1½"	P1@.120	Ea	1.08	4.47	—	5.55
2"	P1@.125	Ea	1.85	4.65	—	6.50
3"	P1@.175	Ea	4.65	6.51	—	11.16
4"	P1@.250	Ea	10.60	9.31	—	19.91
6"	P1@.325	Ea	60.00	12.10	—	72.10
ABS DWV closet bend with solvent-weld joints						
3" x 4"	P1@.250	Ea	15.80	9.31	—	25.11
ABS DWV closet flange with solvent-weld joints						
4"	P1@.125	Ea	7.10	4.65	—	11.75

ABS, DWV with Solvent-Weld Joints

Description	Craft@Hrs	Unit	Material $	Labor $	Equipment $	Total $

ABS DWV P-trap with solvent-weld joints

1½"	P1@.150	Ea	3.27	5.58	—	8.85
2"	P1@.190	Ea	8.07	7.07	—	15.14
3"	P1@.260	Ea	19.20	9.68	—	28.88
4"	P1@.350	Ea	37.10	13.00	—	50.10

ABS DWV sanitary tee (TY) with solvent-weld joints

1½"	P1@.150	Ea	1.57	5.58	—	7.15
2"	P1@.190	Ea	3.62	7.07	—	10.69
3"	P1@.260	Ea	6.97	9.68	—	16.65
4"	P1@.350	Ea	14.70	13.00	—	27.70
6"	P1@.450	Ea	67.20	16.70	—	83.90

ABS DWV reducing sanitary tee (TY) with solvent-weld joints

2" x 1½"	P1@.160	Ea	2.62	5.96	—	8.58
3" x 1½"	P1@.210	Ea	4.47	7.82	—	12.29
3" x 2"	P1@.235	Ea	4.99	8.75	—	13.74
4" x 2"	P1@.300	Ea	14.40	11.20	—	25.60
4" x 3"	P1@.325	Ea	17.10	12.10	—	29.20

ABS DWV double sanitary tee (TY) with solvent-weld joints

1½"	P1@.210	Ea	4.79	7.82	—	12.61
2"	P1@.250	Ea	7.22	9.31	—	16.53
3"	P1@.350	Ea	11.40	13.00	—	24.40
4"	P1@.475	Ea	45.00	17.70	—	62.70

ABS DWV combination wye and 1/8 bend with solvent-weld joints

1½"	P1@.150	Ea	9.46	5.58	—	15.04
2"	P1@.190	Ea	12.30	7.07	—	19.37
3"	P1@.260	Ea	16.50	9.68	—	26.18
4"	P1@.350	Ea	28.90	13.00	—	41.90

ABS DWV combination reducing wye and 1/8 bend with solvent-weld joints

2" x 1½"	P1@.160	Ea	13.70	5.96	—	19.66
3" x 1½"	P1@.210	Ea	20.90	7.82	—	28.72
3" x 2"	P1@.235	Ea	14.60	8.75	—	23.35
4" x 2"	P1@.300	Ea	26.40	11.20	—	37.60
4" x 3"	P1@.325	Ea	31.20	12.10	—	43.30

ABS, DWV with Solvent-Weld Joints

Description	Craft@Hrs	Unit	Material $	Labor $	Equipment $	Total $
ABS DWV bushing with solvent-weld joints						
2" x 1½"	P1@.062	Ea	.82	2.31	—	3.13
3" x 1½"	P1@.088	Ea	4.06	3.28	—	7.34
3" x 2"	P1@.088	Ea	2.90	3.28	—	6.18
4" x 2"	P1@.125	Ea	5.87	4.65	—	10.52
4" x 3"	P1@.125	Ea	5.87	4.65	—	10.52
6" x 4"	P1@.163	Ea	29.20	6.07	—	35.27
ABS DWV FIP adapter						
1½"	P1@.080	Ea	1.23	2.98	—	4.21
2"	P1@.085	Ea	2.28	3.16	—	5.44
3"	P1@.155	Ea	5.39	5.77	—	11.16
4"	P1@.220	Ea	10.30	8.19	—	18.49
6"	P1@.285	Ea	39.50	10.60	—	50.10
ABS DWV test cap						
1½"	P1@.030	Ea	.62	1.12	—	1.74
2"	P1@.040	Ea	1.00	1.49	—	2.49
3"	P1@.050	Ea	1.58	1.86	—	3.44
4"	P1@.090	Ea	2.48	3.35	—	5.83
ABS DWV cleanout plug						
1½"	P1@.090	Ea	2.88	3.35	—	6.23
2"	P1@.120	Ea	4.22	4.47	—	8.69
3"	P1@.240	Ea	7.51	8.93	—	16.44
4"	P1@.320	Ea	16.20	11.90	—	28.10
6"	P1@.450	Ea	51.90	16.70	—	68.60
ABS DWV cleanout tee with solvent-weld joints						
1½"	P1@.120	Ea	4.46	4.47	—	8.93
2"	P1@.125	Ea	7.22	4.65	—	11.87
3"	P1@.175	Ea	12.00	6.51	—	18.51
4"	P1@.250	Ea	24.40	9.31	—	33.71
ABS DWV wye with solvent-weld joints						
1½"	P1@.150	Ea	2.09	5.58	—	7.67
2"	P1@.190	Ea	3.89	7.07	—	10.96
3"	P1@.260	Ea	7.22	9.68	—	16.90
4"	P1@.350	Ea	16.70	13.00	—	29.70
6"	P1@.450	Ea	70.30	16.70	—	87.00
ABS DWV reducing wye with solvent-weld joints						
2" x 1½"	P1@.160	Ea	3.62	5.96	—	9.58
3" x 1½"	P1@.210	Ea	5.04	7.82	—	12.86
3" x 2"	P1@.235	Ea	5.64	8.75	—	14.39
4" x 2"	P1@.300	Ea	12.10	11.20	—	23.30
4" x 3"	P1@.325	Ea	15.20	12.10	—	27.30
6" x 4"	P1@.400	Ea	58.30	14.90	—	73.20

ABS, DWV with Solvent-Weld Joints

Description	Craft@Hrs	Unit	Material $	Labor $	Equipment $	Total $

ABS DWV reducer with solvent-weld joints

2" x 1½"	P1@.125	Ea	.82	4.65	—	5.47
3" x 1½"	P1@.150	Ea	4.06	5.58	—	9.64
3" x 2"	P1@.150	Ea	2.90	5.58	—	8.48
4" x 2"	P1@.185	Ea	5.87	6.89	—	12.76
4" x 3"	P1@.210	Ea	5.80	7.82	—	13.62

ABS DWV coupling with solvent-weld joints

1½"	P1@.120	Ea	.69	4.47	—	5.16
2"	P1@.125	Ea	1.02	4.65	—	5.67
3"	P1@.175	Ea	2.38	6.51	—	8.89
4"	P1@.250	Ea	4.33	9.31	—	13.64
6"	P1@.325	Ea	25.60	12.10	—	37.70

Hanger with swivel assembly

1½"	P1@.300	Ea	5.43	11.20	—	16.63
2"	P1@.300	Ea	5.66	11.20	—	16.86
3"	P1@.350	Ea	9.53	13.00	—	22.53
4"	P1@.350	Ea	10.50	13.00	—	23.50
6"	P1@.450	Ea	17.20	16.70	—	33.90

Riser clamp

1½"	P1@.110	Ea	5.98	4.09	—	10.07
2"	P1@.115	Ea	6.33	4.28	—	10.61
3"	P1@.120	Ea	7.21	4.47	—	11.68
4"	P1@.125	Ea	9.19	4.65	—	13.84
6"	P1@.130	Ea	15.90	4.84	—	20.74

Galvanized steel pipe sleeves

2"	P1@.130	Ea	7.11	4.84	—	11.95
2½"	P1@.150	Ea	7.25	5.58	—	12.83
3"	P1@.180	Ea	7.40	6.70	—	14.10
4"	P1@.220	Ea	8.43	8.19	—	16.62
5"	P1@.250	Ea	10.50	9.31	—	19.81
6"	P1@.270	Ea	11.30	10.00	—	21.30

Lead roof pipe flashing

1½"	P1@.250	Ea	32.70	9.31	—	42.01
2"	P1@.250	Ea	43.00	9.31	—	52.31
3"	P1@.250	Ea	49.90	9.31	—	59.21
4"	P1@.250	Ea	54.70	9.31	—	64.01
6"	P1@.350	Ea	70.00	13.00	—	83.00

PVC, DWV with Solvent-Weld Joints

PVC (Polyvinyl Chloride) meets ASTM – D3034, D3212, F1336, F679 & CSA 182.2 (CSA – Canadian Standards Association – similar to UL – Underwriters Laboratories). The applications for PVC DWV have increased in number due to the unique flame-spread characteristics of their compounds, as well as the development of new fire stop devices for use with combustible piping systems.

PVC will not rust, pit or degrade when exposed to moisture and is extremely resistant to a broad range of corrosive agents.

PVC DWV is suitable for sewer, vent and storm water drainage applications in both below- and above-grade applications.

When penetrating a vertical or horizontal fire separation (fire-rated wall or floor) with combustible piping, certified fire stop devices must be used in compliance with local and national codes.

PVC DWV is not permitted in a ceiling space used as a return air plenum and is not permitted in a vertical shaft. PVC DWV has a flame-spread rating of 15.

This section has been arranged to save the estimator's time by including all normally-used system components such as pipe, fittings, hanger assemblies and riser clamps under one heading. The cost estimates in this section are based on the conditions, limitations and wage rates described in the section "How to Use This Book" beginning on page 5.

Description	Craft@Hrs	Unit	Material $	Labor $	Equipment $	Total $

PVC DWV pipe with solvent-weld joints

1½"	P1@.021	LF	.93	.78	—	1.71
2"	P1@.026	LF	1.49	.97	—	2.46
3"	P1@.032	LF	3.17	1.19	—	4.36
4"	P1@.042	LF	3.90	1.56	—	5.46
6"	P1@.058	LF	7.01	2.16	—	9.17

PVC DWV 90-degree elbow with solvent-weld joints

1½"	P1@.126	Ea	1.63	4.69	—	6.32
2"	P1@.131	Ea	2.45	4.88	—	7.33
3"	P1@.184	Ea	6.43	6.85	—	13.28
4"	P1@.263	Ea	12.70	9.79	—	22.49
6"	P1@.341	Ea	53.40	12.70	—	66.10

PVC DWV 90-degree street elbow with solvent-weld joints

1½"	P1@.063	Ea	1.71	2.34	—	4.05
2"	P1@.068	Ea	2.92	2.53	—	5.45
3"	P1@.092	Ea	8.54	3.42	—	11.96
4"	P1@.131	Ea	20.60	4.88	—	25.48
6"	P1@.171	Ea	75.60	6.36	—	81.96

PVC DWV 45-degree elbow with solvent-weld joints

1½"	P1@.126	Ea	1.37	4.69	—	6.06
2"	P1@.131	Ea	1.50	4.88	—	6.38
3"	P1@.184	Ea	3.77	6.85	—	10.62
4"	P1@.263	Ea	9.52	9.79	—	19.31
6"	P1@.341	Ea	57.90	12.70	—	70.60

PVC, DWV with Solvent-Weld Joints

Description	Craft@Hrs	Unit	Material $	Labor $	Equipment $	Total $

PVC DWV wye with solvent-weld joints

1½"	P1@.158	Ea	2.53	5.88	—	8.41
2"	P1@.200	Ea	5.18	7.44	—	12.62
3"	P1@.273	Ea	8.53	10.20	—	18.73
4"	P1@.368	Ea	14.00	13.70	—	27.70
6"	P1@.473	Ea	54.80	17.60	—	72.40

PVC DWV reducing wye with solvent-weld joints

2" x 1½"	P1@.168	Ea	4.48	6.25	—	10.73
3" x 1½"	P1@.221	Ea	5.15	8.23	—	13.38
3" x 2"	P1@.247	Ea	7.38	9.19	—	16.57
4" x 2"	P1@.315	Ea	14.30	11.70	—	26.00
4" x 3"	P1@.341	Ea	16.90	12.70	—	29.60
6" x 3"	P1@.473	Ea	62.50	17.60	—	80.10
6" x 4"	P1@.473	Ea	66.50	17.60	—	84.10

PVC DWV sanitary tee (TY) with solvent-weld joints

1½"	P1@.158	Ea	1.80	5.88	—	7.68
2"	P1@.200	Ea	3.06	7.44	—	10.50
3"	P1@.273	Ea	6.57	10.20	—	16.77
4"	P1@.368	Ea	12.60	13.70	—	26.30
6"	P1@.473	Ea	68.90	17.60	—	86.50

PVC DWV reducing sanitary tee (TY) with solvent-weld joints

2" x 1½"	P1@.168	Ea	2.38	6.25	—	8.63
3" x 1½"	P1@.221	Ea	4.10	8.23	—	12.33
3" x 2"	P1@.247	Ea	5.50	9.19	—	14.69
4" x 2"	P1@.315	Ea	12.00	11.70	—	23.70
4" x 3"	P1@.341	Ea	13.90	12.70	—	26.60
6" x 3"	P1@.473	Ea	52.80	17.60	—	70.40
6" x 4"	P1@.473	Ea	52.80	17.60	—	70.40

PVC DWV double wye with solvent-weld joints

2" x 1½"	P1@.263	Ea	6.73	9.79	—	16.52
3" x 2"	P1@.368	Ea	13.70	13.70	—	27.40
4" x 3"	P1@.473	Ea	44.60	17.60	—	62.20
6" x 4"	P1@.578	Ea	53.90	21.50	—	75.40

PVC DWV closet flange with solvent-weld joints, adjustable with test plate

4" x 3"	P1@.131	Ea	11.60	4.88	—	16.48

PVC DWV closet flange with solvent-weld joints, non-adjustable with test plate

4" x 3"	P1@.131	Ea	8.83	4.88	—	13.71

PVC, DWV with Solvent-Weld Joints

Description	Craft@Hrs	Unit	Material $	Labor $	Equipment $	Total $
PVC DWV closet flange with solvent-weld joints, non-adjustable without test plate						
4" x 3"	P1@.131	Ea	8.01	4.88	—	12.89
PVC DWV closet flange with solvent-weld joints, adjustable, offset without test plate						
4" x 3"	P1@.170	Ea	17.20	6.33	—	23.53
PVC DWV P-trap with solvent-weld joints						
1½"	P1@.160	Ea	4.10	5.96	—	10.06
2"	P1@.200	Ea	9.18	7.44	—	16.62
3"	P1@.285	Ea	21.50	10.60	—	32.10
4"	P1@.370	Ea	42.30	13.80	—	56.10
6"	P1@.580	Ea	163.00	21.60	—	184.60
PVC DWV bushing with solvent-weld joints						
2" x 1½"	P1@.062	Ea	.77	2.31	—	3.08
3" x 2"	P1@.088	Ea	2.28	3.28	—	5.56
4" x 3"	P1@.125	Ea	4.77	4.65	—	9.42
6" x 4"	P1@.163	Ea	24.60	6.07	—	30.67
PVC DWV FIP adapter						
1½"	P1@.080	Ea	1.12	2.98	—	4.10
2"	P1@.085	Ea	1.83	3.16	—	4.99
3"	P1@.155	Ea	4.43	5.77	—	10.20
4"	P1@.220	Ea	8.66	8.19	—	16.85
6"	P1@.285	Ea	38.90	10.60	—	49.50
PVC DWV MIP adapter						
1½"	P1@.080	Ea	1.25	2.98	—	4.23
2"	P1@.085	Ea	1.90	3.16	—	5.06
3"	P1@.155	Ea	3.94	5.77	—	9.71
4"	P1@.220	Ea	10.60	8.19	—	18.79
6"	P1@.285	Ea	19.30	10.60	—	29.90
PVC DWV coupling with solvent-weld joints						
1½"	P1@.120	Ea	.60	4.47	—	5.07
2"	P1@.125	Ea	.86	4.65	—	5.51
3"	P1@.175	Ea	2.02	6.51	—	8.53
4"	P1@.250	Ea	4.23	9.31	—	13.54
6"	P1@.325	Ea	20.60	12.10	—	32.70

PVC, DWV with Solvent-Weld Joints

Description	Craft@Hrs	Unit	Material $	Labor $	Equipment $	Total $

PVC DWV reducer with solvent-weld joints

2" x 1½"	P1@.125	Ea	1.28	4.65	—	5.93
3" x 2"	P1@.150	Ea	3.85	5.58	—	9.43
4" x 3"	P1@.185	Ea	5.03	6.89	—	11.92
6" x 4"	P1@.210	Ea	50.90	7.82	—	58.72

PVC DWV in-line cleanout with solvent-weld joints

1½"	P1@.120	Ea	5.22	4.47	—	9.69
2"	P1@.125	Ea	8.20	4.65	—	12.85
3"	P1@.175	Ea	12.00	6.51	—	18.51
4"	P1@.250	Ea	25.10	9.31	—	34.41
6" x 4"	P1@.250	Ea	85.30	9.31	—	94.61
6"	P1@.250	Ea	85.30	9.31	—	94.61

PVC DWV end cleanout

1½"	P1@.090	Ea	2.30	3.35	—	5.65
2"	P1@.120	Ea	3.43	4.47	—	7.90
3"	P1@.240	Ea	6.95	8.93	—	15.88
4"	P1@.320	Ea	12.90	11.90	—	24.80
6"	P1@.450	Ea	45.20	16.70	—	61.90

PVC DWV fitting end cleanout

1½"	P1@.090	Ea	2.18	3.35	—	5.53
2"	P1@.120	Ea	3.97	4.47	—	8.44
3"	P1@.240	Ea	4.63	8.93	—	13.56
4"	P1@.320	Ea	9.87	11.90	—	21.77
6"	P1@.450	Ea	51.50	16.70	—	68.20

Hanger with swivel assembly

1½"	P1@.300	Ea	6.91	11.20	—	18.11
2"	P1@.300	Ea	7.24	11.20	—	18.44
3"	P1@.350	Ea	12.30	13.00	—	25.30
4"	P1@.350	Ea	13.30	13.00	—	26.30
6"	P1@.450	Ea	21.70	16.70	—	38.40

Riser clamp

1½"	P1@.110	Ea	5.98	4.09	—	10.07
2"	P1@.115	Ea	6.33	4.28	—	10.61
3"	P1@.120	Ea	7.21	4.47	—	11.68
4"	P1@.125	Ea	9.19	4.65	—	13.84
6"	P1@.130	Ea	15.90	4.84	—	20.74

PVC, DWV with Solvent-Weld Joints

Description	Craft@Hrs	Unit	Material $	Labor $	Equipment $	Total $
Galvanized steel pipe sleeves						
2"	P1@.130	Ea	7.11	4.84	—	11.95
2½"	P1@.150	Ea	7.25	5.58	—	12.83
3"	P1@.180	Ea	7.40	6.70	—	14.10
4"	P1@.220	Ea	8.43	8.19	—	16.62
5"	P1@.250	Ea	10.50	9.31	—	19.81
6"	P1@.270	Ea	11.30	10.00	—	21.30
Lead roof pipe flashing						
1½"	P1@.400	Ea	34.20	14.90	—	49.10
2"	P1@.450	Ea	45.10	16.70	—	61.80
3"	P1@.500	Ea	52.10	18.60	—	70.70
4"	P1@.550	Ea	57.60	20.50	—	78.10

PVC, DWV with Gasketed Bell and Spigot Joints

PVC (Polyvinyl Chloride) is polymerized vinyl chloride produced from acetylene and anhydrous hydrochloric acid.

PVC sewer pipe (ASTM D-3034) 10' lay length, is for underground installation and is used primarily to convey sewage and rain water. It is of the bell & spigot type with joints made water-tight by means of an elastomeric gasket (O-ring) factory-installed in the hub end of the pipe and fittings.

This section includes pipe and all commonly-used fittings. Additional items can be found under "Plumbing and Piping Specialties." The cost estimates in this section are based on the conditions, limitations, and wage rates described in the section "How to Use This Book" beginning on page 5.

Description	Craft@Hrs	Unit	Material $	Labor $	Equipment $	Total $

PVC SDR35 sewer pipe with bell and spigot gasketed joints

4"	P1@.050	LF	4.12	1.86	—	5.98
6"	P1@.070	LF	9.06	2.61	—	11.67
8"	P1@.090	LF	14.30	3.35	—	17.65
10"	P1@.110	LF	21.40	4.09	—	25.49
12"	P1@.140	LF	27.00	5.21	—	32.21
15"	P1@.180	LF	39.80	6.70	—	46.50
18"	P1@.230	LF	62.10	8.56	—	70.66

PVC sewer pipe 1/16 bend B x B with bell and spigot gasketed joints

4"	P1@.320	Ea	20.00	11.90	—	31.90
6"	P1@.400	Ea	39.10	14.90	—	54.00
8"	P1@.480	Ea	114.00	17.90	—	131.90
10"	P1@.800	Ea	311.00	29.80	—	340.80
12"	P1@1.06	Ea	410.00	39.50	—	449.50
15"	P1@1.34	Ea	1,080.00	49.90	—	1,129.90
18"	P1@1.60	Ea	1,560.00	59.60	—	1,619.60

PVC sewer pipe 1/16 bend B x S with bell and spigot gasketed joints

4"	P1@.320	Ea	18.80	11.90	—	30.70
6"	P1@.400	Ea	37.20	14.90	—	52.10
8"	P1@.480	Ea	119.00	17.90	—	136.90
10"	P1@.800	Ea	307.00	29.80	—	336.80
12"	P1@1.06	Ea	397.00	39.50	—	436.50
15"	P1@1.34	Ea	865.00	49.90	—	914.90
18"	P1@1.60	Ea	1,330.00	59.60	—	1,389.60

PVC sewer pipe 1/8 bend B x B with bell and spigot gasketed joints

4"	P1@.320	Ea	20.10	11.90	—	32.00
6"	P1@.400	Ea	40.30	14.90	—	55.20
8"	P1@.480	Ea	108.00	17.90	—	125.90
10"	P1@.800	Ea	292.00	29.80	—	321.80
12"	P1@1.06	Ea	431.00	39.50	—	470.50
15"	P1@1.34	Ea	953.00	49.90	—	1,002.90
18"	P1@1.60	Ea	1,520.00	59.60	—	1,579.60

PVC, DWV with Gasketed Bell and Spigot Joints

Description	Craft@Hrs	Unit	Material $	Labor $	Equipment $	Total $
PVC sewer pipe 1/8 bend B x S with bell and spigot gasketed joints						
4"	P1@.320	Ea	18.00	11.90	—	29.90
6"	P1@.400	Ea	36.00	14.90	—	50.90
8"	P1@.480	Ea	108.00	17.90	—	125.90
10"	P1@.800	Ea	288.00	29.80	—	317.80
12"	P1@1.06	Ea	417.00	39.50	—	456.50
15"	P1@1.34	Ea	747.00	49.90	—	796.90
18"	P1@1.60	Ea	1,260.00	59.60	—	1,319.60
PVC sewer pipe ¼ bend B x B with bell and spigot gasketed joints						
4"	P1@.320	Ea	28.70	11.90	—	40.60
6"	P1@.400	Ea	53.20	14.90	—	68.10
8"	P1@.480	Ea	143.00	17.90	—	160.90
10"	P1@.800	Ea	494.00	29.80	—	523.80
12"	P1@1.06	Ea	633.00	39.50	—	672.50
15"	P1@1.34	Ea	1,350.00	49.90	—	1,399.90
18"	P1@1.60	Ea	2,260.00	59.60	—	2,319.60
PVC sewer pipe ¼ bend B x S with bell and spigot gasketed joints						
4"	P1@.320	Ea	26.20	11.90	—	38.10
6"	P1@.400	Ea	55.90	14.90	—	70.80
8"	P1@.480	Ea	152.00	17.90	—	169.90
10"	P1@.800	Ea	489.00	29.80	—	518.80
12"	P1@1.06	Ea	616.00	39.50	—	655.50
15"	P1@1.34	Ea	1,200.00	49.90	—	1,249.90
18"	P1@1.60	Ea	2,110.00	59.60	—	2,169.60
PVC sewer pipe coupling B x B with bell and spigot gasketed joints						
4"	P1@.320	Ea	24.60	11.90	—	36.50
6"	P1@.400	Ea	49.40	14.90	—	64.30
8"	P1@.480	Ea	84.10	17.90	—	102.00
10"	P1@.800	Ea	184.00	29.80	—	213.80
12"	P1@1.06	Ea	272.00	39.50	—	311.50
15"	P1@1.34	Ea	559.00	49.90	—	608.90
18"	P1@1.60	Ea	1,090.00	59.60	—	1,149.60
PVC sewer pipe wye B x B x B with bell and spigot gasketed joints						
4"	P1@.480	Ea	33.40	17.90	—	51.30
6"	P1@.600	Ea	76.60	22.30	—	98.90
8"	P1@.720	Ea	212.00	26.80	—	238.80
10"	P1@1.20	Ea	616.00	44.70	—	660.70
12"	P1@1.59	Ea	874.00	59.20	—	933.20
15"	P1@2.01	Ea	1,520.00	74.80	—	1,594.80
18"	P1@2.40	Ea	2,680.00	89.30	—	2,769.30

PVC, DWV with Gasketed Bell and Spigot Joints

Description	Craft@Hrs	Unit	Material $	Labor $	Equipment $	Total $
PVC sewer pipe wye B x S x B with bell and spigot gasketed joints						
10"	P1@1.20	Ea	608.00	44.70	—	652.70
12"	P1@1.59	Ea	865.00	59.20	—	924.20
15"	P1@2.01	Ea	1,390.00	74.80	—	1,464.80
18"	P1@2.40	Ea	2,460.00	89.30	—	2,549.30
PVC sewer pipe reducing wye B x B x B with bell and spigot gasketed joints						
6" x 4"	P1@.560	Ea	68.10	20.80	—	88.90
8" x 4"	P1@.640	Ea	102.00	23.80	—	125.80
8" x 6"	P1@.680	Ea	120.00	25.30	—	145.30
10" x 4"	P1@.960	Ea	307.00	35.70	—	342.70
10" x 6"	P1@1.00	Ea	311.00	37.20	—	348.20
10" x 8"	P1@1.04	Ea	494.00	38.70	—	532.70
12" x 4"	P1@1.22	Ea	444.00	45.40	—	489.40
12" x 6"	P1@1.26	Ea	457.00	46.90	—	503.90
12" x 8"	P1@1.30	Ea	681.00	48.40	—	729.40
12" x 10"	P1@1.46	Ea	830.00	54.30	—	884.30
15" x 4"	P1@1.50	Ea	670.00	55.80	—	725.80
15" x 6"	P1@1.54	Ea	792.00	57.30	—	849.30
15" x 8"	P1@1.58	Ea	896.00	58.80	—	954.80
15" x 10"	P1@1.74	Ea	1,070.00	64.80	—	1,134.80
15" x 12"	P1@1.87	Ea	1,190.00	69.60	—	1,259.60
18" x 4"	P1@1.76	Ea	1,520.00	65.50	—	1,585.50
18" x 6"	P1@1.80	Ea	1,540.00	67.00	—	1,607.00
18" x 8"	P1@1.84	Ea	1,640.00	68.50	—	1,708.50
18" x 10"	P1@2.00	Ea	1,800.00	74.40	—	1,874.40
18" x 12"	P1@2.13	Ea	1,940.00	79.30	—	2,019.30
18" x 15"	P1@2.27	Ea	2,070.00	84.50	—	2,154.50
PVC sewer pipe reducing wye B x S x B with bell and spigot gasketed joints						
10" x 4"	P1@.960	Ea	299.00	35.70	—	334.70
10" x 6"	P1@1.00	Ea	315.00	37.20	—	352.20
10" x 8"	P1@1.04	Ea	507.00	38.70	—	545.70
12" x 4"	P1@1.22	Ea	450.00	45.40	—	495.40
12" x 6"	P1@1.26	Ea	457.00	46.90	—	503.90
12" x 8"	P1@1.30	Ea	672.00	48.40	—	720.40
12" x 10"	P1@1.46	Ea	816.00	54.30	—	870.30
15" x 4"	P1@1.50	Ea	684.00	55.80	—	739.80
15" x 6"	P1@1.54	Ea	774.00	57.30	—	831.30
15" x 8"	P1@1.58	Ea	910.00	58.80	—	968.80
15" x 10"	P1@1.74	Ea	1,030.00	64.80	—	1,094.80
15" x 12"	P1@1.87	Ea	1,200.00	69.60	—	1,269.60
18" x 4"	P1@1.76	Ea	1,420.00	65.50	—	1,485.50
18" x 6"	P1@1.80	Ea	1,540.00	67.00	—	1,607.00
18" x 8"	P1@1.84	Ea	1,640.00	68.50	—	1,708.50
18" x 10"	P1@2.00	Ea	1,800.00	74.40	—	1,874.40
18" x 12"	P1@2.13	Ea	1,940.00	79.30	—	2,019.30
18" x 15"	P1@2.27	Ea	2,070.00	84.50	—	2,154.50

PVC, DWV with Gasketed Bell and Spigot Joints

Description	Craft@Hrs	Unit	Material $	Labor $	Equipment $	Total $

PVC sewer pipe tee-wye B x B x B with bell and spigot gasketed joints

4"	P1@.480	Ea	37.20	17.90	—	55.10
6"	P1@.600	Ea	94.40	22.30	—	116.70
6" x 4"	P1@.560	Ea	80.50	20.80	—	101.30
8" x 4"	P1@.640	Ea	102.00	23.80	—	125.80
8" x 6"	P1@.680	Ea	117.00	25.30	—	142.30

PVC sewer pipe tee B x S x B with bell and spigot gasketed joints

10"	P1@1.20	Ea	514.00	44.70	—	558.70
12"	P1@1.59	Ea	712.00	59.20	—	771.20
15"	P1@2.01	Ea	1,130.00	74.80	—	1,204.80
18"	P1@2.40	Ea	2,130.00	89.30	—	2,219.30

PVC sewer pipe reducing tee B x S x B with bell and spigot gasketed joints

10" x 4"	P1@.960	Ea	326.00	35.70	—	361.70
10" x 6"	P1@1.00	Ea	340.00	37.20	—	377.20
10" x 8"	P1@1.04	Ea	532.00	38.70	—	570.70
12" x 4"	P1@1.22	Ea	400.00	45.40	—	445.40
12" x 6"	P1@1.26	Ea	410.00	46.90	—	456.90
12" x 8"	P1@1.30	Ea	562.00	48.40	—	610.40
12" x 10"	P1@1.46	Ea	672.00	54.30	—	726.30
15" x 4"	P1@1.50	Ea	648.00	55.80	—	703.80
15" x 6"	P1@1.54	Ea	681.00	57.30	—	738.30
15" x 8"	P1@1.58	Ea	698.00	58.80	—	756.80
15" x 10"	P1@1.74	Ea	816.00	64.80	—	880.80
15" x 12"	P1@1.87	Ea	896.00	69.60	—	965.60
18" x 4"	P1@1.76	Ea	1,650.00	65.50	—	1,715.50
18" x 6"	P1@1.80	Ea	1,680.00	67.00	—	1,747.00
18" x 8"	P1@1.84	Ea	1,780.00	68.50	—	1,848.50
18" x 10"	P1@2.00	Ea	1,820.00	74.40	—	1,894.40
18" x 12"	P1@2.13	Ea	1,880.00	79.30	—	1,959.30
18" x 15"	P1@2.27	Ea	1,880.00	84.50	—	1,964.50

PVC, DWV with Gasketed Bell and Spigot Joints

Description	Craft@Hrs	Unit	Material $	Labor $	Equipment $	Total $

PVC sewer pipe reducer S x B with bell and spigot gasketed joints

Description	Craft@Hrs	Unit	Material $	Labor $	Equipment $	Total $
6" x 4"	P1@.360	Ea	36.00	13.40	—	49.40
8" x 4"	P1@.400	Ea	98.50	14.90	—	113.40
8" x 6"	P1@.440	Ea	110.00	16.40	—	126.40
10" x 4"	P1@.560	Ea	258.00	20.80	—	278.80
10" x 6"	P1@.600	Ea	262.00	22.30	—	284.30
10" x 8"	P1@.640	Ea	317.00	23.80	—	340.80
12" x 4"	P1@.690	Ea	340.00	25.70	—	365.70
12" x 6"	P1@.730	Ea	348.00	27.20	—	375.20
12" x 8"	P1@.770	Ea	400.00	28.70	—	428.70
12" x 10"	P1@.930	Ea	444.00	34.60	—	478.60
15" x 4"	P1@.830	Ea	571.00	30.90	—	601.90
15" x 6"	P1@.870	Ea	593.00	32.40	—	625.40
15" x 8"	P1@.910	Ea	672.00	33.90	—	705.90
15" x 10"	P1@1.07	Ea	672.00	39.80	—	711.80
15" x 12"	P1@1.20	Ea	774.00	44.70	—	818.70
18" x 8"	P1@1.04	Ea	830.00	38.70	—	868.70
18" x 10"	P1@1.20	Ea	841.00	44.70	—	885.70
18" x 12"	P1@1.33	Ea	845.00	49.50	—	894.50
18" x 15"	P1@1.47	Ea	865.00	54.70	—	919.70

PVC sewer pipe adapter B x S with bell and spigot gasketed joints

Description	Craft@Hrs	Unit	Material $	Labor $	Equipment $	Total $
4"	P1@.320	Ea	21.10	11.90	—	33.00
6"	P1@.400	Ea	39.10	14.90	—	54.00
8"	P1@.480	Ea	100.00	17.90	—	117.90
10"	P1@.800	Ea	123.00	29.80	—	152.80
12"	P1@1.06	Ea	188.00	39.50	—	227.50
15"	P1@1.34	Ea	287.00	49.90	—	336.90
18"	P1@1.60	Ea	628.00	59.60	—	687.60

PVC sewer pipe cap with bell and spigot gasketed joints

Description	Craft@Hrs	Unit	Material $	Labor $	Equipment $	Total $
4"	P1@.160	Ea	12.80	5.96	—	18.76
6"	P1@.200	Ea	24.00	7.44	—	31.44
8"	P1@.240	Ea	64.80	8.93	—	73.73
10"	P1@.400	Ea	202.00	14.90	—	216.90
12"	P1@.530	Ea	307.00	19.70	—	326.70
15"	P1@.670	Ea	494.00	24.90	—	518.90
18"	P1@.800	Ea	712.00	29.80	—	741.80

PVC sewer pipe test plug with bell and spigot gasketed joints

Description	Craft@Hrs	Unit	Material $	Labor $	Equipment $	Total $
4"	P1@.160	Ea	9.69	5.96	—	15.65
6"	P1@.200	Ea	14.80	7.44	—	22.24
8"	P1@.240	Ea	53.20	8.93	—	62.13
10"	P1@.400	Ea	177.00	14.90	—	191.90
12"	P1@.530	Ea	211.00	19.70	—	230.70
15"	P1@.670	Ea	417.00	24.90	—	441.90
18"	P1@.800	Ea	578.00	29.80	—	607.80

Polypropylene, Schedule 40, with Heat-Fusioned Joints

Polypropylene is used almost exclusively for acid and laboratory drain, waste and vent (DWV) systems.

Joints are made by applying low-voltage current to electrical resistance coils imbedded in propylene collars, which are slipped over the ends of the pipe, then inserted into the hubs of the fittings. Compression clamps are placed around the assemblies and tightened to compress the joints prior to applying electrical current from the power unit to fuse the joint.

This section has been arranged to save the estimator's time by including all normally-used system components such as pipe, fittings, hanger assemblies, and riser clamps under one heading. The cost estimates in this section are based on the conditions, limitations and wage rates described in the section "How to Use This Book" beginning on page 5.

Description	Craft@Hrs	Unit	Material $	Labor $	Equipment $	Total $

Schedule 40 DWV polypropylene pipe with heat-fusioned joints

Size	Craft@Hrs	Unit	Material $	Labor $	Equipment $	Total $
1½"	P1@.050	LF	5.80	1.86	—	7.66
2"	P1@.055	LF	7.88	2.05	—	9.93
3"	P1@.060	LF	14.00	2.23	—	16.23
4"	P1@.080	LF	20.20	2.98	—	23.18
6"	P1@.110	LF	36.30	4.09	—	40.39

Schedule 40 DWV polypropylene 1/8 bend with heat-fusioned joints

Size	Craft@Hrs	Unit	Material $	Labor $	Equipment $	Total $
1½"	P1@.350	Ea	25.90	13.00	—	38.90
2"	P1@.400	Ea	31.60	14.90	—	46.50
3"	P1@.450	Ea	57.60	16.70	—	74.30
4"	P1@.500	Ea	65.20	18.60	—	83.80
6"	P1@.750	Ea	179.00	27.90	—	206.90

Schedule 40 DWV polypropylene ¼ bend with heat-fusioned joints

Size	Craft@Hrs	Unit	Material $	Labor $	Equipment $	Total $
1½"	P1@.350	Ea	26.10	13.00	—	39.10
2"	P1@.400	Ea	32.20	14.90	—	47.10
3"	P1@.450	Ea	55.30	16.70	—	72.00
4"	P1@.500	Ea	87.70	18.60	—	106.30
6"	P1@.750	Ea	216.00	27.90	—	243.90

Schedule 40 DWV polypropylene long sweep ¼ bend with heat-fusioned joints

Size	Craft@Hrs	Unit	Material $	Labor $	Equipment $	Total $
1½"	P1@.350	Ea	26.50	13.00	—	39.50
2"	P1@.400	Ea	37.20	14.90	—	52.10
3"	P1@.450	Ea	63.30	16.70	—	80.00
4"	P1@.500	Ea	90.50	18.60	—	109.10

Schedule 40 DWV polypropylene P-trap with heat-fusioned joints

Size	Craft@Hrs	Unit	Material $	Labor $	Equipment $	Total $
1½"	P1@.500	Ea	51.20	18.60	—	69.80
2"	P1@.560	Ea	71.20	20.80	—	92.00
3"	P1@.630	Ea	129.00	23.40	—	152.40
4"	P1@.700	Ea	231.00	26.10	—	257.10

Polypropylene, Schedule 40, with Heat-Fusioned Joints

Description	Craft@Hrs	Unit	Material $	Labor $	Equipment $	Total $

Schedule 40 DWV polypropylene sanitary tee (TY) with heat-fusioned joints

1½"	P1@.500	Ea	32.20	18.60	—	50.80
2"	P1@.560	Ea	39.20	20.80	—	60.00
3"	P1@.630	Ea	64.90	23.40	—	88.30
4"	P1@.700	Ea	116.00	26.10	—	142.10

Schedule 40 DWV polypropylene reducing sanitary tee (TY) with heat-fusioned joints

2" x 1½"	P1@.510	Ea	42.90	19.00	—	61.90
3" x 1½"	P1@.530	Ea	74.40	19.70	—	94.10
3" x 2"	P1@.580	Ea	79.40	21.60	—	101.00
4" x 2"	P1@.650	Ea	114.00	24.20	—	138.20
4" x 3"	P1@.700	Ea	119.00	26.10	—	145.10
6" x 4"	P1@.850	Ea	216.00	31.60	—	247.60

Schedule 40 DWV polypropylene combination wye and 1/8 bend with heat-fusioned joints

1½"	P1@.500	Ea	47.50	18.60	—	66.10
2"	P1@.560	Ea	63.60	20.80	—	84.40
3"	P1@.630	Ea	101.00	23.40	—	124.40
4"	P1@.700	Ea	142.00	26.10	—	168.10

Schedule 40 DWV polypropylene combination reducing wye and 1/8 bend with heat-fusioned joints

2" x 1½"	P1@.500	Ea	57.00	18.60	—	75.60
3" x 1½"	P1@.520	Ea	84.00	19.40	—	103.40
3" x 2"	P1@.570	Ea	91.10	21.20	—	112.30
4" x 2"	P1@.580	Ea	124.00	21.60	—	145.60
4" x 3"	P1@.630	Ea	133.00	23.40	—	156.40

Schedule 40 DWV polypropylene wye with heat-fusioned joints

1½"	P1@.500	Ea	36.50	18.60	—	55.10
2"	P1@.560	Ea	51.80	20.80	—	72.60
3"	P1@.630	Ea	91.10	23.40	—	114.50
4"	P1@.700	Ea	133.00	26.10	—	159.10
6"	P1@1.05	Ea	336.00	39.10	—	375.10

Polypropylene, Schedule 40, with Heat-Fusioned Joints

Description	Craft@Hrs	Unit	Material $	Labor $	Equipment $	Total $

Schedule 40 DWV polypropylene reducing wye with heat-fusioned joints

Description	Craft@Hrs	Unit	Material $	Labor $	Equipment $	Total $
2" x 1½"	P1@.510	Ea	50.70	19.00	—	69.70
3" x 2"	P1@.580	Ea	84.60	21.60	—	106.20
4" x 2"	P1@.600	Ea	121.00	22.30	—	143.30
4" x 3"	P1@.650	Ea	125.00	24.20	—	149.20
6" x 2"	P1@.900	Ea	213.00	33.50	—	246.50
6" x 3"	P1@.900	Ea	216.00	33.50	—	249.50
6" x 4"	P1@.950	Ea	219.00	35.40	—	254.40

Schedule 40 DWV polypropylene female adapter with heat-fusioned joints

Description	Craft@Hrs	Unit	Material $	Labor $	Equipment $	Total $
1½"	P1@.280	Ea	17.70	10.40	—	28.10
2"	P1@.330	Ea	23.70	12.30	—	36.00
3"	P1@.350	Ea	39.90	13.00	—	52.90
4"	P1@.400	Ea	76.20	14.90	—	91.10

Schedule 40 DWV polypropylene male adapter with heat-fusioned joints

Description	Craft@Hrs	Unit	Material $	Labor $	Equipment $	Total $
1½"	P1@.280	Ea	17.40	10.40	—	27.80
2"	P1@.330	Ea	21.10	12.30	—	33.40
3"	P1@.350	Ea	35.00	13.00	—	48.00
4"	P1@.400	Ea	72.30	14.90	—	87.20

Schedule 40 DWV polypropylene cleanout adapter with heat-fusioned joints

Description	Craft@Hrs	Unit	Material $	Labor $	Equipment $	Total $
1½"	P1@.210	Ea	18.50	7.82	—	26.32
2"	P1@.280	Ea	21.90	10.40	—	32.30
3"	P1@.430	Ea	40.60	16.00	—	56.60
4"	P1@.640	Ea	69.60	23.80	—	93.40
6"	P1@.850	Ea	120.00	31.60	—	151.60

Schedule 40 DWV polypropylene plug with heat-fusioned joints

Description	Craft@Hrs	Unit	Material $	Labor $	Equipment $	Total $
1½"	P1@.060	Ea	6.41	2.23	—	8.64
2"	P1@.065	Ea	7.00	2.42	—	9.42
3"	P1@.070	Ea	12.20	2.61	—	14.81
4"	P1@.090	Ea	13.80	3.35	—	17.15

Schedule 40 DWV polypropylene coupling with heat-fusioned joints

Description	Craft@Hrs	Unit	Material $	Labor $	Equipment $	Total $
1½"	P1@.350	Ea	22.60	13.00	—	35.60
2"	P1@.400	Ea	27.80	14.90	—	42.70
3"	P1@.450	Ea	35.60	16.70	—	52.30
4"	P1@.500	Ea	50.70	18.60	—	69.30
6"	P1@.750	Ea	81.00	27.90	—	108.90

Polypropylene, Schedule 40, with Heat-Fusioned Joints

Description	Craft@Hrs	Unit	Material $	Labor $	Equipment $	Total $

Schedule 40 DWV polypropylene reducer with heat-fusioned joints

2" x 1½"	P1@.360	Ea	23.70	13.40	—	37.10
3" x 2"	P1@.410	Ea	27.20	15.30	—	42.50
4" x 2"	P1@.430	Ea	60.90	16.00	—	76.90
4" x 3"	P1@.450	Ea	62.60	16.70	—	79.30

Schedule 40 DWV polypropylene flange with heat-fusioned joints

2"	P1@.350	Ea	78.30	13.00	—	91.30
3"	P1@.380	Ea	162.00	14.10	—	176.10
4"	P1@.430	Ea	211.00	16.00	—	227.00
6"	P1@.630	Ea	277.00	23.40	—	300.40

Schedule 40 DWV polypropylene bolt and gasket set

2"	P1@.500	Ea	7.71	18.60	—	26.31
2½"	P1@.650	Ea	8.16	24.20	—	32.36
3"	P1@.750	Ea	8.58	27.90	—	36.48
4"	P1@1.00	Ea	15.40	37.20	—	52.60

Hanger with swivel assembly

1½"	P1@.300	Ea	5.43	11.20	—	16.63
2"	P1@.300	Ea	5.66	11.20	—	16.86
3"	P1@.350	Ea	9.53	13.00	—	22.53
4"	P1@.350	Ea	10.50	13.00	—	23.50
6"	P1@.450	Ea	17.20	16.70	—	33.90

Riser clamp

1½"	P1@.110	Ea	5.98	4.09	—	10.07
2"	P1@.115	Ea	6.33	4.28	—	10.61
3"	P1@.120	Ea	7.21	4.47	—	11.68
4"	P1@.125	Ea	9.19	4.65	—	13.84
6"	P1@.200	Ea	15.90	7.44	—	23.34

Pipe sleeve

2"	P1@.130	Ea	8.00	4.84	—	12.84
2½"	P1@.150	Ea	8.11	5.58	—	13.69
3"	P1@.180	Ea	8.32	6.70	—	15.02
4"	P1@.220	Ea	9.45	8.19	—	17.64
5"	P1@.250	Ea	11.70	9.31	—	21.01
6"	P1@.270	Ea	12.80	10.00	—	22.80
8"	P1@.270	Ea	14.60	10.00	—	24.60

Lead roof pipe flashing

1½"	P1@.250	Ea	34.40	9.31	—	43.71
2"	P1@.250	Ea	45.60	9.31	—	54.91
3"	P1@.250	Ea	52.40	9.31	—	61.71
4"	P1@.250	Ea	58.00	9.31	—	67.31

Floor, Area, Roof and Planter Drains

Description	Craft@Hrs	Unit	Material $	Labor $	Equipment $	Total $

Floor drain, cast iron, nickel-bronze grate. Add for rough-in (P-trap, fittings, hangers, etc.).

3" x 2"	P1@.500	Ea	155.00	18.60	—	173.60
5" x 3"	P1@.500	Ea	191.00	18.60	—	209.60
6" x 4"	P1@.500	Ea	246.00	18.60	—	264.60
8" x 6"	P1@.500	Ea	326.00	18.60	—	344.60

Area drain, cast iron grate. Add for rough-in (P-trap, fittings, hangers, etc.).

10" x 3"	P1@.650	Ea	326.00	24.20	—	350.20
12" x 4"	P1@.650	Ea	357.00	24.20	—	381.20
14" x 6"	P1@.700	Ea	384.00	26.10	—	410.10

Roof and overflow drain, cast iron, with plastic dome, flow control weir and deck clamp. Add for rough-in (pipe, fittings, hangers, etc.).

8" x 3"	P1@1.50	Ea	440.00	55.80	—	495.80
10" x 4"	P1@1.50	Ea	465.00	55.80	—	520.80
12" x 6"	P1@1.65	Ea	465.00	61.40	—	526.40

Planter drain, cast iron. Add for rough-in (pipe, fittings, hangers, etc.).

2"	P1@1.00	Ea	84.40	37.20	—	121.60
3"	P1@1.10	Ea	97.30	40.90	—	138.20
4"	P1@1.15	Ea	118.00	42.80	—	160.80
6"	P1@1.25	Ea	172.00	46.50	—	218.50

Planter drain, plastic. Add for rough-in (pipe, fittings, hangers, etc.).

4"	P1@.800	Ea	25.80	29.80	—	55.60
5"	P1@1.00	Ea	27.40	37.20	—	64.60
6"	P1@1.05	Ea	44.80	39.10	—	83.90
8"	P1@1.10	Ea	68.30	40.90	—	109.20

Cleanouts

Description	Craft@Hrs	Unit	Material $	Labor $	Equipment $	Total $

Cleanout, in-line, (Barrett) cast iron, (MJ) brass plug. Add for MJ couplings.

2"	P1@.350	Ea	71.50	13.00	—	84.50
3"	P1@.350	Ea	91.50	13.00	—	104.50
4"	P1@.400	Ea	130.00	14.90	—	144.90
6"	P1@.400	Ea	455.00	14.90	—	469.90

Cleanout, in-line (Barrett), ABS/PVC

2"	P1@.300	Ea	22.00	11.20	—	33.20
3"	P1@.300	Ea	35.40	11.20	—	46.60
4"	P1@.350	Ea	64.10	13.00	—	77.10
6"	P1@.350	Ea	250.00	13.00	—	263.00
8"	P1@.400	Ea	485.00	14.90	—	499.90
10"	P1@.550	Ea	598.00	20.50	—	618.50

Cleanout, end of line (Malcolm), cast iron body, brass top. Add for MJ coupling.

2" raised	P1@.250	Ea	25.70	9.31	—	35.01
2" recessed	P1@.250	Ea	64.20	9.31	—	73.51
3" raised	P1@.250	Ea	30.00	9.31	—	39.31
3" recessed	P1@.300	Ea	94.30	11.20	—	105.50
4" raised	P1@.300	Ea	40.10	11.20	—	51.30
4" recessed	P1@.350	Ea	123.00	13.00	—	136.00
6" raised	P1@.350	Ea	127.00	13.00	—	140.00
6" recessed	P1@.400	Ea	235.00	14.90	—	249.90
8" raised	P1@.450	Ea	273.00	16.70	—	289.70

Cleanout, end-of-line (Malcolm), ABS/PVC

2"	P1@.250	Ea	11.70	9.31	—	21.01
3"	P1@.250	Ea	19.60	9.31	—	28.91
4"	P1@.300	Ea	35.40	11.20	—	46.60
6"	P1@.300	Ea	148.00	11.20	—	159.20
8"	P1@.400	Ea	269.00	14.90	—	283.90
10"	P1@.450	Ea	306.00	16.70	—	322.70
12"	P1@.550	Ea	358.00	20.50	—	378.50

Fire Protection Sprinklers

Description	Unit	Material $	Labor $	Equipment $	Total $

Complete wet sprinkler system (square foot costs). System includes subcontractors' overhead and profit. Cost includes design drawings, zone and supervision valves, sprinkler heads, pipe, and connection to water supply service. Make additional allowances if a fire pump is required. Costs are based on ordinary hazard coverage (110 sf per head). Coverage density can vary depending on room sizes and/or hazard rating. Use these costs for preliminary estimates.

Description	Unit	Material $	Labor $	Equipment $	Total $
Exposed piping to 5,000	SF	—	—	—	4.92
Exposed piping 5,000-15,000	SF	—	—	—	4.14
Exposed piping over 15,000	SF	—	—	—	3.71
Concealed piping to 5,000	SF	—	—	—	4.58
Concealed piping 5,000-15,000	SF	—	—	—	3.96
Concealed piping over 15,000	SF	—	—	—	3.10
Add for dry system	%	—	—	—	20.0

Complete wet sprinkler system (per head costs). Cost includes sprinkler heads, sprinkler mains, branch piping, and supports. Make additional allowances for zone and supervision valves, alarms, connection to water service and fire pump when required. Costs are based on ordinary hazard 155 degree heads. Use these costs for preliminary estimates.

Description	Unit	Material $	Labor $	Equipment $	Total $
Upright heads to 5,000 SF	Ea	—	—	—	344.00
Upright heads 5,000 to 15,000 SF	Ea	—	—	—	328.00
Upright heads over 15,000 SF	Ea	—	—	—	312.00
Pendent heads to 5,000 SF	Ea	—	—	—	520.00
Pendent heads 5,000 to 15,000 SF	Ea	—	—	—	434.00
Pendent heads over 15,000 SF	Ea	—	—	—	367.00

Description	Craft@Hrs	Unit	Material $	Labor $	Equipment $	Total $

Supervision (zone) valves, flanged. Add for alarm trim if required.

Description	Craft@Hrs	Unit	Material $	Labor $	Equipment $	Total $
2½" OS&Y gate valve	SL@1.70	Ea	392.00	62.80	—	454.80
3" OS&Y gate vlv	SL@1.80	Ea	415.00	66.50	—	481.50
4" OS&Y gate vlv	SL@2.20	Ea	479.00	81.30	—	560.30
6" OS&Y gate vlv	SL@2.80	Ea	735.00	103.00	—	838.00
8" OS&Y gate vlv	SL@3.50	Ea	1,170.00	129.00	—	1,299.00
2½" butterfly valve (lug)	SL@.600	Ea	596.00	22.20	—	618.20
3" butterfly vlv	SL@.700	Ea	602.00	25.90	—	627.90
4" butterfly vlv	SL@.850	Ea	617.00	31.40	—	648.40
6" butterfly vlv	SL@1.15	Ea	855.00	42.50	—	897.50
8" butterfly vlv	SL@1.85	Ea	1,290.00	68.40	—	1,358.40

Fire Protection Sprinklers

Description	Craft@Hrs	Unit	Material $	Labor $	Equipment $	Total $

Supervision (zone) valves, grooved. Add for alarm trim if required

Description	Craft@Hrs	Unit	Material $	Labor $	Equipment $	Total $
4" butterfly valve	SL@.550	Ea	460.00	20.30	—	480.30
6" butterfly valve	SL@.850	Ea	598.00	31.40	—	629.40
8" butterfly valve	SL@1.25	Ea	905.00	46.20	—	951.20

Alarm valves (flanged or grooved)

Description	Craft@Hrs	Unit	Material $	Labor $	Equipment $	Total $
4" alarm valve	SL@3.50	Ea	629.00	129.00	—	758.00
6" alarm valve	SL@4.00	Ea	774.00	148.00	—	922.00
8" alarm valve	SL@4.75	Ea	1,130.00	176.00	—	1,306.00
4" alarm vlv trim	SL@1.25	Ea	421.00	46.20	—	467.20
6" alarm vlv trim	SL@1.25	Ea	421.00	46.20	—	467.20
8" alarm vlv trim	SL@1.25	Ea	421.00	46.20	—	467.20
4" alarm vlv pkg	SL@4.75	Ea	1,970.00	176.00	—	2,146.00
6" alarm vlv pkg	SL@5.50	Ea	2,190.00	203.00	—	2,393.00
8" alarm vlv pkg	SL@6.00	Ea	2,660.00	222.00	—	2,882.00

Dry valves and trim (deluge/pre-action system)

Description	Craft@Hrs	Unit	Material $	Labor $	Equipment $	Total $
2" dry valve	SL@2.25	Ea	1,180.00	83.20	—	1,263.20
3" dry valve	SL@2.75	Ea	1,230.00	102.00	—	1,332.00
4" dry valve	SL@3.05	Ea	1,370.00	113.00	—	1,483.00
6" dry valve	SL@3.65	Ea	1,770.00	135.00	—	1,905.00
2" dry valve trim	SL@1.25	Ea	576.00	46.20	—	622.20
3" dry valve trim	SL@1.25	Ea	576.00	46.20	—	622.20
4" dry valve trim	SL@1.25	Ea	576.00	46.20	—	622.20
6" dry valve trim	SL@1.25	Ea	576.00	46.20	—	622.20

Check valves, grooved

Description	Craft@Hrs	Unit	Material $	Labor $	Equipment $	Total $
2½"	SL@.450	Ea	225.00	16.60	—	241.60
3"	SL@.500	Ea	248.00	18.50	—	266.50
4"	SL@.550	Ea	222.00	20.30	—	242.30
6"	SL@.850	Ea	437.00	31.40	—	468.40

Check valves, wafer-type, flanged

Description	Craft@Hrs	Unit	Material $	Labor $	Equipment $	Total $
2½"	SL@1.65	Ea	280.00	61.00	—	341.00
3"	SL@1.70	Ea	291.00	62.80	—	353.80
4"	SL@2.10	Ea	315.00	77.60	—	392.60
6"	SL@2.60	Ea	514.00	96.10	—	610.10
8"	SL@3.20	Ea	730.00	118.00	—	848.00

Double-check detector valve assembly (flanged)

Description	Craft@Hrs	Unit	Material $	Labor $	Equipment $	Total $
2½"	SL@4.00	Ea	2,920.00	148.00	—	3,068.00
3"	SL@4.35	Ea	3,140.00	161.00	—	3,301.00
4"	SL@4.95	Ea	3,410.00	183.00	—	3,593.00
6"	SL@6.25	Ea	5,340.00	231.00	—	5,571.00
8"	SL@6.95	Ea	9,620.00	257.00	—	9,877.00

Fire Protection Sprinklers

Description	Craft@Hrs	Unit	Material $	Labor $	Equipment $	Total $

Switches: flow, pressure, monitor, supervisory

Description	Craft@Hrs	Unit	Material $	Labor $	Equipment $	Total $
Flow switch	SL@1.00	Ea	195.00	37.00	—	232.00
Pressure switch	SL@1.00	Ea	147.00	37.00	—	184.00
Monitor switch	SL@1.00	Ea	142.00	37.00	—	179.00
Supervisory sw.	SL@1.00	Ea	130.00	37.00	—	167.00

Sprinkler heads, 155 degree

Description	Craft@Hrs	Unit	Material $	Labor $	Equipment $	Total $
Pendent, brass	SL@.350	Ea	8.53	12.90	—	21.43
Pendent, chrome	SL@.350	Ea	9.13	12.90	—	22.03
Upright, brass	SL@.350	Ea	8.53	12.90	—	21.43
Upright, chrome	SL@.350	Ea	21.40	12.90	—	34.30
Sidewall, brass	SL@.350	Ea	11.10	12.90	—	24.00
Sidewall, chrome	SL@.350	Ea	11.80	12.90	—	24.70

Sprinkler heads, dry, 165 degree

Description	Craft@Hrs	Unit	Material $	Labor $	Equipment $	Total $
Dry pendent	SL@.450	Ea	85.10	16.60	—	101.70
Dry sidewall	SL@.450	Ea	118.00	16.60	—	134.60

Sprinkler heads, 200 degree

Description	Craft@Hrs	Unit	Material $	Labor $	Equipment $	Total $
Pendent, brass	SL@.350	Ea	8.53	12.90	—	21.43
Pendent, chrome	SL@.350	Ea	9.13	12.90	—	22.03
Upright, brass	SL@.350	Ea	8.53	12.90	—	21.43
Upright, chrome	SL@.350	Ea	21.40	12.90	—	34.30
Sidewall, brass	SL@.350	Ea	11.10	12.90	—	24.00
Sidewall, chrome	SL@.350	Ea	11.80	12.90	—	24.70

Sprinkler heads, 286 degree

Description	Craft@Hrs	Unit	Material $	Labor $	Equipment $	Total $
Pendent, brass	SL@.350	Ea	8.53	12.90	—	21.43
Pendent, chrome	SL@.350	Ea	9.13	12.90	—	22.03
Upright, brass	SL@.350	Ea	8.53	12.90	—	21.43
Upright, chrome	SL@.350	Ea	21.40	12.90	—	34.30
Sidewall, brass	SL@.350	Ea	11.10	12.90	—	24.00
Sidewall, chrome	SL@.350	Ea	11.80	12.90	—	24.70

Sprinkler heads, 360 degree

Description	Craft@Hrs	Unit	Material $	Labor $	Equipment $	Total $
Pendent, brass	SL@.350	Ea	5.75	12.90	—	18.65
Upright, brass	SL@.350	Ea	5.75	12.90	—	18.65

Sprinkler heads, 400 degree

Description	Craft@Hrs	Unit	Material $	Labor $	Equipment $	Total $
Pendent, brass	SL@.350	Ea	25.30	12.90	—	38.20
Upright, brass	SL@.350	Ea	25.30	12.90	—	38.20

Fire Protection Equipment

Description	Craft@Hrs	Unit	Material $	Labor $	Equipment $	Total $

Fire hose cabinet, complete assembly. Includes recessed steel cabinet, 2½" and 1½" angle valves, 100' fire hose, hose rack and fog nozzle.

Description	Craft@Hrs	Unit	Material $	Labor $	Equipment $	Total $
FHC, chrome trim	SL@1.95	Ea	409.00	72.10	—	481.10
FHC, brass trim	SL@1.95	Ea	400.00	72.10	—	472.10
FHC, plastic trim	SL@1.95	Ea	388.00	72.10	—	460.10

Fire hose cabinet components

Description	Craft@Hrs	Unit	Material $	Labor $	Equipment $	Total $
Recessed cabinet	SL@1.00	Ea	181.00	37.00	—	218.00
Surface cabinet	SL@.850	Ea	210.00	31.40	—	241.40
Cabinet glass	SL@.350	Ea	13.30	12.90	—	26.20
100' fire hose	SL@.250	Ea	170.00	9.24	—	179.24
75' hose rack	SL@.350	Ea	52.50	12.90	—	65.40
100' hose rack	SL@.350	Ea	71.10	12.90	—	84.00
1½" brass fire hose angle valve	SL@.350	Ea	85.30	12.90	—	98.20
1½" chrome fire hose angle valve	SL@.350	Ea	142.00	12.90	—	154.90
2½" brass fire hose angle valve	SL@.450	Ea	164.00	16.60	—	180.60
2½" chrome fire hose angle valve	SL@.450	Ea	64.70	16.60	—	81.30
Brass fog nozzle	SL@.150	Ea	63.70	5.54	—	69.24
Plastic fog nozzle	SL@.150	Ea	17.60	5.54	—	23.14
Chrome fog nozzle	SL@.150	Ea	65.30	5.54	—	70.84

Fire Protection Equipment

Description	Craft@Hrs	Unit	Material $	Labor $	Equipment $	Total $
Electric fire pump, skid-mounted with controller and jockey pump						
15 KW, 500 gpm	—	Ea	—	—	—	49,200.00
22 KW, 500 gpm	—	Ea	—	—	—	52,700.00
30 KW, 750 gpm	—	Ea	—	—	—	55,700.00
38 KW, 1,000 gpm	—	Ea	—	—	—	61,700.00
56 KW, 1,000 gpm	—	Ea	—	—	—	62,600.00
Diesel fire pump, skid-mounted with controller and jockey pump						
250 gpm	—	Ea	—	—	—	73,500.00
500 gpm	—	Ea	—	—	—	76,600.00
750 gpm	—	Ea	—	—	—	78,000.00
1,000 gpm	—	Ea	—	—	—	79,500.00
Excess pressure pump						
1/4 hp	SL@1.35	Ea	555.00	49.90	—	604.90
1/3 hp	SL@1.35	Ea	557.00	49.90	—	606.90
1/2 hp	SL@1.50	Ea	578.00	55.40	—	633.40
Siamese connection, flush, fire dept. connection						
4" brass	SL@4.00	Ea	767.00	148.00	—	915.00
4" chrome	SL@4.00	Ea	827.00	148.00	—	975.00
6" brass	SL@4.50	Ea	890.00	166.00	—	1,056.00
6" chrome	SL@4.50	Ea	906.00	166.00	—	1,072.00
Siamese connection, surface, fire dept. connection						
4" brass	SL@2.85	Ea	182.00	105.00	—	287.00
6" brass	SL@3.25	Ea	343.00	120.00	—	463.00
Siamese connection, sidewalk, fire dept. connection						
4" brass	SL@3.25	Ea	693.00	120.00	—	813.00
6" brass	SL@3.75	Ea	980.00	139.00	—	1,119.00

Fire Protection Equipment

Description	Craft@Hrs	Unit	Material $	Labor $	Equipment $	Total $
Water motor gong						
Gong	SL@1.50	Ea	341.00	55.40	—	396.40
Multi-purpose fire extinguisher						
2 lb.	SL@.500	Ea	40.30	18.50	—	58.80
5 lb.	SL@.500	Ea	56.70	18.50	—	75.20
10 lb.	SL@.500	Ea	81.20	18.50	—	99.70
20 lb.	SL@.500	Ea	171.00	18.50	—	189.50
Fire hydrant with spool						
6"	P1@5.00	Ea	1,570.00	186.00	—	1,756.00
Indicator post						
6"	P1@6.50	Ea	903.00	242.00	—	1,145.00

Fire Protection Sprinkler Pipe and Fittings (Roll-Grooved)

Description	Craft@Hrs	Unit	Material $	Labor $	Equipment $	Total $

Black steel pipe (A53), installed horizontal, roll-grooved, Schedule 40. Standard weight, including hangers every 10' and a fitting every 33'. Use these figures for preliminary estimates.

2" (50mm)	SL@.160	LF	7.86	5.91	—	13.77
2½" (65mm)	SL@.180	LF	12.60	6.65	—	19.25
3" (75mm)	SL@.210	LF	16.20	7.76	—	23.96
4" (10cm)	SL@.300	LF	21.80	11.10	—	32.90
6" (15cm)	SK@.450	LF	43.60	17.90	.29	61.79
8" (20cm)	SK@.630	LF	64.80	25.00	.41	90.21
10" (25cm)	SK@.750	LF	126.00	29.80	.49	156.29
12" (30cm)	SK@.800	LF	165.00	31.70	.52	197.22

Black steel pipe (A53), installed riser, roll-grooved, Schedule 40. Standard weight, including a riser clamp every other floor and a tee at every floor. Make additional allowances for sleeving or coring as required. Use these figures for preliminary estimates.

2" (50mm)	SL@.110	LF	8.13	4.07	—	12.20
2½" (65mm)	SL@.120	LF	13.10	4.44	—	17.54
3" (75mm)	SL@.150	LF	15.60	5.54	—	21.14
4" (10cm)	SL@.190	LF	22.10	7.02	—	29.12
6" (15cm)	SK@.300	LF	42.20	11.90	.20	54.30
8" (20cm)	SK@.400	LF	65.60	15.90	.26	81.76
10" (25cm)	SK@.510	LF	102.00	20.20	.33	122.53
12" (30cm)	SK@.560	LF	148.00	22.20	.36	170.56

Black steel pipe (A53), installed horizontal, roll-grooved, thin wall, Schedule 10. Light weight, including hangers every 10' and a fitting every 33'. Use these figures for preliminary estimates.

2" (50mm)	SL@.160	LF	6.53	5.91	—	12.44
2½" (65mm)	SL@.180	LF	9.53	6.65	—	16.18
3" (75mm)	SL@.210	LF	11.10	7.76	—	18.86
4" (10cm)	SL@.300	LF	16.10	11.10	—	27.20
6" (15cm)	SK@.450	LF	31.00	17.90	.29	49.19
8" (20cm)	SK@.630	LF	57.50	25.00	.41	82.91

Black steel pipe (A53), installed riser, roll-grooved, thin wall, Schedule 10. Light weight, including a riser clamp every other floor and a tee at every floor. Make additional allowances for sleeving or coring as required. Use these figures for preliminary estimates.

2" (50mm)	SL@.110	LF	7.33	4.07	—	11.40
2½" (65mm)	SL@.120	LF	10.60	4.44	—	15.04
3" (75mm)	SL@.150	LF	12.40	5.54	—	17.94
4" (10cm)	SL@.190	LF	17.40	7.02	—	24.42
6" (15cm)	SK@.300	LF	33.30	11.90	.20	45.40
8" (20cm)	SK@.400	LF	64.80	15.90	.26	80.96

Fire Protection Sprinkler Pipe and Fittings (Roll-Grooved)

Description	Craft@Hrs	Unit	Material $	Labor $	Equipment $	Total $

Black steel pipe only (A53), roll-grooved, Schedule 40. Standard weight, no hangers or fittings.

2" (50mm)	SL@.070	LF	3.47	2.59	—	6.06
2½" (65mm)	SL@.070	LF	7.42	2.59	—	10.01
3" (75mm)	SL@.090	LF	8.26	3.33	—	11.59
4" (10cm)	SL@.130	LF	11.60	4.80	—	16.40
6" (15cm)	SK@.160	LF	17.10	6.35	.10	23.55
8" (20cm)	SK@.210	LF	25.90	8.33	.14	34.37
10" (25cm)	SK@.280	LF	37.40	11.10	.18	48.68
12" (30cm)	SK@.340	LF	63.10	13.50	.22	76.82
14" (36cm)	SK@.400	LF	72.80	15.90	.26	88.96
16" (41cm)	SK@.550	LF	84.60	21.80	.36	106.76

Black steel pipe only, (A53) roll-grooved, thin wall, Schedule 10. Light weight, no hangers or fittings.

2" (50mm)	SL@.070	LF	2.60	2.59	—	5.19
2½" (65mm)	SL@.070	LF	5.50	2.59	—	8.09
3" (75mm)	SL@.090	LF	6.09	3.33	—	9.42
4" (10cm)	SL@.130	LF	8.73	4.80	—	13.53
6" (15cm)	SK@.160	LF	12.60	6.35	.10	19.05
8" (20cm)	SK@.210	LF	21.60	8.33	.14	30.07
10" (25cm)	SK@.280	LF	31.40	11.10	.18	42.68
12" (30cm)	SK@.340	LF	53.10	13.50	.22	66.82
14" (36cm)	SK@.400	LF	61.50	15.90	.26	77.66
16" (41cm)	SK@.550	LF	71.50	21.80	.36	93.66

90-degree elbow, roll-grooved, Style #10 (Victaulic)

2"	SL@.410	Ea	29.90	15.20	—	45.10
2½"	SL@.430	Ea	29.90	15.90	—	45.80
3"	SL@.500	Ea	39.40	18.50	—	57.90
4"	SL@.690	Ea	61.60	25.50	—	87.10
6"	SL@1.19	Ea	134.00	44.00	—	178.00
8"	SL@1.69	Ea	229.00	62.50	—	291.50
10"	SL@2.23	Ea	441.00	82.40	—	523.40
12"	SL@2.48	Ea	695.00	91.70	—	786.70

45-degree elbow, roll-grooved, Style #11 (Victaulic)

2"	SL@.410	Ea	29.90	15.20	—	45.10
2½"	SL@.430	Ea	29.90	15.90	—	45.80
3"	SL@.500	Ea	39.40	18.50	—	57.90
4"	SL@.690	Ea	61.60	25.50	—	87.10
6"	SL@1.19	Ea	134.00	44.00	—	178.00
8"	SL@1.69	Ea	229.00	62.50	—	291.50
10"	SL@2.23	Ea	441.00	82.40	—	523.40
12"	SL@2.48	Ea	695.00	91.70	—	786.70

Fire Protection Sprinkler Pipe and Fittings (Roll-Grooved)

Description	Craft@Hrs	Unit	Material $	Labor $	Equipment $	Total $
22½-degree elbow, roll-grooved, Style #12 (Victaulic)						
2"	SL@.410	Ea	35.60	15.20	—	50.80
2½"	SL@.430	Ea	35.60	15.90	—	51.50
3"	SL@.500	Ea	37.20	18.50	—	55.70
4"	SL@.690	Ea	70.20	25.50	—	95.70
6"	SL@1.19	Ea	152.00	44.00	—	196.00
8"	SL@1.69	Ea	313.00	62.50	—	375.50
10"	SL@2.23	Ea	443.00	82.40	—	525.40
12"	SL@2.48	Ea	473.00	91.70	—	564.70
Tee, roll-grooved, Style #20 (Victaulic)						
2"	SL@.510	Ea	57.70	18.80	—	76.50
2½"	SL@.540	Ea	57.70	20.00	—	77.70
3"	SL@.630	Ea	69.60	23.30	—	92.90
4"	SL@.860	Ea	97.00	31.80	—	128.80
6"	SL@1.49	Ea	224.00	55.10	—	279.10
8"	SL@2.11	Ea	379.00	78.00	—	457.00
10"	SL@2.79	Ea	670.00	103.00	—	773.00
12"	SL@3.10	Ea	785.00	115.00	—	900.00
Reducing tee, roll-grooved, Style #25 (Victaulic)						
2"	SL@.510	Ea	57.70	18.80	—	76.50
2½"	SL@.540	Ea	75.80	20.00	—	95.80
3"	SL@.630	Ea	68.60	23.30	—	91.90
4"	SL@.860	Ea	116.00	31.80	—	147.80
6"	SL@1.49	Ea	216.00	55.10	—	271.10
8"	SL@2.11	Ea	393.00	78.00	—	471.00
10"	SL@2.79	Ea	520.00	103.00	—	623.00
12"	SL@3.10	Ea	575.00	115.00	—	690.00
Reducer, roll-grooved, Style #50 (Victaulic)						
2"	SL@.410	Ea	19.20	15.20	—	34.40
2½"	SL@.430	Ea	19.20	15.90	—	35.10
3"	SL@.500	Ea	26.70	18.50	—	45.20
4"	SL@.690	Ea	44.70	25.50	—	70.20
6"	SL@1.19	Ea	65.00	44.00	—	109.00
8"	SL@1.69	Ea	114.00	62.50	—	176.50
10"	SL@2.23	Ea	238.00	82.40	—	320.40
Cap, roll-grooved, Style #60 (Victaulic)						
2"	SL@.270	Ea	16.90	9.98	—	26.88
2½"	SL@.280	Ea	16.90	10.30	—	27.20
3"	SL@.330	Ea	21.20	12.20	—	33.40
4"	SL@.450	Ea	24.50	16.60	—	41.10
6"	SL@.770	Ea	47.80	28.50	—	76.30
8"	SL@1.10	Ea	77.30	40.70	—	118.00
10"	SL@1.61	Ea	330.00	59.50	—	389.50

Fire Protection Sprinkler Fittings (Roll-Grooved)

Description	Craft@Hrs	Unit	Material $	Labor $	Equipment $	Total $

Mechanical tee (saddle tee), roll-grooved, Style #920 (Victaulic)

Description	Craft@Hrs	Unit	Material $	Labor $	Equipment $	Total $
3"	SL@.750	Ea	58.20	27.70	—	85.90
4"	SL@1.00	Ea	66.10	37.00	—	103.10
6"	SL@1.30	Ea	78.60	48.00	—	126.60
8"	SL@1.50	Ea	124.00	55.40	—	179.40

Coupling, roll-grooved, 800 PSI, Style #77 (Victaulic). Labor included with fittings.

Description	Craft@Hrs	Unit	Material $	Labor $	Equipment $	Total $
3"	—	Ea	26.40	—	—	26.40
4"	—	Ea	43.70	—	—	43.70
6"	—	Ea	71.70	—	—	71.70
8"	—	Ea	97.80	—	—	97.80
10"	—	Ea	132.00	—	—	132.00
12"	—	Ea	176.00	—	—	176.00

Coupling, roll-grooved, light weight, Style #75 (Victaulic). Labor included with fittings.

Description	Craft@Hrs	Unit	Material $	Labor $	Equipment $	Total $
3"	—	Ea	21.30	—	—	21.30
4"	—	Ea	23.00	—	—	23.00
6"	—	Ea	41.10	—	—	41.10
8"	—	Ea	63.00	—	—	63.00

Coupling, roll-grooved, zeroflex, Style #07 (Victaulic). Labor included with fittings.

Description	Craft@Hrs	Unit	Material $	Labor $	Equipment $	Total $
3"	—	Ea	24.60	—	—	24.60
4"	—	Ea	32.30	—	—	32.30
6"	—	Ea	47.50	—	—	47.50
8"	—	Ea	72.30	—	—	72.30
10"	—	Ea	106.00	—	—	106.00
12"	—	Ea	147.00	—	—	147.00

Flange, roll-grooved, Style #741 (Victaulic)

Description	Craft@Hrs	Unit	Material $	Labor $	Equipment $	Total $
3"	SL@.500	Ea	92.80	18.50	—	111.30
4"	SL@.690	Ea	126.00	25.50	—	151.50
6"	SL@1.19	Ea	153.00	44.00	—	197.00
8"	SL@1.69	Ea	178.00	62.50	—	240.50
10"	SL@2.23	Ea	265.00	82.40	—	347.40
12"	SL@2.48	Ea	361.00	91.70	—	452.70

Fire Protection Branch Pipe & Fittings

Description	Craft@Hrs	Unit	Material $	Labor $	Equipment $	Total $

Schedule 40 carbon steel pipe, threaded and coupled, pipe only. Add for hangers, fittings etc.

Description	Craft@Hrs	Unit	Material $	Labor $	Equipment $	Total $
1"	P1@.090	LF	2.74	3.35	—	6.09
1¼"	P1@.100	LF	3.58	3.72	—	7.30
1½"	P1@.110	LF	4.26	4.09	—	8.35
2"	P1@.120	LF	5.75	4.47	—	10.22

125# cast iron 90-degree ell, threaded joints

Description	Craft@Hrs	Unit	Material $	Labor $	Equipment $	Total $
1" x ¾"	P1@.130	Ea	4.85	4.84	—	9.69
1"	P1@.180	Ea	4.50	6.70	—	11.20
1¼"	P1@.240	Ea	6.37	8.93	—	15.30
1½"	P1@.300	Ea	8.87	11.20	—	20.07
2"	P1@.380	Ea	13.50	14.10	—	27.60

125# cast iron 45-degree ell, threaded joints

Description	Craft@Hrs	Unit	Material $	Labor $	Equipment $	Total $
1"	P1@.180	Ea	5.90	6.70	—	12.60
1¼"	P1@.240	Ea	8.40	8.93	—	17.33
1½"	P1@.300	Ea	10.80	11.20	—	22.00
2"	P1@.380	Ea	14.60	14.10	—	28.70

125# cast iron tee, threaded joints

Description	Craft@Hrs	Unit	Material $	Labor $	Equipment $	Total $
1"	P1@.230	Ea	6.20	8.56	—	14.76
1¼"	P1@.310	Ea	11.20	11.50	—	22.70
1½"	P1@.390	Ea	14.40	14.50	—	28.90
2"	P1@.490	Ea	19.70	18.20	—	37.90

125# cast iron reducing tee, threaded joints

Description	Craft@Hrs	Unit	Material $	Labor $	Equipment $	Total $
1" x ¾"	P1@.220	Ea	7.91	8.19	—	16.10
1¼" x ¾"	P1@.290	Ea	10.10	10.80	—	20.90
1¼" x 1"	P1@.290	Ea	11.00	10.80	—	21.80
1½" x ¾"	P1@.370	Ea	14.00	13.80	—	27.80
1½" x 1"	P1@.370	Ea	15.30	13.80	—	29.10
1½" x 1¼"	P1@.370	Ea	16.90	13.80	—	30.70
2" x 1"	P1@.460	Ea	21.90	17.10	—	39.00
2" x 1¼"	P1@.460	Ea	23.50	17.10	—	40.60
2" x 1½"	P1@.460	Ea	25.30	17.10	—	42.40

125# cast iron reducer, threaded joints

Description	Craft@Hrs	Unit	Material $	Labor $	Equipment $	Total $
1" x ¾"	P1@.200	Ea	6.20	7.44	—	13.64
1¼" x 1"	P1@.210	Ea	7.46	7.82	—	15.28
1½" x 1"	P1@.270	Ea	9.16	10.00	—	19.16
1½" x 1¼"	P1@.270	Ea	9.67	10.00	—	19.67
2" x 1¼"	P1@.340	Ea	14.50	12.70	—	27.20
2" x 1½"	P1@.340	Ea	15.20	12.70	—	27.90

Fire Protection Branch Fittings

Description	Craft@Hrs	Unit	Material $	Labor $	Equipment $	Total $

125# cast iron cross, threaded joints

1"	P1@.360	Ea	14.50	13.40	—	27.90
1¼"	P1@.480	Ea	23.80	17.90	—	41.70
1½"	P1@.600	Ea	29.00	22.30	—	51.30
2"	P1@.700	Ea	47.80	26.10	—	73.90

125# cast iron cap, threaded joint

¾"	P1@.100	Ea	3.05	3.72	—	6.77
1"	P1@.140	Ea	3.69	5.21	—	8.90
1¼"	P1@.180	Ea	4.71	6.70	—	11.41
1½"	P1@.230	Ea	5.97	8.56	—	14.53
2"	P1@.290	Ea	9.40	10.80	—	20.20

125# cast iron plug, threaded joint

¾"	P1@.100	Ea	1.98	3.72	—	5.70
1"	P1@.140	Ea	2.63	5.21	—	7.84
1¼"	P1@.180	Ea	4.09	6.70	—	10.79
1½"	P1@.230	Ea	5.19	8.56	—	13.75
2"	P1@.290	Ea	6.43	10.80	—	17.23

125# cast iron coupling, threaded joints

1"	P1@.180	Ea	5.38	6.70	—	12.08
1¼"	P1@.240	Ea	6.63	8.93	—	15.56
1½"	P1@.300	Ea	8.33	11.20	—	19.53
2"	P1@.380	Ea	11.20	14.10	—	25.30

Fire Protection Sprinkler Branch Fittings

Description	Craft@Hrs	Unit	Material $	Labor $	Equipment $	Total $

Steel pipe nipples — 3" long, standard right-hand thread

Description	Craft@Hrs	Unit	Material $	Labor $	Equipment $	Total $
1"	P1@.025	Ea	2.08	.93	—	3.01
1¼"	P1@.025	Ea	2.67	.93	—	3.60
1½"	P1@.025	Ea	3.18	.93	—	4.11
2"	P1@.030	Ea	3.96	1.12	—	5.08

Steel pipe nipples — 4" long, standard right-hand thread

Description	Craft@Hrs	Unit	Material $	Labor $	Equipment $	Total $
1"	P1@.025	Ea	2.48	.93	—	3.41
1¼"	P1@.030	Ea	3.10	1.12	—	4.22
1½"	P1@.030	Ea	3.82	1.12	—	4.94
2"	P1@.030	Ea	4.92	1.12	—	6.04

Steel pipe nipples — 6" long, standard right-hand thread

Description	Craft@Hrs	Unit	Material $	Labor $	Equipment $	Total $
1"	P1@.030	Ea	3.18	1.12	—	4.30
1¼"	P1@.030	Ea	3.99	1.12	—	5.11
1½"	P1@.030	Ea	4.85	1.12	—	5.97
2"	P1@.035	Ea	6.52	1.30	—	7.82

Fire Protection Sprinkler Pipe and Fittings (CPVC)

Description	Craft@Hrs	Unit	Material $	Labor $	Equipment $	Total $

Chlorinated Polyvinyl Chloride pipe (CPVC), CPVC pipe only, solvent weld. No hangers or fittings.

¾"	SL@.055	LF	3.54	2.03	—	5.57
1"	SL@.055	LF	5.17	2.03	—	7.20
1¼"	SL@.065	LF	8.09	2.40	—	10.49
1½"	SL@.075	LF	11.40	2.77	—	14.17
2"	SL@.085	LF	16.80	3.14	—	19.94
2½"	SL@.110	LF	26.10	4.07	—	30.17
3"	SL@.125	LF	39.70	4.62	—	44.32

CPVC 90-degree elbow, solvent weld

¾"	SL@.100	Ea	4.10	3.70	—	7.80
1"	SL@.120	Ea	6.18	4.44	—	10.62
1¼"	SL@.170	Ea	7.90	6.28	—	14.18
1½"	SL@.180	Ea	11.90	6.65	—	18.55
2"	SL@.200	Ea	13.80	7.39	—	21.19

CPVC 45-degree elbow, solvent weld

¾"	SL@.100	Ea	4.24	3.70	—	7.94
1"	SL@.120	Ea	5.54	4.44	—	9.98
1¼"	SL@.170	Ea	7.29	6.28	—	13.57
1½"	SL@.180	Ea	9.80	6.65	—	16.45
2"	SL@.200	Ea	12.20	7.39	—	19.59

CPVC tee, solvent weld

¾"	SL@.150	Ea	4.53	5.54	—	10.07
1"	SL@.170	Ea	7.71	6.28	—	13.99
1¼"	SL@.230	Ea	12.10	8.50	—	20.60
1½"	SL@.250	Ea	16.00	9.24	—	25.24
2"	SL@.280	Ea	25.30	10.30	—	35.60

CPVC reducing tee, solvent weld

1"	SL@.170	Ea	6.64	6.28	—	12.92
1¼"	SL@.230	Ea	12.10	8.50	—	20.60
1½"	SL@.250	Ea	16.10	9.24	—	25.34
2"	SL@.280	Ea	25.80	10.30	—	36.10
3"	SL@.350	Ea	40.60	12.90	—	53.50

CPVC coupling, solvent weld

¾"	SL@.100	Ea	2.98	3.70	—	6.68
1"	SL@.120	Ea	4.04	4.44	—	8.48
1¼"	SL@.170	Ea	6.38	6.28	—	12.66
1½"	SL@.180	Ea	8.04	6.65	—	14.69
2"	SL@.200	Ea	10.10	7.39	—	17.49
3"	SL@.280	Ea	20.80	10.30	—	31.10

Fire Protection Sprinkler Fittings (CPVC)

Description	Craft@Hrs	Unit	Material $	Labor $	Equipment $	Total $
CPVC head adapter, solvent weld						
¾" x ½"	SL@.150	Ea	6.75	5.54	—	12.29
1" x ½"	SL@.165	Ea	7.21	6.10	—	13.31
CPVC cap, solvent weld						
¾"	SL@.060	Ea	2.21	2.22	—	4.43
1"	SL@.065	Ea	3.09	2.40	—	5.49
1¼"	SL@.075	Ea	5.30	2.77	—	8.07
1½"	SL@.085	Ea	6.79	3.14	—	9.93
2"	SL@.095	Ea	7.91	3.51	—	11.42
CPVC flange, solvent weld						
1"	SL@.165	Ea	7.91	6.10	—	14.01
1¼"	SL@.175	Ea	10.20	6.47	—	16.67
1½"	SL@.185	Ea	15.90	6.84	—	22.74
2"	SL@.200	Ea	23.90	7.39	—	31.29
3"	SL@.225	Ea	39.80	8.32	—	48.12
CPVC female adapter, solvent weld						
1"	SL@.065	Ea	19.60	2.40	—	22.00
1¼"	SL@.075	Ea	32.60	2.77	—	35.37
1½"	SL@.085	Ea	41.80	3.14	—	44.94
2"	SL@.095	Ea	54.00	3.51	—	57.51

Commercial Boilers

Boilers for industrial, commercial and institutional applications. ASME "H" or "U" stamped, LFUE (Levelized Fuel Utilization Efficiency) certified 90% or better. U.S. Green Building council certified. Cleaver-Brooks or equal. Shipped factory-assembled and skid-mounted. Includes the boiler shell and structure, I-beam skids, combustion chamber, fire tubes, crown sheet, tube sheets (for fire tube), water tubes, mud drum(s), steam drum (for water tube boiler), finned water tubes and check valves (for hot water generators), refractory brick lining, explosion doors, combustion air inlets and stack outlet. Add the cost of the fuel train piping, electrical wiring, refractory, stack, feedwater system, combustion controls, water treatment, expansion tanks (if hot water) or condensate return (if steam) and system start-up. See the sections following packaged boilers. Costs are based on boiler horsepower rating (BHP). Add the cost of commissioning, insurance inspection, permit, permit inspection, hydraulic test, calibration and lockout sealing of emissions monitor, and final run test. Equipment cost assumes a 17-ton hydraulic truck-mounted crane (at $130 per hour). LEED (Leadership in Energy & Environmental Design) certification requires 88% boiler efficiency, recording controls, compliance with SCAQMD 1146.2 emission standards and zone recording thermostats. A LEED-certified boiler with a U.S. Green Building Council rating will add 4% to 15% to the cost of the boiler, as identified in the cost tables that follow.

Description	Craft@Hrs	Unit	Material $	Labor $	Equipment $	Total $

Hot water or steam cast iron gas boiler, 30 PSIG. Costs shown are based on boiler horsepower rating (BHP). On-site assembly. Includes allowances for associated appurtenances and pipe connections.

Description	Craft@Hrs	Unit	Material $	Labor $	Equipment $	Total $
4.5 BHP	SN@8.50	Ea	5,280.00	333.00	144.00	5,757.00
5.9 BHP	SN@9.50	Ea	6,590.00	372.00	161.00	7,123.00
9.0 BHP	SN@10.0	Ea	9,820.00	392.00	169.00	10,381.00
15.0 BHP	SN@12.0	Ea	15,300.00	470.00	203.00	15,973.00
20.0 BHP	SN@14.0	Ea	20,300.00	549.00	237.00	21,086.00
30.0 BHP	SN@16.0	Ea	26,100.00	627.00	270.00	26,997.00
45.0 BHP	SN@18.0	Ea	37,900.00	705.00	304.00	38,909.00
60.0 BHP	SN@20.0	Ea	49,200.00	784.00	338.00	50,322.00
Add for LEED-certified boiler	—	%	4.0	—	—	—
Add for LEED inspection	—	Ea	—	—	—	2,060.00
Add for central monitor/recorder	—	Ea	1,770.00	—	—	1,770.00
Add for recording thermostat	—	Ea	412.00	—	—	412.00

Commercial Boilers

Description	Craft@Hrs	Unit	Material $	Labor $	Equipment $	Total $

Hot water Scotch marine firetube gas boiler, 30 PSIG. Costs shown are based on boiler horsepower rating (BHP). Set in place only. Make additional allowances for associated appurtenances and pipe connections.

Description	Craft@Hrs	Unit	Material $	Labor $	Equipment $	Total $
30 BHP	SN@10.0	Ea	37,700.00	392.00	169.00	38,261.00
40 BHP	SN@10.0	Ea	39,900.00	392.00	169.00	40,461.00
50 BHP	SN@10.0	Ea	41,600.00	392.00	169.00	42,161.00
60 BHP	SN@12.0	Ea	46,200.00	470.00	203.00	46,873.00
70 BHP	SN@12.0	Ea	49,900.00	470.00	203.00	50,573.00
80 BHP	SN@14.0	Ea	53,100.00	549.00	237.00	53,886.00
100 BHP	SN@14.0	Ea	63,300.00	549.00	237.00	64,086.00
125 BHP	SN@16.0	Ea	67,900.00	627.00	270.00	68,797.00
150 BHP	SN@18.0	Ea	72,900.00	705.00	304.00	73,909.00
200 BHP	SN@18.0	Ea	85,900.00	705.00	304.00	86,909.00
250 BHP	SN@26.0	Ea	94,600.00	1,020.00	439.00	96,059.00
300 BHP	SN@26.0	Ea	102,000.00	1,020.00	439.00	103,459.00
350 BHP	SN@26.0	Ea	116,000.00	1,020.00	439.00	117,459.00
400 BHP	SN@32.0	Ea	119,000.00	1,250.00	541.00	120,791.00
500 BHP	SN@32.0	Ea	141,000.00	1,250.00	541.00	142,791.00
600 BHP	SN@32.0	Ea	159,000.00	1,250.00	541.00	160,791.00
750 BHP	SN@40.0	Ea	166,000.00	1,570.00	676.00	168,246.00
Add for LEED-certified boiler	—	%	4.0	—	—	—
Add for LEED inspection	—	Ea	—	—	—	2,060.00
Add for central monitor/recorder	—	Ea	4,120.00	—	—	4,120.00
Add for recording thermostat	—	Ea	412.00	—	—	412.00

Commercial Boilers

Description	Craft@Hrs	Unit	Material $	Labor $	Equipment $	Total $

Packaged Scotch marine firetube oil or gas fired boiler, 150 PSIG.
Costs shown are based on boiler horsepower rating (BHP). Set in place only.
Make additional allowances for associated appurtenances and pipe connections.

Description	Craft@Hrs	Unit	Material $	Labor $	Equipment $	Total $
60 BHP	SN@12.0	Ea	53,100.00	470.00	203.00	53,773.00
70 BHP	SN@12.0	Ea	57,100.00	470.00	203.00	57,773.00
80 BHP	SN@14.0	Ea	61,200.00	549.00	237.00	61,986.00
100 BHP	SN@14.0	Ea	73,400.00	549.00	237.00	74,186.00
125 BHP	SN@16.0	Ea	77,800.00	627.00	270.00	78,697.00
150 BHP	SN@18.0	Ea	83,100.00	705.00	304.00	84,109.00
200 BHP	SN@18.0	Ea	110,000.00	705.00	304.00	111,009.00
250 BHP	SN@26.0	Ea	117,000.00	1,020.00	439.00	118,459.00
300 BHP	SN@26.0	Ea	132,000.00	1,020.00	439.00	133,459.00
350 BHP	SN@26.0	Ea	136,000.00	1,020.00	439.00	137,459.00
400 BHP	SN@32.0	Ea	163,000.00	1,250.00	541.00	164,791.00
500 BHP	SN@32.0	Ea	180,000.00	1,250.00	541.00	181,791.00
600 BHP	SN@32.0	Ea	185,000.00	1,250.00	541.00	186,791.00
700 BHP	SN@40.0	Ea	191,000.00	1,570.00	676.00	193,246.00
Add for LEED-certified boiler	—	%	4.0	—	—	—
Add for LEED inspection	—	Ea	—	—	—	2,060.00
Add for central monitor/ recorder	—	Ea	4,120.00	—	—	4,120.00
Add for recording thermostat	—	Ea	412.00	—	—	412.00

Packaged hot water firebox oil or gas fired boiler. Costs shown are based on boiler horsepower rating (BHP). Set in place only. Make additional allowances for associated appurtenances and pipe connections.

Description	Craft@Hrs	Unit	Material $	Labor $	Equipment $	Total $
13 BHP	SN@8.00	Ea	20,000.00	314.00	135.00	20,449.00
16 BHP	SN@8.00	Ea	20,300.00	314.00	135.00	20,749.00
19 BHP	SN@8.00	Ea	20,900.00	314.00	135.00	21,349.00
22 BHP	SN@8.00	Ea	21,500.00	314.00	135.00	21,949.00
28 BHP	SN@8.00	Ea	22,500.00	314.00	135.00	22,949.00
34 BHP	SN@8.00	Ea	25,800.00	314.00	135.00	26,249.00
40 BHP	SN@8.00	Ea	28,500.00	314.00	135.00	28,949.00
46 BHP	SN@8.00	Ea	32,500.00	314.00	135.00	32,949.00
52 BHP	SN@8.00	Ea	38,000.00	314.00	135.00	38,449.00
61 BHP	SN@8.00	Ea	41,900.00	314.00	135.00	42,349.00
70 BHP	SN@8.00	Ea	44,800.00	314.00	135.00	45,249.00
80 BHP	SN@8.00	Ea	46,200.00	314.00	135.00	46,649.00
100 BHP	SN@8.00	Ea	52,200.00	314.00	135.00	52,649.00
125 BHP	SN@16.0	Ea	62,000.00	627.00	270.00	62,897.00
150 BHP	SN@16.0	Ea	66,600.00	627.00	270.00	67,497.00
Add for LEED-certified boiler	—	%	4.0	—	—	—
Add for LEED inspection	—	Ea	—	—	—	2,060.00
Add for central monitor/ recorder	—	Ea	2,940.00	—	—	2,940.00
Add for recording thermostat	—	Ea	412.00	—	—	412.00

Commercial Boilers

Description	Craft@Hrs	Unit	Material $	Labor $	Equipment $	Total $

Packaged watertube industrial high pressure boilers. Costs shown are based on boiler horsepower rating (BHP). Set in place only. Make additional allowances for associated appurtenances and pipe connections.

Description	Craft@Hrs	Unit	Material $	Labor $	Equipment $	Total $
480 BHP	SN@32.0	Ea	377,000.00	1,250.00	541.00	378,791.00
526 BHP	SN@32.0	Ea	385,000.00	1,250.00	541.00	386,791.00
572 BHP	SN@32.0	Ea	407,000.00	1,250.00	541.00	408,791.00
618 BHP	SN@32.0	Ea	415,000.00	1,250.00	541.00	416,791.00
664 BHP	SN@32.0	Ea	433,000.00	1,250.00	541.00	434,791.00
710 BHP	SN@64.0	Ea	439,000.00	2,510.00	1,080.00	442,590.00
747 BHP	SN@64.0	Ea	446,000.00	2,510.00	1,080.00	449,590.00
792 BHP	SN@64.0	Ea	460,000.00	2,510.00	1,080.00	463,590.00
838 BHP	SN@64.0	Ea	474,000.00	2,510.00	1,080.00	477,590.00
884 BHP	SN@64.0	Ea	488,000.00	2,510.00	1,080.00	491,590.00
930 BHP	SN@64.0	Ea	494,000.00	2,510.00	1,080.00	497,590.00
1,000 BHP	SN@64.0	Ea	503,000.00	2,510.00	1,080.00	506,590.00
Add for LEED-certified boiler	—	%	4.0	—	—	—
Add for LEED inspection	—	Ea	—	—	—	2,060.00
Add for central monitor/ recorder	—	Ea	5,060.00	—	—	5,060.00
Add for recording thermostat	—	Ea	412.00	—	—	412.00

Gas fired hot water steel boilers, 12 PSIG Mid-efficiency (82%) atmospheric draft, standing pilot light, c/w primary loop circulating pump. Labor includes setting in place and connecting pump and controls. Make additional allowances for distribution piping connections, safety devices, flue, gas connections, wiring, etc.

Description	Craft@Hrs	Unit	Material $	Labor $	Equipment $	Total $
50,000 Btu	P1@4.00	Ea	1,430.00	149.00	67.60	1,646.60
70,000 Btu	P1@4.25	Ea	1,640.00	158.00	71.50	1,869.50
105,000 Btu	P1@4.50	Ea	1,800.00	167.00	76.10	2,043.10
140,000 Btu	P1@4.75	Ea	2,070.00	177.00	80.30	2,327.30
175,000 Btu	P1@5.00	Ea	2,300.00	186.00	84.50	2,570.50
210,000 Btu	P1@5.50	Ea	2,920.00	205.00	93.00	3,218.00
245,000 Btu	P1@7.00	Ea	3,250.00	261.00	118.00	3,629.00
Electronic spark ignition Add	—	Ea	307.00	—	—	307.00
Add for LEED-certified boiler	—	%	15.0	—	—	—
Add for LEED inspection	—	Ea	—	—	—	2,060.00
Add for central monitor/ recorder	—	Ea	1,290.00	—	—	1,290.00
Add for recording thermostat	—	Ea	177.00	—	—	177.00

Commercial Boilers

Description	Craft@Hrs	Unit	Material $	Labor $	Equipment $	Total $

Gas fired hot water cast iron boilers, 12 PSIG Mid-efficiency (82%) direct vent, forced draft electronic spark ignition, c/w primary loop circulating pump. Labor includes setting in place and connecting pump and controls. Make additional allowances for distribution piping connections, safety devices, flue, gas connections, wiring, etc.

Description	Craft@Hrs	Unit	Material $	Labor $	Equipment $	Total $
50,000 Btu	P1@4.00	Ea	1,790.00	149.00	67.60	2,006.60
65,000 Btu	P1@4.25	Ea	2,010.00	158.00	71.50	2,239.50
100,000 Btu	P1@4.50	Ea	2,190.00	167.00	76.10	2,433.10
130,000 Btu	P1@4.75	Ea	2,430.00	177.00	80.30	2,687.30
170,000 Btu	P1@5.00	Ea	2,810.00	186.00	84.50	3,080.50
200,000 Btu	P1@5.50	Ea	3,090.00	205.00	93.00	3,388.00
235,000 Btu	P1@7.00	Ea	3,410.00	261.00	118.00	3,789.00
Add for LEED-certified boiler	—	%	15.0	—	—	—
Add for LEED inspection	—	Ea	—	—	—	2,060.00
Add for central monitor/ recorder	—	Ea	1,290.00	—	—	1,290.00
Add for recording thermostat	—	Ea	177.00	—	—	177.00

Description	Craft@Hrs	Unit	Material $	Labor $	Total $

Steam heating boilers for residential and light commercial applications. Parker or equal 15 PSIG Spiral watertube type, with ASTM SA-53 grade tubing, full American Society of Mechanical Engineers "M", "H" and "S" ratings and Underwriters Laboratories certifications. Specifically designed for "twinning" and multiple unit additions and retrofits. Packaged type units with ASME rated pressure relief valve, low water cutoff valve, main and check valves, pressure gauge, and pre-wired Fireye/McDonnell Miller combustion and steam regulator controls. Also includes pre-assembled FM/IRI approved natural gas combustion train w/Fisher or equal diaphragm regulator valve, check trim and vent line fittings or pre-assembled FM/IRI fuel oil combustion train with oil preheater, valved recirculation loop, filter check trim and fuel oil pump. Does not include sheet metal stack or vents, installation permits and approvals, or inspection costs. (BHP = boiler horse power)

Description	Craft@Hrs	Unit	Material $	Labor $	Total $
37,500 Btu (1 BHP)	P1@3.00	Ea	1,960.00	112.00	2,072.00
50,000 Btu (1.5 BHP)	P1@3.00	Ea	2,380.00	112.00	2,492.00
100,000 Btu (2.0 BHP)	P1@4.00	Ea	2,910.00	149.00	3,059.00
150,000 Btu (2.5 BHP)	P1@4.25	Ea	3,670.00	158.00	3,828.00
200,000 Btu (3.0 BHP)	P1@5.00	Ea	3,890.00	186.00	4,076.00
Add for LEED-certified boiler	—	%	15.0	—	—
Add for LEED inspection	—	Ea	—	—	2,060.00
Add for central monitor/ recorder	—	Ea	1,350.00	—	1,350.00
Add for recording thermostat	—	Ea	177.00	—	177.00

Commercial Boiler Connections

Description	Craft@Hrs	Unit	Material $	Labor $	Equipment $	Total $

Commercial hot water boiler connection. Includes pipe, fittings, water feed and pressure relief (drains), pipe insulation, valves, gauges, vents and thermometers to connect boiler to distribution system.

Description	Craft@Hrs	Unit	Material $	Labor $	Equipment $	Total $
2½" supply	SN@18.0	Ea	2,260.00	705.00	11.70	2,976.70
3" supply	SN@20.5	Ea	2,840.00	803.00	13.40	3,656.40
4" supply	SN@26.0	Ea	4,100.00	1,020.00	16.90	5,136.90
6" supply	SN@32.0	Ea	6,030.00	1,250.00	20.90	7,300.90
8" supply	SN@39.5	Ea	8,040.00	1,550.00	25.70	9,615.70
10" supply	SN@48.0	Ea	11,400.00	1,880.00	31.30	13,311.30
12" supply	SN@54.0	Ea	14,300.00	2,120.00	35.20	16,455.20

Commercial steam boiler connection. Includes pipe, fittings, water feed and pressure relief (drains), pipe insulation, valves, gauges, vents and thermometers to connect boiler to distribution system.

Description	Craft@Hrs	Unit	Material $	Labor $	Equipment $	Total $
2½" supply	SN@34.0	Ea	4,340.00	1,330.00	22.20	5,692.20
3" supply	SN@38.0	Ea	4,560.00	1,490.00	24.80	6,074.80
4" supply	SN@43.0	Ea	5,310.00	1,690.00	28.00	7,028.00
6" supply	SN@53.0	Ea	7,280.00	2,080.00	34.50	9,394.50
8" supply	SN@62.0	Ea	9,370.00	2,430.00	40.40	11,840.40
10" supply	SN@70.0	Ea	12,800.00	2,740.00	45.60	15,585.60
12" supply	SN@82.0	Ea	16,000.00	3,210.00	53.40	19,263.40

Additional costs for industrial, commercial and institutional packaged boilers. See more detailed installation costs and boiler trim costs in the sections that follow.

Description	Craft@Hrs	Unit	Material $	Labor $	Equipment $	Total $
Form and pour an interior slab	CF@.156	SF	9.46	5.60	—	15.06
Place vibration pad and anchors	CF@.100	SF	5.75	3.59	—	9.34
Boltdown of boiler	P1@3.00	Ea	—	112.00	112.00	111.00
Fuel train and feedwater lines	P1@.200	LF	11.50	7.44	—	18.94
Mount interior boiler drain	P1@.500	LF	8.94	18.60	—	27.54
Bore pipe hole in concrete wall	P1@.250	Ea	—	9.31	—	9.31
Bore stack vent in concrete wall	P1@.250	Ea	—	9.31	—	9.31
Mount and edge-seal stack	SN@.100	LF	9.67	3.92	—	13.59
Mount circulating pump	P1@.450	Ea	11.50	16.70	—	28.20
Expansion tank or condensate return lines	P1@.250	LF	3.70	9.31	—	13.01
Electrical wiring	BE@.150	LF	2.19	6.15	—	8.34
Combustion controls	BE@.200	LF	4.04	8.21	—	12.25
Install flue gas monitoring system	BE@.200	LF	3.99	8.21	—	12.20

Commercial Boiler Connections

Description	Craft@Hrs	Unit	Material $	Labor $	Equipment $	Total $

Additional costs for steam heating boilers for residential and light commercial applications. See more detailed installation costs and boiler trim costs in the sections that follow.

Description	Craft@Hrs	Unit	Material $	Labor $	Equipment $	Total $
Form and pour a 4' x 4' interior slab	CF@2.50	Ea	150.00	89.70	—	239.70
Place 4' x 4' x 1/2" vibration pads	CF@.750	Ea	43.90	26.90	—	70.80
Boltdown of boiler	P1@1.00	Ea	—	37.20	—	37.20
Fuel train and feedwater lines	P1@2.50	Ea	—	93.10	—	93.10
Mount interior boiler drain	P1@.500	LF	8.83	18.60	—	27.43
Bore piping hole in basement wall	P1@.250	Ea	—	9.31	—	9.31
Bore stack vent in concrete wall	P1@.250	Ea	—	9.31	—	9.31
Mount and edge-seal stack	P1@.500	Ea	45.80	18.60	—	64.40
Mount circulating pump	P1@.450	Ea	—	16.70	—	16.70
Install expansion tank						
2.1 gallon	P1@.300	Ea	40.20	11.20	—	51.40
4.5 gallon	P1@.300	Ea	67.50	11.20	—	78.70

Commercial Boiler Components and Accessories

Costs for Combustion Trains for Boilers. Factory Mutual, Industrial Risk Insurers, and Underwriters' Laboratories certified, U.S. Green Building Council rated, LFUE (Levelized Fuel Utilization Efficiency) rated at 90% or better. Includes the burner, burner head, burner mount flange, associated refractory and gasketing, main fuel valve, draft damper and fan (for a power burner), or burner face plate (for an atmospheric or induced-draft burner). Note: BHP multiplied by 33.5 is the approximate required 1,000 Btu per hour input (MBH input). Cost of fuel line relief vent piping and trip valve to outside of the building are not required in all jurisdictions and are not assumed in these costs.

Description	Craft@Hrs	Unit	Material $	Labor $	Equipment $	Total $

Combustion train for Scotch marine firetube boilers. Costs shown are based on boiler horsepower rating (BHP).

Description	Craft@Hrs	Unit	Material $	Labor $	Equipment $	Total $
20 BHP	SN@8.00	Ea	3,030.00	314.00	5.21	3,349.21
30 BHP	SN@8.00	Ea	4,090.00	314.00	5.21	4,409.21
40 BHP	SN@8.00	Ea	5,620.00	314.00	5.21	5,939.21
50 BHP	SN@8.00	Ea	6,960.00	314.00	5.21	7,279.21
60 BHP	SN@8.00	Ea	7,740.00	314.00	5.21	8,059.21
70 BHP	SN@8.00	Ea	8,470.00	314.00	5.21	8,789.21
80 BHP	SN@8.00	Ea	8,630.00	314.00	5.21	8,949.21
100 BHP	SN@8.00	Ea	9,430.00	314.00	5.21	9,749.21
125 BHP	SN@8.00	Ea	9,790.00	314.00	5.21	10,109.21
150 BHP	SN@8.00	Ea	10,800.00	314.00	5.21	11,119.21
200 BHP	SN@8.00	Ea	11,200.00	314.00	5.21	11,519.21
250 BHP	SN@8.00	Ea	12,500.00	314.00	5.21	12,819.21
300 BHP	SN@8.00	Ea	14,100.00	314.00	135.00	14,549.00
350 BHP	SN@8.00	Ea	14,300.00	314.00	135.00	14,749.00
400 BHP	SN@8.00	Ea	14,600.00	314.00	135.00	15,049.00
500 BHP	SN@16.0	Ea	15,800.00	627.00	270.00	16,697.00
600 BHP	SN@16.0	Ea	17,200.00	627.00	270.00	18,097.00
750 BHP	SN@32.0	Ea	18,200.00	1,250.00	541.00	19,991.00
800 BHP	SN@32.0	Ea	18,900.00	1,250.00	541.00	20,691.00

Combustion train for package firebox boilers. Costs shown are based on boiler horsepower rating (BHP).

Description	Craft@Hrs	Unit	Material $	Labor $	Equipment $	Total $
13.4 BHP	SN@8.00	Ea	1,550.00	314.00	135.00	1,999.00
16.4 BHP	SN@8.00	Ea	2,070.00	314.00	135.00	2,519.00
17.4 BHP	SN@8.00	Ea	2,450.00	314.00	135.00	2,899.00
22.4 BHP	SN@8.00	Ea	2,950.00	314.00	135.00	3,399.00
28.4 BHP	SN@8.00	Ea	3,110.00	314.00	135.00	3,559.00
34.4 BHP	SN@8.00	Ea	3,440.00	314.00	135.00	3,889.00
40.3 BHP	SN@8.00	Ea	3,740.00	314.00	135.00	4,189.00
46.3 BHP	SN@8.00	Ea	4,110.00	314.00	135.00	4,559.00
52.3 BHP	SN@8.00	Ea	4,590.00	314.00	135.00	5,039.00
61.1 BHP	SN@8.00	Ea	4,990.00	314.00	135.00	5,439.00
70.0 BHP	SN@8.00	Ea	5,100.00	314.00	135.00	5,549.00
80.0 BHP	SN@8.00	Ea	5,310.00	314.00	135.00	5,759.00
100.0 BHP	SN@8.00	Ea	5,620.00	314.00	135.00	6,069.00
125.0 BHP	SN@8.00	Ea	6,010.00	314.00	135.00	6,459.00
150.0 BHP	SN@8.00	Ea	6,510.00	314.00	135.00	6,959.00

Commercial Boiler Components and Accessories

Description	Craft@Hrs	Unit	Material $	Labor $	Equipment $	Total $

Natural gas fuel train piping for firetube and watertube boilers. Meets requirements of Underwriters' Laboratories. Shipped pre-assembled. Add the cost of venting and piping to a gas meter. Costs shown are based on boiler horsepower rating (BHP).

Description	Craft@Hrs	Unit	Material $	Labor $	Equipment $	Total $
20 BHP	SN@8.00	Ea	791.00	314.00	135.00	1,240.00
30 BHP	SN@8.00	Ea	1,080.00	314.00	135.00	1,529.00
40 BHP	SN@8.00	Ea	1,660.00	314.00	135.00	2,109.00
50 BHP	SN@8.00	Ea	1,960.00	314.00	135.00	2,409.00
60 BHP	SN@8.00	Ea	2,150.00	314.00	135.00	2,599.00
70 BHP	SN@8.00	Ea	2,280.00	314.00	135.00	2,729.00
80 BHP	SN@8.00	Ea	2,480.00	314.00	135.00	2,929.00
100 BHP	SN@8.00	Ea	2,870.00	314.00	135.00	3,319.00
125 BHP	SN@8.00	Ea	3,420.00	314.00	135.00	3,869.00
150 BHP	SN@8.00	Ea	4,050.00	314.00	135.00	4,499.00
200 BHP	SN@8.00	Ea	4,860.00	314.00	135.00	5,309.00
250 BHP	SN@8.00	Ea	6,010.00	314.00	135.00	6,459.00
300 BHP	SN@8.00	Ea	6,330.00	314.00	135.00	6,779.00
350 BHP	SN@8.00	Ea	6,780.00	314.00	135.00	7,229.00
400 BHP	SN@8.00	Ea	7,290.00	314.00	135.00	7,739.00
500 BHP	SN@16.0	Ea	8,190.00	627.00	270.00	9,087.00
600 BHP	SN@16.0	Ea	9,430.00	627.00	270.00	10,327.00
750 BHP	SN@16.0	Ea	10,200.00	627.00	270.00	11,097.00
800 BHP	SN@16.0	Ea	10,600.00	627.00	270.00	11,497.00

Natural gas fuel train piping for package firebox boilers.
Costs shown are based on boiler horsepower rating (BHP).

Description	Craft@Hrs	Unit	Material $	Labor $	Equipment $	Total $
13 BHP	SN@8.00	Ea	1,050.00	314.00	135.00	1,499.00
16 BHP	SN@8.00	Ea	1,220.00	314.00	135.00	1,669.00
19 BHP	SN@8.00	Ea	1,300.00	314.00	135.00	1,749.00
22 BHP	SN@8.00	Ea	1,430.00	314.00	135.00	1,879.00
28 BHP	SN@8.00	Ea	1,500.00	314.00	135.00	1,949.00
34 BHP	SN@8.00	Ea	1,660.00	314.00	135.00	2,109.00
40 BHP	SN@8.00	Ea	1,960.00	314.00	135.00	2,409.00
45 BHP	SN@8.00	Ea	2,070.00	314.00	135.00	2,519.00
50 BHP	SN@8.00	Ea	2,480.00	314.00	135.00	2,929.00
60 BHP	SN@8.00	Ea	2,820.00	314.00	135.00	3,269.00
70 BHP	SN@8.00	Ea	3,160.00	314.00	135.00	3,609.00
80 BHP	SN@8.00	Ea	3,320.00	314.00	135.00	3,769.00
100 BHP	SN@8.00	Ea	3,970.00	314.00	135.00	4,419.00
125 BHP	SN@8.00	Ea	4,100.00	314.00	135.00	4,549.00
150 BHP	SN@8.00	Ea	4,500.00	314.00	135.00	4,949.00
Add for code approval:						
Factory Mutual	—	%	7.0	—	—	—
Industrial Risk	—	%	10.0	—	—	—

Commercial Boiler Components and Accessories

Oil Fuel Train Piping. Costs shown include recirculating pump, electric oil preheater, regulating valve, check valves, isolating pump stop cocks, oil line pressure relief valve and pressure gauges. Add the cost of concrete pump pad, electrical connections, relief line piping back to the oil tank and tank connecting piping. Add for number 4-6 fuel oil as shown.

Description	Craft@Hrs	Unit	Material $	Labor $	Equipment $	Total $

Oil fuel train piping for boilers. Costs shown are based on boiler horsepower rating (BHP) using number 2 fuel oil.

Description	Craft@Hrs	Unit	Material $	Labor $	Equipment $	Total $
20 to 30 BHP	SN@4.00#	Ea	1,680.00	157.00	67.60	1,904.60
40 to 60 BHP	SN@4.00#	Ea	1,890.00	157.00	67.60	2,114.60
70 to 80 BHP	SN@4.00#	Ea	2,060.00	157.00	67.60	2,284.60
100 BHP	SN@8.00#	Ea	2,640.00	314.00	135.00	3,089.00
125 to 150 BHP	SN@8.00#	Ea	3,420.00	314.00	135.00	3,869.00
200 BHP	SN@8.00#	Ea	4,510.00	314.00	135.00	4,959.00
250 BHP	SN@8.00#	Ea	5,340.00	314.00	135.00	5,789.00
300 BHP	SN@16.0	Ea	5,710.00	627.00	270.00	6,607.00
350 BHP	SN@16.0	Ea	6,380.00	627.00	270.00	7,277.00
400 BHP	SN@16.0	Ea	7,660.00	627.00	270.00	8,557.00
500 BHP	SN@32.0	Ea	7,750.00	1,250.00	541.00	9,541.00
600 BHP	SN@32.0	Ea	8,130.00	1,250.00	541.00	9,921.00
750 to 800 BHP	SN@32.0	Ea	8,880.00	1,250.00	541.00	10,671.00
900 to 1,000 BHP	SN@32.0	Ea	10,400.00	1,250.00	541.00	12,191.00
Add for #4-6 oil	—	%	10.0	—	—	—
Add for tie-in with:						
dual-fuel burner	—	%	5.0	—	—	—

Commercial Boiler Components and Accessories

Dual-fuel boiler burners. For a new installation or retrofit for Scotch marine firetube, firebox or package watertube boilers. Rated at the best available control technology (BACT) for natural gas. Complies with SCAQMD rules 1146.2, 222 and 219 and the Clean Air Act section 307(d). Includes control of CO_2 emissions as well as NOx, SOx, VMs. Cleaver Brooks or equal. Factory Mutual, Industrial Risk Insurers, and Underwriters' Laboratories certified, U.S. Green Building Council rated, LFUE (Levelized Fuel Utilization Efficiency) rated at 90% or better. Includes the burner, burner head, burner mount flange, associated refractory and gasketing, main natural gas and oil fuel valves, draft damper and fan (for a power burner), or burner face plate (for an atmospheric or induced-draft burner) and the natural gas fuel train. Add the cost of an oil fuel train to complete dual-fuel bid. The control panel includes the control circuit, burner mounted control panel, motor starter(s), panel signal lights and full modulation with manual potentiometer, flame safeguard controls, flue gas recirculator air inlet butterfly valve and relay actuator, combustion air proving switch and high fire air interlock switch. Note: BHP multiplied by 33.5 is the approximate required 1,000 Btu per hour input (MBH input). Add the installation costs below and the cost of natural gas fuel line relief vent piping and a trip valve to outside of the building if required.

Description	Craft@Hrs	Unit	Material $	Labor $	Equipment $	Total $

Dual-fuel boiler burners. Costs shown are based on boiler horsepower rating (BHP).

Description	Craft@Hrs	Unit	Material $	Labor $	Equipment $	Total $
20 BHP	SN@8.00	Ea	5,390.00	314.00	—	5,704.00
30 BHP	SN@8.00	Ea	6,260.00	314.00	—	6,574.00
40 BHP	SN@8.00	Ea	7,730.00	314.00	—	8,044.00
50 BHP	SN@8.00	Ea	8,830.00	314.00	—	9,144.00
60 BHP	SN@8.00	Ea	10,000.00	314.00	—	10,314.00
70 BHP	SN@8.00	Ea	11,000.00	314.00	—	11,314.00
80 BHP	SN@8.00	Ea	12,600.00	314.00	—	12,914.00
100 BHP	SN@8.00	Ea	14,800.00	314.00	—	15,114.00
125 BHP	SN@8.00	Ea	16,700.00	314.00	—	17,014.00
150 BHP	SN@8.00	Ea	17,500.00	314.00	—	17,814.00
200 BHP	SN@8.00	Ea	22,200.00	314.00	—	22,514.00
250 BHP	SN@8.00	Ea	28,400.00	314.00	—	28,714.00
300 BHP	SN@8.00	Ea	33,600.00	314.00	—	33,914.00
350 BHP	SN@8.00	Ea	37,300.00	314.00	—	37,614.00
400 BHP	SN@8.00	Ea	39,600.00	314.00	—	39,914.00
500 BHP	SN@16.0	Ea	40,600.00	627.00	—	41,227.00
600 BHP	SN@16.0	Ea	45,600.00	627.00	—	46,227.00
750 BHP	SN@32.0	Ea	49,000.00	1,250.00	—	50,250.00
800 BHP	SN@32.0	Ea	49,700.00	1,250.00	—	50,950.00

Commercial Boiler Connections

Description	Craft@Hrs	Unit	Material $	Labor $	Equipment $	Total $
Additional costs for installing dual-fuel boiler burners						
Form and pour 4' x 4' x 2" interior slab	CF@2.50	Ea	150.00	89.70	—	239.70
Bore fuel line hole through wall or pad	P1@.250	Ea	—	9.31	—	9.31
Bore flue gas hole through wall or pad	P1@.250	Ea	—	9.31	—	9.31
Place burner and control panel enclosure	ER@8.00	Ea	—	320.00	—	320.00
Mount, level and grout burner, duct and control panel	P1@.450	Ea	473.00	16.70	—	489.70
Piping for fuel lines and flue gas recirculator	P1@.500	LF	8.94	18.60	—	27.54
Electrical wiring	BE@.300	LF	3.89	12.30	—	16.19
Wire control panel	BE@.300	LF	1.81	12.30	—	14.11
Ground fault protection wiring	BE@.300	LF	5.70	12.30	—	18.00
Commission and test	BE@8.00	Ea	—	328.00	—	328.00

Costs for Electrical Service for Boilers, Subcontract. Costs shown include typical main power tie-in and fusing, circuit breaker panel, and main bus. Assumes controls, fan, blower switchgear and control console are pre-wired electrical service. These costs include the electrical subcontractor's overhead and profit.

Description	Craft@Hrs	Unit	Material $	Labor $	Total $
Electrical service for firetube and watertube boilers. Costs shown are based on boiler horsepower rating (BHP).					
20 to 100 BHP	—	Ea	—	—	1,480.00
125 to 200 BHP	—	Ea	—	—	1,790.00
250 to 350 BHP	—	Ea	—	—	2,030.00
400 to 500 BHP	—	Ea	—	—	2,330.00
500 to 600 BHP	—	Ea	—	—	3,360.00
750 to 800 BHP	—	Ea	—	—	6,060.00

Electrical service for package firebox boilers. Costs shown are based on boiler horsepower rating (BHP).

Description	Craft@Hrs	Unit	Material $	Labor $	Total $
13 to 70 BHP	—	Ea	—	—	1,440.00
80 to 125 BHP	—	Ea	—	—	1,470.00
150 BHP	—	Ea	—	—	1,810.00

Commercial Boiler Components and Accessories

Costs for Refractory for Boilers, Subcontract. For package Scotch marine firetube, firebox and watertube boilers. Most new boilers are shipped from the factory with the refractory brick installed. If refractory brick is installed on the job site to repair shipping damage or to create firedoor seals, use the figures below. These costs include the refractory subcontractor's overhead and profit. Costs shown are per 100 pounds of refractory used.

Description	Craft@Hrs	Unit	Material $	Labor $	Equipment $	Total $
Repairs and seals	—	100 lb	—	—	—	248.00

Costs for Boiler Feedwater Pumps. Costs shown are for the pump only. Add the cost of piping, pump actuating valves and concrete pump base.

Feedwater pumps for firetube and watertube boilers.
Costs shown are based on boiler horsepower rating (BHP).

20 to 40 BHP	SN@8.00	Ea	11,200.00	314.00	135.00	11,649.00
50 to 80 BHP	SN@8.00	Ea	13,500.00	314.00	135.00	13,949.00
100 BHP	SN@8.00	Ea	15,000.00	314.00	135.00	15,449.00
125 to 200 BHP	SN@16.0	Ea	15,400.00	627.00	270.00	16,297.00
250 to 400 BHP	SN@32.0	Ea	17,200.00	1,250.00	541.00	18,991.00
500 BHP	SN@32.0	Ea	18,200.00	1,250.00	541.00	19,991.00
600 BHP	SN@32.0	Ea	18,700.00	1,250.00	541.00	20,491.00
750 BHP	SN@32.0	Ea	19,200.00	1,250.00	541.00	20,991.00
800 BHP	SN@32.0	Ea	26,000.00	1,250.00	541.00	27,791.00

Feedwater pumps for package firebox boilers.
Costs shown are based on boiler horsepower rating (BHP).

13 to 35 BHP	SN@8.00	Ea	6,030.00	314.00	135.00	6,479.00
40 to 45 BHP	SN@8.00	Ea	7,090.00	314.00	135.00	7,539.00
50 to 60 BHP	SN@8.00	Ea	8,880.00	314.00	135.00	9,329.00
70 to 80 BHP	SN@8.00	Ea	10,500.00	314.00	135.00	10,949.00
100 BHP	SN@8.00	Ea	13,500.00	314.00	135.00	13,949.00
125 BHP	SN@16.0	Ea	13,500.00	627.00	270.00	14,397.00
150 BHP	SN@16.0	Ea	15,000.00	627.00	270.00	15,897.00

Costs for Natural Gas Combustion Controls for Boilers. Instrumentation Society of America, Underwriters' Laboratories, Factory Mutual, and Industrial Risk Insurers certified. Fireye or equal. Includes automatic recycling or continuous pilot, spark igniter and coil (for electric ignition), timer to control the purge cycle, pilot, main burner (hi-lo-off) or main burner (proportional control), flame safeguard and sensor and combustion control panel.

Natural gas combustion controls for firetube and watertube boilers.
Costs shown are based on boiler horsepower rating (BHP).

20 to 80 BHP	SN@4.00	Ea	4,590.00	157.00	67.60	4,814.60
100 BHP	SN@4.00	Ea	5,100.00	157.00	67.60	5,324.60
125 to 200 BHP	SN@8.00	Ea	5,100.00	314.00	135.00	5,549.00
250 BHP	SN@16.0	Ea	5,100.00	627.00	270.00	5,997.00
300 BHP	SN@16.0	Ea	6,120.00	627.00	270.00	7,017.00
350 or 400 BHP	SN@16.0	Ea	7,170.00	627.00	270.00	8,067.00
500 BHP	SN@32.0	Ea	7,170.00	1,250.00	541.00	8,961.00
600 BHP	SN@32.0	Ea	8,560.00	1,250.00	541.00	10,351.00
750 BHP	SN@64.0	Ea	9,980.00	2,510.00	1,080.00	13,570.00
800 BHP	SN@64.0	Ea	11,500.00	2,510.00	1,080.00	15,090.00

Commercial Boiler Components and Accessories

Description	Craft@Hrs	Unit	Material $	Labor $	Equipment $	Total $

Natural gas combustion controls for firebox boilers. Costs shown are based on boiler horsepower rating (BHP).

13.4-80 BHP	SN@4.00	Ea	4,410.00	157.00	67.60	4,634.60
100.0 BHP	SN@4.00	Ea	5,100.00	157.00	67.60	5,324.60
126.9 BHP	SN@8.00	Ea	5,100.00	314.00	135.00	5,549.00
150.0 BHP	SN@8.00	Ea	9,510.00	314.00	135.00	9,959.00
Add if burner controls are for:						
#2 oil-fired	—	%	—	—	—	10.0
Heavy oil-fired	—	%	—	—	—	20.0
Dual fuel burners	—	%	—	—	—	70.0

Costs for Water Softening Systems for Use with Boilers. Costs shown include the metering valve, tank, pump, and zeolite media. Add the cost of piping to connect the water softening system to the main feedwater tank and pump.

Water softening system for firetube and watertube boilers. Costs shown are based on boiler horsepower rating (BHP).

20 to 80 BHP	SN@4.00	Ea	7,350.00	157.00	67.60	7,574.60
100 BHP	SN@4.00	Ea	8,620.00	157.00	67.60	8,844.60
125 BHP	SN@8.00	Ea	1,000.00	314.00	135.00	1,449.00
150 or 200 BHP	SN@8.00	Ea	12,800.00	314.00	135.00	13,249.00
250 BHP	SN@8.00	Ea	15,000.00	314.00	135.00	15,449.00
300 or 350 BHP	SN@16.0	Ea	19,200.00	627.00	270.00	20,097.00
400 or 500 BHP	SN@16.0	Ea	24,100.00	627.00	270.00	24,997.00
600 BHP	SN@32.0	Ea	29,100.00	1,250.00	541.00	30,891.00
750 or 800 BHP	SN@64.0	Ea	33,900.00	2,510.00	1,080.00	37,490.00

Water softening system for package firebox boilers. Costs shown are based on boiler horsepower rating (BHP).

13.4 to 80 BHP	SN@4.00	Ea	8,640.00	157.00	67.60	8,864.60
100.0 BHP	SN@4.00	Ea	8,640.00	157.00	67.60	8,864.60
126.9 BHP	SN@8.00	Ea	10,200.00	314.00	135.00	10,649.00
150.0 BHP	SN@8.00	Ea	11,700.00	314.00	135.00	12,149.00

Costs for Start-up, Shakedown, and Calibration of Boilers. Costs shown include minor changes and adjustments to meet requirements of the boiler inspector, adjustment of combustion efficiency at the burner by a control technician, pump and metering calibration testing and steam or hot water metering on start-up.

Start-up, shakedown & calibration of firetube and watertube boilers. Costs shown are based on boiler horsepower rating (BHP.) If a factory technician performs the start-up, add this cost at a minimum of $800 (8 hours at $100.00 per hour).

13.4 to 100 BHP	S2@4.00	Ea	—	146.00	—	146.00
125 to 200 BHP	S2@8.00	Ea	—	292.00	—	292.00
250 to 400 BHP	S2@16.0	Ea	—	585.00	—	585.00
500 to 600 BHP	S2@24.0	Ea	—	877.00	—	877.00
750 to 800 BHP	S2@30.0	Ea	—	1,100.00	—	1,100.00

Commercial Boiler Components and Accessories

Costs for Stack Economizer Units for Boilers. Costs shown include piping connection to existing boiler, economizer fin tube coils, feedwater pump, check valves, thermostatic regulating valve, stack dampers, pressure relief valve, stack mount flanges and insulating enclosure, but no architectural modifications.

Description	Craft@Hrs	Unit	Material $	Labor $	Equipment $	Total $

Stack waste heat recovery system for boilers. Costs shown are based on boiler horsepower rating (BHP).

Description	Craft@Hrs	Unit	Material $	Labor $	Equipment $	Total $
500 BHP	SN@40.0	Ea	43,000.00	1,570.00	676.00	45,246.00
750 BHP	SN@48.0	Ea	53,500.00	1,880.00	811.00	56,191.00
800 BHP	SN@48.0	Ea	57,700.00	1,880.00	811.00	60,391.00
1,000 BHP	SN@56.0	Ea	72,000.00	2,190.00	947.00	75,137.00

Pollution control modules for gas-fired industrial package watertube boilers. N0x, S0x, VM (volatile materials) and particulates emissions reduction module. Free-standing unit. Verantis, Croll-Reynolds or equal. Designed to meet or exceed 2010 California Code, Tier IV, EPA and European ISO standards. Either vertical counter-flow or packed-bed type. For multi-dwelling, process and industrial applications. By boiler rating in millions of Btu's per hour (MMBtu/hr). See installation and efficiency optimizing control costs below. Oil-fired boilers will add about 15% to the installed cost, including a second stage reduction module, second stage pump, sulfur extraction, venturi particulate separator and sludge tank. To calculate the annual lease rate for a complete installation, divide the total installed cost by five. Then add 2% per year to cover vendor-supplied maintenance.

Description	Craft@Hrs	Unit	Material $	Labor $	Equipment $	Total $
1.0 MMBtu/Hr	—	Ea	39,900.00	—	—	39,900.00
1.7 MMBtu/Hr	—	Ea	71,000.00	—	—	71,000.00
3.4 MMBtu/Hr	—	Ea	126,000.00	—	—	126,000.00
4.1 MMBtu/Hr	—	Ea	131,000.00	—	—	131,000.00
5.0 MMBtu/Hr	—	Ea	142,000.00	—	—	142,000.00
6.7 MMBtu/Hr	—	Ea	152,000.00	—	—	152,000.00
13.4 MMBtu/Hr	—	Ea	158,000.00	—	—	158,000.00
20.1 MMBtu/Hr	—	Ea	176,000.00	—	—	176,000.00
26.8 MMBtu/Hr	—	Ea	186,000.00	—	—	186,000.00
33.5 MMBtu/Hr	—	Ea	203,000.00	—	—	203,000.00

Installation costs for pollution control modules for gas-fired industrial package watertube boilers.

Description	Craft@Hrs	Unit	Material $	Labor $	Equipment $	Total $
Form and pour foundation	ER@.040	SF	2.42	1.60	—	4.02
Position main housing, fans and pumps	ER@4.00	Ea	—	160.00	—	160.00
Tie-down and leveling	ER@4.00	Ea	581.00	160.00	—	741.00
Mount piping system frame supports	ER@6.00	Ea	1.76	240.00	—	241.76
Rig and mount plenum and ducts	SN@5.00	Ea	—	196.00	—	196.00
Rig and mount tank for emission reduction, fluid and housing inlet metering valve	ER@4.00	Ea	—	160.00	—	160.00

Commercial Boiler Components and Accessories

Description	Craft@Hrs	Unit	Material $	Labor $	Equipment $	Total $

Installation costs for pollution control modules for gas-fired industrial package watertube boilers – (continued).

Description	Craft@Hrs	Unit	Material $	Labor $	Equipment $	Total $
Weld systems piping	ER@8.00	Ea	2.45	320.00	—	322.45
Wire electrical system	BE@8.00	Ea	2.09	328.00	—	330.09
Wire controls system	BE@16.0	Ea	2.21	656.00	—	658.21
Level and balance rotating equipment	P1@8.00	Ea	234.00	298.00	—	532.00
Calibration and test	ER@2.00	Ea	—	79.90	—	79.90
Witnessed run trials (add the permit fee)	P1@4.00	Ea	—	149.00	—	149.00

Efficiency-optimizing centralized boiler burner controls. Retrofit or new construction. Preferred Instruments or equal. For use with emissions monitoring and digital reporting pollution control equipment installed for a gas-fired industrial boiler. Complies with the Clean Air Act section 307(d), California and Massachusetts boiler codes and NFPA 85. Includes a boiler-mounted control panel with chassis, servo motor controls for burner fuel line valves, LCD touchpad control panel, ultra-violet, infra-red or self-checking flame scanner, operator interface panel and interface to a remote PC (not included). Controls variable speed drives for burner forced-draft fans, induced draft fans and feedwater pumps.

Description	Craft@Hrs	Unit	Material $	Labor $	Equipment $	Total $
Mount automation module, less wiring	BE@.250	Ea	1,290.00	10.30	—	1,300.30
Mount flame safety sensors less wiring	BE@.250	Ea	205.00	10.30	—	215.30
Servomotor mounting less wiring	BE@.250	Ea	546.00	10.30	—	556.30
Mount operator interface terminal with power supply less wiring	BE@.250	Ea	93.40	10.30	—	103.70
Operator interface terminal bridge, less wiring	BE@.250	Ea	4,440.00	10.30	—	4,450.30
Bore control wiring hole through basement wall	CF@.250	Ea	—	8.97	—	8.97
Twisted pair Cat 6, 1 GBPS wire 4 pair	BE@.008	LF	.56	.33	—	.89
Panel wiring to emissions monitoring array	BE@.250	LF	2.94	10.30	—	13.24
Sensor wiring to controller	BE@.300	LF	2.68	12.30	—	14.98
Electric servomotor wiring to controller	BE@.400	LF	2.45	16.40	—	18.85
Mount and wire PC module	BE@2.00	LF	23.50	82.10	—	105.60
Ground fault protection wiring	BE@1.25	LF	146.00	51.30	—	197.30
Startup, programming and system test	BE@3.00	Ea	—	123.00	—	123.00
Commissioning and final test	BE@8.00	Ea	—	328.00	—	328.00

Commercial Boiler Components and Accessories

Costs for Condensate Receiver and Pumping Units for Boilers. Costs shown include a stainless steel factory assembled tank and pump with motor, float switch, condensate inlet, feedwater tank, connection, level gauge, thermometer, solenoid valve for direct boiler feed. No concrete mounting pad required.

Description	Craft@Hrs	Unit	Material $	Labor $	Equipment $	Total $

Condensate receiver and pumping unit for boilers. Unit sizing and costs are based on storage for one minute at the designed steam return rate. Set in place only. Make additional allowances for pipe and electrical connections.

Description	Craft@Hrs	Unit	Material $	Labor $	Equipment $	Total $
750 lbs per hour	SN@4.00	Ea	5,600.00	157.00	2.61	5,759.61
1,500 lbs per hour	SN@4.00	Ea	5,600.00	157.00	2.61	5,759.61
3,000 lbs per hour	SN@4.00	Ea	7,120.00	157.00	2.61	7,279.61
5,000 lbs per hour	SN@6.00	Ea	7,850.00	235.00	3.91	8,088.91
6,250 lbs per hour	SN@6.00	Ea	8,750.00	235.00	3.91	8,988.91

Costs for Deaerator/Condenser Units for Boilers. Pressurized jet spray type. Costs shown include an ASME storage receiver, stand, jet spray deaerator head, makeup control valve, makeup controller, pressure control valve, level controls, pressure and temperature controls, pumped condensate inlet, pumps and motor, steam inlet and system accessories. Add for a concrete mounting pad and piping to the condensate return line (inlet) and piping to the boiler feedwater system (outlet).

Deaerator/condenser unit for boilers. Costs shown are based on boiler horsepower rating (BHP).

Description	Craft@Hrs	Unit	Material $	Labor $	Equipment $	Total $
100 BHP	SN@8.00	Ea	58,600.00	314.00	5.21	58,919.21
200 BHP	SN@12.0	Ea	60,900.00	470.00	7.82	61,377.82
300 BHP	SN@14.0	Ea	64,600.00	549.00	9.12	65,158.12
400 BHP	SN@16.0	Ea	73,200.00	627.00	10.40	73,837.40
600 BHP	SN@24.0	Ea	78,500.00	941.00	15.60	79,456.60
800 BHP	SN@36.0	Ea	85,800.00	1,410.00	23.50	87,233.50
1,000 BHP	SN@48.0	Ea	100,000.00	1,880.00	31.30	101,911.30

Rolairtrol type air separators. Priced by capacity in gallons per minute (GPM) and by pipe connection size. Includes hanging separator but no connection, (2" is a screwed connection, 2½" and up are flanged connections) using small tools.

Description	Craft@Hrs	Unit	Material $	Labor $	Equipment $	Total $
56 GPM, 2"	S2@2.50	Ea	1,050.00	91.40	1.63	1,143.03
90 GPM, 2½"	S2@2.50	Ea	1,150.00	91.40	1.63	1,243.03
170 GPM, 3"	S2@3.00	Ea	1,500.00	110.00	1.96	1,611.96
300 GPM, 4"	S2@3.50	Ea	2,260.00	128.00	2.28	2,390.28
500 GPM, 6"	S2@4.75	Ea	2,950.00	174.00	3.10	3,127.10
700 GPM, 6"	S2@4.75	Ea	3,810.00	174.00	3.10	3,987.10
1,300 GPM, 8"	SN@6.00	Ea	5,120.00	235.00	3.91	5,358.91
2,000 GPM, 10"	SN@6.00	Ea	7,160.00	235.00	3.91	7,398.91
2,750 GPM, 12"	SN@12.0	Ea	9,290.00	470.00	7.82	9,767.82

Commercial Boiler Components and Accessories

Description	Craft@Hrs	Unit	Material $	Labor $	Equipment $	Total $

Costs for Chemical Feed Duplex Pump/ Packaged Units for Boilers. Costs shown include the chemical tanks, electronic metering, valve and stand. Add the cost of piping to the feedwater tank.

Chemical feed duplex pump for boilers. Costs shown are based on boiler horsepower rating (BHP).

Description	Craft@Hrs	Unit	Material $	Labor $	Equipment $	Total $
20 to 100 BHP	SN@3.00	Ea	2,150.00	118.00	50.70	2,318.70
25 to 200 BHP	SN@3.00	Ea	2,820.00	118.00	50.70	2,988.70
200 to 400 BHP	SN@6.00	Ea	2,820.00	235.00	101.00	3,156.00
500 to 600 BHP	SN@6.00	Ea	6,400.00	235.00	101.00	6,736.00
700 to 800 BHP#	SN@8.00	Ea	7,150.00	314.00	135.00	7,599.00
1,000 BHP	SN@10.0	Ea	9,670.00	392.00	169.00	10,231.00

Chemical feed simple in-line system for hot water heating system.
Costs shown include 3 gallon pot feeder, two 3/4" isolation valves, 10' of ¾" pipe and two ¾" threadolets. Chemicals are drawn in through suction created by hot water system circulation pumps.

Description	Craft@Hrs	Unit	Material $	Labor $	Equipment $	Total $
50 to 20 BHP	SN@3.00	Ea	499.00	118.00	50.70	667.70

Packaged Boiler Feedwater Systems for Boilers. Costs shown include duplex pumps with heater assembly, thermal lining, pressure gauges, makeup feeder valve, ASME feedwater tank and stand, single phase electric motor and level gauge. Add the costs for a concrete mounting pad and piping to the boilers.

Packaged boiler feedwater systems. Costs shown are based on boiler horsepower rating (BHP).

Description	Craft@Hrs	Unit	Material $	Labor $	Equipment $	Total $
15 to 50 BHP	SN@5.00	Ea	15,000.00	196.00	84.50	15,280.50
60 to 80 BHP	SN@5.00	Ea	16,600.00	196.00	84.50	16,880.50
100 BHP	SN@6.00	Ea	17,000.00	235.00	101.00	17,336.00
125 to 150 BHP	SN@6.00	Ea	18,200.00	235.00	101.00	18,536.00
200 BHP	SN@8.00	Ea	18,500.00	314.00	135.00	18,949.00
250 BHP	SN@8.00	Ea	19,300.00	314.00	135.00	19,749.00
300 BHP	SN@8.00	Ea	20,600.00	314.00	135.00	21,049.00
400 BHP	SN@8.00	Ea	21,200.00	314.00	135.00	21,649.00
500 BHP	SN@8.00	Ea	25,800.00	314.00	135.00	26,249.00
600 BHP	SN@12.0	Ea	26,900.00	470.00	203.00	27,573.00
750 BHP	SN@12.0	Ea	26,800.00	470.00	203.00	27,473.00

Continuous Blowdown Heat Recovery System. Includes a flash receiver (50 PSI ASME rated tank), plate and frame heat exchanger, pneumatic level controller and valve, air filter regulator, water gauge, pressure gauge, backflush piping, pressure relief valve connected to mud drum and drain. Add for concrete pad mount, piping and electrical wiring. Add the cost of automatic controls from the following section.

Continuous blowdown heat recovery systems. Costs shown by pounds per hour (lbs per hr) of rated makeup capacity.

Description	Craft@Hrs	Unit	Material $	Labor $	Equipment $	Total $
5,000 lbs per hr	SN@8.00	Ea	50,400.00	314.00	135.00	50,849.00
10,000 lbs per hr	SN@8.00	Ea	99,800.00	314.00	135.00	100,249.00
25,000 lbs per hr	SN@8.00	Ea	127,000.00	314.00	135.00	127,449.00
50,000 lbs per hr	SN@8.00	Ea	202,000.00	314.00	135.00	202,449.00

Commercial Boiler Components and Accessories

Boiler blowdown waste heat recovery system controls. Typical costs in rated pounds per hour for high-efficiency commercial or industrial controls for a manually-operated boiler blowdown waste heat recovery system. Wilson Equipment, Spirax Sarco, or equal. Meets EPA efficiency specifications and liquid waste control requirements. For use with package industrial watertube, firebox and firetube boilers up to 250 PSIG and up to 800 boiler horsepower. Includes timer, total dissolved solids sensor, digital control minipanel, valve, valve actuator and T-type security key for manual actuation. For new construction, add the appropriately sized blowdown waste heat recovery system shown above.

Description	Craft@Hrs	Unit	Material $	Labor $	Equipment $	Total $
5,000 lbs per hr	SN@8.00	Ea	12,900.00	314.00	—	13,214.00
10,000 lbs per hr	SN@8.00	Ea	17,700.00	314.00	—	18,014.00
25,000 lbs per hr	SN@8.00	Ea	30,900.00	314.00	—	31,214.00
50,000 lbs per hr	SN@8.00	Ea	37,400.00	314.00	—	37,714.00
Add for piping for blowdown valve	P1@.500	LF	9.05	18.60	—	27.65
Add for wiring the valve actuator	BE@.300	LF	3.71	12.30	—	16.01
Add for control panel wiring	BE@.300	LF	1.77	12.30	—	14.07
Add for ground fault protection wiring	BE@.300	LF	5.83	12.30	—	18.13
Add for commissioning and tests	BE@8.00	Ea	—	328.00	—	328.00

Boiler Stacks. Costs shown assume a 20-foot stack height and include rigging, breeching, connection of stack sections, guy wires where required, vendor-supplied stack supports and insulation. Add the cost of a concrete base.

Boiler stack. Costs shown assume a 20-foot stack height.

10" diameter	SN@12.0	Ea	3,590.00	470.00	203.00	4,263.00
12" diameter	SN@12.0	Ea	4,480.00	470.00	203.00	5,153.00
16" diameter	SN@16.0	Ea	5,350.00	627.00	270.00	6,247.00
20" diameter	SN@20.0#	Ea	6,720.00	784.00	338.00	7,842.00
24" diameter	SN@24.0	Ea	8,880.00	941.00	406.00	10,227.00
30" diameter	SN@32.0	Ea	10,500.00	1,250.00	541.00	12,291.00
56" diameter	SN@40.0	Ea	20,800.00	1,570.00	676.00	23,046.00
Add for each 10' section (or portion thereof) beyond 20' high per 10' sect.	—#	%	—	—	—	15.0

Costs for Tankless Indirect Water Heater. Provides domestic hot water from boiler coil insert. Costs shown include heat exchanger, coil, circulating pump, thermostatically controlled valve, bypass piping and air bleeder valve. Add the cost of an equipment base, electrical wiring, piping and an insulated hot water storage tank.

Tankless indirect water heater. Costs by gallons per hour (GPH).

240 GPH	SN@8.00	Ea	20,900.00	314.00	135.00	21,349.00
360 GPH	SN@8.00	Ea	26,200.00	314.00	135.00	26,649.00
480 GPH	SN@10.0	Ea	26,900.00	392.00	169.00	27,461.00
840 GPH	SN@12.0	Ea	61,000.00	470.00	203.00	61,673.00
920 GPH	SN@16.0	Ea	68,000.00	627.00	270.00	68,897.00
1,240 GPH	SN@20.0	Ea	76,200.00	784.00	338.00	77,322.00

Commercial Boiler Components and Accessories

Costs for Boiler Trim. Costs shown include the pressure relief valve, level gauge (for steam systems), level controller & low water cutoff, pressure and temperature gauge, feedwater valve & actuator, main valve, blowdown valve, cleanout manholes and gaskets.

Description	Craft@Hrs	Unit	Material $	Labor $	Equipment $	Total $
Boiler trim. Costs shown are per boiler horsepower rating (BHP).						
20 to 100 BHP	SN@4.00	Ea	3,340.00	157.00	2.61	3,499.61
100 to 200 BHP	SN@6.00	Ea	3,760.00	235.00	3.91	3,998.91
200 to 400 BHP	SN@8.00	Ea	4,630.00	314.00	5.21	4,949.21
400 to 600 BHP	SN@8.00	Ea	7,080.00	314.00	5.21	7,399.21
600 to 800 BHP	SN@8.00	Ea	8,350.00	314.00	5.21	8,669.21
800 to 1,000 BHP	SN@10.0	Ea	8,360.00	392.00	6.52	8,758.52

Heat reclaimer condensate return & feedwater system, stainless steel non-siphon, including flash tank with level control and vacuum breaker, overflow relief drain, simplex centrifugal pump, pre-assembled piping to feedwater tank, structural steel stand and frame, check trim and pre-wired controls. Also provides potable hot water as part of the condensed steam return cycle. Capacity rating is in GPH (gallons per hour of condensate)

Description	Craft@Hrs	Unit	Material $	Labor $	Total $
33 gal. ½ HP pump	P1@3.00	Ea	992.00	112.00	1,104.00
46 gal. ¾ HP pump	P1@4.00	Ea	1,220.00	149.00	1,369.00

Vacuum breakers, rough brass, threaded, to be mounted on finned radiation tube inlet valve fitting

½" vacuum breaker	P1@.250	Ea	46.30	9.31	55.61

Steam traps, float valve Sarco type or equal w/drain fitting, brazed

¾" with check valve	P1@.250	Ea	44.10	9.31	53.41
1.0" with check valve	P1@.450	Ea	70.60	16.70	87.30

Baseboard fin tube radiation, per linear foot, 15 PSIG test thinwall copper tube with aluminum fins, braze fitted wall mounted

¾" element with 8" cover	P1@.280	LF	11.20	10.40	21.60
1" element with 8" cover	P1@.350	LF	18.30	13.00	31.30
Add for corners, fillers and caps, average per foot	—	%	—	15.0	—

Solenoid valves, zone thermostatically-controlled steam electric, 15 PSIG rating, braze fitting. Thermostatic Honeywell-type or equal solenoid relay valves

¾", brazed	P1@.250	Ea	14.60	9.31	23.91
1", brazed	P1@.250	Ea	14.60	9.31	23.91

Commercial Boiler Components and Accessories

Description	Craft@Hrs	Unit	Material $	Labor $	Equipment $	Total $

Additional costs for boiler pumps, condenser units, feedwater systems, heat recovery and oil fuel train piping.

Description	Craft@Hrs	Unit	Material $	Labor $	Equipment $	Total $
Form and pour concrete pad	CF@1.56	SF	9.36	56.00	—	65.36
Place vibration pads	CF@.750	Ea	44.90	26.90	—	71.80
Mount interior drain	P1@.500	LF	8.83	18.60	—	27.43
Bore hole through wall or pad	P1@.250	Ea	—	9.31	—	9.31
Piping to load (typical)	P1@.500	LF	8.87	18.60	—	27.47
Bore burner stack vent in exterior wall	P1@.250	Ea	—	9.31	—	9.31
Mounting and edge-sealing stack	P1@.500	Ea	45.40	18.60	—	64.00

Emissions sensing and recording equipment. For stack gas combustion monitoring of boilers. Monitors and records NOx, SOx, CO, H²S, volatile materials and particulates emissions to meet 2011 EPA greenhouse gas regulations. CO_2 monitoring and recording is required on boilers producing 25,000 tons or more a year of carbon emissions. Signal Group Limited or equal. Includes a stack-mounted pulse type infra-red single-beam light source with solid-state detector. Sensitivity range is variable between .1 ppm and 1,000 ppm. Includes a hot-wire radiation source, rotating shutter with reference target for continuous unmanned recalibration, 240 x 64 pixel LCD display with remote panel-mounted solid-state digital analyzer and lockout type inspector-only access recording circuitry. Cost includes calibration and testing by a factory technician. Add the cost of the permit, inspection, test and documentation.

Description	Craft@Hrs	Unit	Material $	Labor $	Equipment $	Total $
Analyzer and recorder installation to remote panel	BE@3.00	Ea	29,700.00	123.00	—	29,823.00
Infrared sensor unit installation to stack	BE@2.00	Ea	9,310.00	82.10	—	9,392.10
Wiring of sensor to analyzer and recorder	BE@8.00	Ea	581.00	328.00	—	909.00
Circuit testing and grounding	BE@2.00	Ea	—	82.10	—	82.10

Centrifugal Pumps and Pump Connections

Description	Craft@Hrs	Unit	Material $	Labor $	Equipment $	Total $

Base-mounted, bronze fitted centrifugal pump, set in place only. Make additional allowances for pipe connection assembly.

Description	Craft@Hrs	Unit	Material $	Labor $	Equipment $	Total $
½ HP	P1@1.50	Ea	1,430.00	55.80	—	1,485.80
¾ HP	P1@1.75	Ea	1,820.00	65.10	—	1,885.10
1 HP	P1@2.00	Ea	2,230.00	74.40	—	2,304.40
1½ HP	P1@2.40	Ea	2,700.00	89.30	—	2,789.30
2 HP	P1@2.70	Ea	3,320.00	100.00	—	3,420.00
3 HP	P1@3.00	Ea	3,960.00	112.00	—	4,072.00
5 HP	P1@3.50	Ea	4,520.00	130.00	—	4,650.00
7½ HP	P1@4.00	Ea	5,010.00	149.00	—	5,159.00
10 HP	P1@4.50	Ea	7,310.00	167.00	—	7,477.00
15 HP	ER@5.00	Ea	8,460.00	200.00	3.26	8,663.26
20 HP	ER@6.00	Ea	9,780.00	240.00	3.91	10,023.91
25 HP	ER@7.00	Ea	10,200.00	280.00	4.56	10,484.56
30 HP	ER@7.40	Ea	12,100.00	296.00	4.82	12,400.82
40 HP	ER@7.70	Ea	12,000.00	308.00	5.02	12,313.02
50 HP	ER@8.00	Ea	14,600.00	320.00	5.21	14,925.21

Base-mounted centrifugal pump connection assembly. Includes pipe, fittings, pipe insulation, valves, strainers, suction guides and gauges.

Description	Craft@Hrs	Unit	Material $	Labor $	Equipment $	Total $
2½" supply	P1@12.0	Ea	2,460.00	447.00	—	2,907.00
3" supply	P1@13.5	Ea	2,850.00	502.00	—	3,352.00
4" supply	P1@16.0	Ea	3,910.00	596.00	—	4,506.00
6" supply	P1@22.0	Ea	6,470.00	819.00	—	7,289.00
8" supply	P1@28.0	Ea	9,380.00	1,040.00	—	10,420.00
10" supply	P1@35.0	Ea	15,500.00	1,300.00	—	16,800.00
12" supply	P1@42.0	Ea	20,600.00	1,560.00	—	22,160.00

Cast iron line-mounted centrifugal pump, including connection and supports.

Description	Craft@Hrs	Unit	Material $	Labor $	Equipment $	Total $
1/6 HP	P1@1.25	Ea	370.00	46.50	—	416.50
1/4 HP	P1@1.40	Ea	473.00	52.10	—	525.10
1/3 HP	P1@1.50	Ea	584.00	55.80	—	639.80
1/2 HP	P1@1.70	Ea	747.00	63.30	—	810.30
3/4 HP	P1@1.80	Ea	1,010.00	67.00	—	1,077.00
1 HP	P1@2.00	Ea	1,200.00	74.40	—	1,274.40

Heat Exchangers and Connections

Description	Craft@Hrs	Unit	Material $	Labor $	Equipment $	Total $

2-pass shell and tube type heat exchanger, set in place only. Make additional allowances for pipe connection assembly.

Description	Craft@Hrs	Unit	Material $	Labor $	Equipment $	Total $
4" dia. x 30"	SN@2.00	Ea	1,770.00	78.40	1.30	1,849.70
4" dia. x 60"	SN@2.50	Ea	2,010.00	98.00	1.63	2,109.63
6" dia. x 30"	SN@3.25	Ea	2,460.00	127.00	2.12	2,589.12
6" dia. x 60"	SN@3.50	Ea	3,060.00	137.00	2.28	3,199.28
8" dia. x 30"	SN@4.25	Ea	3,800.00	167.00	2.77	3,969.77
8" dia. x 60"	SN@4.50	Ea	4,900.00	176.00	2.93	5,078.93
10" dia. x 30"	SN@5.50	Ea	5,630.00	216.00	3.58	5,849.58
10" dia. x 60"	SN@5.75	Ea	7,440.00	225.00	3.75	7,668.75
10" dia. x 90"	SN@6.00	Ea	9,270.00	235.00	3.91	9,508.91
12" dia. x 60"	SN@6.50	Ea	9,750.00	255.00	4.24	10,009.24
12" dia. x 90"	SN@7.00	Ea	12,500.00	274.00	4.45	12,778.45
12" dia. x 120"	SN@7.25	Ea	15,200.00	284.00	4.72	15,488.72
14" dia. x 60"	SN@7.50	Ea	14,400.00	294.00	4.89	14,698.89
14" dia. x 90"	SN@8.00	Ea	17,900.00	314.00	5.21	18,219.21
14" dia. x 120"	SN@8.25	Ea	21,600.00	323.00	5.38	21,928.38
16" dia. x 60"	SN@8.50	Ea	17,800.00	333.00	5.54	18,138.54
16" dia. x 90"	SN@9.00	Ea	22,200.00	353.00	5.87	22,558.87
16" dia. x 120"	SN@9.25	Ea	25,100.00	363.00	6.03	25,469.03

Heat exchanger connection assembly, hot water to hot water. Includes pipe, fittings, pipe insulation, valves, thermometers gauges and vents.

Description	Craft@Hrs	Unit	Material $	Labor $	Equipment $	Total $
1½" supply	P1@32.0	Ea	3,910.00	1,190.00	—	5,100.00
2" supply	P1@38.0	Ea	4,820.00	1,410.00	—	6,230.00
2½" supply	P1@50.0	Ea	6,280.00	1,860.00	—	8,140.00
3" supply	P1@55.0	Ea	6,600.00	2,050.00	—	8,650.00
4" supply	P1@62.0	Ea	8,030.00	2,310.00	—	10,340.00
6" supply	P1@85.0	Ea	13,000.00	3,160.00	5,030.00	21,190.00
8" supply	P1@102.	Ea	18,400.00	3,800.00	6,030.00	28,230.00
10" supply	P1@130.	Ea	24,100.00	4,840.00	7,690.00	36,630.00

Heat exchanger connection assembly, steam to hot water. Includes pipe, fittings, pipe insulation, valves, thermometers gauges and vents.

Description	Craft@Hrs	Unit	Material $	Labor $	Equipment $	Total $
1½" supply	P1@22.0	Ea	4,480.00	819.00	—	5,299.00
2" supply	P1@26.0	Ea	5,230.00	968.00	—	6,198.00
2½" supply	P1@41.0	Ea	6,490.00	1,530.00	—	8,020.00
3" supply	P1@44.0	Ea	7,520.00	1,640.00	—	9,160.00
4" supply	P1@54.5	Ea	8,570.00	2,030.00	—	10,600.00
6" supply	P1@67.0	Ea	13,600.00	2,490.00	3,960.00	20,050.00
8" supply	P1@80.0	Ea	19,200.00	2,980.00	4,730.00	26,910.00

Fan Coil Units and Connections

Description	Craft@Hrs	Unit	Material $	Labor $	Equipment $	Total $

Ceiling-hung horizontal, 2 pipe hydronic fan coil unit with cabinet, set in place only. Make additional allowances for pipe connection assembly.

200 CFM	P1@2.00	Ea	970.00	74.40	—	1,044.40
400 CFM	P1@2.50	Ea	1,140.00	93.10	—	1,233.10
600 CFM	P1@3.00	Ea	1,350.00	112.00	—	1,462.00
800 CFM	P1@3.25	Ea	1,710.00	121.00	—	1,831.00
1,000 CFM	P1@3.50	Ea	2,140.00	130.00	—	2,270.00
Deduct for "less cabinet"	—	%	-20.0	—	—	—
Add for "floor-mounted"	—	%	10.0	—	—	—

Vertical, 4 pipe hydronic fan coil unit, set in place only. Make additional allowances for pipe connection assembly.

200 CFM	P1@2.15	Ea	1,090.00	80.00	—	1,170.00
400 CFM	P1@2.75	Ea	1,280.00	102.00	—	1,382.00
600 CFM	P1@3.25	Ea	1,510.00	121.00	—	1,631.00
800 CFM	P1@3.45	Ea	1,870.00	128.00	—	1,998.00
1,000 CFM	P1@3.60	Ea	2,340.00	134.00	—	2,474.00

2 pipe fan coil unit connection. Coil connection includes type L copper pipe and wrought fittings, pipe insulation, condensate drain, isolation valves and 2-way control valve.

½" supply	P1@2.50	Ea	158.00	93.10	—	251.10
¾" supply	P1@2.80	Ea	191.00	104.00	—	295.00
1" supply	P1@3.40	Ea	289.00	127.00	—	416.00

4 pipe fan coil unit connection. Coil connection includes type L copper pipe and wrought fittings, pipe insulation, condensate drain, isolation valves and 3-way control valve.

½" supply	P1@2.75	Ea	191.00	102.00	—	293.00
¾" supply	P1@2.95	Ea	221.00	110.00	—	331.00

Reheat Coils and Connections

Description	Craft@Hrs	Unit	Material $	Labor $	Equipment $	Total $

Duct-mounted flanged hot water reheat coil, set in place only. Make additional allowances for pipe connection assembly.

12" x 6"	SN@1.00	Ea	374.00	39.20	—	413.20
12" x 8"	SN@1.00	Ea	393.00	39.20	—	432.20
12" x 10"	SN@1.00	Ea	409.00	39.20	—	448.20
12" x 12"	SN@1.00	Ea	447.00	39.20	—	486.20
18" x 6"	SN@1.25	Ea	473.00	49.00	—	522.00
18" x 12"	SN@1.25	Ea	487.00	49.00	—	536.00
18" x 18"	SN@1.25	Ea	523.00	49.00	—	572.00
24" x 12"	SN@1.50	Ea	564.00	58.80	—	622.80
24" x 18"	SN@1.50	Ea	589.00	58.80	—	647.80
24" x 24"	SN@1.75	Ea	663.00	68.60	—	731.60

Duct-mounted hot water reheat coil connection assembly. Includes pipe, fittings, pipe insulation, valves and strainers.

½" steel	P1@2.25	Ea	191.00	83.70	—	274.70
¾" steel	P1@2.35	Ea	227.00	87.50	—	314.50
1" steel	P1@2.65	Ea	297.00	98.60	—	395.60
1¼" steel	P1@3.00	Ea	359.00	112.00	—	471.00
1½" steel	P1@3.25	Ea	383.00	121.00	—	504.00
2" steel	P1@3.75	Ea	427.00	140.00	—	567.00

Duct-mounted slip-in electric reheat coil, set in place only. Make additional allowances for electrical connection.

12" x 6"	SN@1.10	Ea	833.00	43.10	—	876.10
12" x 8"	SN@1.20	Ea	949.00	47.00	—	996.00
12" x 10"	SN@1.30	Ea	1,140.00	50.90	—	1,190.90
12" x 12"	SN@1.40	Ea	1,260.00	54.90	—	1,314.90
18" x 6"	SN@1.50	Ea	1,400.00	58.80	—	1,458.80
18" x 12"	SN@1.60	Ea	1,500.00	62.70	—	1,562.70
18" x 18"	SN@1.70	Ea	1,600.00	66.60	—	1,666.60
24" x 12"	SN@1.90	Ea	1,720.00	74.50	—	1,794.50
24" x 18"	SN@2.10	Ea	1,810.00	82.30	—	1,892.30
24" x 24"	SN@2.40	Ea	2,210.00	94.10	—	2,304.10

Unit Heaters and Connections

Description	Craft@Hrs	Unit	Material $	Labor $	Equipment $	Total $

Gas-fired unit heater, hang in place only. Make additional allowances for gas connection assembly.

Description	Craft@Hrs	Unit	Material $	Labor $	Equipment $	Total $
60mbh, 200cfm	P1@2.00	Ea	1,750.00	74.40	—	1,824.40
100mbh, 400cfm	P1@2.50	Ea	2,030.00	93.10	—	2,123.10
150mbh, 600cfm	P1@3.00	Ea	2,480.00	112.00	—	2,592.00
175mbh, 800cfm	P1@3.25	Ea	2,670.00	121.00	—	2,791.00
200mbh, 1,000cfm	P1@3.50	Ea	2,870.00	130.00	—	3,000.00
250mbh, 1,500cfm	P1@3.75	Ea	3,370.00	140.00	—	3,510.00
300mbh, 2,000cfm	P1@4.00	Ea	3,750.00	149.00	—	3,899.00

Hot water unit heater, hang in place only. Make additional allowances for pipe connection assembly.

Description	Craft@Hrs	Unit	Material $	Labor $	Equipment $	Total $
8mbh, 400cfm	P1@2.00	Ea	461.00	74.40	4.56	539.96
12mbh, 400cfm	P1@2.00	Ea	593.00	74.40	4.56	671.96
16mbh, 400cfm	P1@2.00	Ea	626.00	74.40	4.56	704.96
30mbh, 700cfm	P1@3.00	Ea	837.00	112.00	6.84	955.84
43mbh, 1,040cfm	P1@3.50	Ea	1,030.00	130.00	7.98	1,167.98
57mbh, 1,370cfm	P1@4.00	Ea	1,100.00	149.00	9.12	1,258.12
68mbh, 1,715cfm	P1@4.00	Ea	1,750.00	149.00	9.12	1,908.12
105mbh, 2,360cfm	P1@5.00	Ea	2,020.00	186.00	2,206.00	2,215.10
123mbh, 2,900cfm	P1@6.00	Ea	2,590.00	223.00	2,813.00	2,824.40
140mbh, 3,530cfm	P1@6.00	Ea	3,000.00	223.00	3,223.00	3,234.40
156mbh, 4,250cfm	P1@6.00	Ea	3,150.00	223.00	3,373.00	3,384.40
210mbh, 4,450cfm	P1@8.00	Ea	5,590.00	298.00	5,888.00	5,902.80
123mbh, 5,200cfm	P1@6.00	Ea	4,610.00	223.00	4,833.00	4,844.40
257mbh, 5,350cfm	P1@8.00	Ea	9,460.00	298.00	9,758.00	9,772.80

Hot water unit heater connection. Includes Schedule 40 threaded and coupled supply and return piping with malleable iron fittings, pipe insulation, valves and vents Make additional allowances for electrical connection.

Description	Craft@Hrs	Unit	Material $	Labor $	Equipment $	Total $
¾" supply	P1@2.50	Ea	267.00	93.10	—	360.10
1" supply	P1@3.00	Ea	313.00	112.00	—	425.00
1¼" supply	P1@3.30	Ea	372.00	123.00	—	495.00
1½" supply	P1@3.75	Ea	433.00	140.00	—	573.00
2" supply	P1@4.10	Ea	560.00	153.00	—	713.00

Gas-fired unit heater flue connection

Description	Craft@Hrs	Unit	Material $	Labor $	Equipment $	Total $
2" B vent	SN@.090	LF	5.81	3.53	—	9.34
3" B vent	SN@.100	LF	7.20	3.92	—	11.12
4" B vent	SN@.110	LF	9.63	4.31	—	13.94
6" B vent	SN@.130	LF	10.90	5.09	—	15.99

Gas-fired unit heater fuel connection. Assembly includes 8' of black steel pipe, malleable fittings, plug valve and regulator.

Description	Craft@Hrs	Unit	Material $	Labor $	Equipment $	Total $
¾" gas conn.	P1@1.30	Ea	97.10	48.40	—	145.50
1" gas conn.	P1@1.60	Ea	105.00	59.60	—	164.60
1¼" gas conn.	P1@2.10	Ea	188.00	78.20	—	266.20
1½" gas conn.	P1@2.50	Ea	217.00	93.10	—	310.10

Chillers and Chiller Connections

Description	Craft@Hrs	Unit	Material $	Labor $	Equipment $	Total $

Centrifugal water-cooled chiller, set in place only. Unit comes factory assembled complete. Make additional allowances for cooling tower pipe connections.

Description	Craft@Hrs	Unit	Material $	Labor $	Equipment $	Total $
100 tons	SN@20.0	Ea	107,000.00	784.00	338.00	108,122.00
150 tons	SN@20.0	Ea	162,000.00	784.00	338.00	163,122.00
200 tons	SN@20.0	Ea	174,000.00	784.00	338.00	175,122.00
250 tons	SN@20.0	Ea	205,000.00	784.00	338.00	206,122.00
300 tons	SN@25.0	Ea	219,000.00	980.00	423.00	220,403.00
350 tons	SN@25.0	Ea	226,000.00	980.00	423.00	227,403.00
400 tons	SN@25.0	Ea	267,000.00	980.00	423.00	268,403.00
500 tons	SN@25.0	Ea	334,000.00	980.00	423.00	335,403.00

Reciprocating water-cooled chiller, set in place only. Unit comes factory assembled complete. Make additional allowances for cooling tower pipe connections.

Description	Craft@Hrs	Unit	Material $	Labor $	Equipment $	Total $
20 tons	SN@12.0	Ea	32,800.00	470.00	203.00	33,473.00
25 tons	SN@12.0	Ea	37,500.00	470.00	203.00	38,173.00
30 tons	SN@12.0	Ea	40,800.00	470.00	203.00	41,473.00
40 tons	SN@12.0	Ea	49,100.00	470.00	203.00	49,773.00
50 tons	SN@16.0	Ea	69,300.00	627.00	270.00	70,197.00
60 tons	SN@16.0	Ea	89,100.00	627.00	270.00	89,997.00
75 tons	SN@16.0	Ea	95,700.00	627.00	270.00	96,597.00
100 tons	SN@20.0	Ea	102,000.00	784.00	338.00	103,122.00
125 tons	SN@20.0	Ea	124,000.00	784.00	338.00	125,122.00
150 tons	SN@20.0	Ea	141,000.00	784.00	338.00	142,122.00

Water-cooled chiller connection assembly. Includes condenser and chilled water pipe, fittings, pipe insulation, valves, thermometers and gauges.

Description	Craft@Hrs	Unit	Material $	Labor $	Equipment $	Total $
2½" chiller conn.	SN@36.0	Ea	3,060.00	1,410.00	—	4,470.00
3" chiller conn.	SN@38.0	Ea	3,370.00	1,490.00	—	4,860.00
4" chiller conn.	SN@51.0	Ea	4,490.00	2,000.00	—	6,490.00
6" chiller conn.	SN@65.0	Ea	7,530.00	2,550.00	1,100.00	11,180.00
8" chiller conn.	SN@77.0	Ea	11,500.00	3,020.00	1,300.00	15,820.00
10" chiller conn.	SN@100.	Ea	21,000.00	3,920.00	1,690.00	26,610.00
12" chiller conn.	SN@125.	Ea	24,300.00	4,900.00	2,110.00	31,310.00

Condensing Units and Cooling Towers

Description	Craft@Hrs	Unit	Material $	Labor $	Equipment $	Total $

Air-cooled condensing unit, set in place only. Unit comes factory assembled complete. Make additional allowances for pipe and electrical connections.

Description	Craft@Hrs	Unit	Material $	Labor $	Equipment $	Total $
3 tons	SN@2.00	Ea	4,300.00	78.40	33.80	4,412.20
5 tons	SN@3.00	Ea	7,320.00	118.00	50.70	7,488.70
7½ tons	SN@4.00	Ea	9,480.00	157.00	67.60	9,704.60
10 tons	SN@4.50	Ea	12,600.00	176.00	76.10	12,852.10
12½ tons	SN@5.00	Ea	14,800.00	196.00	84.50	15,080.50
15 tons	SN@5.50	Ea	17,200.00	216.00	93.00	17,509.00
20 tons	SN@6.00	Ea	22,600.00	235.00	101.00	22,936.00
25 tons	SN@7.00	Ea	28,500.00	274.00	118.00	28,892.00
30 tons	SN@8.00	Ea	34,100.00	314.00	135.00	34,549.00
40 tons	SN@9.00	Ea	44,500.00	353.00	152.00	45,005.00
50 tons	SN@10.0	Ea	56,000.00	392.00	169.00	56,561.00

Galvanized steel cooling tower. Assemble and set in place only. Add for pipe connections.

Description	Craft@Hrs	Unit	Material $	Labor $	Equipment $	Total $
10 tons	SN@16.0	Ea	4,350.00	627.00	270.00	5,247.00
15 tons	SN@16.0	Ea	4,930.00	627.00	270.00	5,827.00
20 tons	SN@18.0	Ea	7,740.00	705.00	304.00	8,749.00
25 tons	SN@20.0	Ea	9,550.00	784.00	338.00	10,672.00
30 tons	SN@25.0	Ea	10,900.00	980.00	423.00	12,303.00
40 tons	SN@30.0	Ea	13,900.00	1,180.00	507.00	15,587.00
50 tons	SN@35.0	Ea	15,300.00	1,370.00	592.00	17,262.00
60 tons	SN@40.0	Ea	16,200.00	1,570.00	676.00	18,446.00
80 tons	SN@45.0	Ea	17,500.00	1,760.00	760.00	20,020.00
100 tons	SN@50.0	Ea	20,800.00	1,960.00	845.00	23,605.00
125 tons	SN@60.0	Ea	25,300.00	2,350.00	1,010.00	28,660.00
150 tons	SN@65.0	Ea	29,200.00	2,550.00	1,100.00	32,850.00
175 tons	SN@70.0	Ea	32,900.00	2,740.00	1,180.00	36,820.00
200 tons	SN@70.0	Ea	36,300.00	2,740.00	1,180.00	40,220.00
300 tons	SN@75.0	Ea	52,200.00	2,940.00	1,270.00	56,410.00
400 tons	SN@80.0	Ea	66,100.00	3,140.00	1,350.00	70,590.00
500 tons	SN@80.0	Ea	79,200.00	3,140.00	1,350.00	83,690.00

Redwood or fir induced-draft cooling tower. Assemble and set in place only. Add for pipe connections.

Description	Craft@Hrs	Unit	Material $	Labor $	Equipment $	Total $
100 tons	SN@75.0	Ea	13,300.00	2,940.00	1,270.00	17,510.00
200 tons	SN@80.0	Ea	22,500.00	3,140.00	1,350.00	26,990.00
300 tons	SN@85.0	Ea	32,400.00	3,330.00	1,440.00	37,170.00
400 tons	SN@90.0	Ea	37,000.00	3,530.00	1,520.00	42,050.00
500 tons	SN@90.0	Ea	43,100.00	3,530.00	1,520.00	48,150.00

Cooling Towers and Cooling Tower Connections

Description	Craft@Hrs	Unit	Material $	Labor $	Equipment $	Total $

Redwood or fir forced-draft cooling tower. Assemble and set in place only. Add for pipe connections.

Description	Craft@Hrs	Unit	Material $	Labor $	Equipment $	Total $
100 tons	SN@75.0	Ea	21,500.00	2,940.00	1,270.00	25,710.00
200 tons	SN@80.0	Ea	41,300.00	3,140.00	1,350.00	45,790.00
300 tons	SN@85.0	Ea	48,100.00	3,330.00	1,440.00	52,870.00
400 tons	SN@90.0	Ea	63,700.00	3,530.00	1,520.00	68,750.00
500 tons	SN@90.0	Ea	77,400.00	3,530.00	1,520.00	82,450.00

Cooling tower connection assembly. Includes condenser water pipe, fittings, valves, thermometers and gauges.

Description	Craft@Hrs	Unit	Material $	Labor $	Equipment $	Total $
2½" connection	SN@42.0	Ea	5,590.00	1,650.00	—	7,240.00
3" connection	SN@47.0	Ea	6,340.00	1,840.00	—	8,180.00
4" connection	SN@62.0	Ea	9,130.00	2,430.00	—	11,560.00
6" connection	SN@96.0	Ea	14,400.00	3,760.00	62.60	18,222.60
8" connection	SN@118.	Ea	20,600.00	4,620.00	76.90	25,296.90
10" connection	SN@150.	Ea	28,500.00	5,880.00	97.80	34,477.80
12" connection	SN@215.	Ea	34,700.00	8,430.00	140.00	43,270.00
14" connection	SN@260.	Ea	39,800.00	10,200.00	169.00	50,169.00

Carbon Steel, Schedule 40 with 150# Fittings & Butt-Welded Joints

Schedule 40 carbon steel (ASTM A-53 and A-120) pipe with 150 pound carbon steel welding fittings is commonly used for heating hot water, chilled water, glycol and low pressure steam systems where operating pressures don't exceed 250 PSIG.

Schedule 40 (ASTM A-53B SML) steel pipe with carbon steel welding fittings can also be used for refrigerant piping systems.

The cost estimates in this section are based on the conditions, limitations and wage rates described in the section "How to Use This Book" beginning on page 5.

Equipment cost, where shown, is $4.45 per hour for a 2-ton chain hoist and $6.02 per hour for 200 amp gas welder, trailer mounted.

Description	Craft@Hrs	Unit	Material $	Labor $	Equipment $	Total $

Schedule 40 carbon steel butt-welded horizontal pipe assembly.
Horizontally hung in a building. Assembly includes fittings and hanger assemblies. Based on a reducing tee and a 45-degree elbow every 16 feet for ½" pipe. A reducing tee and a 45-degree elbow every 50 feet for 12" pipe. Hangers spaced to meet plumbing code. *(Distance between fittings in the pipe assembly increases as pipe diameter increases.)* Use these figures for preliminary estimates.

Description	Craft@Hrs	Unit	Material $	Labor $	Equipment $	Total $
½"	P1@.144	LF	2.05	5.36	.46	7.87
¾"	P1@.170	LF	2.22	6.33	.54	9.09
1"	P1@.206	LF	2.73	7.67	.65	11.05
1¼"	P1@.208	LF	3.63	7.74	.66	12.03
1½"	P1@.238	LF	3.89	8.86	.75	13.50
2"	P1@.286	LF	5.14	10.60	.90	16.64
2½"	P1@.344	LF	9.43	12.80	1.09	23.32
3"	P1@.390	LF	11.80	14.50	1.23	27.53
4"	P1@.507	LF	16.40	18.90	1.60	36.90
6"	ER@.712	LF	26.30	28.50	1.11	55.91
8"	ER@.847	LF	45.50	33.80	1.32	80.62
10"	ER@1.02	LF	66.30	40.80	1.59	108.69
12"	ER@1.25	LF	84.90	50.00	1.94	136.84

Schedule 40 carbon steel butt-welded vertical pipe assembly. Riser assembly, including fittings and riser clamps. Based on a reducing tee, and a riser clamp on every floor. Use these figures for preliminary estimates.

Description	Craft@Hrs	Unit	Material $	Labor $	Equipment $	Total $
½"	P1@.104	LF	1.56	3.87	.33	5.76
¾"	P1@.128	LF	1.75	4.76	.41	6.92
1"	P1@.162	LF	2.24	6.03	.51	8.78
1¼"	P1@.184	LF	3.89	6.85	.58	11.32
1½"	P1@.204	LF	4.35	7.59	.65	12.59
2"	P1@.247	LF	5.78	9.19	.78	15.75
2½"	P1@.304	LF	9.84	11.30	.96	22.10
3"	P1@.351	LF	11.90	13.10	1.11	26.11
4"	P1@.505	LF	15.70	18.80	1.60	36.10
6"	ER@.737	LF	28.20	29.50	1.15	58.85
8"	ER@.878	LF	51.00	35.10	1.37	87.47
10"	ER@1.07	LF	77.60	42.80	1.66	122.06
12"	ER@1.32	LF	102.00	52.70	2.05	156.75

Carbon Steel, Schedule 40 with 150# Fittings & Butt-Welded Joints

Description	Craft@Hrs	Unit	Material $	Labor $	Equipment $	Total $

Schedule 40 carbon steel pipe, plain end, butt-welded joints, pipe only

½"	P1@.070	LF	.94	2.61	.22	3.77
¾"	P1@.080	LF	1.07	2.98	.25	4.30
1"	P1@.100	LF	1.58	3.72	.32	5.62
1¼"	P1@.110	LF	2.05	4.09	.35	6.49
1½"	P1@.120	LF	2.46	4.47	.38	7.31
2"	P1@.140	LF	3.34	5.21	.44	8.99
2½"	P1@.170	LF	5.96	6.33	.54	12.83
3"	P1@.190	LF	7.83	7.07	.60	15.50
4"	P1@.300	LF	11.10	11.20	.95	23.25
6"	ER@.420	LF	19.30	16.80	.65	36.75
8"	ER@.510	LF	25.10	20.40	.79	46.29
10"	ER@.600	LF	35.70	24.00	.93	60.63
12"	ER@.760	LF	43.00	30.40	1.18	74.58

Schedule 40 carbon steel 45-degree ell, butt-welded joints

½"	P1@.220	Ea	2.77	8.19	.70	11.66
¾"	P1@.330	Ea	2.77	12.30	1.04	16.11
1"	P1@.440	Ea	2.77	16.40	1.39	20.56
1¼"	P1@.560	Ea	2.92	20.80	1.77	25.49
1½"	P1@.710	Ea	2.92	26.40	2.25	31.57
2"	P1@.890	Ea	3.03	33.10	2.82	38.95
2½"	P1@1.11	Ea	5.45	41.30	3.51	50.26
3"	P1@1.33	Ea	6.59	49.50	4.21	60.30
4"	P1@1.78	Ea	9.71	66.30	5.63	81.64
6"	ER@2.67	Ea	25.10	107.00	4.15	136.25
8"	ER@3.20	Ea	59.30	128.00	4.98	192.28
10"	ER@4.00	Ea	90.80	160.00	6.22	257.02
12"	ER@4.80	Ea	133.00	192.00	7.47	332.47

Schedule 40 carbon steel 90-degree ell, butt-welded joints

½"	P1@.220	Ea	2.70	8.19	.70	11.59
¾"	P1@.330	Ea	2.70	12.30	1.04	16.04
1"	P1@.440	Ea	2.70	16.40	1.39	20.49
1¼"	P1@.560	Ea	2.88	20.80	1.77	25.45
1½"	P1@.670	Ea	2.88	24.90	2.25	30.03
2"	P1@.890	Ea	3.15	33.10	2.68	38.93
2½"	P1@1.11	Ea	6.25	41.30	3.51	51.06
3"	P1@1.33	Ea	8.38	49.50	4.21	62.09
4"	P1@1.78	Ea	13.50	66.30	5.63	85.43
6"	ER@2.67	Ea	38.30	107.00	4.15	149.45
8"	ER@3.20	Ea	87.10	128.00	4.98	220.08
10"	ER@4.00	Ea	156.00	160.00	6.22	322.22
12"	ER@4.80	Ea	231.00	192.00	7.47	430.47

Carbon Steel, Schedule 40 with 150# Fittings & Butt-Welded Joints

Description	Craft@Hrs	Unit	Material $	Labor $	Equipment $	Total $

Schedule 40 carbon steel tee, butt-welded joints

Description	Craft@Hrs	Unit	Material $	Labor $	Equipment $	Total $
½"	P1@.330	Ea	8.56	12.30	1.04	21.90
¾"	P1@.500	Ea	8.56	18.60	1.58	28.74
1"	P1@.670	Ea	8.56	24.90	2.12	35.58
1¼"	P1@.830	Ea	9.52	30.90	2.63	43.05
1½"	P1@1.00	Ea	10.00	37.20	3.16	50.36
2"	P1@1.33	Ea	11.00	49.50	4.21	64.71
2½"	P1@1.67	Ea	19.10	62.20	5.28	86.58
3"	P1@2.00	Ea	21.30	74.40	6.33	102.03
4"	P1@2.67	Ea	29.10	99.40	8.45	136.95
6"	ER@4.00	Ea	53.30	160.00	6.22	219.52
8"	ER@4.80	Ea	124.00	192.00	7.47	323.47
10"	ER@6.00	Ea	213.00	240.00	9.33	462.33
12"	ER@7.20	Ea	319.00	288.00	11.20	618.20

Schedule 40 carbon steel reducing tee, butt-welded joints

Description	Craft@Hrs	Unit	Material $	Labor $	Equipment $	Total $
1½" x ½"	P1@.940	Ea	13.00	35.00	2.97	50.97
1½" x ¾"	P1@.940	Ea	13.00	35.00	2.97	50.97
1½" x 1"	P1@.940	Ea	13.00	35.00	2.97	50.97
1½" x 1¼"	P1@.940	Ea	13.00	35.00	2.97	50.97
2" x ¾"	P1@1.22	Ea	17.50	45.40	2.97	65.87
2" x 1"	P1@1.22	Ea	17.50	45.40	3.86	66.76
2" x 1¼"	P1@1.22	Ea	17.50	45.40	3.86	66.76
2" x 1½"	P1@1.22	Ea	17.50	45.40	3.86	66.76
2½" x 1"	P1@1.56	Ea	27.00	58.10	4.94	90.04
2½" x 1¼"	P1@1.56	Ea	27.00	58.10	4.94	90.04
2½" x 1½"	P1@1.56	Ea	27.00	58.10	4.94	90.04
2½" x 2"	P1@1.56	Ea	27.00	58.10	4.94	90.04
3" x 1¼"	P1@1.89	Ea	27.80	70.30	5.98	104.08
3" x 2"	P1@1.89	Ea	27.80	70.30	5.98	104.08
3" x 2½"	P1@1.89	Ea	27.80	70.30	5.98	104.08
4" x 1½"	P1@2.44	Ea	29.60	90.80	7.72	128.12
4" x 2"	P1@2.44	Ea	29.60	90.80	7.72	128.12
4" x 2½"	P1@2.44	Ea	29.60	90.80	7.72	128.12
4" x 3"	P1@2.44	Ea	29.60	90.80	7.72	128.12
6" x 2½"	ER@3.78	Ea	67.00	151.00	5.88	223.88
6" x 3"	ER@3.78	Ea	67.00	151.00	5.88	223.88
6" x 4"	ER@3.78	Ea	67.00	151.00	5.88	223.88
8" x 3"	ER@4.40	Ea	165.00	176.00	6.85	347.85
8" x 4"	ER@4.40	Ea	165.00	176.00	6.85	347.85
8" x 6"	ER@4.40	Ea	165.00	176.00	6.85	347.85
10" x 4"	ER@5.60	Ea	280.00	224.00	8.71	512.71
10" x 6"	ER@5.60	Ea	280.00	224.00	8.71	512.71
10" x 8"	ER@5.60	Ea	280.00	224.00	8.71	512.71
12" x 6"	ER@6.80	Ea	415.00	272.00	10.60	697.60
12" x 8"	ER@6.80	Ea	415.00	272.00	10.60	697.60
12" x 10"	ER@6.80	Ea	415.00	272.00	10.60	697.60

Carbon Steel, Schedule 40 with 150# Fittings & Butt-Welded Joints

Description	Craft@Hrs	Unit	Material $	Labor $	Equipment $	Total $
Schedule 40 carbon steel concentric reducer, butt-welded joints						
¾"	P1@.295	Ea	5.78	11.00	.93	17.71
1"	P1@.395	Ea	6.91	14.70	1.25	22.86
1¼"	P1@.500	Ea	7.31	18.60	1.58	27.49
1½"	P1@.600	Ea	8.14	22.30	1.90	32.34
2"	P1@.800	Ea	7.59	29.80	2.53	39.92
2½"	P1@1.00	Ea	15.00	37.20	3.16	55.36
3"	P1@1.20	Ea	12.00	44.70	3.80	60.50
4"	P1@1.60	Ea	17.10	59.60	5.06	81.76
6"	ER@2.40	Ea	31.50	95.90	3.73	131.13
8"	ER@2.88	Ea	60.00	115.00	4.48	179.48
10"	ER@3.60	Ea	91.90	144.00	5.60	241.50
12"	ER@4.30	Ea	141.00	172.00	6.99	319.99
Schedule 40 carbon steel eccentric reducer, butt-welded joints						
¾"	P1@.295	Ea	7.10	11.00	.93	19.03
1"	P1@.395	Ea	9.59	14.70	1.25	25.54
1¼"	P1@.500	Ea	11.10	18.60	1.58	31.28
1½"	P1@.600	Ea	11.70	22.30	1.90	35.90
2"	P1@.800	Ea	9.08	29.80	2.53	41.41
2½"	P1@1.00	Ea	18.30	37.20	3.16	58.66
3"	P1@1.20	Ea	16.10	44.70	3.80	64.60
4"	P1@1.60	Ea	22.90	59.60	5.06	87.56
6"	ER@2.40	Ea	45.50	95.90	3.73	145.13
8"	ER@2.88	Ea	83.80	115.00	4.48	203.28
10"	ER@3.60	Ea	144.00	144.00	5.60	293.60
12"	ER@4.30	Ea	220.00	172.00	6.69	398.69
Schedule 40 carbon steel cap, butt-welded joints						
½"	P1@.150	Ea	6.56	5.58	.47	12.61
¾"	P1@.230	Ea	7.02	8.56	.73	16.31
1"	P1@.300	Ea	7.50	11.20	.95	19.65
1¼"	P1@.390	Ea	9.36	14.50	1.23	25.09
1½"	P1@.470	Ea	9.36	17.50	1.49	28.35
2"	P1@.620	Ea	9.36	23.10	1.96	34.42
2½"	P1@.770	Ea	14.50	28.70	2.44	45.64
3"	P1@.930	Ea	13.90	34.60	2.94	51.44
4"	P1@1.25	Ea	17.80	46.50	3.96	68.26
6"	ER@1.87	Ea	20.60	74.70	2.91	98.21
8"	ER@2.24	Ea	37.30	89.50	3.48	130.28
10"	ER@2.80	Ea	62.90	112.00	4.36	179.26
12"	ER@3.36	Ea	78.90	134.00	5.23	218.13

Carbon Steel, Schedule 40 with 150# Fittings & Butt-Welded Joints

Description	Craft@Hrs	Unit	Material $	Labor $	Equipment $	Total $
Schedule 40 carbon steel union, butt-welded joints						
½"	P1@.220	Ea	4.73	8.19	.70	13.62
¾"	P1@.330	Ea	6.10	12.30	1.04	19.44
1"	P1@.440	Ea	8.43	16.40	1.39	26.22
1¼"	P1@.560	Ea	17.80	20.80	1.77	40.37
1½"	P1@.670	Ea	19.40	24.90	2.12	46.42
2"	P1@.890	Ea	26.10	33.10	2.82	62.02
2½"	P1@1.11	Ea	87.10	41.30	3.51	131.91
3"	P1@1.33	Ea	103.00	49.50	4.21	156.71
Carbon steel weldolet, Schedule 40						
½"	P1@.330	Ea	6.58	12.30	1.04	19.92
¾"	P1@.500	Ea	7.26	18.60	1.58	27.44
1"	P1@.670	Ea	8.10	24.90	2.12	35.12
1¼"	P1@.830	Ea	12.30	30.90	2.63	45.83
1½"	P1@1.00	Ea	12.30	37.20	3.16	52.66
2"	P1@1.33	Ea	13.40	49.50	4.21	67.11
2½"	P1@1.67	Ea	45.80	62.20	5.28	113.28
3"	P1@2.00	Ea	49.40	74.40	6.33	130.13
4"	P1@2.67	Ea	64.30	99.40	8.45	172.15
6"	P1@4.00	Ea	261.00	149.00	12.70	422.70
8"	P1@4.80	Ea	499.00	179.00	15.20	693.20
10"	P1@6.00	Ea	685.00	223.00	19.00	927.00
12"	P1@7.20	Ea	838.00	268.00	22.80	1,128.80
Carbon steel threadolet, Schedule 40						
¾"	P1@.330	Ea	4.38	12.30	1.04	17.72
1"	P1@.440	Ea	5.19	16.40	1.39	22.98
1¼"	P1@.560	Ea	8.97	20.80	1.77	31.54
1½"	P1@.670	Ea	10.10	24.90	2.12	37.12
2"	P1@.890	Ea	11.70	33.10	2.82	47.62
2½"	P1@1.11	Ea	56.40	41.30	3.51	101.21
3"	P1@1.33	Ea	61.80	49.50	4.21	115.51
4"	P1@1.78	Ea	121.00	66.30	5.63	192.93
150# forged steel companion flange, weld neck						
2½"	P1@.610	Ea	32.50	22.70	1.93	57.13
3"	P1@.730	Ea	30.30	27.20	2.31	59.81
4"	P1@.980	Ea	41.90	36.50	3.10	81.50
6"	ER@1.47	Ea	67.40	58.70	2.29	128.39
8"	ER@1.77	Ea	136.00	70.70	2.75	209.45
10"	ER@2.20	Ea	224.00	87.90	3.42	315.32
12"	ER@2.64	Ea	325.00	105.00	4.11	434.11

Carbon Steel, Schedule 40 with 150# Fittings & Butt-Welded Joints

Description	Craft@Hrs	Unit	Material $	Labor $	Equipment $	Total $

Class 125 bronze body gate valve, threaded ends

½"	P1@.210	Ea	19.60	7.82	—	27.42
¾"	P1@.250	Ea	24.50	9.31	—	33.81
1"	P1@.300	Ea	34.40	11.20	—	45.60
1¼"	P1@.400	Ea	43.40	14.90	—	58.30
1½"	P1@.450	Ea	58.60	16.70	—	75.30
2"	P1@.500	Ea	98.00	18.60	—	116.60
2½"	P1@.750	Ea	161.00	27.90	—	188.90
3"	P1@.950	Ea	232.00	35.40	—	267.40

Class 125 iron body gate valve, flanged ends

2"	P1@.500	Ea	375.00	18.60	—	393.60
2½"	P1@.600	Ea	506.00	22.30	—	528.30
3"	P1@.750	Ea	550.00	27.90	—	577.90
4"	P1@1.35	Ea	805.00	50.20	—	855.20
6"	ER@2.50	Ea	1,530.00	99.90	1.63	1,631.53
8"	ER@3.00	Ea	2,520.00	120.00	1.96	2,641.96
10"	ER@4.00	Ea	4,130.00	160.00	2.61	4,292.61
12"	ER@4.50	Ea	5,710.00	180.00	2.93	5,892.93

Class 125 bronze body globe valve, threaded ends

½"	P1@.210	Ea	36.50	7.82	—	44.32
¾"	P1@.250	Ea	48.40	9.31	—	57.71
1"	P1@.300	Ea	69.40	11.20	—	80.60
1¼"	P1@.400	Ea	97.70	14.90	—	112.60
1½"	P1@.450	Ea	131.00	16.70	—	147.70
2"	P1@.500	Ea	214.00	18.60	—	232.60

Class 125 iron body globe valve, flanged ends

2½"	P1@.600	Ea	425.00	22.30	—	447.30
3"	P1@.750	Ea	509.00	27.90	—	536.90
4"	P1@1.35	Ea	684.00	50.20	—	734.20
6"	ER@2.50	Ea	1,210.00	99.90	1.63	1,311.53
8"	ER@3.00	Ea	1,840.00	120.00	1.96	1,961.96
10"	ER@4.00	Ea	2,810.00	160.00	2.61	2,972.61

200 PSIG iron body butterfly valve, lug-type, lever operated

2"	P1@.450	Ea	168.00	16.70	—	184.70
2½"	P1@.450	Ea	175.00	16.70	—	191.70
3"	P1@.550	Ea	183.00	20.50	—	203.50
4"	P1@.550	Ea	228.00	20.50	—	248.50
6"	ER@.800	Ea	368.00	32.00	.52	400.52
8"	ER@.800	Ea	503.00	32.00	.52	535.52
10"	ER@.900	Ea	699.00	36.00	.59	735.59
12"	ER@1.00	Ea	913.00	40.00	.65	953.65

Carbon Steel, Schedule 40 with 150# Fittings & Butt-Welded Joints

Description	Craft@Hrs	Unit	Material $	Labor $	Equipment $	Total $

200 PSIG iron body butterfly valve, wafer-type, lever operated

Description	Craft@Hrs	Unit	Material $	Labor $	Equipment $	Total $
2"	P1@.450	Ea	153.00	16.70	—	169.70
2½"	P1@.450	Ea	156.00	16.70	—	172.70
3"	P1@.550	Ea	168.00	20.50	—	188.50
4"	P1@.550	Ea	204.00	20.50	—	224.50
6"	ER@.800	Ea	341.00	32.00	.52	373.52
8"	ER@.800	Ea	472.00	32.00	.52	504.52
10"	ER@.900	Ea	660.00	36.00	.59	696.59
12"	ER@1.00	Ea	867.00	40.00	.65	907.65

Class 125 bronze body swing check valve, threaded

Description	Craft@Hrs	Unit	Material $	Labor $	Equipment $	Total $
½"	P1@.210	Ea	20.80	7.82	—	28.62
¾"	P1@.250	Ea	29.70	9.31	—	39.01
1"	P1@.300	Ea	38.80	11.20	—	50.00
1¼"	P1@.400	Ea	55.40	14.90	—	70.30
1½"	P1@.450	Ea	78.10	16.70	—	94.80
2"	P1@.500	Ea	129.00	18.60	—	147.60

Class 125 iron body swing check valve, flanged joints

Description	Craft@Hrs	Unit	Material $	Labor $	Equipment $	Total $
2"	P1@.500	Ea	182.00	18.60	—	200.60
2½"	P1@.600	Ea	230.00	22.30	—	252.30
3"	P1@.750	Ea	287.00	27.90	—	314.90
4"	P1@1.35	Ea	421.00	50.20	—	471.20
6"	ER@2.50	Ea	817.00	99.90	1.63	918.53
8"	ER@3.00	Ea	1,470.00	120.00	1.96	1,591.96
10"	ER@4.00	Ea	2,390.00	160.00	2.61	2,552.61
12"	ER@4.50	Ea	3,280.00	180.00	2.93	3,462.93

Class 125 iron body silent check valve, flanged joints

Description	Craft@Hrs	Unit	Material $	Labor $	Equipment $	Total $
2"	P1@.500	Ea	133.00	18.60	—	151.60
2½"	P1@.600	Ea	153.00	22.30	—	175.30
3"	P1@.750	Ea	173.00	27.90	—	200.90
4"	P1@1.35	Ea	226.00	50.20	—	276.20
6"	ER@2.50	Ea	406.00	99.90	1.63	507.53
8"	ER@3.00	Ea	732.00	120.00	1.96	853.96
10"	ER@4.00	Ea	1,130.00	160.00	2.61	1,292.61

Class 125 bronze body strainer, threaded ends

Description	Craft@Hrs	Unit	Material $	Labor $	Equipment $	Total $
½"	P1@.210	Ea	30.20	7.82	—	38.02
¾"	P1@.250	Ea	39.80	9.31	—	49.11
1"	P1@.300	Ea	48.70	11.20	—	59.90
1¼"	P1@.400	Ea	68.10	14.90	—	83.00
1½"	P1@.450	Ea	102.00	16.70	—	118.70
2"	P1@.500	Ea	178.00	18.60	—	196.60

Carbon Steel, Schedule 40 with 150# Fittings & Butt-Welded Joints

Description	Craft@Hrs	Unit	Material $	Labor $	Equipment $	Total $
Class 125 iron body strainer, flanged						
2"	P1@.500	Ea	136.00	18.60	—	154.60
2½"	P1@.600	Ea	152.00	22.30	—	174.30
3"	P1@.750	Ea	176.00	27.90	—	203.90
4"	P1@1.35	Ea	298.00	50.20	—	348.20
6"	ER@2.50	Ea	606.00	99.90	1.63	707.53
8"	ER@3.00	Ea	1,020.00	120.00	1.96	1,141.96

Description	Craft@Hrs	Unit	Material $	Labor $	Equipment $	Total $
Installation of 2-way control valve, threaded joints						
½"	P1@.210	Ea	—	7.82	—	7.82
¾"	P1@.250	Ea	—	9.31	—	9.31
1"	P1@.300	Ea	—	11.20	—	11.20
1¼"	P1@.400	Ea	—	14.90	—	14.90
1½"	P1@.450	Ea	—	16.70	—	16.70
2"	P1@.500	Ea	—	18.60	—	18.60
2½"	P1@.830	Ea	—	30.90	—	30.90
3"	P1@.990	Ea	—	36.80	—	36.80

Description	Craft@Hrs	Unit	Material $	Labor $	Equipment $	Total $
Installation of 2-way control valve, flanged joints						
2"	P1@.500	Ea	—	18.60	—	18.60
2½"	P1@.600	Ea	—	22.30	—	22.30
3"	P1@.750	Ea	—	27.90	—	27.90
4"	P1@1.35	Ea	—	50.20	—	50.20
6"	P1@2.50	Ea	—	93.10	—	93.10
8"	P1@3.00	Ea	—	112.00	—	112.00
10"	P1@4.00	Ea	—	149.00	—	149.00
12"	P1@4.50	Ea	—	167.00	—	167.00

Description	Craft@Hrs	Unit	Material $	Labor $	Equipment $	Total $
Installation of 3-way control valve, threaded joints						
½"	P1@.260	Ea	—	9.68	—	9.68
¾"	P1@.365	Ea	—	13.60	—	13.60
1"	P1@.475	Ea	—	17.70	—	17.70
1¼"	P1@.575	Ea	—	21.40	—	21.40
1½"	P1@.680	Ea	—	25.30	—	25.30
2"	P1@.910	Ea	—	33.90	—	33.90
2½"	P1@1.12	Ea	—	41.70	—	41.70
3"	P1@1.33	Ea	—	49.50	—	49.50

Description	Craft@Hrs	Unit	Material $	Labor $	Equipment $	Total $
Installation of 3-way control valve, flanged joints						
2"	P1@.910	Ea	—	33.90	—	33.90
2½"	P1@1.12	Ea	—	41.70	—	41.70
3"	P1@1.33	Ea	—	49.50	—	49.50
4"	P1@2.00	Ea	—	74.40	—	74.40
6"	P1@3.70	Ea	—	138.00	—	138.00
8"	P1@4.40	Ea	—	164.00	—	164.00
10"	P1@5.90	Ea	—	220.00	—	220.00
12"	P1@6.50	Ea	—	242.00	—	242.00

Carbon Steel, Schedule 40 with 150# Fittings & Butt-Welded Joints

Description	Craft@Hrs	Unit	Material $	Labor $	Equipment $	Total $
Bolt and gasket sets						
2"	P1@.500	Ea	4.22	18.60	—	22.82
2½"	P1@.650	Ea	4.93	24.20	—	29.13
3"	P1@.750	Ea	8.33	27.90	—	36.23
4"	P1@1.00	Ea	13.90	37.20	—	51.10
6"	P1@1.20	Ea	23.20	44.70	—	67.90
8"	P1@1.25	Ea	26.00	46.50	—	72.50
10"	P1@1.70	Ea	44.70	63.30	—	108.00
12"	P1@2.20	Ea	50.40	81.90	—	132.30
Thermometer with well						
7"	P1@.250	Ea	188.00	9.31	—	197.31
9"	P1@.250	Ea	251.00	9.31	—	260.31
Dial-type pressure gauge						
2½"	P1@.200	Ea	38.30	7.44	—	45.74
3½"	P1@.200	Ea	50.40	7.44	—	57.84
Pressure/temperature tap						
Tap	P1@.150	Ea	17.60	5.58	—	23.18
Hanger with swivel assembly						
2"	P1@.300	Ea	6.63	11.20	—	17.83
2½"	P1@.350	Ea	8.58	13.00	—	21.58
3"	P1@.350	Ea	10.60	13.00	—	23.60
4"	P1@.350	Ea	16.40	13.00	—	29.40
5"	P1@.450	Ea	18.40	16.70	—	35.10
6"	P1@.450	Ea	23.90	16.70	—	40.60
8"	P1@.450	Ea	31.80	16.70	—	48.50
10"	P1@.550	Ea	43.70	20.50	—	64.20
12"	P1@.600	Ea	57.90	22.30	—	80.20
Riser clamp						
2"	P1@.115	Ea	6.19	4.28	—	10.47
2½"	P1@.120	Ea	6.50	4.47	—	10.97
3"	P1@.120	Ea	7.04	4.47	—	11.51
4"	P1@.125	Ea	8.97	4.65	—	13.62
5"	P1@.180	Ea	12.90	6.70	—	19.60
6"	P1@.200	Ea	15.50	7.44	—	22.94
8"	P1@.200	Ea	25.20	7.44	—	32.64
10"	P1@.250	Ea	37.50	9.31	—	46.81
12"	P1@.250	Ea	44.50	9.31	—	53.81

Carbon Steel, Schedule 40 with 150# M.I. Fittings & Threaded Joints

Schedule 40 carbon steel (ASTM A-53 and A-120) pipe with 150 pound malleable iron threaded fittings is commonly used for heating hot water, chilled water, glycol and low pressure steam systems where operating pressures do not exceed 125 PSIG and temperatures are below 250 degrees F., for pipe sizes 2" and smaller. For pipe sizes 2½" and larger, wrought steel fittings with either welded or grooved joints should be used.

Schedule 40 (ASTM A-53B SML) steel pipe with carbon steel welding fittings can also be used for refrigerant piping systems.

This section has been arranged to save the estimator's time by including all normally-used system components such as pipe, fittings, valves, hanger assemblies, riser clamps and miscellaneous items under one heading. Additional items can be found under "Plumbing and Piping Specialties." The cost estimates in this section are based on the conditions, limitations and wage rates described in the section "How to Use This Book" beginning on page 5.

Description	Craft@Hrs	Unit	Material $	Labor $	Equipment $	Total $

Schedule 40 carbon steel threaded horizontal pipe assembly.
Horizontally hung in a building. Assembly includes fittings and hanger assemblies. Based on a reducing tee and a 45-degree elbow every 16 feet for ½" pipe. A reducing tee and a 45-degree elbow every 30 feet for 4" pipe. Hangers spaced to meet plumbing code. *(Distance between fittings in the pipe assembly increases as pipe diameter increases.)* Use these figures for preliminary estimates.

Size	Craft@Hrs	Unit	Material $	Labor $	Equipment $	Total $
½"	P1@.119	LF	1.85	4.43	—	6.28
¾"	P1@.132	LF	2.16	4.91	—	7.07
1"	P1@.144	LF	2.98	5.36	—	8.34
1¼"	P1@.160	LF	3.89	5.96	—	9.85
1½"	P1@.184	LF	4.32	6.85	—	11.17
2"	P1@.200	LF	6.54	7.44	—	13.98
2½"	P1@.236	LF	12.90	8.78	—	21.68
3"	P1@.271	LF	17.70	10.10	—	27.80
4"	P1@.398	LF	26.40	14.80	—	41.20

Schedule 40 carbon steel threaded vertical pipe assembly.
Riser assembly, including fittings and riser clamps. Based on a reducing tee, and a riser clamp on every floor. Use these figures for preliminary estimates.

Size	Craft@Hrs	Unit	Material $	Labor $	Equipment $	Total $
½"	P1@.082	LF	1.41	3.05	—	4.46
¾"	P1@.096	LF	1.79	3.57	—	5.36
1"	P1@.121	LF	2.54	4.50	—	7.04
1¼"	P1@.138	LF	3.63	5.14	—	8.77
1½"	P1@.156	LF	4.13	5.81	—	9.94
2"	P1@.173	LF	6.15	6.44	—	12.59
2½"	P1@.200	LF	12.60	7.44	—	20.04
3"	P1@.236	LF	18.50	8.78	—	27.28
4"	P1@.354	LF	24.30	13.20	—	37.50

Carbon Steel, Schedule 40 with 150# M.I. Fittings & Threaded Joints

Description	Craft@Hrs	Unit	Material $	Labor $	Equipment $	Total $

Schedule 40 carbon steel pipe, threaded and coupled, pipe only

½"	P1@.060	LF	1.32	2.23	—	3.55
¾"	P1@.070	LF	1.47	2.61	—	4.08
1"	P1@.090	LF	2.23	3.35	—	5.58
1¼"	P1@.100	LF	2.90	3.72	—	6.62
1½"	P1@.110	LF	3.45	4.09	—	7.54
2"	P1@.120	LF	5.11	4.47	—	9.58
2½"	P1@.130	LF	7.85	4.84	—	12.69
3"	P1@.150	LF	10.40	5.58	—	15.98
4"	P1@.260	LF	14.90	9.68	—	24.58

150# malleable iron 45-degree ell, threaded joints

½"	P1@.120	Ea	1.86	4.47	—	6.33
¾"	P1@.130	Ea	2.26	4.84	—	7.10
1"	P1@.180	Ea	2.88	6.70	—	9.58
1¼"	P1@.240	Ea	5.10	8.93	—	14.03
1½"	P1@.300	Ea	6.30	11.20	—	17.50
2"	P1@.380	Ea	9.41	14.10	—	23.51
2½"	P1@.510	Ea	27.20	19.00	—	46.20
3"	P1@.640	Ea	35.30	23.80	—	59.10
4"	P1@.850	Ea	68.80	31.60	—	100.40

150# malleable iron 90-degree ell, threaded joints

½"	P1@.120	Ea	1.15	4.47	—	5.62
¾"	P1@.130	Ea	1.41	4.84	—	6.25
1"	P1@.180	Ea	2.41	6.70	—	9.11
1¼"	P1@.240	Ea	3.93	8.93	—	12.86
1½"	P1@.300	Ea	5.21	11.20	—	16.41
2"	P1@.380	Ea	8.87	14.10	—	22.97
2½"	P1@.510	Ea	19.40	19.00	—	38.40
3"	P1@.640	Ea	28.80	23.80	—	52.60
4"	P1@.850	Ea	67.30	31.60	—	98.90

150# malleable iron tee, threaded joints

½"	P1@.180	Ea	1.53	6.70	—	8.23
¾"	P1@.190	Ea	2.21	7.07	—	9.28
1"	P1@.230	Ea	3.75	8.56	—	12.31
1¼"	P1@.310	Ea	6.14	11.50	—	17.64
1½"	P1@.390	Ea	7.54	14.50	—	22.04
2"	P1@.490	Ea	12.80	18.20	—	31.00
2½"	P1@.660	Ea	27.10	24.60	—	51.70
3"	P1@.830	Ea	39.60	30.90	—	70.50
4"	P1@1.10	Ea	94.90	40.90	—	135.80

Carbon Steel, Schedule 40 with 150# M.I. Fittings & Threaded Joints

Description	Craft@Hrs	Unit	Material $	Labor $	Equipment $	Total $

150# malleable iron reducing tee, threaded joints

Description	Craft@Hrs	Unit	Material $	Labor $	Equipment $	Total $
¾" x ½"	P1@.180	Ea	3.27	6.70	—	9.97
1" x ½"	P1@.220	Ea	4.16	8.19	—	12.35
1" x ¾"	P1@.220	Ea	4.16	8.19	—	12.35
1¼" x ½"	P1@.290	Ea	7.54	10.80	—	18.34
1¼" x ¾"	P1@.290	Ea	6.64	10.80	—	17.44
1¼" x 1"	P1@.290	Ea	7.19	10.80	—	17.99
1½" x ½"	P1@.370	Ea	8.85	13.80	—	22.65
1½" x ¾"	P1@.370	Ea	8.35	13.80	—	22.15
1½" x 1"	P1@.370	Ea	8.35	13.80	—	22.15
1½" x 1¼"	P1@.370	Ea	10.30	13.80	—	24.10
2" x 1"	P1@.460	Ea	13.10	17.10	—	30.20
2" x 1¼"	P1@.460	Ea	14.20	17.10	—	31.30
2" x 1½"	P1@.460	Ea	14.20	17.10	—	31.30
4" x 1½"	P1@1.05	Ea	98.30	39.10	—	137.40
4" x 2"	P1@1.05	Ea	98.30	39.10	—	137.40
4" x 3"	P1@1.05	Ea	98.30	39.10	—	137.40

150# malleable iron reducer, threaded joints

Description	Craft@Hrs	Unit	Material $	Labor $	Equipment $	Total $
¾" x ½"	P1@.130	Ea	2.07	4.84	—	6.91
1" x ½"	P1@.160	Ea	3.24	5.96	—	9.20
1" x ¾"	P1@.160	Ea	3.24	5.96	—	9.20
1¼" x ¾"	P1@.210	Ea	3.24	7.82	—	11.06
1¼" x 1"	P1@.210	Ea	4.02	7.82	—	11.84
1½" x 1"	P1@.270	Ea	5.89	10.00	—	15.89
1½" x 1¼"	P1@.270	Ea	5.10	10.00	—	15.10
2" x 1¼"	P1@.340	Ea	8.43	12.70	—	21.13
2" x 1½"	P1@.340	Ea	7.40	12.70	—	20.10
3" x 2"	P1@.530	Ea	22.00	19.70	—	41.70
3" x 2½"	P1@.530	Ea	25.70	19.70	—	45.40
4" x 3"	P1@.750	Ea	25.70	27.90	—	53.60

150# malleable iron cross, threaded joints

Description	Craft@Hrs	Unit	Material $	Labor $	Equipment $	Total $
½"	P1@.280	Ea	5.58	10.40	—	15.98
¾"	P1@.320	Ea	6.81	11.90	—	18.71
1"	P1@.360	Ea	8.35	13.40	—	21.75
1¼"	P1@.480	Ea	13.40	17.90	—	31.30
1½"	P1@.600	Ea	16.60	22.30	—	38.90
2"	P1@.700	Ea	27.60	26.10	—	53.70
2½"	P1@1.02	Ea	50.90	38.00	—	88.90
3"	P1@1.28	Ea	70.50	47.60	—	118.10

Carbon Steel, Schedule 40 with 150# M.I. Fittings & Threaded Joints

Description	Craft@Hrs	Unit	Material $	Labor $	Equipment $	Total $
150# malleable iron cap, threaded joint						
½"	P1@.090	Ea	1.16	3.35	—	4.51
¾"	P1@.100	Ea	1.55	3.72	—	5.27
1"	P1@.140	Ea	1.91	5.21	—	7.12
1¼"	P1@.180	Ea	2.47	6.70	—	9.17
1½"	P1@.230	Ea	3.36	8.56	—	11.92
2"	P1@.290	Ea	4.93	10.80	—	15.73
2½"	P1@.380	Ea	11.00	14.10	—	25.10
3"	P1@.480	Ea	16.30	17.90	—	34.20
4"	P1@.640	Ea	27.70	23.80	—	51.50
150# malleable iron plug, threaded joint						
½"	P1@.090	Ea	.91	3.35	—	4.26
¾"	P1@.100	Ea	.98	3.72	—	4.70
1"	P1@.140	Ea	.98	5.21	—	6.19
1¼"	P1@.180	Ea	1.56	6.70	—	8.26
1½"	P1@.230	Ea	2.24	8.56	—	10.80
2"	P1@.290	Ea	2.61	10.80	—	13.41
2½"	P1@.380	Ea	6.32	14.10	—	20.42
3"	P1@.480	Ea	7.35	17.90	—	25.25
4"	P1@.640	Ea	13.40	23.80	—	37.20
150# malleable iron union, threaded joints						
½"	P1@.140	Ea	5.06	5.21	—	10.27
¾"	P1@.150	Ea	5.80	5.58	—	11.38
1"	P1@.210	Ea	7.55	7.82	—	15.37
1¼"	P1@.280	Ea	10.80	10.40	—	21.20
1½"	P1@.360	Ea	13.40	13.40	—	26.80
2"	P1@.450	Ea	15.60	16.70	—	32.30
2½"	P1@.610	Ea	46.50	22.70	—	69.20
3"	P1@.760	Ea	55.90	28.30	—	84.20
150# malleable iron coupling, threaded joints						
½"	P1@.120	Ea	1.55	4.47	—	6.02
¾"	P1@.130	Ea	1.84	4.84	—	6.68
1"	P1@.180	Ea	2.76	6.70	—	9.46
1¼"	P1@.240	Ea	3.56	8.93	—	12.49
1½"	P1@.300	Ea	4.75	11.20	—	15.95
2"	P1@.380	Ea	6.90	14.10	—	21.00

Carbon Steel, Schedule 40 with 150# M.I. Fittings & Threaded Joints

Description	Craft@Hrs	Unit	Material $	Labor $	Equipment $	Total $

Steel pipe nipples — Close nipple, standard right-hand thread

½"	P1@.025	Ea	.60	.93	—	1.53
¾"	P1@.025	Ea	.71	.93	—	1.64
1"	P1@.025	Ea	1.10	.93	—	2.03
1¼"	P1@.025	Ea	1.35	.93	—	2.28
1½"	P1@.025	Ea	1.62	.93	—	2.55
2"	P1@.025	Ea	2.08	.93	—	3.01
2½"	P1@.030	Ea	8.11	1.12	—	9.23
3"	P1@.030	Ea	8.78	1.12	—	9.90
4"	P1@.040	Ea	14.00	1.49	—	15.49
6"	P1@.050	Ea	70.70	1.86	—	72.56
8"	P1@.065	Ea	124.00	2.42	—	126.42

Steel pipe nipples — 1½" long, standard right-hand thread

½"	P1@.025	Ea	.63	.93	—	1.56
¾"	P1@.025	Ea	.72	.93	—	1.65

Steel pipe nipples — 2" long, standard right-hand thread

½"	P1@.025	Ea	.63	.93	—	1.56
¾"	P1@.025	Ea	.72	.93	—	1.65
1"	P1@.025	Ea	1.10	.93	—	2.03
1¼"	P1@.025	Ea	1.35	.93	—	2.28
1½"	P1@.025	Ea	1.62	.93	—	2.55

Steel pipe nipples — 2½" long, standard right-hand thread

½"	P1@.025	Ea	.64	.93	—	1.57
¾"	P1@.025	Ea	.76	.93	—	1.69
1"	P1@.025	Ea	1.25	.93	—	2.18
1¼"	P1@.025	Ea	1.60	.93	—	2.53
1½"	P1@.025	Ea	1.98	.93	—	2.91
2"	P1@.025	Ea	2.34	.93	—	3.27

Steel pipe nipples — 3" long, standard right-hand thread

½"	P1@.025	Ea	.75	.93	—	1.68
¾"	P1@.025	Ea	.89	.93	—	1.82
1"	P1@.025	Ea	1.30	.93	—	2.23
1¼"	P1@.025	Ea	1.69	.93	—	2.62
1½"	P1@.025	Ea	2.02	.93	—	2.95
2"	P1@.030	Ea	2.51	1.12	—	3.63
2½"	P1@.030	Ea	8.42	1.12	—	9.54
3"	P1@.035	Ea	9.49	1.30	—	10.79

Carbon Steel, Schedule 40 with 150# M.I. Fittings & Threaded Joints

Description	Craft@Hrs	Unit	Material $	Labor $	Equipment $	Total $
Steel pipe nipples — 3½" long, standard right-hand thread						
½"	P1@.025	Ea	.86	.93	—	1.79
¾"	P1@.025	Ea	1.03	.93	—	1.96
1"	P1@.025	Ea	1.49	.93	—	2.42
1¼"	P1@.025	Ea	1.90	.93	—	2.83
1½"	P1@.030	Ea	2.40	1.12	—	3.52
2"	P1@.030	Ea	2.96	1.12	—	4.08
2½"	P1@.035	Ea	11.20	1.30	—	12.50
3"	P1@.035	Ea	12.90	1.30	—	14.20
4"	P1@.040	Ea	17.70	1.49	—	19.19
Steel pipe nipples — 4" long, standard right-hand thread						
½"	P1@.025	Ea	.92	.93	—	1.85
¾"	P1@.025	Ea	1.10	.93	—	2.03
1"	P1@.025	Ea	1.58	.93	—	2.51
1¼"	P1@.030	Ea	1.95	1.12	—	3.07
1½"	P1@.030	Ea	2.47	1.12	—	3.59
2"	P1@.030	Ea	3.12	1.12	—	4.24
2½"	P1@.035	Ea	9.69	1.30	—	10.99
3"	P1@.040	Ea	10.80	1.49	—	12.29
4"	P1@.040	Ea	15.80	1.49	—	17.29
Steel pipe nipples — 4½" long, standard right-hand thread						
½"	P1@.025	Ea	1.10	.93	—	2.03
¾"	P1@.025	Ea	1.24	.93	—	2.17
1"	P1@.025	Ea	1.69	.93	—	2.62
1¼"	P1@.030	Ea	2.66	1.12	—	3.78
1½"	P1@.030	Ea	3.21	1.12	—	4.33
2"	P1@.030	Ea	3.45	1.12	—	4.57
2½"	P1@.035	Ea	12.40	1.30	—	13.70
3"	P1@.040	Ea	15.50	1.49	—	16.99
4"	P1@.050	Ea	19.50	1.86	—	21.36
6"	P1@.070	Ea	85.00	2.61	—	87.61
Steel pipe nipples — 5" long, standard right-hand thread						
½"	P1@.025	Ea	1.12	.93	—	2.05
¾"	P1@.025	Ea	1.24	.93	—	2.17
1"	P1@.030	Ea	1.84	1.12	—	2.96
1¼"	P1@.030	Ea	2.38	1.12	—	3.50
1½"	P1@.030	Ea	2.85	1.12	—	3.97
2"	P1@.035	Ea	3.58	1.30	—	4.88
2½"	P1@.040	Ea	13.10	1.49	—	14.59
3"	P1@.045	Ea	13.60	1.67	—	15.27
4"	P1@.050	Ea	21.30	1.86	—	23.16
6"	P1@.075	Ea	86.20	2.79	—	88.99
8"	P1@.100	Ea	144.00	3.72	—	147.72

Carbon Steel, Schedule 40 with 150# M.I. Fittings & Threaded Joints

Description	Craft@Hrs	Unit	Material $	Labor $	Equipment $	Total $

Steel pipe nipples — 6" long, standard right-hand thread

Description	Craft@Hrs	Unit	Material $	Labor $	Equipment $	Total $
½"	P1@.025	Ea	1.24	.93	—	2.17
¾"	P1@.025	Ea	1.51	.93	—	2.44
1"	P1@.030	Ea	2.02	1.12	—	3.14
1¼"	P1@.030	Ea	2.52	1.12	—	3.64
1½"	P1@.030	Ea	3.04	1.12	—	4.16
2"	P1@.035	Ea	4.09	1.30	—	5.39
2½"	P1@.040	Ea	11.80	1.49	—	13.29
3"	P1@.045	Ea	13.10	1.67	—	14.77
4"	P1@.060	Ea	19.50	2.23	—	21.73
6"	P1@.085	Ea	96.50	3.16	—	99.66
8"	P1@.120	Ea	154.00	4.47	—	158.47

Steel pipe nipples — 8" long, standard right-hand thread

Description	Craft@Hrs	Unit	Material $	Labor $	Equipment $	Total $
2½"	P1@.050	Ea	15.40	1.86	—	17.26
3"	P1@.055	Ea	18.90	2.05	—	20.95
4"	P1@.070	Ea	30.30	2.61	—	32.91
6"	P1@.100	Ea	101.00	3.72	—	104.72
8"	P1@.150	Ea	173.00	5.58	—	178.58

Steel pipe nipples — 10" long, standard right-hand thread

Description	Craft@Hrs	Unit	Material $	Labor $	Equipment $	Total $
2½"	P1@.055	Ea	18.60	2.05	—	20.65
3"	P1@.060	Ea	22.50	2.23	—	24.73
4"	P1@.080	Ea	36.70	2.98	—	39.68
6"	P1@.115	Ea	113.00	4.28	—	117.28
8"	P1@.165	Ea	194.00	6.14	—	200.14

Steel pipe nipples — 12" long, standard right-hand thread

Description	Craft@Hrs	Unit	Material $	Labor $	Equipment $	Total $
2½"	P1@.060	Ea	18.00	2.23	—	20.23
3"	P1@.065	Ea	22.10	2.42	—	24.52
4"	P1@.080	Ea	35.90	2.98	—	38.88
6"	P1@.130	Ea	129.00	4.84	—	133.84
8"	P1@.190	Ea	213.00	7.07	—	220.07

Class 125 bronze body gate valve, threaded ends

Description	Craft@Hrs	Unit	Material $	Labor $	Equipment $	Total $
½"	P1@.210	Ea	19.60	7.82	—	27.42
¾"	P1@.250	Ea	24.50	9.31	—	33.81
1"	P1@.300	Ea	34.40	11.20	—	45.60
1¼"	P1@.400	Ea	43.40	14.90	—	58.30
1½"	P1@.450	Ea	58.60	16.70	—	75.30
2"	P1@.500	Ea	98.00	18.60	—	116.60
2½"	P1@.750	Ea	161.00	27.90	—	188.90
3"	P1@.950	Ea	231.00	35.40	—	266.40

Carbon Steel, Schedule 40 with 150# M.I. Fittings & Threaded Joints

Description	Craft@Hrs	Unit	Material $	Labor $	Equipment $	Total $

Class 125 iron body gate valve, flanged ends

2"	P1@.500	Ea	335.00	18.60	—	353.60
2½"	P1@.600	Ea	506.00	22.30	—	528.30
3"	P1@.750	Ea	550.00	27.90	—	577.90
4"	P1@1.35	Ea	805.00	50.20	—	855.20

Class 125 bronze body globe valve, threaded ends

½"	P1@.210	Ea	36.50	7.82	—	44.32
¾"	P1@.250	Ea	48.40	9.31	—	57.71
1"	P1@.300	Ea	69.40	11.20	—	80.60
1¼"	P1@.400	Ea	97.70	14.90	—	112.60
1½"	P1@.450	Ea	131.00	16.70	—	147.70
2"	P1@.500	Ea	214.00	18.60	—	232.60

Class 125 iron body globe valve, flanged ends

2"	P1@.500	Ea	273.00	18.60	—	291.60
2½"	P1@.600	Ea	425.00	22.30	—	447.30
3"	P1@.750	Ea	509.00	27.90	—	536.90
4"	P1@1.35	Ea	684.00	50.20	—	734.20

200 PSIG iron body butterfly valve, lug-type, lever operated

2"	P1@.450	Ea	168.00	16.70	—	184.70
2½"	P1@.450	Ea	175.00	16.70	—	191.70
3"	P1@.550	Ea	183.00	20.50	—	203.50
4"	P1@.550	Ea	228.00	20.50	—	248.50

200 PSIG iron body butterfly valve, wafer-type, lever operated

2"	P1@.450	Ea	153.00	16.70	—	169.70
2½"	P1@.450	Ea	156.00	16.70	—	172.70
3"	P1@.550	Ea	168.00	20.50	—	188.50
4"	P1@.550	Ea	204.00	20.50	—	224.50

Class 125 bronze body 2-piece ball valve, threaded ends

½"	P1@.210	Ea	10.40	7.82	—	18.22
¾"	P1@.250	Ea	13.80	9.31	—	23.11
1"	P1@.300	Ea	24.10	11.20	—	35.30
1¼"	P1@.400	Ea	40.10	14.90	—	55.00
1½"	P1@.450	Ea	54.10	16.70	—	70.80
2"	P1@.500	Ea	68.70	18.60	—	87.30
3"	P1@.625	Ea	470.00	23.30	—	493.30
4"	P1@.690	Ea	615.00	25.70	—	640.70

Carbon Steel, Schedule 40 with 150# M.I. Fittings & Threaded Joints

Description	Craft@Hrs	Unit	Material $	Labor $	Equipment $	Total $

Class 125 bronze body swing check valve, threaded

½"	P1@.210	Ea	20.80	7.82	—	28.62
¾"	P1@.250	Ea	29.70	9.31	—	39.01
1"	P1@.300	Ea	38.80	11.20	—	50.00
1¼"	P1@.400	Ea	55.40	14.90	—	70.30
1½"	P1@.450	Ea	78.10	16.70	—	94.80
2"	P1@.500	Ea	129.00	18.60	—	147.60

Class 125 iron body swing check valve, flanged ends

2"	P1@.500	Ea	182.00	18.60	—	200.60
2½"	P1@.600	Ea	230.00	22.30	—	252.30
3"	P1@.750	Ea	287.00	27.90	—	314.90
4"	P1@1.35	Ea	421.00	50.20	—	471.20

Class 125 iron body silent check valve, wafer-type

2"	P1@.500	Ea	133.00	18.60	—	151.60
2½"	P1@.600	Ea	148.00	22.30	—	170.30
3"	P1@.750	Ea	170.00	27.90	—	197.90
4"	P1@1.35	Ea	227.00	50.20	—	277.20

Class 125 bronze body strainer, threaded ends

½"	P1@.210	Ea	29.80	7.82	—	37.62
¾"	P1@.250	Ea	39.20	9.31	—	48.51
1"	P1@.300	Ea	48.00	11.20	—	59.20
1¼"	P1@.400	Ea	67.10	14.90	—	82.00
1½"	P1@.450	Ea	100.00	16.70	—	116.70
2"	P1@.500	Ea	175.00	18.60	—	193.60

Class 125 iron body strainer, flanged ends

2"	P1@.500	Ea	134.00	18.60	—	152.60
2½"	P1@.600	Ea	150.00	22.30	—	172.30
3"	P1@.750	Ea	173.00	27.90	—	200.90
4"	P1@1.35	Ea	518.00	50.20	—	568.20

Installation of 2-way control valve, threaded joints

½"	P1@.210	Ea	—	7.82	—	7.82
¾"	P1@.275	Ea	—	10.20	—	10.20
1"	P1@.350	Ea	—	13.00	—	13.00
1¼"	P1@.430	Ea	—	16.00	—	16.00
1½"	P1@.505	Ea	—	18.80	—	18.80
2"	P1@.675	Ea	—	25.10	—	25.10
2½"	P1@.830	Ea	—	30.90	—	30.90
3"	P1@.990	Ea	—	36.80	—	36.80

Carbon Steel, Schedule 40 with 150# M.I. Fittings & Threaded Joints

Description	Craft@Hrs	Unit	Material $	Labor $	Equipment $	Total $
Installation of 3-way control valve, threaded joints						
½"	P1@.260	Ea	—	9.68	—	9.68
¾"	P1@.365	Ea	—	13.60	—	13.60
1"	P1@.475	Ea	—	17.70	—	17.70
1¼"	P1@.575	Ea	—	21.40	—	21.40
1½"	P1@.680	Ea	—	25.30	—	25.30
2"	P1@.910	Ea	—	33.90	—	33.90
2½"	P1@1.12	Ea	—	41.70	—	41.70
3"	P1@1.33	Ea	—	49.50	—	49.50
Companion flange, threaded						
2"	P1@.290	Ea	42.30	10.80	—	53.10
2½"	P1@.380	Ea	65.10	14.10	—	79.20
3"	P1@.460	Ea	70.40	17.10	—	87.50
4"	P1@.600	Ea	96.80	22.30	—	119.10
Bolt and gasket sets						
2"	P1@.500	Ea	4.26	18.60	—	22.86
2½"	P1@.650	Ea	4.98	24.20	—	29.18
3"	P1@.750	Ea	8.42	27.90	—	36.32
4"	P1@1.00	Ea	14.00	37.20	—	51.20
Thermometer with well						
7"	P1@.250	Ea	187.00	9.31	—	196.31
9"	P1@.250	Ea	193.00	9.31	—	202.31
Dial-type pressure gauge						
2½"	P1@.200	Ea	39.10	7.44	—	46.54
3½"	P1@.200	Ea	51.50	7.44	—	58.94
Pressure/temperature tap						
Tap	P1@.150	Ea	18.00	5.58	—	23.58
Hanger with swivel assembly						
½"	P1@.250	Ea	4.77	9.31	—	14.08
¾"	P1@.250	Ea	4.98	9.31	—	14.29
1"	P1@.250	Ea	5.24	9.31	—	14.55
1¼"	P1@.300	Ea	5.48	11.20	—	16.68
1½"	P1@.300	Ea	6.33	11.20	—	17.53
2"	P1@.300	Ea	6.94	11.20	—	18.14
2½"	P1@.350	Ea	8.99	13.00	—	21.99
3"	P1@.350	Ea	11.10	13.00	—	24.10
4"	P1@.350	Ea	17.50	13.00	—	30.50

Carbon Steel, Schedule 40 with 150# M.I. Fittings & Threaded Joints

Description	Craft@Hrs	Unit	Material $	Labor $	Equipment $	Total $
Riser clamp						
½"	P1@.100	Ea	3.20	3.72	—	6.92
¾"	P1@.100	Ea	4.69	3.72	—	8.41
1"	P1@.100	Ea	4.75	3.72	—	8.47
1¼"	P1@.105	Ea	5.73	3.91	—	9.64
1½"	P1@.110	Ea	6.04	4.09	—	10.13
2"	P1@.115	Ea	6.39	4.28	—	10.67
2½"	P1@.120	Ea	6.74	4.47	—	11.21
3"	P1@.120	Ea	7.29	4.47	—	11.76
4"	P1@.125	Ea	9.29	4.65	—	13.94

Carbon Steel, Schedule 5 with Pressfit Fittings

Schedule 5 carbon steel (ASTM A-53, A135 and A-795) pipe with pressfit fittings. The pressfit jointing method offers economy, speed and reliability for joining ¾" to 2" diameter pipe for fire protection, heating and cooling water systems. *(Use roll-grooved pipe and fittings for 2½" and larger pipe diameters.)*

The pressfit system for carbon steel is BOCA approved. Listed by SBCCI PST and ESI and is UL/ULC listed and FM approved for 175 PSI (1,200 kPa) on fire protection services and is rated to 300 PSI (2,065 kPa) for heating water systems.

The pressfit system requires no special preparation of the pipe ends before assembly. Pipe should be square cut (+/–0.030") and de-burred. Pressfit system products are designed only for use on approved Schedule 5 carbon steel pipe.

O-ring selection as specified by the mechanical design engineer. Type O O-rings are rated for application temperatures of +20 degrees F to +300 degrees F. Type O O-rings are recommended for many oxidizing acids, petroleum oils, halogenated hydrocarbons, lubricants, hydraulic fluids, organic liquids and air with hydrocarbons within the specified temperature range. Not recommended for steam services.

Standard type E O-rings are rated for application temperatures of –30 degrees F to +230 degrees F. Type E O-rings are recommended for hot water service within the specified temperature range plus a variety of dilute acids, oil free air and many chemical services.

Type T O-rings are rated for application temperatures of –20 degrees F to +180 degrees F. Type T O-rings are recommended for petroleum products, air with oil vapors, vegetable or mineral oils within the specified temperature range. Not recommended for hot water services over 150 degrees F., or steam, or hot dry air over 140 degrees F.

A specialized hand-held electric pressfit tool is required to make the pipe/fitting connections. Equipment cost, where shown, is $1.95 per hour for this electric pressfit tool.

Description	Craft@Hrs	Unit	Material $	Labor $	Equipment $	Total $

Schedule 5 carbon steel pipe, plain end, pressfit joints, pipe only. No hangers or fittings.

Description	Craft@Hrs	Unit	Material $	Labor $	Equipment $	Total $
¾"	P1@.030	Ea	2.14	1.12	.03	3.29
1"	P1@.040	LF	2.92	1.49	.04	4.45
1¼"	P1@.040	LF	3.72	1.49	.04	5.25
1½"	P1@.050	LF	4.44	1.86	.05	6.35
2"	P1@.050	LF	5.33	1.86	.05	7.24

45-degree pressfit ell, with Type O O-ring (P x P)

Description	Craft@Hrs	Unit	Material $	Labor $	Equipment $	Total $
¾"	P1@.170	Ea	32.80	6.33	.17	39.30
1"	P1@.180	Ea	39.40	6.70	.18	46.28
1¼"	P1@.190	Ea	54.20	7.07	.18	61.45
1½"	P1@.200	Ea	69.00	7.44	.19	76.63
2"	P1@.200	Ea	93.50	7.44	.19	101.13

Carbon Steel, Schedule 5 with Pressfit Fittings

Description	Craft@Hrs	Unit	Material $	Labor $	Equipment $	Total $
90-degree standard radius pressfit ell, with Type O O-ring (P x P)						
¾"	P1@.170	Ea	34.40	6.33	.17	40.90
1"	P1@.180	Ea	38.80	6.70	.18	45.68
1¼"	P1@.190	Ea	54.20	7.07	.18	61.45
1½"	P1@.200	Ea	66.30	7.44	.19	73.93
2"	P1@.200	Ea	91.70	7.44	.19	99.33
90-degree short radius pressfit ell, with Type O O-ring (P x P)						
¾"	P1@.170	Ea	39.40	6.33	.17	45.90
1"	P1@.180	Ea	47.60	6.70	.18	54.48
1¼"	P1@.190	Ea	72.40	7.07	.18	79.65
1½"	P1@.200	Ea	89.50	7.44	.19	97.13
2"	P1@.200	Ea	117.00	7.44	.19	124.63
90-degree standard radius reducing pressfit x FIP ell, with Type O O-ring (P x F)						
1" x ½"	P1@.185	Ea	44.40	6.89	.18	51.47
1" x ¾"	P1@.185	Ea	44.90	6.89	.18	51.97
1¼" x ½"	P1@.195	Ea	59.20	7.26	.19	66.65
1¼" x ½"	P1@.195	Ea	60.90	7.26	.19	68.35
1½" x ½"	P1@.215	Ea	74.40	8.00	.21	82.61
1½" x ¾"	P1@.215	Ea	75.00	8.00	.21	83.21
90-degree short radius reducing pressfit x FIP ell, with Type O O-ring (P x F)						
1" x ¾"	P1@.185	Ea	40.40	6.89	.18	47.47
¾" x 1"	P1@.185	Ea	41.00	6.89	.18	48.07
1¼" x ¾"	P1@.195	Ea	56.70	7.26	.19	64.15
1½" x ¾"	P1@.215	Ea	72.00	8.00	.21	80.21
Standard coupling, pressfit, with Type O O-ring (P x P)						
¾"	P1@.170	Ea	25.70	6.33	.17	32.20
1"	P1@.180	Ea	33.40	6.70	.18	40.28
1¼"	P1@.190	Ea	46.20	7.07	.18	53.45
1½"	P1@.200	Ea	56.80	7.44	.19	64.43
2"	P1@.200	Ea	76.20	7.44	.19	83.83

Carbon Steel, Schedule 5 with Pressfit Fittings

Description	Craft@Hrs	Unit	Material $	Labor $	Equipment $	Total $
Slip coupling, pressfit, with Type O O-ring (P x P)						
¾"	P1@.170	Ea	34.60	6.33	.17	41.10
1"	P1@.180	Ea	44.40	6.70	.18	51.28
1¼"	P1@.190	Ea	59.20	7.07	.18	66.45
1½"	P1@.200	Ea	73.30	7.44	.19	80.93
2"	P1@.200	Ea	97.40	7.44	.19	105.03
Tee, pressfit, with Type O O-ring (P x P)						
¾"	P1@.190	Ea	39.30	7.07	.18	46.55
1"	P1@.220	Ea	46.40	8.19	.21	54.80
1¼"	P1@.240	Ea	61.30	8.93	.23	70.46
1½"	P1@.255	Ea	74.80	9.49	.25	84.54
2"	P1@.270	Ea	102.00	10.00	.26	112.26
Reducing tee, pressfit, with Type O O-ring (P x P x P)						
1" x ¾"	P1@.220	Ea	31.10	8.19	.21	39.50
1¼" x ¾"	P1@.240	Ea	63.00	8.93	.23	72.16
1¼" x 1"	P1@.240	Ea	75.00	8.93	.23	84.16
1½" x ¾"	P1@.255	Ea	73.30	9.49	.25	83.04
1½" x 1"	P1@.255	Ea	84.80	9.49	.25	94.54
2" x ¾"	P1@.270	Ea	96.80	10.00	.26	107.06
2" x 1"	P1@.270	Ea	111.00	10.00	.26	121.26
2" x 1½"	P1@.270	Ea	117.00	10.00	.26	127.26
Male threaded pressfit adapter with Type O O-ring (P x M)						
¾" x ½"	P1@.180	Ea	23.90	6.70	.18	30.78
¾" x ¾"	P1@.190	Ea	25.70	7.07	.18	32.95
¾" x 1"	P1@.190	Ea	28.40	7.07	.18	35.65
1" x ¾"	P1@.190	Ea	33.30	7.07	.18	40.55
1" x 1"	P1@.190	Ea	35.60	7.07	.18	42.85
1¼" x 1¼"	P1@.200	Ea	49.10	7.44	.19	56.73
1½" x 1½"	P1@.215	Ea	61.50	8.00	.21	69.71
2" x 2"	P1@.225	Ea	79.70	8.37	.22	88.29
Female threaded pressfit adapter with Type O O-ring (P x F)						
¾" x ½"	P1@.180	Ea	23.90	6.70	.18	30.78
¾" x ¾"	P1@.190	Ea	25.70	7.07	.18	32.95
1" x ½"	P1@.190	Ea	26.60	7.07	.18	33.85
1" x ¾"	P1@.190	Ea	35.60	7.07	.18	42.85
1" x 1"	P1@.200	Ea	36.50	7.44	.19	44.13

Carbon Steel, Schedule 80 with 300# Fittings & Butt-Welded Joints

Schedule 80 carbon steel (ASTM A-53 and A-120) pipe with 300 pound carbon steel welding fittings is commonly used for steam condensate and re-circulating water systems where operating pressures do not exceed 700 PSIG.

The cost estimates in this section are based on the conditions, limitations and wage rates described in the section "How to Use This Book" beginning on page 5.

Equipment cost, where shown, is $4.45 per hour for a 2-ton chain hoist and $6.02 per hour for 200 amp gas welder, trailer mounted.

Description	Craft@Hrs	Unit	Material $	Labor $	Equipment $	Total $

Schedule 80 carbon steel butt-welded horizontal pipe assembly.
Horizontally hung in a building. Assembly includes fittings and hanger assemblies. Based on a reducing tee and a 45-degree elbow every 16 feet for ½" pipe. A reducing tee and a 45-degree elbow every 50 feet for 12" pipe. Hangers spaced to meet plumbing code. *(Distance between fittings in the pipe assembly increases as pipe diameter increases.)* Use these figures for preliminary estimates.

Description	Craft@Hrs	Unit	Material $	Labor $	Equipment $	Total $
½"	P1@.183	LF	4.89	6.81	.58	12.28
¾"	P1@.203	LF	5.47	7.56	.64	13.67
1"	P1@.241	LF	6.25	8.97	.76	15.98
1¼"	P1@.246	LF	7.01	9.16	.78	16.95
1½"	P1@.271	LF	7.60	10.10	.86	18.56
2"	P1@.350	LF	9.90	13.00	1.11	24.01
2½"	P1@.437	LF	14.70	16.30	1.38	32.38
3"	P1@.532	LF	19.20	19.80	1.68	40.68
4"	P1@.601	LF	27.70	22.40	1.90	52.00
6"	ER@.833	LF	38.80	33.30	1.30	73.40
8"	ER@.994	LF	60.00	39.70	1.55	101.25
10"	ER@1.20	LF	102.00	48.00	1.87	151.87
12"	ER@1.43	LF	119.00	57.10	2.22	178.32

Schedule 80 carbon steel butt-welded vertical pipe assembly.
Riser assembly, including fittings and riser clamps. Based on a reducing tee, and a riser clamp on every floor. Use these figures for preliminary estimates.

Description	Craft@Hrs	Unit	Material $	Labor $	Equipment $	Total $
½"	P1@.153	LF	4.82	5.69	.48	10.99
¾"	P1@.163	LF	5.41	6.07	.52	12.00
1"	P1@.196	LF	6.19	7.30	.62	14.11
1¼"	P1@.229	LF	7.72	8.52	.72	16.96
1½"	P1@.240	LF	8.24	8.93	.76	17.93
2"	P1@.310	LF	10.30	11.50	.98	22.78
2½"	P1@.395	LF	15.00	14.70	1.25	30.95
3"	P1@.516	LF	19.30	19.20	1.63	40.13
4"	P1@.610	LF	27.70	22.70	1.93	52.33
6"	ER@.877	LF	41.00	35.00	1.36	77.36
8"	ER@1.04	LF	65.40	41.60	1.62	108.62
10"	ER@1.28	LF	109.00	51.10	1.99	162.09
12"	ER@1.58	LF	135.00	63.10	2.46	200.56

Carbon Steel, Schedule 80 with 300# Fittings & Butt-Welded Joints

Description	Craft@Hrs	Unit	Material $	Labor $	Equipment $	Total $
Schedule 80 carbon steel pipe, plain end, butt-welded						
½"	P1@.080	LF	1.95	2.98	.25	5.18
¾"	P1@.090	LF	2.42	3.35	.28	6.05
1"	P1@.110	LF	3.33	4.09	.35	7.77
1¼"	P1@.120	LF	4.43	4.47	.38	9.28
1½"	P1@.130	LF	5.33	4.84	.41	10.58
2"	P1@.170	LF	7.40	6.33	.54	14.27
2½"	P1@.220	LF	11.40	8.19	.70	20.29
3"	P1@.280	LF	15.00	10.40	.89	26.29
4"	P1@.340	LF	22.10	12.70	1.08	35.88
6"	ER@.460	LF	33.40	18.40	.72	52.52
8"	ER@.560	LF	50.20	22.40	.87	73.47
10"	ER@.660	LF	87.10	26.40	1.03	114.53
12"	ER@.840	LF	100.00	33.60	1.31	134.91
Schedule 80 carbon steel 45-degree ell, butt-welded						
½"	P1@.300	Ea	7.08	11.20	.95	19.23
¾"	P1@.440	Ea	7.08	16.40	1.39	24.87
1"	P1@.590	Ea	7.08	22.00	1.87	30.95
1¼"	P1@.740	Ea	7.80	27.50	2.34	37.64
1½"	P1@.890	Ea	8.12	33.10	2.82	44.04
2"	P1@1.18	Ea	7.62	43.90	3.73	55.25
2½"	P1@1.49	Ea	10.40	55.50	4.71	70.61
3"	P1@1.77	Ea	13.10	65.90	5.60	84.60
4"	P1@2.37	Ea	19.90	88.20	7.50	115.60
6"	ER@3.55	Ea	55.70	142.00	5.52	203.22
8"	ER@4.26	Ea	90.20	170.00	6.63	266.83
10"	ER@5.32	Ea	144.00	213.00	8.28	365.28
12"	ER@6.38	Ea	217.00	255.00	9.93	481.93
Schedule 80 carbon steel 90-degree ell, butt-welded						
½"	P1@.300	Ea	7.15	11.20	.95	19.30
¾"	P1@.440	Ea	7.15	16.40	1.39	24.94
1"	P1@.590	Ea	7.15	22.00	1.87	31.02
1¼"	P1@.740	Ea	8.22	27.50	2.34	38.06
1½"	P1@.890	Ea	8.81	33.10	2.82	44.73
2"	P1@1.18	Ea	13.20	43.90	3.73	60.83
2½"	P1@1.49	Ea	13.20	55.50	4.71	73.41
3"	P1@1.77	Ea	17.30	65.90	5.60	88.80
4"	P1@2.37	Ea	41.10	88.20	7.50	136.80
6"	ER@3.55	Ea	78.20	142.00	5.52	225.72
8"	ER@4.26	Ea	143.00	170.00	6.63	319.63
10"	ER@5.32	Ea	261.00	213.00	8.28	482.28
12"	ER@6.38	Ea	367.00	255.00	9.93	631.93

Carbon Steel, Schedule 80 with 300# Fittings & Butt-Welded Joints

Description	Craft@Hrs	Unit	Material $	Labor $	Equipment $	Total $

Schedule 80 carbon steel tee, butt-welded

Description	Craft@Hrs	Unit	Material $	Labor $	Equipment $	Total $
½"	P1@.440	Ea	24.00	16.40	1.39	41.79
¾"	P1@.670	Ea	24.00	24.90	2.12	51.02
1"	P1@.890	Ea	24.00	33.10	2.82	59.92
1¼"	P1@1.11	Ea	24.00	41.30	3.51	68.81
1½"	P1@1.33	Ea	27.10	49.50	4.21	80.81
2"	P1@1.77	Ea	23.20	65.90	5.60	94.70
2½"	P1@2.22	Ea	55.90	82.60	7.02	145.52
3"	P1@2.66	Ea	61.40	99.00	8.42	168.82
4"	P1@3.55	Ea	64.80	132.00	11.20	208.00
6"	ER@5.32	Ea	104.00	213.00	8.28	325.28
8"	ER@6.38	Ea	209.00	255.00	9.93	473.93
10"	ER@7.98	Ea	337.00	319.00	12.40	668.40
12"	ER@9.58	Ea	487.00	383.00	14.90	884.90

Carbon Steel, Schedule 80 with 300# Fittings & Butt-Welded Joints

Description	Craft@Hrs	Unit	Material $	Labor $	Equipment $	Total $
Schedule 80 carbon steel reducing tee, butt-welded						
1" x ¾"	P1@.810	Ea	29.30	30.10	2.56	61.96
1¼" x ½"	P1@1.04	Ea	29.30	38.70	3.29	71.29
1¼" x ¾"	P1@1.04	Ea	29.30	38.70	3.29	71.29
1¼" x 1"	P1@1.04	Ea	29.30	38.70	3.29	71.29
1½" x ½"	P1@1.26	Ea	29.30	46.90	3.99	80.19
1½" x ¾"	P1@1.26	Ea	29.30	46.90	3.99	80.19
1½" x 1"	P1@1.26	Ea	29.30	46.90	3.99	80.19
1½" x 1¼"	P1@1.26	Ea	29.30	46.90	3.99	80.19
2" x ¾"	P1@1.63	Ea	31.80	60.70	5.16	97.66
2" x 1"	P1@1.63	Ea	31.80	60.70	5.16	97.66
2" x 1¼"	P1@1.63	Ea	31.80	60.70	5.16	97.66
2" x 1½"	P1@1.63	Ea	31.80	60.70	5.16	97.66
2½" x 1"	P1@2.07	Ea	44.10	77.00	6.55	127.65
2½" x 1¼"	P1@2.07	Ea	44.10	77.00	6.55	127.65
2½" x 1½"	P1@2.07	Ea	44.10	77.00	6.55	127.65
2½" x 2"	P1@2.07	Ea	44.10	77.00	6.55	127.65
3" x 1¼"	P1@2.51	Ea	44.10	93.40	7.94	145.44
3" x 2"	P1@2.51	Ea	44.10	93.40	7.94	145.44
3" x 2½"	P1@2.51	Ea	44.10	93.40	7.94	145.44
4" x 1½"	P1@3.25	Ea	69.20	121.00	10.30	200.50
4" x 2"	P1@3.25	Ea	69.20	121.00	10.30	200.50
4" x 2½"	P1@3.25	Ea	69.20	121.00	10.30	200.50
4" x 3"	P1@3.25	Ea	69.20	121.00	10.30	200.50
6" x 2½"	ER@5.03	Ea	128.00	201.00	7.83	336.83
6" x 3"	ER@5.03	Ea	128.00	201.00	7.83	336.83
6" x 4"	ER@5.03	Ea	128.00	201.00	7.83	336.83
8" x 3"	ER@5.85	Ea	227.00	234.00	9.10	470.10
8" x 4"	ER@5.85	Ea	227.00	234.00	9.10	470.10
8" x 6"	ER@5.85	Ea	227.00	234.00	9.10	470.10
10" x 4"	ER@7.45	Ea	361.00	298.00	11.60	670.60
10" x 6"	ER@7.45	Ea	361.00	298.00	11.60	670.60
10" x 8"	ER@7.45	Ea	361.00	298.00	11.60	670.60
12" x 6"	ER@9.04	Ea	663.00	361.00	14.10	1,038.10
12" x 8"	ER@9.04	Ea	663.00	361.00	14.10	1,038.10
12" x 10"	ER@9.04	Ea	663.00	361.00	14.10	1,038.10
Schedule 80 carbon steel concentric reducer, butt-welded						
¾"	P1@.360	Ea	14.20	13.40	1.14	28.74
1"	P1@.500	Ea	15.30	18.60	1.58	35.48
1¼"	P1@.665	Ea	13.20	24.80	2.10	40.10
1½"	P1@.800	Ea	14.10	29.80	2.53	46.43
2"	P1@1.00	Ea	12.10	37.20	3.16	52.46
2½"	P1@1.33	Ea	16.10	49.50	4.21	69.81
3"	P1@1.63	Ea	15.30	60.70	5.16	81.16
4"	P1@2.00	Ea	20.20	74.40	6.33	100.93
6"	ER@2.96	Ea	44.10	118.00	2.68	164.78
8"	ER@3.90	Ea	65.30	156.00	3.53	224.83
10"	ER@4.79	Ea	90.20	191.00	4.33	285.53
12"	ER@5.85	Ea	132.00	234.00	5.29	371.29

Carbon Steel, Schedule 80 with 300# Fittings & Butt-Welded Joints

Description	Craft@Hrs	Unit	Material $	Labor $	Equipment $	Total $

Schedule 80 carbon steel eccentric reducer, butt-welded

Description	Craft@Hrs	Unit	Material $	Labor $	Equipment $	Total $
¾"	P1@.360	Ea	17.50	13.40	1.14	32.04
1"	P1@.500	Ea	22.70	18.60	1.58	42.88
1¼"	P1@.665	Ea	18.30	24.80	2.10	45.20
1½"	P1@.800	Ea	21.70	29.80	2.53	54.03
2"	P1@1.00	Ea	17.30	37.20	3.16	57.66
2½"	P1@1.33	Ea	20.90	49.50	4.21	74.61
3"	P1@1.63	Ea	20.20	60.70	5.16	86.06
4"	P1@2.00	Ea	28.40	74.40	6.33	109.13
6"	ER@2.96	Ea	64.90	118.00	2.68	185.58
8"	ER@3.90	Ea	96.80	156.00	3.53	256.33
10"	ER@4.79	Ea	169.00	191.00	4.33	364.33
12"	ER@5.85	Ea	263.00	234.00	5.29	502.29

Schedule 80 carbon steel cap, butt-welded

Description	Craft@Hrs	Unit	Material $	Labor $	Equipment $	Total $
½"	P1@.180	Ea	14.10	6.70	.57	21.37
¾"	P1@.270	Ea	19.70	10.00	.85	30.55
1"	P1@.360	Ea	9.12	13.40	1.14	23.66
1¼"	P1@.460	Ea	9.12	17.10	1.46	27.68
1½"	P1@.560	Ea	9.98	20.80	1.77	32.55
2"	P1@.740	Ea	8.65	27.50	2.34	38.49
2½"	P1@.920	Ea	9.68	34.20	2.91	46.79
3"	P1@1.11	Ea	10.30	41.30	3.51	55.11
4"	P1@1.50	Ea	13.80	55.80	4.75	74.35
6"	ER@2.24	Ea	25.70	89.50	2.03	117.23
8"	ER@2.68	Ea	39.70	107.00	2.42	149.12
10"	ER@3.36	Ea	59.10	134.00	3.04	196.14
12"	ER@4.00	Ea	78.00	160.00	3.62	241.62

Schedule 80 carbon steel union, butt-welded

Description	Craft@Hrs	Unit	Material $	Labor $	Equipment $	Total $
½"	P1@.300	Ea	11.90	11.20	.95	24.05
¾"	P1@.440	Ea	14.20	16.40	1.39	31.99
1"	P1@.590	Ea	18.90	22.00	1.87	42.77
1¼"	P1@.740	Ea	31.80	27.50	2.34	61.64
1½"	P1@.890	Ea	34.40	33.10	2.82	70.32
2"	P1@1.18	Ea	46.00	43.90	3.73	93.63
2½"	P1@1.49	Ea	100.00	55.50	4.71	160.21
3"	P1@1.77	Ea	124.00	65.90	5.60	195.50

Carbon steel weldolet

Description	Craft@Hrs	Unit	Material $	Labor $	Equipment $	Total $
½"	P1@.440	Ea	13.00	16.40	1.39	30.79
¾"	P1@.670	Ea	13.30	24.90	2.12	40.32
1"	P1@.890	Ea	14.40	33.10	2.82	50.32
1¼"	P1@1.11	Ea	18.90	41.30	3.51	63.71
1½"	P1@1.33	Ea	18.90	49.50	4.21	72.61
2"	P1@1.77	Ea	19.30	65.90	5.60	90.80
2½"	P1@2.22	Ea	44.00	82.60	7.02	133.62
3"	P1@2.66	Ea	45.60	99.00	8.42	153.02
4"	P1@3.55	Ea	101.00	132.00	11.20	244.20

Carbon Steel, Schedule 80 with 300# Fittings & Butt-Welded Joints

Description	Craft@Hrs	Unit	Material $	Labor $	Equipment $	Total $
Carbon steel threadolet						
½"	P1@.300	Ea	9.51	11.20	.95	21.66
¾"	P1@.440	Ea	10.50	16.40	1.39	28.29
1"	P1@.590	Ea	13.00	22.00	1.87	36.87
1¼"	P1@.740	Ea	58.30	27.50	2.34	88.14
1½"	P1@.890	Ea	58.30	33.10	2.82	94.22
2"	P1@1.18	Ea	76.80	43.90	3.73	124.43
300# forged steel slip-on companion flange, welding-type						
2½"	P1@.810	Ea	32.40	30.10	2.56	65.06
3"	P1@.980	Ea	32.40	36.50	3.10	72.00
4"	P1@1.30	Ea	47.80	48.40	4.11	100.31
6"	ER@1.95	Ea	86.70	77.90	1.76	166.36
8"	ER@2.35	Ea	122.00	93.90	2.12	218.02
10"	ER@2.90	Ea	202.00	116.00	2.62	320.62
12"	ER@3.50	Ea	323.00	140.00	3.16	466.16
Class 300 bronze body gate valve, threaded joints						
½"	P1@.210	Ea	45.90	7.82	—	53.72
¾"	P1@.250	Ea	59.90	9.31	—	69.21
1"	P1@.300	Ea	72.60	11.20	—	83.80
1¼"	P1@.400	Ea	99.00	14.90	—	113.90
1½"	P1@.450	Ea	125.00	16.70	—	141.70
2"	P1@.500	Ea	184.00	18.60	—	202.60
2½"	P1@.750	Ea	376.00	27.90	—	403.90
3"	P1@.950	Ea	533.00	35.40	—	568.40
Class 250 iron body gate valve, flanged						
2½"	P1@.600	Ea	114.00	22.30	—	136.30
3"	P1@.750	Ea	825.00	27.90	—	852.90
4"	P1@1.35	Ea	1,210.00	50.20	—	1,260.20
6"	ER@2.50	Ea	1,620.00	99.90	1.63	1,721.53
8"	ER@3.00	Ea	5,560.00	120.00	1.96	5,681.96
10"	ER@4.00	Ea	9,070.00	160.00	2.61	9,232.61
12"	ER@4.50	Ea	14,800.00	180.00	2.93	14,982.93
Class 300 bronze body globe valve, threaded						
½"	P1@.210	Ea	68.40	7.82	—	76.22
¾"	P1@.250	Ea	86.80	9.31	—	96.11
1"	P1@.300	Ea	124.00	11.20	—	135.20
1¼"	P1@.400	Ea	180.00	14.90	—	194.90
1½"	P1@.450	Ea	239.00	16.70	—	255.70
2"	P1@.500	Ea	331.00	18.60	—	349.60

Carbon Steel, Schedule 80 with 300# Fittings & Butt-Welded Joints

Description	Craft@Hrs	Unit	Material $	Labor $	Equipment $	Total $

Class 250 iron body globe valve, flanged

2"	P1@.500	Ea	601.00	18.60	—	619.60
2½"	P1@.600	Ea	746.00	22.30	—	768.30
3"	P1@.750	Ea	889.00	27.90	—	916.90
4"	P1@1.35	Ea	1,230.00	50.20	—	1,280.20
6"	ER@2.50	Ea	2,420.00	99.90	1.63	2,521.53
8"	ER@3.00	Ea	5,350.00	120.00	1.96	5,471.96

200 PSIG iron body butterfly valve, lug-type, lever operated

2"	P1@.450	Ea	168.00	16.70	—	184.70
2½"	P1@.450	Ea	175.00	16.70	—	191.70
3"	P1@.550	Ea	183.00	20.50	—	203.50
4"	P1@.550	Ea	228.00	20.50	—	248.50
6"	ER@.800	Ea	368.00	32.00	.52	400.52
8"	ER@.800	Ea	503.00	32.00	.52	535.52
10"	ER@.900	Ea	699.00	36.00	.59	735.59
12"	ER@1.00	Ea	913.00	40.00	.65	953.65

200 PSIG iron body butterfly valve, wafer-type, lever operated

2"	P1@.450	Ea	153.00	16.70	—	169.70
2½"	P1@.450	Ea	156.00	16.70	—	172.70
3"	P1@.550	Ea	168.00	20.50	—	188.50
4"	P1@.550	Ea	204.00	20.50	—	224.50
6"	ER@.800	Ea	341.00	32.00	.52	373.52
8"	ER@.800	Ea	472.00	32.00	.52	504.52
10"	ER@.900	Ea	660.00	36.00	.59	696.59
12"	ER@1.00	Ea	867.00	40.00	.65	907.65

Class 300 bronze body swing check valve, threaded

½"	P1@.210	Ea	142.00	7.82	—	149.82
¾"	P1@.250	Ea	165.00	9.31	—	174.31
1"	P1@.300	Ea	256.00	11.20	—	267.20
1¼"	P1@.400	Ea	293.00	14.90	—	307.90
1½"	P1@.450	Ea	418.00	16.70	—	434.70
2"	P1@.500	Ea	585.00	18.60	—	603.60

Class 250 iron body swing check valve, flanged

2"	P1@.500	Ea	469.00	18.60	—	487.60
2½"	P1@.600	Ea	547.00	22.30	—	569.30
3"	P1@.750	Ea	679.00	27.90	—	706.90
4"	P1@1.35	Ea	833.00	50.20	—	883.20
6"	ER@2.50	Ea	1,610.00	99.90	1.63	1,711.53
8"	ER@3.00	Ea	3,020.00	120.00	1.96	3,141.96

Carbon Steel, Schedule 80 with 300# Fittings & Butt-Welded Joints

Description	Craft@Hrs	Unit	Material $	Labor $	Equipment $	Total $
Class 250 iron body silent check valve, flanged						
2"	P1@.500	Ea	196.00	18.60	—	214.60
2½"	P1@.600	Ea	226.00	22.30	—	248.30
3"	P1@.750	Ea	251.00	27.90	—	278.90
4"	P1@1.35	Ea	341.00	50.20	—	391.20
6"	ER@2.50	Ea	599.00	99.90	1.63	700.53
8"	ER@3.00	Ea	1,090.00	120.00	1.96	1,211.96
10"	ER@4.00	Ea	1,730.00	160.00	2.61	1,892.61
Class 250 bronze body strainer, threaded						
½"	P1@.210	Ea	40.20	7.82	—	48.02
¾"	P1@.250	Ea	54.30	9.31	—	63.61
1"	P1@.300	Ea	69.00	11.20	—	80.20
1¼"	P1@.400	Ea	97.70	14.90	—	112.60
1½"	P1@.450	Ea	128.00	16.70	—	144.70
2"	P1@.500	Ea	217.00	18.60	—	235.60
Class 250 iron body strainer, flanged						
2"	P1@.500	Ea	232.00	18.60	—	250.60
2½"	P1@.600	Ea	254.00	22.30	—	276.30
3"	P1@.750	Ea	296.00	27.90	—	323.90
4"	P1@1.35	Ea	526.00	50.20	—	576.20
6"	ER@2.50	Ea	842.00	99.90	1.63	943.53
8"	ER@3.00	Ea	1,540.00	120.00	1.96	1,661.96
Installation of 2-way control valve, threaded joints						
½"	P1@.210	Ea	—	7.82	—	7.82
¾"	P1@.250	Ea	—	9.31	—	9.31
1"	P1@.300	Ea	—	11.20	—	11.20
1¼"	P1@.400	Ea	—	14.90	—	14.90
1½"	P1@.450	Ea	—	16.70	—	16.70
2"	P1@.500	Ea	—	18.60	—	18.60
2½"	P1@.830	Ea	—	30.90	—	30.90
3"	P1@.990	Ea	—	36.80	—	36.80
Installation of 2-way control valve, flanged joints						
2"	P1@.500	Ea	—	18.60	—	18.60
2½"	P1@.600	Ea	—	22.30	—	22.30
3"	P1@.750	Ea	—	27.90	—	27.90
4"	P1@1.35	Ea	—	50.20	—	50.20
6"	P1@2.50	Ea	—	93.10	—	93.10
8"	P1@3.00	Ea	—	112.00	—	112.00
10"	P1@4.00	Ea	—	149.00	—	149.00
12"	P1@4.50	Ea	—	167.00	—	167.00

Carbon Steel, Schedule 80 with 300# Fittings & Butt-Welded Joints

Description	Craft@Hrs	Unit	Material $	Labor $	Equipment $	Total $

Installation of 3-way control valve, threaded joints

½"	P1@.260	Ea	—	9.68	—	9.68
¾"	P1@.365	Ea	—	13.60	—	13.60
1"	P1@.475	Ea	—	17.70	—	17.70
1¼"	P1@.575	Ea	—	21.40	—	21.40
1½"	P1@.680	Ea	—	25.30	—	25.30
2"	P1@.910	Ea	—	33.90	—	33.90
2½"	P1@1.12	Ea	—	41.70	—	41.70
3"	P1@1.33	Ea	—	49.50	—	49.50

Installation of 3-way control valve, flanged joints

2"	P1@.910	Ea	—	33.90	—	33.90
2½"	P1@1.12	Ea	—	41.70	—	41.70
3"	P1@1.33	Ea	—	49.50	—	49.50
4"	P1@2.00	Ea	—	74.40	—	74.40
6"	P1@3.70	Ea	—	138.00	—	138.00
8"	P1@4.40	Ea	—	164.00	—	164.00
10"	P1@5.90	Ea	—	220.00	—	220.00
12"	P1@6.50	Ea	—	242.00	—	242.00

Bolt and gasket set

2"	P1@.500	Ea	4.26	18.60	—	22.86
2½"	P1@.650	Ea	4.98	24.20	—	29.18
3"	P1@.750	Ea	8.42	27.90	—	36.32
4"	P1@1.00	Ea	14.00	37.20	—	51.20
6"	P1@1.20	Ea	23.40	44.70	—	68.10
8"	P1@1.25	Ea	26.30	46.50	—	72.80
10"	P1@1.70	Ea	45.20	63.30	—	108.50
12"	P1@2.20	Ea	50.90	81.90	—	132.80

Thermometer with well

7"	P1@.250	Ea	187.00	9.31	—	196.31
9"	P1@.250	Ea	193.00	9.31	—	202.31

Dial-type pressure gauge

2½"	P1@.200	Ea	39.10	7.44	—	46.54
3½"	P1@.200	Ea	51.50	7.44	—	58.94

Pressure/temperature tap

Tap	P1@.150	Ea	18.00	5.58	—	23.58

Carbon Steel, Schedule 80 with 300# Fittings & Butt-Welded Joints

Description	Craft@Hrs	Unit	Material $	Labor $	Equipment $	Total $
Hanger with swivel assembly						
½"	P1@.250	Ea	6.45	9.31	—	15.76
¾"	P1@.250	Ea	7.09	9.31	—	16.40
1"	P1@.250	Ea	7.67	9.31	—	16.98
1½"	P1@.300	Ea	8.03	11.20	—	19.23
2"	P1@.300	Ea	10.50	11.20	—	21.70
2½"	P1@.350	Ea	13.40	13.00	—	26.40
3"	P1@.350	Ea	17.00	13.00	—	30.00
4"	P1@.350	Ea	29.00	13.00	—	42.00
5"	P1@.450	Ea	33.90	16.70	—	50.60
6"	P1@.450	Ea	37.70	16.70	—	54.40
8"	P1@.450	Ea	49.80	16.70	—	66.50
10"	P1@.550	Ea	68.20	20.50	—	88.70
12"	P1@.600	Ea	90.20	22.30	—	112.50
Riser clamp						
½"	P1@.100	Ea	3.20	3.72	—	6.92
¾"	P1@.100	Ea	4.69	3.72	—	8.41
1"	P1@.100	Ea	4.75	3.72	—	8.47
1¼"	P1@.105	Ea	5.73	3.91	—	9.64
1½"	P1@.110	Ea	6.04	4.09	—	10.13
2"	P1@.115	Ea	6.39	4.28	—	10.67
2½"	P1@.120	Ea	6.74	4.47	—	11.21
3"	P1@.120	Ea	7.29	4.47	—	11.76
4"	P1@.125	Ea	9.29	4.65	—	13.94
5"	P1@.180	Ea	13.30	6.70	—	20.00
6"	P1@.200	Ea	16.00	7.44	—	23.44
8"	P1@.200	Ea	26.10	7.44	—	33.54
10"	P1@.250	Ea	38.90	9.31	—	48.21
12"	P1@.250	Ea	46.10	9.31	—	55.41

Carbon Steel, Schedule 80 with 300# M.I. Fittings & Threaded Joints

Schedule 80 carbon steel (ASTM A-53 and A-120) pipe with 300 pound malleable iron threaded fittings is commonly used for heating hot water, chilled water, steam and steam condensate systems where operating pressures do not exceed 250 PSIG for pipe sized 2" and smaller. For pipe sizes 2½" and larger wrought steel fittings either welded or grooved should be used.

The cost estimates in this section are based on the conditions, limitations and wage rates described in the section "How to Use This Book" beginning on page 5.

Equipment cost, where shown, is $4.45 per hour for a 2-ton chain hoist.

Description	Craft@Hrs	Unit	Material $	Labor $	Equipment $	Total $

Schedule 80 carbon steel threaded horizontal pipe assembly.
Horizontally hung in a building. Assembly includes fittings and hanger assemblies. Based on a reducing tee and a 45-degree elbow every 16 feet for ½" pipe and a reducing tee and a 45-degree elbow every 30 feet for 4" pipe. Hangers spaced to meet plumbing code. *(Distance between fittings in the pipe assembly increases as pipe diameter increases.)* Use these figures for preliminary estimates.

Description	Craft@Hrs	Unit	Material $	Labor $	Equipment $	Total $
½"	P1@.121	LF	4.05	4.50	—	8.55
¾"	P1@.125	LF	4.63	4.65	—	9.28
1"	P1@.149	LF	5.80	5.55	—	11.35
1¼"	P1@.163	LF	6.67	6.07	—	12.74
1½"	P1@.186	LF	6.91	6.92	—	13.83
2"	P1@.210	LF	8.47	7.82	—	16.29
2½"	P1@.297	LF	11.20	11.10	—	22.30
3"	P1@.340	LF	24.00	12.70	—	36.70
4"	P1@.409	LF	36.30	15.20	—	51.50

Schedule 80 carbon steel threaded vertical pipe assembly.
Riser assembly, including fittings and riser clamps. Based on a reducing tee, and a riser clamp on every floor. Use these figures for preliminary estimates.

Description	Craft@Hrs	Unit	Material $	Labor $	Equipment $	Total $
½"	P1@.094	LF	3.33	3.50	—	6.83
¾"	P1@.104	LF	3.80	3.87	—	7.67
1"	P1@.124	LF	5.00	4.62	—	9.62
1¼"	P1@.143	LF	5.84	5.32	—	11.16
1½"	P1@.161	LF	6.44	5.99	—	12.43
2"	P1@.179	LF	7.68	6.66	—	14.34
2½"	P1@.257	LF	10.80	9.57	—	20.37
3"	P1@.296	LF	32.30	11.00	—	43.30
4"	P1@.373	LF	53.60	13.90	—	67.50

Carbon Steel, Schedule 80 with 300# M.I. Fittings & Threaded Joints

Description	Craft@Hrs	Unit	Material $	Labor $	Equipment $	Total $

Schedule 80 carbon steel pipe, threaded and coupled joints

Description	Craft@Hrs	Unit	Material $	Labor $	Equipment $	Total $
½"	P1@.070	LF	2.04	2.61	—	4.65
¾"	P1@.080	LF	2.57	2.98	—	5.55
1"	P1@.100	LF	3.49	3.72	—	7.21
1¼"	P1@.110	LF	4.60	4.09	—	8.69
1½"	P1@.120	LF	5.43	4.47	—	9.90
2"	P1@.130	LF	7.41	4.84	—	12.25
2½"	P1@.190	LF	11.60	7.07	—	18.67
3"	P1@.220	LF	15.40	8.19	—	23.59
4"	P1@.290	LF	22.10	10.80	—	32.90

300# malleable iron 45-degree ell, threaded joints

Description	Craft@Hrs	Unit	Material $	Labor $	Equipment $	Total $
½"	P1@.132	Ea	9.71	4.91	—	14.62
¾"	P1@.143	Ea	10.80	5.32	—	16.12
1"	P1@.198	Ea	12.00	7.37	—	19.37
1¼"	P1@.264	Ea	18.50	9.83	—	28.33
1½"	P1@.330	Ea	24.10	12.30	—	36.40
2"	P1@.418	Ea	35.90	15.60	—	51.50
2½"	P1@.561	Ea	56.70	20.90	—	77.60
3"	P1@.704	Ea	75.50	26.20	—	101.70
4"	P1@.935	Ea	145.00	34.80	—	179.80

300# malleable iron 90-degree ell, threaded joints

Description	Craft@Hrs	Unit	Material $	Labor $	Equipment $	Total $
½"	P1@.132	Ea	7.47	4.91	—	12.38
¾"	P1@.143	Ea	8.65	5.32	—	13.97
1"	P1@.198	Ea	11.00	7.37	—	18.37
1¼"	P1@.264	Ea	15.40	9.83	—	25.23
1½"	P1@.330	Ea	18.50	12.30	—	30.80
2"	P1@.418	Ea	25.80	15.60	—	41.40
2½"	P1@.561	Ea	49.00	20.90	—	69.90
3"	P1@.704	Ea	67.00	26.20	—	93.20
4"	P1@.935	Ea	134.00	34.80	—	168.80

Carbon Steel, Schedule 80 with 300# M.I. Fittings & Threaded Joints

Description	Craft@Hrs	Unit	Material $	Labor $	Equipment $	Total $

300# malleable iron tee, threaded joints

Description	Craft@Hrs	Unit	Material $	Labor $	Equipment $	Total $
½"	P1@.198	Ea	10.50	7.37	—	17.87
¾"	P1@.209	Ea	11.60	7.78	—	19.38
1"	P1@.253	Ea	13.90	9.42	—	23.32
1¼"	P1@.341	Ea	18.50	12.70	—	31.20
1½"	P1@.429	Ea	22.80	16.00	—	38.80
2"	P1@.539	Ea	33.50	20.10	—	53.60
2½"	P1@.726	Ea	63.80	27.00	—	90.80
3"	P1@.913	Ea	107.00	34.00	—	141.00
4"	P1@1.21	Ea	182.00	45.00	—	227.00

300# malleable iron reducing tee, threaded joints

Description	Craft@Hrs	Unit	Material $	Labor $	Equipment $	Total $
¾" x ½"	P1@.205	Ea	14.40	7.63	—	22.03
1" x ½"	P1@.205	Ea	18.20	7.63	—	25.83
1" x ¾"	P1@.205	Ea	18.20	7.63	—	25.83
1¼" x ½"	P1@.310	Ea	23.40	11.50	—	34.90
1¼" x ¾"	P1@.310	Ea	23.40	11.50	—	34.90
1¼" x 1"	P1@.310	Ea	23.40	11.50	—	34.90
1½" x ½"	P1@.400	Ea	28.90	14.90	—	43.80
1½" x ¾"	P1@.400	Ea	28.90	14.90	—	43.80
1½" x 1"	P1@.400	Ea	28.90	14.90	—	43.80
1½" x 1¼"	P1@.400	Ea	28.90	14.90	—	43.80
2" x 1"	P1@.500	Ea	42.90	18.60	—	61.50
2" x 1¼"	P1@.500	Ea	42.90	18.60	—	61.50
2" x 1½"	P1@.500	Ea	42.90	18.60	—	61.50
3" x 1¼"	P1@.830	Ea	223.00	30.90	—	253.90
3" x 1½"	P1@.830	Ea	223.00	30.90	—	253.90
3" x 2"	P1@.830	Ea	223.00	30.90	—	253.90

300# malleable iron reducer, threaded joints

Description	Craft@Hrs	Unit	Material $	Labor $	Equipment $	Total $
¾" x ½"	P1@.135	Ea	8.58	5.02	—	13.60
1" x ½"	P1@.170	Ea	10.30	6.33	—	16.63
1" x ¾"	P1@.170	Ea	10.30	6.33	—	16.63
1¼" x ¾"	P1@.220	Ea	13.00	8.19	—	21.19
1¼" x 1"	P1@.220	Ea	13.00	8.19	—	21.19
1½" x 1"	P1@.285	Ea	17.80	10.60	—	28.40
1½" x 1¼"	P1@.285	Ea	17.80	10.60	—	28.40
2" x 1¼"	P1@.360	Ea	25.70	13.40	—	39.10
2" x 1½"	P1@.360	Ea	25.70	13.40	—	39.10
3" x 2"	P1@.550	Ea	62.40	20.50	—	82.90
3" x 2½"	P1@.550	Ea	62.40	20.50	—	82.90

Carbon Steel, Schedule 80 with 300# M.I. Fittings & Threaded Joints

Description	Craft@Hrs	Unit	Material $	Labor $	Equipment $	Total $
300# malleable iron cross, threaded joints						
½"	P1@.310	Ea	25.20	11.50	—	36.70
¾"	P1@.350	Ea	27.50	13.00	—	40.50
1"	P1@.400	Ea	31.20	14.90	—	46.10
1¼"	P1@.530	Ea	38.50	19.70	—	58.20
1½"	P1@.660	Ea	50.30	24.60	—	74.90
2"	P1@.770	Ea	72.30	28.70	—	101.00
300# malleable iron cap, threaded joint						
½"	P1@.100	Ea	5.03	3.72	—	8.75
¾"	P1@.110	Ea	6.70	4.09	—	10.79
1"	P1@.155	Ea	8.67	5.77	—	14.44
1¼"	P1@.200	Ea	11.80	7.44	—	19.24
1½"	P1@.250	Ea	14.00	9.31	—	23.31
2"	P1@.320	Ea	19.30	11.90	—	31.20
3"	P1@.530	Ea	40.20	19.70	—	59.90
4"	P1@.705	Ea	73.90	26.20	—	100.10
300# malleable iron plug, threaded joint						
½"	P1@.090	Ea	1.24	3.35	—	4.59
¾"	P1@.100	Ea	1.69	3.72	—	5.41
1"	P1@.140	Ea	2.02	5.21	—	7.23
1¼"	P1@.180	Ea	2.17	6.70	—	8.87
1½"	P1@.230	Ea	2.77	8.56	—	11.33
2"	P1@.290	Ea	4.29	10.80	—	15.09
2½"	P1@.380	Ea	6.06	14.10	—	20.16
3"	P1@.480	Ea	10.10	17.90	—	28.00
4"	P1@.640	Ea	28.90	23.80	—	52.70
300# malleable iron union, threaded joints						
½"	P1@.150	Ea	9.26	5.58	—	14.84
¾"	P1@.165	Ea	10.40	6.14	—	16.54
1"	P1@.230	Ea	13.70	8.56	—	22.26
1¼"	P1@.310	Ea	20.90	11.50	—	32.40
1½"	P1@.400	Ea	22.70	14.90	—	37.60
2"	P1@.500	Ea	28.90	18.60	—	47.50
2½"	P1@.670	Ea	63.80	24.90	—	88.70
3"	P1@.840	Ea	107.00	31.30	—	138.30
4"	P1@.960	Ea	300.00	35.70	—	335.70

Carbon Steel, Schedule 80 with 300# M.I. Fittings & Threaded Joints

Description	Craft@Hrs	Unit	Material $	Labor $	Equipment $	Total $

300# malleable iron coupling, threaded joints

½"	P1@.132	Ea	5.62	4.91	—	10.53
¾"	P1@.143	Ea	6.57	5.32	—	11.89
1"	P1@.198	Ea	7.73	7.37	—	15.10
1¼"	P1@.264	Ea	9.26	9.83	—	19.09
1½"	P1@.330	Ea	13.90	12.30	—	26.20
2"	P1@.418	Ea	19.30	15.60	—	34.90
2½"	P1@.561	Ea	24.80	20.90	—	45.70
3"	P1@.704	Ea	40.20	26.20	—	66.40
4"	P1@.935	Ea	96.10	34.80	—	130.90

Class 300 bronze body gate valve, threaded joints

½"	P1@.210	Ea	45.90	7.82	—	53.72
¾"	P1@.250	Ea	59.60	9.31	—	68.91
1"	P1@.300	Ea	72.50	11.20	—	83.70
1¼"	P1@.400	Ea	98.80	14.90	—	113.70
1½"	P1@.450	Ea	125.00	16.70	—	141.70
2"	P1@.500	Ea	183.00	18.60	—	201.60

Class 250 iron body gate valve, flanged joints

2"	P1@.500	Ea	114.00	18.60	—	132.60
2½"	P1@.600	Ea	825.00	22.30	—	847.30
3"	P1@.750	Ea	1,040.00	27.90	—	1,067.90
4"	P1@1.35	Ea	1,780.00	50.20	—	1,830.20

Class 300 bronze body globe valve, threaded joints

½"	P1@.210	Ea	68.40	7.82	—	76.22
¾"	P1@.250	Ea	86.80	9.31	—	96.11
1"	P1@.300	Ea	124.00	11.20	—	135.20
1¼"	P1@.400	Ea	180.00	14.90	—	194.90
1½"	P1@.450	Ea	239.00	16.70	—	255.70
2"	P1@.500	Ea	331.00	18.60	—	349.60

Class 250 iron body globe valve, flanged joints

2"	P1@.500	Ea	601.00	18.60	—	619.60
2½"	P1@.600	Ea	746.00	22.30	—	768.30
3"	P1@.750	Ea	889.00	27.90	—	916.90
4"	P1@1.35	Ea	1,230.00	50.20	—	1,280.20

200 PSIG iron body butterfly valve, lug-type, lever operated

2"	P1@.450	Ea	168.00	16.70	—	184.70
2½"	P1@.450	Ea	175.00	16.70	—	191.70
3"	P1@.550	Ea	183.00	20.50	—	203.50
4"	P1@.550	Ea	228.00	20.50	—	248.50

Carbon Steel, Schedule 80 with 300# M.I. Fittings & Threaded Joints

Description	Craft@Hrs	Unit	Material $	Labor $	Equipment $	Total $

200 PSIG iron body butterfly valve, wafer-type, lever operated

2"	P1@.450	Ea	153.00	16.70	—	169.70
2½"	P1@.450	Ea	156.00	16.70	—	172.70
3"	P1@.550	Ea	168.00	20.50	—	188.50
4"	P1@.550	Ea	204.00	20.50	—	224.50

Class 150 bronze body 2-piece ball valve, threaded joints

½"	P1@.210	Ea	15.70	7.82	—	23.52
¾"	P1@.250	Ea	25.30	9.31	—	34.61
1"	P1@.300	Ea	32.10	11.20	—	43.30
1¼"	P1@.400	Ea	40.10	14.90	—	55.00
1½"	P1@.450	Ea	54.10	16.70	—	70.80
2"	P1@.500	Ea	68.70	18.60	—	87.30
3"	P1@.625	Ea	470.00	23.30	—	493.30
4"	P1@.690	Ea	615.00	25.70	—	640.70

Class 300 bronze body swing check valve, threaded joints

½"	P1@.210	Ea	142.00	7.82	—	149.82
¾"	P1@.250	Ea	165.00	9.31	—	174.31
1"	P1@.300	Ea	256.00	11.20	—	267.20
1¼"	P1@.400	Ea	293.00	14.90	—	307.90
1½"	P1@.450	Ea	418.00	16.70	—	434.70
2"	P1@.500	Ea	585.00	18.60	—	603.60

Class 250 iron body swing check valve, flanged joints

2"	P1@.500	Ea	469.00	18.60	—	487.60
2½"	P1@.600	Ea	547.00	22.30	—	569.30
3"	P1@.750	Ea	679.00	27.90	—	706.90
4"	P1@1.35	Ea	833.00	50.20	—	883.20

Class 250 iron body silent check valve, wafer-type

2"	P1@.500	Ea	196.00	18.60	—	214.60
2½"	P1@.600	Ea	226.00	22.30	—	248.30
3"	P1@.750	Ea	251.00	27.90	—	278.90
4"	P1@1.35	Ea	341.00	50.20	—	391.20

Class 250 bronze body wye strainer, threaded joints

½"	P1@.210	Ea	40.20	7.82	—	48.02
¾"	P1@.250	Ea	54.30	9.31	—	63.61
1"	P1@.300	Ea	69.00	11.20	—	80.20
1¼"	P1@.400	Ea	97.70	14.90	—	112.60
1½"	P1@.450	Ea	128.00	16.70	—	144.70
2"	P1@.500	Ea	217.00	18.60	—	235.60

Carbon Steel, Schedule 80 with 300# M.I. Fittings & Threaded Joints

Description	Craft@Hrs	Unit	Material $	Labor $	Equipment $	Total $

Class 250 iron body strainer, flanged ends

2"	P1@.500	Ea	232.00	18.60	—	250.60
2½"	P1@.600	Ea	254.00	22.30	—	276.30
3"	P1@.750	Ea	296.00	27.90	—	323.90
4"	P1@1.35	Ea	526.00	50.20	—	576.20

Installation of 2-way control valve, threaded joints

½"	P1@.210	Ea	—	7.82	—	7.82
¾"	P1@.275	Ea	—	10.20	—	10.20
1"	P1@.350	Ea	—	13.00	—	13.00
1¼"	P1@.430	Ea	—	16.00	—	16.00
1½"	P1@.505	Ea	—	18.80	—	18.80
2"	P1@.675	Ea	—	25.10	—	25.10
2½"	P1@.830	Ea	—	30.90	—	30.90
3"	P1@.990	Ea	—	36.80	—	36.80

Installation of 3-way control valve, threaded joints

½"	P1@.260	Ea	—	9.68	—	9.68
¾"	P1@.365	Ea	—	13.60	—	13.60
1"	P1@.475	Ea	—	17.70	—	17.70
1¼"	P1@.575	Ea	—	21.40	—	21.40
1½"	P1@.680	Ea	—	25.30	—	25.30
2"	P1@.910	Ea	—	33.90	—	33.90
2½"	P1@1.12	Ea	—	41.70	—	41.70
3"	P1@1.33	Ea	—	49.50	—	49.50

Threaded companion flange, 300# malleable iron

2"	P1@.290	Ea	104.00	10.80	—	114.80
2½"	P1@.380	Ea	143.00	14.10	—	157.10
3"	P1@.460	Ea	136.00	17.10	—	153.10
4"	P1@.600	Ea	251.00	22.30	—	273.30

Bolt and gasket sets

2"	P1@.500	Ea	4.26	18.60	—	22.86
2½"	P1@.650	Ea	4.98	24.20	—	29.18
3"	P1@.750	Ea	8.45	27.90	—	36.35
4"	P1@1.00	Ea	14.10	37.20	—	51.30

Carbon Steel, Schedule 80 with 300# M.I. Fittings & Threaded Joints

Description	Craft@Hrs	Unit	Material $	Labor $	Equipment $	Total $
Thermometer with well						
7"	P1@.250	Ea	187.00	9.31	—	196.31
9"	P1@.250	Ea	193.00	9.31	—	202.31
Dial-type pressure gauge						
2½"	P1@.200	Ea	39.10	7.44	—	46.54
3½"	P1@.200	Ea	51.50	7.44	—	58.94
Pressure/temperature tap						
Tap	P1@.150	Ea	18.00	5.58	—	23.58
Hanger with swivel assembly						
½"	P1@.250	Ea	4.77	9.31	—	14.08
¾"	P1@.250	Ea	4.98	9.31	—	14.29
1"	P1@.250	Ea	5.24	9.31	—	14.55
1¼"	P1@.300	Ea	5.48	11.20	—	16.68
1½"	P1@.300	Ea	6.33	11.20	—	17.53
2"	P1@.300	Ea	6.94	11.20	—	18.14
2½"	P1@.350	Ea	8.99	13.00	—	21.99
3"	P1@.350	Ea	11.10	13.00	—	24.10
4"	P1@.350	Ea	17.20	13.00	—	30.20
Riser clamp						
½"	P1@.100	Ea	3.20	3.72	—	6.92
¾"	P1@.100	Ea	4.69	3.72	—	8.41
1"	P1@.100	Ea	4.75	3.72	—	8.47
1¼"	P1@.105	Ea	5.73	3.91	—	9.64
1½"	P1@.110	Ea	6.04	4.09	—	10.13
2"	P1@.115	Ea	6.39	4.28	—	10.67
2½"	P1@.120	Ea	6.74	4.47	—	11.21
3"	P1@.120	Ea	7.29	4.47	—	11.76
4"	P1@.125	Ea	9.29	4.65	—	13.94

Carbon Steel, Schedule 160 with 3,000-6,000# Fittings

Schedule 160 carbon steel (ASTM A-106) pipe with steel fittings is commonly used for steam, steam condensate and recirculating systems.

The cost estimates in this section are based on the conditions, limitations and wage rates described in the section "How to Use This Book" beginning on page 5.

Description	Craft@Hrs	Unit	Material $	Labor $	Equipment $	Total $

Schedule 160 carbon steel threaded horizontal pipe assembly.
Horizontally hung in a building. Assembly includes fittings and hanger assemblies. Based on a reducing tee and a 45-degree elbow every 16 feet for ½" pipe. A reducing tee and a 45-degree elbow every 30 feet for 4" pipe. Hangers spaced to meet plumbing code. *(Distance between fittings in the pipe assembly increases as pipe diameter increases.)* Use these figures for preliminary estimates.

Description	Craft@Hrs	Unit	Material $	Labor $	Equipment $	Total $
½"	P1@.140	LF	5.91	5.21	—	11.12
¾"	P1@.145	LF	6.92	5.40	—	12.32
1"	P1@.172	LF	9.49	6.40	—	15.89
1¼"	P1@.184	LF	10.40	6.85	—	17.25
1½"	P1@.207	LF	12.00	7.70	—	19.70
2"	P1@.272	LF	18.20	10.10	—	28.30
2½"	P1@.388	LF	27.80	14.40	—	42.20
3"	P1@.434	LF	41.50	16.20	—	57.70
4"	P1@.530	LF	68.10	19.70	—	87.80

Schedule 160 carbon steel socket weld horizontal pipe assembly.
Horizontally hung in a building. Assembly includes fittings and hanger assemblies. Based on a reducing tee and a 45-degree elbow every 16 feet for ½" pipe. A reducing tee and a 45-degree elbow every 30 feet for 4" pipe Hangers spaced to meet plumbing code. *(Distance between fittings in the pipe assembly increases as pipe diameter increases.)* Use these figures for preliminary estimates.

Description	Craft@Hrs	Unit	Material $	Labor $	Equipment $	Total $
½"	P1@.175	LF	5.75	6.51	.55	12.81
¾"	P1@.206	LF	6.65	7.67	.65	14.97
1"	P1@.247	LF	9.15	9.19	.78	19.12
1¼"	P1@.250	LF	10.20	9.31	.79	20.30
1½"	P1@.284	LF	11.00	10.60	.90	22.50
2"	P1@.375	LF	17.30	14.00	1.19	32.49
2½"	P1@.503	LF	26.00	18.70	1.59	46.29
3"	P1@.568	LF	40.40	21.10	1.80	63.30
4"	P1@.709	LF	66.70	26.40	2.24	95.34

Carbon Steel, Schedule 160 with 3,000-6,000# Fittings

Description	Craft@Hrs	Unit	Material $	Labor $	Equipment $	Total $

Schedule 160 carbon steel threaded vertical pipe assembly. Riser assembly, including fittings and riser clamps. Based on a reducing tee, and a riser clamp on every floor. Use these figures for preliminary estimates.

Description	Craft@Hrs	Unit	Material $	Labor $	Equipment $	Total $
½"	P1@.114	LF	5.20	4.24	—	9.44
¾"	P1@.125	LF	6.21	4.65	—	10.86
1"	P1@.148	LF	8.60	5.51	—	14.11
1¼"	P1@.165	LF	10.50	6.14	—	16.64
1½"	P1@.183	LF	11.60	6.81	—	18.41
2"	P1@.242	LF	17.40	9.01	—	26.41
2½"	P1@.348	LF	26.00	13.00	—	39.00
3"	P1@.393	LF	40.00	14.60	—	54.60
4"	P1@.487	LF	64.00	18.10	—	82.10

Schedule 160 carbon steel socket weld vertical pipe assembly. Riser assembly, including fittings and riser clamps. Based on a reducing tee, and a riser clamp on every floor. Use these figures for preliminary estimates.

Description	Craft@Hrs	Unit	Material $	Labor $	Equipment $	Total $
½"	P1@.133	LF	5.06	4.95	.42	10.43
¾"	P1@.162	LF	5.98	6.03	.51	12.52
1"	P1@.199	LF	8.40	7.41	.63	16.44
1¼"	P1@.227	LF	9.78	8.45	.72	18.95
1½"	P1@.255	LF	10.80	9.49	.81	21.10
2"	P1@.341	LF	16.50	12.70	1.08	30.28
2½"	P1@.467	LF	24.10	17.40	1.48	42.98
3"	P1@.532	LF	40.40	19.80	1.68	61.88
4"	P1@.674	LF	63.10	25.10	2.13	90.33

Schedule 160 carbon steel plain end pipe

Description	Craft@Hrs	Unit	Material $	Labor $	Equipment $	Total $
½"	P1@.090	LF	4.79	3.35	—	8.14
¾"	P1@.100	LF	5.57	3.72	—	9.29
1"	P1@.120	LF	7.92	4.47	—	12.39
1¼"	P1@.130	LF	9.30	4.84	—	14.14
1½"	P1@.140	LF	14.50	5.21	—	19.71
2"	P1@.190	LF	11.90	7.07	—	18.97
3"	P1@.310	LF	23.30	11.50	—	34.80
4"	P1@.380	LF	36.10	14.10	—	50.20

Carbon Steel, Schedule 160 with 3,000-6,000# Fittings

Description	Craft@Hrs	Unit	Material $	Labor $	Equipment $	Total $

3,000# carbon steel 45-degree ell, threaded joints

Description	Craft@Hrs	Unit	Material $	Labor $	Equipment $	Total $
½"	P1@.132	Ea	9.33	4.91	—	14.24
¾"	P1@.143	Ea	10.60	5.32	—	15.92
1"	P1@.198	Ea	14.70	7.37	—	22.07
1¼"	P1@.264	Ea	20.30	9.83	—	30.13
1½"	P1@.330	Ea	30.70	12.30	—	43.00
2"	P1@.418	Ea	41.80	15.60	—	57.40
2½"	P1@.561	Ea	104.00	20.90	—	124.90
3"	P1@.704	Ea	168.00	26.20	—	194.20
4"	P1@.935	Ea	310.00	34.80	—	344.80

3,000# carbon steel 45-degree ell, socket-welded joints

Description	Craft@Hrs	Unit	Material $	Labor $	Equipment $	Total $
½"	P1@.300	Ea	7.56	11.20	.95	19.71
¾"	P1@.440	Ea	9.15	16.40	1.39	26.94
1"	P1@.590	Ea	11.90	22.00	1.87	35.77
1¼"	P1@.740	Ea	16.30	27.50	2.34	46.14
1½"	P1@.890	Ea	19.80	33.10	2.82	55.72
2"	P1@1.18	Ea	34.20	43.90	3.73	81.83
2½"	P1@1.49	Ea	88.30	55.50	4.71	148.51
3"	P1@1.77	Ea	149.00	65.90	5.60	220.50
4"	P1@2.37	Ea	294.00	88.20	7.50	389.70

6,000# carbon steel 45-degree ell, threaded joints

Description	Craft@Hrs	Unit	Material $	Labor $	Equipment $	Total $
½"	P1@.132	Ea	15.80	4.91	—	20.71
¾"	P1@.143	Ea	21.50	5.32	—	26.82
1"	P1@.198	Ea	24.10	7.37	—	31.47
1¼"	P1@.264	Ea	64.10	9.83	—	73.93
1½"	P1@.330	Ea	64.10	12.30	—	76.40
2"	P1@.418	Ea	89.50	15.60	—	105.10
2½"	P1@.561	Ea	340.00	20.90	—	360.90
3"	P1@.704	Ea	349.00	26.20	—	375.20

6,000# carbon steel 45-degree ell, socket-welded joints

Description	Craft@Hrs	Unit	Material $	Labor $	Equipment $	Total $
½"	P1@.300	Ea	12.20	11.20	.95	24.35
¾"	P1@.440	Ea	14.10	16.40	1.39	31.89
1"	P1@.590	Ea	19.20	22.00	1.87	43.07
1¼"	P1@.740	Ea	68.80	27.50	2.34	98.64
1½"	P1@.890	Ea	46.50	33.10	2.82	82.42
2"	P1@1.18	Ea	57.80	43.90	3.73	105.43
2½"	P1@1.49	Ea	229.00	55.50	4.71	289.21
3"	P1@1.77	Ea	270.00	65.90	5.60	341.50
4"	P1@2.37	Ea	323.00	88.20	7.50	418.70

Carbon Steel, Schedule 160 with 3,000-6,000# Fittings

Description	Craft@Hrs	Unit	Material $	Labor $	Equipment $	Total $
3,000# carbon steel 90-degree ell, threaded joints						
½"	P1@.132	Ea	7.29	4.91	—	12.20
¾"	P1@.143	Ea	8.78	5.32	—	14.10
1"	P1@.198	Ea	12.70	7.37	—	20.07
1¼"	P1@.264	Ea	21.50	9.83	—	31.33
1½"	P1@.330	Ea	32.30	12.30	—	44.60
2"	P1@.418	Ea	38.80	15.60	—	54.40
2½"	P1@.561	Ea	96.30	20.90	—	117.20
3"	P1@.704	Ea	147.00	26.20	—	173.20
4"	P1@.935	Ea	323.00	34.80	—	357.80
3,000# carbon steel 90-degree ell, socket-welded joints						
½"	P1@.300	Ea	7.23	11.20	.95	19.38
¾"	P1@.440	Ea	7.46	16.40	1.39	25.25
1"	P1@.590	Ea	9.60	22.00	1.87	33.47
1¼"	P1@.740	Ea	15.80	27.50	2.34	45.64
1½"	P1@.890	Ea	20.20	33.10	2.82	56.12
2"	P1@1.18	Ea	30.70	43.90	3.73	78.33
2½"	P1@1.49	Ea	73.70	55.50	4.71	133.91
3"	P1@1.77	Ea	121.00	65.90	5.60	192.50
4"	P1@2.37	Ea	357.00	88.20	7.50	452.70
6,000# carbon steel 90-degree ell, threaded joints						
½"	P1@.132	Ea	11.80	4.91	—	16.71
¾"	P1@.143	Ea	15.00	5.32	—	20.32
1"	P1@.198	Ea	19.40	7.37	—	26.77
1¼"	P1@.264	Ea	32.70	9.83	—	42.53
1½"	P1@.330	Ea	50.50	12.30	—	62.80
2"	P1@.418	Ea	88.80	15.60	—	104.40
2½"	P1@.561	Ea	132.00	20.90	—	152.90
3"	P1@.704	Ea	299.00	26.20	—	325.20
6,000# carbon steel 90-degree ell, socket-welded joints						
½"	P1@.300	Ea	10.00	11.20	.95	22.15
¾"	P1@.440	Ea	11.70	16.40	1.39	29.49
1"	P1@.590	Ea	16.30	22.00	1.87	40.17
1¼"	P1@.740	Ea	23.80	27.50	2.34	53.64
1½"	P1@.890	Ea	35.80	33.10	2.82	71.72
2"	P1@1.18	Ea	52.50	43.90	3.73	100.13
2½"	P1@1.49	Ea	109.00	55.50	4.71	169.21
3"	P1@1.77	Ea	193.00	65.90	5.60	264.50
4"	P1@2.37	Ea	425.00	88.20	7.50	520.70

Carbon Steel, Schedule 160 with 3,000-6,000# Fittings

Description	Craft@Hrs	Unit	Material $	Labor $	Equipment $	Total $

3,000# carbon steel tee, threaded joints

Description	Craft@Hrs	Unit	Material $	Labor $	Equipment $	Total $
½"	P1@.198	Ea	9.63	7.37	—	17.00
¾"	P1@.209	Ea	13.80	7.78	—	21.58
1"	P1@.253	Ea	17.50	9.42	—	26.92
1¼"	P1@.341	Ea	29.00	12.70	—	41.70
1½"	P1@.429	Ea	39.50	16.00	—	55.50
2"	P1@.539	Ea	52.60	20.10	—	72.70
2½"	P1@.726	Ea	127.00	27.00	—	154.00
3"	P1@.913	Ea	238.00	34.00	—	272.00
4"	P1@1.21	Ea	394.00	45.00	—	439.00

3,000# carbon steel tee, socket-welded joints

Description	Craft@Hrs	Unit	Material $	Labor $	Equipment $	Total $
½"	P1@.440	Ea	8.30	16.40	1.39	26.09
¾"	P1@.670	Ea	10.50	24.90	2.12	37.52
1"	P1@.890	Ea	14.50	33.10	2.82	50.42
1¼"	P1@1.11	Ea	19.40	41.30	3.51	64.21
1½"	P1@1.33	Ea	30.40	49.50	4.21	84.11
2"	P1@1.77	Ea	42.40	65.90	5.60	113.90
2½"	P1@2.22	Ea	102.00	82.60	7.02	191.62
3"	P1@2.66	Ea	241.00	99.00	8.42	348.42
4"	P1@3.55	Ea	387.00	132.00	11.20	530.20

6,000# carbon steel tee, threaded joints

Description	Craft@Hrs	Unit	Material $	Labor $	Equipment $	Total $
½"	P1@.198	Ea	15.40	7.37	—	22.77
¾"	P1@.209	Ea	20.40	7.78	—	28.18
1"	P1@.253	Ea	34.20	9.42	—	43.62
1¼"	P1@.341	Ea	42.60	12.70	—	55.30
1½"	P1@.429	Ea	58.20	16.00	—	74.20
2"	P1@.539	Ea	114.00	20.10	—	134.10
2½"	P1@.726	Ea	408.00	27.00	—	435.00
3"	P1@.913	Ea	413.00	34.00	—	447.00

6,000# carbon steel tee, socket-welded joints

Description	Craft@Hrs	Unit	Material $	Labor $	Equipment $	Total $
½"	P1@.440	Ea	13.30	16.40	1.39	31.09
¾"	P1@.670	Ea	17.50	24.90	2.12	44.52
1"	P1@.890	Ea	21.90	33.10	2.82	57.82
1¼"	P1@1.11	Ea	62.50	41.30	3.51	107.31
1½"	P1@1.33	Ea	56.00	49.50	4.21	109.71
2"	P1@1.77	Ea	71.00	65.90	5.60	142.50
2½"	P1@2.22	Ea	150.00	82.60	7.02	239.62
3"	P1@2.66	Ea	252.00	99.00	8.42	359.42
4"	P1@3.55	Ea	478.00	132.00	11.20	621.20

Carbon Steel, Schedule 160 with 3,000-6,000# Fittings

Description	Craft@Hrs	Unit	Material $	Labor $	Equipment $	Total $
3,000# carbon steel union, threaded joints						
½"	P1@.150	Ea	10.60	5.58	—	16.18
¾"	P1@.165	Ea	12.70	6.14	—	18.84
1"	P1@.230	Ea	16.70	8.56	—	25.26
1¼"	P1@.310	Ea	28.10	11.50	—	39.60
1½"	P1@.400	Ea	30.70	14.90	—	45.60
2"	P1@.500	Ea	40.80	18.60	—	59.40
2½"	P1@.670	Ea	88.90	24.90	—	113.80
3"	P1@.840	Ea	110.00	31.30	—	141.30
3,000# carbon steel union, socket-welded joints						
½"	P1@.300	Ea	12.70	11.20	.95	24.85
¾"	P1@.440	Ea	14.50	16.40	1.39	32.29
1"	P1@.590	Ea	18.60	22.00	1.87	42.47
1¼"	P1@.740	Ea	29.90	27.50	2.34	59.74
1½"	P1@.890	Ea	31.00	33.10	2.82	66.92
2"	P1@1.18	Ea	43.90	43.90	3.73	91.53
2½"	P1@1.49	Ea	102.00	55.50	4.71	162.21
3"	P1@1.77	Ea	130.00	65.90	5.60	201.50
6,000# carbon steel union, threaded joints						
½"	P1@.150	Ea	24.80	5.58	—	30.38
¾"	P1@.165	Ea	28.10	6.14	—	34.24
1"	P1@.230	Ea	36.90	8.56	—	45.46
1¼"	P1@.310	Ea	60.50	11.50	—	72.00
1½"	P1@.400	Ea	74.70	14.90	—	89.60
2"	P1@.500	Ea	106.00	18.60	—	124.60
6,000# carbon steel union, socket-welded joints						
½"	P1@.300	Ea	25.60	11.20	.95	37.75
¾"	P1@.440	Ea	29.30	16.40	1.39	47.09
1"	P1@.590	Ea	42.70	22.00	1.87	66.57
1¼"	P1@.740	Ea	65.60	27.50	2.34	95.44
1½"	P1@.890	Ea	76.70	33.10	2.82	112.62
2"	P1@1.18	Ea	116.00	43.90	3.73	163.63
3,000# carbon steel reducer, socket-welded joints						
¾"	P1@.360	Ea	6.90	13.40	1.14	21.44
1"	P1@.500	Ea	8.39	18.60	1.58	28.57
1¼"	P1@.665	Ea	9.71	24.80	2.10	36.61
1½"	P1@.800	Ea	12.10	29.80	2.53	44.43
2"	P1@1.00	Ea	16.70	37.20	3.16	57.06
2½"	P1@1.33	Ea	115.00	49.50	4.21	168.71
3"	P1@1.63	Ea	145.00	60.70	5.16	210.86

Carbon Steel, Schedule 160 with 3,000-6,000# Fittings

Description	Craft@Hrs	Unit	Material $	Labor $	Equipment $	Total $

6,000# carbon steel reducer, socket-welded joints

Description	Craft@Hrs	Unit	Material $	Labor $	Equipment $	Total $
¾"	P1@.360	Ea	26.50	13.40	1.14	41.04
1"	P1@.500	Ea	27.40	18.60	1.58	47.58
1¼"	P1@.665	Ea	58.70	24.80	2.10	85.60
1½"	P1@.800	Ea	58.70	29.80	2.53	91.03
2"	P1@1.00	Ea	78.10	37.20	3.16	118.46

3,000# carbon steel cap, socket-welded joint

Description	Craft@Hrs	Unit	Material $	Labor $	Equipment $	Total $
½"	P1@.180	Ea	4.13	6.70	.57	11.40
¾"	P1@.270	Ea	4.79	10.00	.85	15.64
1"	P1@.360	Ea	7.32	13.40	1.14	21.86
1¼"	P1@.460	Ea	8.60	17.10	1.46	27.16
1½"	P1@.560	Ea	11.70	20.80	1.77	34.27
2"	P1@.740	Ea	18.30	27.50	2.34	48.14
2½"	P1@.920	Ea	98.40	34.20	2.91	135.51
3"	P1@1.11	Ea	102.00	41.30	3.51	146.81
4"	P1@1.50	Ea	154.00	55.80	4.75	214.55

6,000# carbon steel cap, socket-welded joint

Description	Craft@Hrs	Unit	Material $	Labor $	Equipment $	Total $
½"	P1@.180	Ea	7.92	6.70	.57	15.19
¾"	P1@.270	Ea	11.50	10.00	.85	22.35
1"	P1@.360	Ea	12.10	13.40	1.14	26.64
1¼"	P1@.460	Ea	28.50	17.10	1.46	47.06
1½"	P1@.560	Ea	36.90	20.80	1.77	59.47
2"	P1@.740	Ea	54.90	27.50	2.34	84.74

3,000# carbon steel weldolet

Description	Craft@Hrs	Unit	Material $	Labor $	Equipment $	Total $
½"	P1@.580	Ea	14.60	21.60	1.84	38.04
¾"	P1@.890	Ea	15.00	33.10	2.82	50.92
1"	P1@1.18	Ea	16.40	43.90	3.73	64.03
1¼"	P1@1.48	Ea	20.40	55.10	4.68	80.18
1½"	P1@1.75	Ea	20.40	65.10	5.54	91.04
2"	P1@2.35	Ea	22.10	87.50	7.44	117.04
2½"	P1@2.90	Ea	50.00	108.00	9.18	167.18
3"	P1@3.50	Ea	51.40	130.00	11.10	192.50
4"	P1@4.75	Ea	63.70	177.00	15.00	255.70

6,000# carbon steel weldolet

Description	Craft@Hrs	Unit	Material $	Labor $	Equipment $	Total $
½"	P1@.690	Ea	63.70	25.70	2.18	91.58
¾"	P1@1.05	Ea	66.40	39.10	3.32	108.82
1"	P1@1.40	Ea	69.70	52.10	4.43	126.23
1¼"	P1@1.75	Ea	83.90	65.10	5.54	154.54
1½"	P1@2.10	Ea	94.10	78.20	6.64	178.94
2"	P1@2.80	Ea	108.00	104.00	8.86	220.86
2½"	P1@3.50	Ea	159.00	130.00	11.10	300.10
3"	P1@4.20	Ea	177.00	156.00	13.30	346.30
4"	P1@5.70	Ea	302.00	212.00	18.00	532.00

Carbon Steel, Schedule 160 with 3,000-6,000# Fittings

Description	Craft@Hrs	Unit	Material $	Labor $	Equipment $	Total $
3,000# carbon steel threadolet						
½"	P1@.400	Ea	11.00	14.90	—	25.90
¾"	P1@.580	Ea	12.20	21.60	—	33.80
1"	P1@.790	Ea	15.00	29.40	—	44.40
1¼"	P1@.980	Ea	67.90	36.50	—	104.40
1½"	P1@1.20	Ea	67.90	44.70	—	112.60
2"	P1@1.60	Ea	89.70	59.60	—	149.30
6,000# carbon steel threadolet						
½"	P1@.480	Ea	16.90	17.90	—	34.80
¾"	P1@.690	Ea	18.50	25.70	—	44.20
1"	P1@.950	Ea	26.40	35.40	—	61.80
1¼"	P1@1.20	Ea	98.40	44.70	—	143.10
1½"	P1@1.45	Ea	98.40	54.00	—	152.40
2"	P1@1.90	Ea	161.00	70.70	—	231.70
Class 300 forged steel companion flange, weld neck						
½"	P1@.240	Ea	20.00	8.93	.76	29.69
¾"	P1@.320	Ea	20.00	11.90	1.01	32.91
1"	P1@.340	Ea	20.00	12.70	1.08	33.78
1¼"	P1@.425	Ea	25.90	15.80	1.34	43.04
1½"	P1@.560	Ea	25.90	20.80	1.77	48.47
2"	P1@.740	Ea	28.30	27.50	2.34	58.14
2½"	P1@.810	Ea	41.50	30.10	2.56	74.16
3"	P1@.980	Ea	42.50	36.50	3.10	82.10
4"	P1@1.30	Ea	62.60	48.40	4.11	115.11
Class 300 forged steel companion flange, slip on						
½"	P1@.240	Ea	15.90	8.93	—	24.83
¾"	P1@.320	Ea	15.90	11.90	—	27.80
1"	P1@.340	Ea	15.90	12.70	—	28.60
1½"	P1@.425	Ea	23.80	15.80	—	39.60
2"	P1@.560	Ea	22.40	20.80	—	43.20
2½"	P1@.740	Ea	38.20	27.50	—	65.70
3"	P1@.810	Ea	38.20	30.10	—	68.30
4"	P1@.980	Ea	56.50	36.50	—	93.00
Class 300 cast steel gate valve, flanged						
2"	P1@.500	Ea	2,080.00	18.60	—	2,098.60
2½"	P1@.600	Ea	2,820.00	22.30	—	2,842.30
3"	P1@.750	Ea	2,820.00	27.90	—	2,847.90
4"	P1@1.35	Ea	3,930.00	50.20	—	3,980.20

Carbon Steel, Schedule 160 with 3,000-6,000# Fittings

Description	Craft@Hrs	Unit	Material $	Labor $	Equipment $	Total $

Class 600 cast steel gate valve, flanged

2"	P1@.500	Ea	3,460.00	18.60	—	3,478.60
2½"	P1@.600	Ea	4,990.00	22.30	—	5,012.30
3"	P1@.750	Ea	4,990.00	27.90	—	5,017.90
4"	P1@1.35	Ea	7,990.00	50.20	—	8,040.20

Class 300 cast steel body globe valve, flanged

2"	P1@.500	Ea	1,450.00	18.60	—	1,468.60
2½"	P1@.600	Ea	1,980.00	22.30	—	2,002.30
3"	P1@.750	Ea	1,980.00	27.90	—	2,007.90
4"	P1@1.35	Ea	2,720.00	50.20	—	2,770.20

Class 600 cast steel body globe valve, flanged

2"	P1@.500	Ea	2,030.00	18.60	—	2,048.60
2½"	P1@.600	Ea	3,080.00	22.30	—	3,102.30
3"	P1@.750	Ea	3,080.00	27.90	—	3,107.90
4"	P1@1.35	Ea	4,970.00	50.20	—	5,020.20

Class 300 cast steel body swing check valve, flanged

2"	P1@.500	Ea	1,370.00	18.60	—	1,388.60
2½"	P1@.600	Ea	2,070.00	22.30	—	2,092.30
3"	P1@.750	Ea	2,070.00	27.90	—	2,097.90
4"	P1@1.35	Ea	2,930.00	50.20	—	2,980.20

Class 600 cast steel body swing check valve, flanged

2"	P1@.500	Ea	2,110.00	18.60	—	2,128.60
2½"	P1@.600	Ea	3,330.00	22.30	—	3,352.30
3"	P1@.750	Ea	3,330.00	27.90	—	3,357.90
4"	P1@1.35	Ea	5,260.00	50.20	—	5,310.20

Installation of cast steel body 2-way control valve, threaded joints

½"	P1@.210	Ea	—	7.82	—	7.82
¾"	P1@.250	Ea	—	9.31	—	9.31
1"	P1@.300	Ea	—	11.20	—	11.20
1¼"	P1@.400	Ea	—	14.90	—	14.90
1½"	P1@.450	Ea	—	16.70	—	16.70
2"	P1@.500	Ea	—	18.60	—	18.60
2½"	P1@.830	Ea	—	30.90	—	30.90
3"	P1@.990	Ea	—	36.80	—	36.80

Installation of cast steel body 2-way control valve, flanged joints

2"	P1@.500	Ea	—	18.60	—	18.60
2½"	P1@.600	Ea	—	22.30	—	22.30
3"	P1@.750	Ea	—	27.90	—	27.90
4"	P1@1.35	Ea	—	50.20	—	50.20

Carbon Steel, Schedule 160 with 3,000-6,000# Fittings

Description	Craft@Hrs	Unit	Material $	Labor $	Equipment $	Total $
Installation of cast steel body 3-way control valve, threaded joints						
½"	P1@.260	Ea	—	9.68	—	9.68
¾"	P1@.365	Ea	—	13.60	—	13.60
1"	P1@.475	Ea	—	17.70	—	17.70
1¼"	P1@.575	Ea	—	21.40	—	21.40
1½"	P1@.680	Ea	—	25.30	—	25.30
2"	P1@.910	Ea	—	33.90	—	33.90
2½"	P1@1.12	Ea	—	41.70	—	41.70
3"	P1@1.33	Ea	—	49.50	—	49.50
Installation of cast steel body 3-way control valve, flanged joints						
2"	P1@.910	Ea	—	33.90	—	33.90
2½"	P1@1.12	Ea	—	41.70	—	41.70
3"	P1@1.33	Ea	—	49.50	—	49.50
4"	P1@2.00	Ea	—	74.40	—	74.40
300# bolt and gasket set, ring face						
2"	P1@.500	Ea	8.45	18.60	—	27.05
2½"	P1@.650	Ea	9.88	24.20	—	34.08
3"	P1@.750	Ea	17.00	27.90	—	44.90
4"	P1@1.00	Ea	27.60	37.20	—	64.80
300# bolt and gasket set, full face						
2"	P1@.500	Ea	17.00	18.60	—	35.60
2½"	P1@.650	Ea	19.80	24.20	—	44.00
3"	P1@.750	Ea	33.90	27.90	—	61.80
4"	P1@1.00	Ea	55.50	37.20	—	92.70
Thermometer with well						
7"	P1@.250	Ea	187.00	9.31	—	196.31
9"	P1@.250	Ea	193.00	9.31	—	202.31
Dial-type pressure gauge						
2½"	P1@.200	Ea	39.10	7.44	—	46.54
3½"	P1@.200	Ea	51.50	7.44	—	58.94
Pressure/temperature tap						
Tap	P1@.150	Ea	18.00	5.58	—	23.58

Carbon Steel, Schedule 160 with 3,000-6,000# Fittings

Description	Craft@Hrs	Unit	Material $	Labor $	Equipment $	Total $
Hanger with swivel assembly						
½"	P1@.250	Ea	6.23	9.31	—	15.54
¾"	P1@.250	Ea	6.84	9.31	—	16.15
1"	P1@.250	Ea	7.17	9.31	—	16.48
1¼"	P1@.300	Ea	7.74	11.20	—	18.94
1½"	P1@.300	Ea	10.10	11.20	—	21.30
2"	P1@.300	Ea	13.00	11.20	—	24.20
2½"	P1@.350	Ea	16.40	13.00	—	29.40
3"	P1@.350	Ea	28.00	13.00	—	41.00
4"	P1@.350	Ea	36.40	13.00	—	49.40
Riser clamp						
½"	P1@.100	Ea	3.08	3.72	—	6.80
¾"	P1@.100	Ea	4.52	3.72	—	8.24
1¼"	P1@.105	Ea	5.53	3.91	—	9.44
1½"	P1@.110	Ea	5.83	4.09	—	9.92
2"	P1@.115	Ea	6.19	4.28	—	10.47
2½"	P1@.120	Ea	6.50	4.47	—	10.97
3"	P1@.120	Ea	7.04	4.47	—	11.51
4"	P1@.125	Ea	8.97	4.65	—	13.62

Carbon Steel, Schedule 40 with Roll-Grooved Joints

Schedule 40 carbon steel (ASTM A-53 and A-120) roll-grooved pipe with factory grooved malleable iron, ductile iron, or steel fittings is commonly used for heating hot water, chilled water and condenser water systems.

Consult the manufacturers for maximum operating temperature/pressure ratings for the various combinations of pipe and fittings and system applications.

The cost estimates in this section are based on the conditions, limitations and wage rates described in the section "How to Use This Book" beginning on page 5.

Equipment cost, where shown, is $4.45 per hour for a 2-ton chain hoist.

Description	Craft@Hrs	Unit	Material $	Labor $	Equipment $	Total $

Schedule 40 carbon steel roll-grooved horizontal pipe assembly.
Horizontally hung in a building. Assembly includes fittings, couplings and hanger assemblies. Based on a reducing tee and a 45-degree elbow every 16 feet for ½" pipe. A reducing tee and a 45-degree elbow every 50 feet for 12" pipe. Hangers spaced to meet plumbing code. *(Distance between fittings in the pipe assembly increases as pipe diameter increases.)* Use these figures for preliminary estimates.

Description	Craft@Hrs	Unit	Material $	Labor $	Equipment $	Total $
2"	P1@.183	LF	9.49	6.81	—	16.30
2½"	P1@.192	LF	12.50	7.15	—	19.65
3"	P1@.220	LF	14.10	8.19	—	22.29
4"	P1@.262	LF	20.40	9.75	—	30.15
6"	ER@.347	LF	37.70	13.90	.23	51.83
8"	ER@.414	LF	55.00	16.50	.27	71.77
10"	ER@.530	LF	81.80	21.20	.35	103.35
12"	ER@.617	LF	146.00	24.70	.40	171.10

Schedule 40 carbon steel roll-grooved vertical pipe assembly.
Riser assembly, including fittings, couplings, and riser clamps. Based on a reducing tee, and a riser clamp on every floor. Use these figures for preliminary estimates.

Description	Craft@Hrs	Unit	Material $	Labor $	Equipment $	Total $
2"	P1@.120	LF	8.31	4.47	—	12.78
2½"	P1@.123	LF	11.80	4.58	—	16.38
3"	P1@.150	LF	12.30	5.58	—	17.88
4"	P1@.209	LF	18.50	7.78	—	26.28
6"	ER@.294	LF	35.90	11.70	.19	47.79
8"	ER@.395	LF	58.90	15.80	.26	74.96
10"	ER@.523	LF	81.80	20.90	.34	103.04
12"	ER@.608	LF	133.00	24.30	.40	157.70

Schedule 40 carbon steel pipe with roll-grooved joints

Description	Craft@Hrs	Unit	Material $	Labor $	Equipment $	Total $
2"	P1@.100	LF	3.76	3.72	—	7.48
2½"	P1@.115	LF	5.90	4.28	—	10.18
3"	P1@.135	LF	8.06	5.02	—	13.08
4"	P1@.190	LF	11.50	7.07	—	18.57
5"	ER@.240	LF	21.60	9.59	.16	31.35
6"	ER@.315	LF	32.80	12.60	.21	45.61
8"	ER@.370	LF	32.80	14.80	.24	47.84
10"	ER@.460	LF	45.90	18.40	.30	64.60
12"	ER@.550	LF	72.50	22.00	.36	94.86

Carbon Steel, Schedule 40 with Roll-Grooved Joints

Description	Craft@Hrs	Unit	Material $	Labor $	Equipment $	Total $

Schedule 40 carbon steel 45-degree ell with roll-grooved joints

2"	P1@.170	Ea	27.60	6.33	—	33.93
2½"	P1@.185	Ea	37.50	6.89	—	44.39
3"	P1@.200	Ea	49.40	7.44	—	56.84
4"	P1@.230	Ea	72.90	8.56	—	81.46
5"	ER@.250	Ea	178.00	9.99	.16	188.15
6"	ER@.350	Ea	208.00	14.00	.23	222.23
8"	ER@.500	Ea	434.00	20.00	.33	454.33
10"	ER@.750	Ea	585.00	30.00	.49	615.49
12"	ER@1.00	Ea	1,020.00	40.00	.65	1,060.65

Schedule 40 carbon steel 90-degree ell with roll-grooved joints

2"	P1@.170	Ea	27.60	6.33	—	33.93
2½"	P1@.185	Ea	37.50	6.89	—	44.39
3"	P1@.200	Ea	49.40	7.44	—	56.84
4"	P1@.230	Ea	72.90	8.56	—	81.46
5"	ER@.250	Ea	178.00	9.99	.16	188.15
6"	ER@.350	Ea	208.00	14.00	.23	222.23
8"	ER@.500	Ea	434.00	20.00	.33	454.33
10"	ER@.750	Ea	673.00	30.00	.49	703.49
12"	ER@1.00	Ea	1,070.00	40.00	.65	1,110.65

Schedule 40 carbon steel tee with roll-grooved joints

2"	P1@.200	Ea	42.40	7.44	—	49.84
2½"	P1@.210	Ea	57.70	7.82	—	65.52
3"	P1@.230	Ea	80.50	8.56	—	89.06
4"	P1@.250	Ea	124.00	9.31	—	133.31
5"	ER@.300	Ea	291.00	12.00	.20	303.20
6"	ER@.500	Ea	338.00	20.00	.33	358.33
8"	ER@.750	Ea	741.00	30.00	.49	771.49
10"	ER@1.00	Ea	1,320.00	40.00	.65	1,360.65
12"	ER@1.50	Ea	1,820.00	59.90	.98	1,880.88

Carbon Steel, Schedule 40 with Roll-Grooved Joints

Description	Craft@Hrs	Unit	Material $	Labor $	Equipment $	Total $

Schedule 40 carbon steel reducing tee with roll-grooved joints

Description	Craft@Hrs	Unit	Material $	Labor $	Equipment $	Total $
3" x 2"	P1@.230	Ea	109.00	8.56	—	117.56
4" x 2"	P1@.240	Ea	148.00	8.93	—	156.93
4" x 2½"	P1@.245	Ea	148.00	9.12	—	157.12
4" x 3"	P1@.250	Ea	148.00	9.31	—	157.31
5" x 2"	P1@.260	Ea	271.00	9.68	—	280.68
5" x 3"	P1@.280	Ea	298.00	10.40	—	308.40
5" x 4"	P1@.300	Ea	320.00	11.20	—	331.20
6" x 3"	ER@.400	Ea	352.00	16.00	.26	368.26
6" x 4"	ER@.450	Ea	352.00	18.00	.29	370.29
6" x 5"	ER@.500	Ea	352.00	20.00	.33	372.33
8" x 3"	ER@.600	Ea	508.00	24.00	.39	532.39
8" x 4"	ER@.650	Ea	741.00	26.00	.42	767.42
8" x 5"	ER@.700	Ea	741.00	28.00	.46	769.46
8" x 6"	ER@.750	Ea	741.00	30.00	.49	771.49
10" x 4"	ER@.800	Ea	790.00	32.00	.52	822.52
10" x 5"	ER@.900	Ea	805.00	36.00	.59	841.59
10" x 6"	ER@1.00	Ea	805.00	40.00	.65	845.65
10" x 8"	ER@1.10	Ea	805.00	44.00	.72	849.72
12" x 6"	ER@1.20	Ea	117.00	48.00	.78	165.78
12" x 8"	ER@1.35	Ea	1,210.00	53.90	.88	1,264.78
12" x 10"	ER@1.50	Ea	1,270.00	59.90	.98	1,330.88

Schedule 40 carbon steel male adapter

Description	Craft@Hrs	Unit	Material $	Labor $	Equipment $	Total $
2"	P1@.170	Ea	18.00	6.33	—	24.33
2½"	P1@.185	Ea	21.30	6.89	—	28.19
3"	P1@.200	Ea	26.60	7.44	—	34.04
4"	P1@.230	Ea	44.50	8.56	—	53.06
5"	ER@.290	Ea	93.70	11.60	.19	105.49
6"	ER@.400	Ea	219.00	16.00	.26	235.26

Schedule 40 carbon steel female adapter

Description	Craft@Hrs	Unit	Material $	Labor $	Equipment $	Total $
2"	P1@.170	Ea	40.70	6.33	—	47.03
3"	P1@.200	Ea	62.90	7.44	—	70.34
4"	P1@.230	Ea	88.10	8.56	—	96.66

Schedule 40 carbon steel reducer, roll-grooved joints

Description	Craft@Hrs	Unit	Material $	Labor $	Equipment $	Total $
2"	P1@.160	Ea	38.00	5.96	—	43.96
3"	P1@.170	Ea	52.70	6.33	—	59.03
4"	P1@.210	Ea	64.20	7.82	—	72.02
5"	ER@.260	Ea	89.30	10.40	.17	99.87
6"	ER@.310	Ea	104.00	12.40	.20	116.60
8"	ER@.450	Ea	271.00	18.00	.29	289.29
10"	ER@.690	Ea	471.00	27.60	.45	499.05
12"	ER@.920	Ea	819.00	36.80	.60	856.40

Carbon Steel, Schedule 40 with Roll-Grooved Joints

Description	Craft@Hrs	Unit	Material $	Labor $	Equipment $	Total $
Schedule 40 carbon steel cap, roll-grooved joint						
2"	P1@.110	Ea	19.70	4.09	—	23.79
2½"	P1@.120	Ea	31.00	4.47	—	35.47
3"	P1@.130	Ea	31.00	4.84	—	35.84
4"	P1@.140	Ea	33.40	5.21	—	38.61
5"	ER@.150	Ea	77.60	5.99	.10	83.69
6"	ER@.170	Ea	80.50	6.79	.11	87.40
8"	ER@.200	Ea	156.00	7.99	.13	164.12
10"	ER@.225	Ea	280.00	8.99	.15	289.14
12"	ER@.250	Ea	453.00	9.99	.16	463.15
Flange with gasket, grooved joint						
2"	P1@.280	Ea	59.10	10.40	—	69.50
2½"	P1@.340	Ea	69.80	12.70	—	82.50
3"	P1@.390	Ea	79.00	14.50	—	93.50
4"	P1@.560	Ea	114.00	20.80	—	134.80
6"	ER@.780	Ea	143.00	31.20	—	174.20
8"	ER@1.10	Ea	168.00	44.00	—	212.00
10"	ER@1.40	Ea	213.00	55.90	—	268.90
12"	ER@1.70	Ea	295.00	67.90	—	362.90
Roll-grooved coupling with gasket						
2"	P1@.300	Ea	30.60	11.20	—	41.80
2½"	P1@.350	Ea	36.80	13.00	—	49.80
3"	P1@.400	Ea	40.70	14.90	—	55.60
4"	P1@.500	Ea	59.10	18.60	—	77.70
5"	ER@.600	Ea	90.60	24.00	—	114.60
6"	ER@.700	Ea	106.00	28.00	—	134.00
8"	ER@.900	Ea	178.00	36.00	—	214.00
10"	ER@1.10	Ea	246.00	44.00	—	290.00
12"	ER@1.35	Ea	280.00	53.90	—	333.90
Class 125 iron body gate valve, flanged ends						
2"	P1@.500	Ea	385.00	18.60	—	403.60
2½"	P1@.600	Ea	520.00	22.30	—	542.30
3"	P1@.750	Ea	564.00	27.90	—	591.90
4"	P1@1.35	Ea	829.00	50.20	—	879.20
6"	ER@2.50	Ea	1,570.00	99.90	1.63	1,671.53
8"	ER@3.00	Ea	2,610.00	120.00	1.96	2,731.96
10"	ER@4.00	Ea	4,250.00	160.00	2.61	4,412.61
12"	ER@4.50	Ea	5,900.00	180.00	2.93	6,082.93

Carbon Steel, Schedule 40 with Roll-Grooved Joints

Description	Craft@Hrs	Unit	Material $	Labor $	Equipment $	Total $

Class 125 iron body globe valve, flanged joints

Description	Craft@Hrs	Unit	Material $	Labor $	Equipment $	Total $
2"	P1@.500	Ea	281.00	18.60	—	299.60
2½"	P1@.600	Ea	436.00	22.30	—	458.30
3"	P1@.750	Ea	523.00	27.90	—	550.90
4"	P1@1.35	Ea	703.00	50.20	—	753.20
6"	ER@2.50	Ea	1,240.00	99.90	1.63	1,341.53
8"	ER@3.00	Ea	1,890.00	120.00	1.96	2,011.96
10"	ER@4.00	Ea	2,880.00	160.00	2.61	3,042.61

200 PSIG iron body butterfly valve, lug-type, lever operated

Description	Craft@Hrs	Unit	Material $	Labor $	Equipment $	Total $
2"	P1@.450	Ea	168.00	16.70	—	184.70
2½"	P1@.450	Ea	175.00	16.70	—	191.70
3"	P1@.550	Ea	183.00	20.50	—	203.50
4"	P1@.550	Ea	228.00	20.50	—	248.50
6"	ER@.800	Ea	368.00	32.00	.52	400.52
8"	ER@.800	Ea	503.00	32.00	.52	535.52
10"	ER@.900	Ea	699.00	36.00	.59	735.59
12"	ER@1.00	Ea	913.00	40.00	.65	953.65

200 PSIG iron body butterfly valve, wafer-type, lever operated

Description	Craft@Hrs	Unit	Material $	Labor $	Equipment $	Total $
2"	P1@.450	Ea	153.00	16.70	—	169.70
2½"	P1@.450	Ea	156.00	16.70	—	172.70
3"	P1@.550	Ea	168.00	20.50	—	188.50
4"	P1@.550	Ea	204.00	20.50	—	224.50
6"	ER@.800	Ea	341.00	32.00	.52	373.52
8"	ER@.800	Ea	472.00	32.00	.52	504.52
10"	ER@.900	Ea	660.00	36.00	.59	696.59
12"	ER@1.00	Ea	867.00	40.00	.65	907.65

Class 125 iron body swing check valve, flanged ends

Description	Craft@Hrs	Unit	Material $	Labor $	Equipment $	Total $
2"	P1@.500	Ea	182.00	18.60	—	200.60
2½"	P1@.600	Ea	230.00	22.30	—	252.30
3"	P1@.750	Ea	287.00	27.90	—	314.90
4"	P1@1.35	Ea	421.00	50.20	—	471.20
6"	ER@2.50	Ea	817.00	99.90	1.63	918.53
8"	ER@3.00	Ea	1,470.00	120.00	1.96	1,591.96
10"	ER@4.00	Ea	2,390.00	160.00	2.61	2,552.61
12"	ER@4.50	Ea	3,280.00	180.00	2.93	3,462.93

Class 125 iron body silent check valve, flanged joints

Description	Craft@Hrs	Unit	Material $	Labor $	Equipment $	Total $
2"	P1@.500	Ea	133.00	18.60	—	151.60
2½"	P1@.600	Ea	153.00	22.30	—	175.30
3"	P1@.750	Ea	173.00	27.90	—	200.90
4"	P1@1.35	Ea	226.00	50.20	—	276.20
5"	ER@2.00	Ea	328.00	79.90	1.30	409.20
6"	ER@2.50	Ea	406.00	99.90	1.63	507.53
8"	ER@3.00	Ea	732.00	120.00	1.96	853.96
10"	ER@4.00	Ea	1,130.00	160.00	2.61	1,292.61

Carbon Steel, Schedule 40 with Roll-Grooved Joints

Description	Craft@Hrs	Unit	Material $	Labor $	Equipment $	Total $

Class 125 iron body strainer, flanged

2"	P1@.500	Ea	136.00	18.60	—	154.60
2½"	P1@.600	Ea	152.00	22.30	—	174.30
3"	P1@.750	Ea	176.00	27.90	—	203.90
4"	P1@1.35	Ea	298.00	50.20	—	348.20
6"	ER@2.50	Ea	606.00	99.90	1.63	707.53
8"	ER@3.00	Ea	100.00	120.00	1.96	221.96

Installation of 2-way control valve, flanged joints

2"	P1@.500	Ea	—	18.60	—	18.60
2½"	P1@.600	Ea	—	22.30	—	22.30
3"	P1@.750	Ea	—	27.90	—	27.90
4"	P1@1.35	Ea	—	50.20	—	50.20
6"	P1@2.50	Ea	—	93.10	—	93.10
8"	P1@3.00	Ea	—	112.00	—	112.00
10"	P1@4.00	Ea	—	149.00	—	149.00
12"	P1@4.50	Ea	—	167.00	—	167.00

Installation of 3-way control valve, flanged joints

2"	P1@.910	Ea	—	33.90	—	33.90
2½"	P1@1.12	Ea	—	41.70	—	41.70
3"	P1@1.33	Ea	—	49.50	—	49.50
4"	P1@2.00	Ea	—	74.40	—	74.40
6"	P1@3.70	Ea	—	138.00	—	138.00
8"	P1@4.40	Ea	—	164.00	—	164.00
10"	P1@5.90	Ea	—	220.00	—	220.00
12"	P1@6.50	Ea	—	242.00	—	242.00

Bolt and gasket sets

2"	P1@.500	Ea	4.26	18.60	—	22.86
2½"	P1@.650	Ea	4.98	24.20	—	29.18
3"	P1@.750	Ea	8.45	27.90	—	36.35
4"	P1@1.00	Ea	14.10	37.20	—	51.30
6"	P1@1.20	Ea	23.40	44.70	—	68.10
8"	P1@1.25	Ea	26.30	46.50	—	72.80
10"	P1@1.70	Ea	45.50	63.30	—	108.80
12"	P1@2.20	Ea	50.90	81.90	—	132.80

Thermometer with well

7"	P1@.250	Ea	185.00	9.31	—	194.31
9"	P1@.250	Ea	190.00	9.31	—	199.31

Dial-type pressure gauge

2½"	P1@.200	Ea	38.70	7.44	—	46.14
3½"	P1@.200	Ea	51.00	7.44	—	58.44

Carbon Steel, Schedule 40 with Roll-Grooved Joints

Description	Craft@Hrs	Unit	Material $	Labor $	Equipment $	Total $
Pressure/temperature tap						
Tap	P1@.150	Ea	17.80	5.58	—	23.38
Hanger with swivel assembly						
2"	P1@.300	Ea	6.70	11.20	—	17.90
2½"	P1@.350	Ea	8.67	13.00	—	21.67
3"	P1@.350	Ea	10.70	13.00	—	23.70
4"	P1@.350	Ea	16.60	13.00	—	29.60
5"	P1@.450	Ea	18.60	16.70	—	35.30
6"	P1@.450	Ea	24.10	16.70	—	40.80
8"	P1@.450	Ea	32.10	16.70	—	48.80
10"	P1@.550	Ea	44.20	20.50	—	64.70
12"	P1@.600	Ea	58.50	22.30	—	80.80
Riser clamp						
2"	P1@.115	Ea	6.19	4.28	—	10.47
2½"	P1@.120	Ea	6.50	4.47	—	10.97
3"	P1@.120	Ea	7.04	4.47	—	11.51
4"	P1@.125	Ea	8.97	4.65	—	13.62
5"	P1@.180	Ea	12.90	6.70	—	19.60
6"	P1@.200	Ea	15.50	7.44	—	22.94
8"	P1@.200	Ea	25.20	7.44	—	32.64
10"	P1@.250	Ea	38.10	9.31	—	47.41
12"	P1@.250	Ea	44.50	9.31	—	53.81

Carbon Steel, Schedule 10 with Roll-Grooved Joints

Schedule 10 carbon steel (ASTM A-53) roll-grooved pipe with factory grooved malleable iron, ductile iron, or steel fittings is commonly used for sprinkler systems, heating hot water, chilled water and condenser water systems.

Consult the manufacturers for maximum operating temperature/pressure ratings for the various combinations of pipe and fittings and system applications.

The cost estimates in this section are based on the conditions, limitations and wage rates described in the section "How to Use This Book" beginning on page 5.

Equipment cost, where shown, is $4.45 per hour for a 2-ton chain hoist.

Description	Craft@Hrs	Unit	Material $	Labor $	Equipment $	Total $

Schedule 10 carbon steel roll-grooved horizontal pipe assembly.
Horizontally hung in a building. Assembly includes fittings, couplings and hanger assemblies. Based on a reducing tee and a 45-degree elbow every 16 feet for ½" pipe. A reducing tee and a 45-degree elbow every 50 feet for 12" pipe. Hangers spaced to meet plumbing code. *(Distance between fittings in the pipe assembly increases as pipe diameter increases.)* Use these figures for preliminary estimates.

Description	Craft@Hrs	Unit	Material $	Labor $	Equipment $	Total $
2"	P1@.178	LF	8.42	6.63	—	15.05
2½"	P1@.188	LF	10.50	7.00	—	17.50
3"	P1@.215	LF	12.20	8.00	—	20.20
4"	P1@.255	LF	17.40	9.49	—	26.89
6"	ER@.339	LF	32.80	13.50	.22	46.52
8"	ER@.404	LF	50.00	16.10	.26	66.36
10"	ER@.517	LF	74.90	20.70	.34	95.94
12"	ER@.602	LF	134.00	24.10	.39	158.49

Schedule 10 carbon steel roll-grooved vertical pipe assembly.
Riser assembly, including fittings, couplings, and riser clamps. Based on a reducing tee, and a riser clamp on every floor. Use these figures for preliminary estimates.

Description	Craft@Hrs	Unit	Material $	Labor $	Equipment $	Total $
2"	P1@.117	LF	8.08	4.35	—	12.43
2½"	P1@.120	LF	10.90	4.47	—	15.37
3"	P1@.146	LF	11.20	5.43	—	16.63
4"	P1@.204	LF	17.30	7.59	—	24.89
6"	ER@.286	LF	34.00	11.40	.19	45.59
8"	ER@.385	LF	59.80	15.40	.25	75.45
10"	ER@.510	LF	83.00	20.40	.33	103.73
12"	ER@.593	LF	134.00	23.70	.39	158.09

Carbon Steel, Schedule 10 with Roll-Grooved Joints

Description	Craft@Hrs	Unit	Material $	Labor $	Equipment $	Total $

Schedule 10 carbon steel pipe, roll-grooved joints

2"	P1@.080	LF	2.46	2.98	—	5.44
2½"	P1@.100	LF	3.35	3.72	—	7.07
3"	P1@.115	LF	4.07	4.28	—	8.35
4"	P1@.160	LF	5.41	5.96	—	11.37
5"	ER@.200	LF	9.24	7.99	.13	17.36
6"	ER@.260	LF	9.74	10.40	.17	20.31
8"	ER@.300	LF	13.40	12.00	.20	25.60
10"	ER@.380	LF	18.30	15.20	.25	33.75
12"	ER@.450	LF	25.90	18.00	.29	44.19

Schedule 10 carbon steel 45-degree ell, roll-grooved joints

2"	P1@.170	Ea	27.60	6.33	—	33.93
2½"	P1@.185	Ea	37.50	6.89	—	44.39
3"	P1@.200	Ea	49.40	7.44	—	56.84
4"	P1@.230	Ea	72.90	8.56	—	81.46
5"	ER@.250	Ea	178.00	9.99	.16	188.15
6"	ER@.350	Ea	208.00	14.00	.23	222.23
8"	ER@.500	Ea	434.00	20.00	.33	454.33
10"	ER@.750	Ea	585.00	30.00	.49	615.49
12"	ER@1.00	Ea	1,020.00	40.00	.65	1,060.65

Schedule 10 carbon steel 90-degree ell, roll-grooved joints

2"	P1@.170	Ea	27.60	6.33	—	33.93
2½"	P1@.185	Ea	37.50	6.89	—	44.39
3"	P1@.200	Ea	49.40	7.44	—	56.84
4"	P1@.230	Ea	72.90	8.56	—	81.46
5"	ER@.250	Ea	178.00	9.99	.16	188.15
6"	ER@.350	Ea	208.00	14.00	.23	222.23
8"	ER@.500	Ea	434.00	20.00	.33	454.33
10"	ER@.750	Ea	673.00	30.00	.49	703.49
12"	ER@1.00	Ea	1,070.00	40.00	.65	1,110.65

Schedule 10 carbon steel tee, roll-grooved joints

2"	P1@.200	Ea	42.40	7.44	—	49.84
2½"	P1@.210	Ea	57.70	7.82	—	65.52
3"	P1@.230	Ea	80.50	8.56	—	89.06
4"	P1@.250	Ea	124.00	9.31	—	133.31
5"	ER@.300	Ea	291.00	12.00	.20	303.20
6"	ER@.500	Ea	338.00	20.00	.33	358.33
8"	ER@.750	Ea	741.00	30.00	.49	771.49
10"	ER@1.00	Ea	1,320.00	40.00	.65	1,360.65
12"	ER@1.50	Ea	1,820.00	59.90	.98	1,880.88

Carbon Steel, Schedule 10 with Roll-Grooved Joints

Description	Craft@Hrs	Unit	Material $	Labor $	Equipment $	Total $

Schedule 10 carbon steel reducing tee, roll-grooved joints

Description	Craft@Hrs	Unit	Material $	Labor $	Equipment $	Total $
3" x 2"	P1@.230	Ea	109.00	8.56	—	117.56
4" x 2"	P1@.240	Ea	148.00	8.93	—	156.93
4" x 2½"	P1@.245	Ea	148.00	9.12	—	157.12
4" x 3"	P1@.250	Ea	148.00	9.31	—	157.31
5" x 2"	P1@.260	Ea	271.00	9.68	—	280.68
5" x 3"	P1@.280	Ea	298.00	10.40	—	308.40
5" x 4"	P1@.300	Ea	320.00	11.20	—	331.20
6" x 3"	ER@.400	Ea	352.00	16.00	.26	368.26
6" x 4"	ER@.450	Ea	352.00	18.00	.29	370.29
6" x 5"	ER@.500	Ea	352.00	20.00	.33	372.33
8" x 3"	ER@.600	Ea	508.00	24.00	.39	532.39
8" x 4"	ER@.650	Ea	741.00	26.00	.42	767.42
8" x 5"	ER@.700	Ea	741.00	28.00	.46	769.46
8" x 6"	ER@.750	Ea	741.00	30.00	.49	771.49
10" x 4"	ER@.800	Ea	790.00	32.00	.52	822.52
10" x 5"	ER@.900	Ea	805.00	36.00	.59	841.59
10" x 6"	ER@1.00	Ea	805.00	40.00	.65	845.65
10" x 8"	ER@1.10	Ea	805.00	44.00	.72	849.72
12" x 6"	ER@1.20	Ea	117.00	48.00	.78	165.78
12" x 8"	ER@1.35	Ea	1,210.00	53.90	.88	1,264.78
12" x 10"	ER@1.50	Ea	1,270.00	59.90	.98	1,330.88

Schedule 10 carbon steel male adapter

Description	Craft@Hrs	Unit	Material $	Labor $	Equipment $	Total $
2"	P1@.170	Ea	18.00	6.33	—	24.33
2½"	P1@.185	Ea	21.30	6.89	—	28.19
3"	P1@.200	Ea	26.60	7.44	—	34.04
4"	P1@.230	Ea	44.50	8.56	—	53.06
5"	ER@.290	Ea	93.70	11.60	—	105.30
6"	ER@.400	Ea	219.00	16.00	—	235.00

Schedule 10 carbon steel female adapter

Description	Craft@Hrs	Unit	Material $	Labor $	Equipment $	Total $
2"	P1@.170	Ea	40.70	6.33	—	47.03
3"	P1@.200	Ea	62.90	7.44	—	70.34
4"	P1@.230	Ea	88.10	8.56	—	96.66

Schedule 10 carbon steel reducer, roll-grooved joints

Description	Craft@Hrs	Unit	Material $	Labor $	Equipment $	Total $
2"	P1@.160	Ea	38.00	5.96	—	43.96
3"	P1@.170	Ea	52.70	6.33	—	59.03
4"	P1@.210	Ea	64.20	7.82	—	72.02
5"	ER@.260	Ea	89.30	10.40	—	99.70
6"	ER@.310	Ea	104.00	12.40	—	116.40
8"	ER@.450	Ea	271.00	18.00	—	289.00
10"	ER@.690	Ea	471.00	27.60	—	498.60
12"	ER@.920	Ea	819.00	36.80	—	855.80

Carbon Steel, Schedule 10 with Roll-Grooved Joints

Description	Craft@Hrs	Unit	Material $	Labor $	Equipment $	Total $

Schedule 10 carbon steel cap, roll-grooved joints

2"	P1@.110	Ea	19.70	4.09	—	23.79
2½"	P1@.120	Ea	31.00	4.47	—	35.47
3"	P1@.130	Ea	31.00	4.84	—	35.84
4"	P1@.140	Ea	33.40	5.21	—	38.61
5"	ER@.150	Ea	77.60	5.99	—	83.59
6"	ER@.170	Ea	80.50	6.79	—	87.29
8"	ER@.200	Ea	156.00	7.99	—	163.99
10"	ER@.225	Ea	280.00	8.99	—	288.99
12"	ER@.250	Ea	453.00	9.99	—	462.99

Flange with gasket, grooved joint

2"	P1@.280	Ea	59.10	10.40	—	69.50
2½"	P1@.340	Ea	69.80	12.70	—	82.50
3"	P1@.390	Ea	79.00	14.50	—	93.50
4"	P1@.560	Ea	114.00	20.80	—	134.80
6"	ER@.780	Ea	143.00	31.20	—	174.20
8"	ER@1.10	Ea	168.00	44.00	—	212.00
10"	ER@1.40	Ea	213.00	55.90	—	268.90
12"	ER@1.70	Ea	295.00	67.90	—	362.90

Roll-grooved coupling with gasket

2"	P1@.300	Ea	30.60	11.20	—	41.80
2½"	P1@.350	Ea	36.80	13.00	—	49.80
3"	P1@.400	Ea	40.70	14.90	—	55.60
4"	P1@.500	Ea	59.10	18.60	—	77.70
5"	ER@.600	Ea	90.60	24.00	—	114.60
6"	ER@.700	Ea	106.00	28.00	—	134.00
8"	ER@.900	Ea	178.00	36.00	—	214.00
10"	ER@1.10	Ea	246.00	44.00	—	290.00
12"	ER@1.35	Ea	280.00	53.90	—	333.90

Class 125 iron body gate valve, flanged joints

2"	P1@.500	Ea	499.00	18.60	—	517.60
2½"	P1@.600	Ea	677.00	22.30	—	699.30
3"	P1@.750	Ea	733.00	27.90	—	760.90
4"	P1@1.35	Ea	1,060.00	50.20	—	1,110.20
5"	ER@2.00	Ea	2,050.00	79.90	1.30	2,131.20
6"	ER@2.50	Ea	2,050.00	99.90	1.63	2,151.53
8"	ER@3.00	Ea	3,350.00	120.00	1.96	3,471.96
10"	ER@4.00	Ea	5,570.00	160.00	2.61	5,732.61
12"	ER@4.50	Ea	7,600.00	180.00	2.93	7,782.93

Carbon Steel, Schedule 10 with Roll-Grooved Joints

Description	Craft@Hrs	Unit	Material $	Labor $	Equipment $	Total $

Class 125 iron body globe valve, flanged joints

2"	P1@.500	Ea	273.00	18.60	—	291.60
2½"	P1@.600	Ea	422.00	22.30	—	444.30
3"	P1@.750	Ea	509.00	27.90	—	536.90
4"	P1@1.35	Ea	680.00	50.20	—	730.20
5"	ER@2.00	Ea	1,210.00	79.90	1.30	1,291.20
6"	ER@2.50	Ea	1,210.00	99.90	1.63	1,311.53
8"	ER@3.00	Ea	1,830.00	120.00	1.96	1,951.96
10"	ER@4.00	Ea	2,810.00	160.00	2.61	2,972.61

200 PSIG iron body butterfly valve, lug-type, lever operated

2"	P1@.450	Ea	168.00	16.70	—	184.70
2½"	P1@.450	Ea	175.00	16.70	—	191.70
3"	P1@.550	Ea	183.00	20.50	—	203.50
4"	P1@.550	Ea	228.00	20.50	—	248.50
5"	ER@.800	Ea	301.00	32.00	.52	333.52
6"	ER@.800	Ea	368.00	32.00	.52	400.52
8"	ER@.800	Ea	503.00	32.00	.52	535.52
10"	ER@.900	Ea	699.00	36.00	.59	735.59
12"	ER@1.00	Ea	913.00	40.00	.65	953.65

200 PSIG iron body butterfly valve, wafer-type, lever operated

2"	P1@.450	Ea	153.00	16.70	—	169.70
2½"	P1@.450	Ea	156.00	16.70	—	172.70
3"	P1@.550	Ea	168.00	20.50	—	188.50
4"	P1@.550	Ea	204.00	20.50	—	224.50
5"	ER@.800	Ea	272.00	32.00	.52	304.52
6"	ER@.800	Ea	341.00	32.00	.52	373.52
8"	ER@.800	Ea	472.00	32.00	.52	504.52
10"	ER@.900	Ea	660.00	36.00	.59	696.59
12"	ER@1.00	Ea	867.00	40.00	.65	907.65

Class 125 iron body swing check valve, flanged joints

2"	P1@.500	Ea	182.00	18.60	—	200.60
2½"	P1@.600	Ea	230.00	22.30	—	252.30
3"	P1@.750	Ea	287.00	27.90	—	314.90
4"	P1@1.35	Ea	421.00	50.20	—	471.20
5"	ER@2.00	Ea	811.00	79.90	1.30	892.20
6"	ER@2.50	Ea	817.00	99.90	1.63	918.53
8"	ER@3.00	Ea	1,470.00	120.00	1.96	1,591.96
10"	ER@4.00	Ea	2,390.00	160.00	2.61	2,552.61
12"	ER@4.50	Ea	3,280.00	180.00	2.93	3,462.93

Carbon Steel, Schedule 10 with Roll-Grooved Joints

Description	Craft@Hrs	Unit	Material $	Labor $	Equipment $	Total $
Class 125 iron body silent check valve, flanged joints						
2"	P1@.500	Ea	133.00	18.60	—	151.60
2½"	P1@.600	Ea	153.00	22.30	—	175.30
3"	P1@.750	Ea	173.00	27.90	—	200.90
4"	P1@1.35	Ea	226.00	50.20	—	276.20
5"	ER@2.00	Ea	328.00	79.90	1.30	409.20
6"	ER@2.50	Ea	406.00	99.90	1.63	507.53
8"	ER@3.00	Ea	732.00	120.00	1.96	853.96
10"	ER@4.00	Ea	1,130.00	160.00	2.61	1,292.61
Class 125 iron body strainer, flanged						
2"	P1@.500	Ea	136.00	18.60	—	154.60
2½"	P1@.600	Ea	152.00	22.30	—	174.30
3"	P1@.750	Ea	175.00	27.90	—	202.90
4"	P1@1.35	Ea	298.00	50.20	—	348.20
5"	ER@2.00	Ea	605.00	79.90	1.30	686.20
6"	ER@2.50	Ea	605.00	99.90	1.63	706.53
8"	ER@3.00	Ea	1,020.00	120.00	1.96	1,141.96
Installation of 2-way control valve, flanged joints						
2"	P1@.500	Ea	—	18.60	—	18.60
2½"	P1@.600	Ea	—	22.30	—	22.30
3"	P1@.750	Ea	—	27.90	—	27.90
4"	P1@1.35	Ea	—	50.20	—	50.20
6"	P1@2.50	Ea	—	93.10	—	93.10
8"	P1@3.00	Ea	—	112.00	—	112.00
10"	P1@4.00	Ea	—	149.00	—	149.00
12"	P1@4.50	Ea	—	167.00	—	167.00
Installation of 3-way control valve, flanged joints						
2"	P1@.910	Ea	—	33.90	—	33.90
2½"	P1@1.12	Ea	—	41.70	—	41.70
3"	P1@1.33	Ea	—	49.50	—	49.50
4"	P1@2.00	Ea	—	74.40	—	74.40
6"	P1@3.70	Ea	—	138.00	—	138.00
8"	P1@4.40	Ea	—	164.00	—	164.00
10"	P1@5.90	Ea	—	220.00	—	220.00
12"	P1@6.50	Ea	—	242.00	—	242.00

Carbon Steel, Schedule 10 with Roll-Grooved Joints

Description	Craft@Hrs	Unit	Material $	Labor $	Equipment $	Total $
Bolt and gasket sets						
2"	P1@.500	Ea	4.26	18.60	—	22.86
2½"	P1@.650	Ea	4.98	24.20	—	29.18
3"	P1@.750	Ea	8.42	27.90	—	36.32
4"	P1@1.00	Ea	14.00	37.20	—	51.20
5"	P1@1.10	Ea	23.40	40.90	—	64.30
6"	P1@1.20	Ea	23.40	44.70	—	68.10
8"	P1@1.25	Ea	26.30	46.50	—	72.80
10"	P1@1.70	Ea	45.20	63.30	—	108.50
12"	P1@2.20	Ea	50.90	81.90	—	132.80
Thermometer with well						
7"	P1@.250	Ea	187.00	9.31	—	196.31
9"	P1@.250	Ea	193.00	9.31	—	202.31
Dial-type pressure gauge						
2½"	P1@.200	Ea	39.10	7.44	—	46.54
3½"	P1@.200	Ea	51.50	7.44	—	58.94
Pressure/temperature tap						
Tap	P1@.150	Ea	18.00	5.58	—	23.58
Hanger with swivel assembly						
2"	P1@.300	Ea	6.94	11.20	—	18.14
2½"	P1@.350	Ea	8.99	13.00	—	21.99
3"	P1@.350	Ea	11.10	13.00	—	24.10
4"	P1@.350	Ea	17.20	13.00	—	30.20
5"	P1@.450	Ea	19.40	16.70	—	36.10
6"	P1@.450	Ea	24.90	16.70	—	41.60
8"	P1@.450	Ea	33.30	16.70	—	50.00
10"	P1@.550	Ea	45.80	20.50	—	66.30
12"	P1@.600	Ea	60.80	22.30	—	83.10
Riser clamp						
2"	P1@.115	Ea	6.39	4.28	—	10.67
2½"	P1@.120	Ea	6.74	4.47	—	11.21
3"	P1@.120	Ea	7.29	4.47	—	11.76
4"	P1@.125	Ea	9.29	4.65	—	13.94
5"	P1@.180	Ea	13.30	6.70	—	20.00
6"	P1@.200	Ea	16.00	7.44	—	23.44
8"	P1@.200	Ea	26.10	7.44	—	33.54
10"	P1@.250	Ea	38.90	9.31	—	48.21
12"	P1@.250	Ea	46.10	9.31	—	55.41

Carbon Steel, Schedule 40 with Cut-Grooved Joints

Schedule 40 carbon steel (ASTM A-53 and A-120) cut-grooved pipe with factory-grooved malleable iron, ductile iron or steel fittings is commonly used for heating hot water, chilled water and condenser water systems.

Consult the manufacturers for maximum operating temperature/pressure ratings for various combinations of pipe and fittings.

This section has been arranged to save the estimator's time by including all normally-used system components such as pipe, fittings, valves, hanger assemblies, riser clamps and miscellaneous items under one heading. Additional items can be found under "Plumbing and Piping Specialties." The cost estimates in this section are based on the conditions, limitations and wage rates described in the section "How to Use This Book" beginning on page 5.

Equipment cost, where shown, is $4.45 per hour for a 2-ton chain hoist.

Description	Craft@Hrs	Unit	Material $	Labor $	Equipment $	Total $

Schedule 40 carbon steel pipe with cut-grooved joints

Description	Craft@Hrs	Unit	Material $	Labor $	Equipment $	Total $
2"	P1@.110	LF	4.19	4.09	—	8.28
2½"	P1@.130	LF	6.58	4.84	—	11.42
3"	P1@.150	LF	8.93	5.58	—	14.51
4"	P1@.210	LF	12.90	7.82	—	20.72
5"	ER@.270	LF	24.20	10.80	.18	35.18
6"	ER@.350	LF	36.50	14.00	.23	50.73
8"	ER@.410	LF	36.50	16.40	.27	53.17
10"	ER@.510	LF	51.20	20.40	.33	71.93
12"	ER@.610	LF	80.80	24.40	.40	105.60

Schedule 40 carbon steel 45-degree ell with cut-grooved joints

Description	Craft@Hrs	Unit	Material $	Labor $	Equipment $	Total $
2"	P1@.170	Ea	27.60	6.33	—	33.93
2½"	P1@.185	Ea	37.70	6.89	—	44.59
3"	P1@.200	Ea	49.40	7.44	—	56.84
4"	P1@.230	Ea	72.90	8.56	—	81.46
5"	ER@.250	Ea	178.00	9.99	.16	188.15
6"	ER@.350	Ea	208.00	14.00	.23	222.23
8"	ER@.500	Ea	434.00	20.00	.33	454.33
10"	ER@.750	Ea	585.00	30.00	.49	615.49
12"	ER@1.00	Ea	1,020.00	40.00	.65	1,060.65

Schedule 40 carbon steel 90-degree ell with cut-grooved joints

Description	Craft@Hrs	Unit	Material $	Labor $	Equipment $	Total $
2"	P1@.170	Ea	27.60	6.33	—	33.93
2½"	P1@.185	Ea	37.70	6.89	—	44.59
3"	P1@.200	Ea	49.40	7.44	—	56.84
4"	P1@.230	Ea	72.90	8.56	—	81.46
5"	ER@.250	Ea	178.00	9.99	.16	188.15
6"	ER@.350	Ea	208.00	14.00	.23	222.23
8"	ER@.500	Ea	434.00	20.00	.33	454.33
10"	ER@.750	Ea	673.00	30.00	.49	703.49
12"	ER@1.00	Ea	1,820.00	40.00	.65	1,860.65

Carbon Steel, Schedule 40 with Cut-Grooved Joints

Description	Craft@Hrs	Unit	Material $	Labor $	Equipment $	Total $

Schedule 40 carbon steel tee with cut-grooved joints

2"	P1@.200	Ea	42.40	7.44	—	49.84
2½"	P1@.210	Ea	57.70	7.82	—	65.52
3"	P1@.230	Ea	80.50	8.56	—	89.06
4"	P1@.250	Ea	124.00	9.31	—	133.31
5"	ER@.300	Ea	291.00	12.00	.20	303.20
6"	ER@.500	Ea	338.00	20.00	.22	358.22
8"	ER@.750	Ea	741.00	30.00	.49	771.49
10"	ER@1.00	Ea	1,320.00	40.00	.65	1,360.65
12"	ER@1.50	Ea	1,820.00	59.90	.98	1,880.88

Schedule 40 carbon steel reducing tee with cut-grooved joints

3" x 2"	P1@.230	Ea	109.00	8.56	—	117.56
4" x 2"	P1@.240	Ea	148.00	8.93	—	156.93
4" x 2½"	P1@.245	Ea	148.00	9.12	—	157.12
4" x 3"	P1@.250	Ea	148.00	9.31	—	157.31
5" x 2"	P1@.260	Ea	271.00	9.68	—	280.68
5" x 3"	P1@.280	Ea	298.00	10.40	—	308.40
5" x 4"	P1@.300	Ea	320.00	11.20	—	331.20
6" x 3"	ER@.400	Ea	352.00	16.00	.26	368.26
6" x 4"	ER@.450	Ea	352.00	18.00	.29	370.29
6" x 5"	ER@.500	Ea	352.00	20.00	.33	372.33
8" x 3"	ER@.600	Ea	508.00	24.00	.39	532.39
8" x 4"	ER@.650	Ea	741.00	26.00	.42	767.42
8" x 5"	ER@.700	Ea	741.00	28.00	.46	769.46
8" x 6"	ER@.750	Ea	741.00	30.00	.49	771.49
10" x 4"	ER@.800	Ea	790.00	32.00	.52	822.52
10" x 5"	ER@.900	Ea	805.00	36.00	.59	841.59
10" x 6"	ER@1.00	Ea	805.00	40.00	.65	845.65
10" x 8"	ER@1.10	Ea	805.00	44.00	.72	849.72
12" x 6"	ER@1.20	Ea	117.00	48.00	.78	165.78
12" x 8"	ER@1.35	Ea	1,210.00	53.90	.88	1,264.78
12" x 10"	ER@1.50	Ea	1,270.00	59.90	.98	1,330.88

Schedule 40 carbon steel male adapter

2"	P1@.170	Ea	18.00	6.33	—	24.33
2½"	P1@.185	Ea	21.30	6.89	—	28.19
3"	P1@.200	Ea	26.60	7.44	—	34.04
4"	P1@.230	Ea	44.20	8.56	—	52.76
5"	ER@.290	Ea	93.70	11.60	—	105.30
6"	ER@.400	Ea	219.00	16.00	—	235.00

Schedule 40 carbon steel female adapter

2"	P1@.170	Ea	40.70	6.33	—	47.03
3"	P1@.200	Ea	62.90	7.44	—	70.34
4"	P1@.230	Ea	88.10	8.56	—	96.66

Carbon Steel, Schedule 40 with Cut-Grooved Joints

Description	Craft@Hrs	Unit	Material $	Labor $	Equipment $	Total $
Schedule 40 carbon steel reducer						
2"	P1@.160	Ea	38.00	5.96	—	43.96
3"	P1@.170	Ea	52.70	6.33	—	59.03
4"	P1@.210	Ea	64.20	7.82	—	72.02
5"	ER@.260	Ea	89.30	10.40	—	99.70
6"	ER@.310	Ea	104.00	12.40	—	116.40
8"	ER@.450	Ea	271.00	18.00	—	289.00
10"	ER@.690	Ea	471.00	27.60	—	498.60
12"	ER@.920	Ea	819.00	36.80	—	855.80
Schedule 40 carbon steel cap						
2"	P1@.110	Ea	19.70	4.09	—	23.79
2½"	P1@.120	Ea	31.00	4.47	—	35.47
3"	P1@.130	Ea	31.00	4.84	—	35.84
4"	P1@.140	Ea	33.40	5.21	—	38.61
5"	ER@.150	Ea	77.60	5.99	—	83.59
6"	ER@.170	Ea	80.50	6.79	—	87.29
8"	ER@.200	Ea	156.00	7.99	—	163.99
10"	ER@.225	Ea	280.00	8.99	—	288.99
12"	ER@.250	Ea	453.00	9.99	—	462.99
Flange with gasket, grooved joint						
2"	P1@.280	Ea	59.10	10.40	—	69.50
2½"	P1@.340	Ea	69.80	12.70	—	82.50
3"	P1@.390	Ea	79.00	14.50	—	93.50
4"	P1@.560	Ea	114.00	20.80	—	134.80
6"	ER@.780	Ea	143.00	31.20	—	174.20
8"	ER@1.10	Ea	168.00	44.00	—	212.00
10"	ER@1.40	Ea	213.00	55.90	—	268.90
12"	ER@1.70	Ea	295.00	67.90	—	362.90
Cut-grooved joint coupling with gasket						
2"	P1@.300	Ea	30.60	11.20	—	41.80
2½"	P1@.350	Ea	36.80	13.00	—	49.80
3"	P1@.400	Ea	40.70	14.90	—	55.60
4"	P1@.500	Ea	59.10	18.60	—	77.70
5"	ER@.600	Ea	90.60	24.00	—	114.60
6"	ER@.700	Ea	106.00	28.00	—	134.00
8"	ER@.900	Ea	178.00	36.00	—	214.00
10"	ER@1.10	Ea	246.00	44.00	—	290.00
12"	ER@1.35	Ea	280.00	53.90	—	333.90

Carbon Steel, Schedule 40 with Cut-Grooved Joints

Description	Craft@Hrs	Unit	Material $	Labor $	Equipment $	Total $

Class 125 iron body gate valve, flanged ends

2"	P1@.500	Ea	535.00	18.60	—	553.60
2½"	P1@.600	Ea	724.00	22.30	—	746.30
3"	P1@.750	Ea	785.00	27.90	—	812.90
4"	P1@1.35	Ea	1,150.00	50.20	—	1,200.20
6"	ER@2.50	Ea	2,220.00	99.90	1.63	2,321.53
8"	ER@3.00	Ea	3,600.00	120.00	1.96	3,721.96
10"	ER@4.00	Ea	5,940.00	160.00	2.61	6,102.61
12"	ER@4.50	Ea	8,140.00	180.00	2.93	8,322.93

Class 125 iron body globe valve, flanged joints

2"	P1@.500	Ea	273.00	18.60	—	291.60
2½"	P1@.600	Ea	425.00	22.30	—	447.30
3"	P1@.750	Ea	509.00	27.90	—	536.90
4"	ER@1.35	Ea	684.00	53.90	—	737.90
6"	ER@2.50	Ea	1,210.00	99.90	1.63	1,311.53
8"	ER@3.00	Ea	1,840.00	120.00	1.96	1,961.96
10"	ER@4.00	Ea	2,810.00	160.00	2.61	2,972.61

200 PSIG iron body butterfly valve, lug-type, lever operated

2"	P1@.450	Ea	168.00	16.70	—	184.70
2½"	P1@.450	Ea	175.00	16.70	—	191.70
3"	P1@.550	Ea	183.00	20.50	—	203.50
4"	P1@.550	Ea	228.00	20.50	—	248.50
6"	ER@.800	Ea	368.00	32.00	.52	400.52
8"	ER@.800	Ea	503.00	32.00	.52	535.52
10"	ER@.900	Ea	699.00	36.00	.59	735.59
12"	ER@1.00	Ea	913.00	40.00	.65	953.65

200 PSIG iron body butterfly valve, wafer-type, lever operated

2"	P1@.450	Ea	153.00	16.70	—	169.70
2½"	P1@.450	Ea	156.00	16.70	—	172.70
3"	P1@.550	Ea	168.00	20.50	—	188.50
4"	P1@.550	Ea	204.00	20.50	—	224.50
6"	ER@.800	Ea	341.00	32.00	.52	373.52
8"	ER@.800	Ea	472.00	32.00	.52	504.52
10"	ER@.900	Ea	660.00	36.00	.59	696.59
12"	ER@1.00	Ea	867.00	40.00	.65	907.65

Class 125 iron body swing check valve, flanged ends

2"	P1@.500	Ea	182.00	18.60	—	200.60
2½"	P1@.600	Ea	230.00	22.30	—	252.30
3"	P1@.750	Ea	287.00	27.90	—	314.90
4"	P1@1.35	Ea	421.00	50.20	—	471.20
6"	ER@2.50	Ea	817.00	99.90	1.63	918.53
8"	ER@3.00	Ea	1,470.00	120.00	1.96	1,591.96
10"	ER@4.00	Ea	2,390.00	160.00	2.61	2,552.61
12"	ER@4.50	Ea	3,280.00	180.00	2.93	3,462.93

Carbon Steel, Schedule 40 with Cut-Grooved Joints

Description	Craft@Hrs	Unit	Material $	Labor $	Equipment $	Total $

Class 125 iron body silent check valve, flanged joints

2"	P1@.500	Ea	133.00	18.60	—	151.60
2½"	P1@.600	Ea	153.00	22.30	—	175.30
3"	P1@.750	Ea	173.00	27.90	—	200.90
4"	P1@1.35	Ea	226.00	50.20	—	276.20
5"	ER@2.00	Ea	328.00	79.90	1.30	409.20
6"	ER@2.50	Ea	406.00	99.90	1.63	507.53
8"	ER@3.00	Ea	732.00	120.00	1.96	853.96
10"	ER@4.00	Ea	1,130.00	160.00	2.61	1,292.61

Class 125 iron body strainer, flanged

2"	P1@.500	Ea	136.00	18.60	—	154.60
2½"	P1@.600	Ea	152.00	22.30	—	174.30
3"	P1@.750	Ea	176.00	27.90	—	203.90
4"	P1@1.35	Ea	298.00	50.20	—	348.20
6"	ER@2.50	Ea	606.00	99.90	1.63	707.53
8"	ER@3.00	Ea	1,020.00	120.00	1.96	1,141.96

Installation of 2-way control valve, flanged joints

2"	P1@.500	Ea	—	18.60	—	18.60
2½"	P1@.600	Ea	—	22.30	—	22.30
3"	P1@.750	Ea	—	27.90	—	27.90
4"	P1@1.35	Ea	—	50.20	—	50.20
6"	P1@2.50	Ea	—	93.10	—	93.10
8"	P1@3.00	Ea	—	112.00	—	112.00
10"	P1@4.00	Ea	—	149.00	—	149.00
12"	P1@4.50	Ea	—	167.00	—	167.00

Installation of 3-way control valve, flanged joints

2"	P1@.910	Ea	—	33.90	—	33.90
2½"	P1@1.12	Ea	—	41.70	—	41.70
3"	P1@1.33	Ea	—	49.50	—	49.50
4"	P1@2.00	Ea	—	74.40	—	74.40
6"	P1@3.70	Ea	—	138.00	—	138.00
8"	P1@4.40	Ea	—	164.00	—	164.00
10"	P1@5.90	Ea	—	220.00	—	220.00
12"	P1@6.50	Ea	—	242.00	—	242.00

Bolt and gasket sets

2"	P1@.500	Ea	4.26	18.60	—	22.86
2½"	P1@.650	Ea	4.98	24.20	—	29.18
3"	P1@.750	Ea	8.42	27.90	—	36.32
4"	P1@1.00	Ea	14.00	37.20	—	51.20
6"	P1@1.20	Ea	23.40	44.70	—	68.10
8"	P1@1.25	Ea	26.30	46.50	—	72.80
10"	P1@1.70	Ea	45.20	63.30	—	108.50
12"	P1@2.20	Ea	50.90	81.90	—	132.80

Carbon Steel, Schedule 40 with Cut-Grooved Joints

Description	Craft@Hrs	Unit	Material $	Labor $	Equipment $	Total $
Thermometer with well						
7"	P1@.250	Ea	187.00	9.31	—	196.31
9"	P1@.250	Ea	193.00	9.31	—	202.31
Dial-type pressure gauge						
2½"	P1@.200	Ea	39.10	7.44	—	46.54
3½"	P1@.200	Ea	51.50	7.44	—	58.94
Pressure/temperature tap						
Tap	P1@.150	Ea	17.90	5.58	—	23.48
Hanger with swivel assembly						
2"	P1@.300	Ea	6.91	11.20	—	18.11
2½"	P1@.350	Ea	8.93	13.00	—	21.93
3"	P1@.350	Ea	11.10	13.00	—	24.10
4"	P1@.350	Ea	17.10	13.00	—	30.10
5"	P1@.450	Ea	19.30	16.70	—	36.00
6"	P1@.450	Ea	24.80	16.70	—	41.50
8"	P1@.450	Ea	33.10	16.70	—	49.80
10"	P1@.550	Ea	45.60	20.50	—	66.10
12"	P1@.600	Ea	60.60	22.30	—	82.90
Riser clamp						
2"	P1@.115	Ea	6.36	4.28	—	10.64
2½"	P1@.120	Ea	6.69	4.47	—	11.16
3"	P1@.120	Ea	7.26	4.47	—	11.73
4"	P1@.125	Ea	9.25	4.65	—	13.90
5"	P1@.180	Ea	13.20	6.70	—	19.90
6"	P1@.200	Ea	16.00	7.44	—	23.44
8"	P1@.200	Ea	26.00	7.44	—	33.44
10"	P1@.250	Ea	38.80	9.31	—	48.11
12"	P1@.250	Ea	45.90	9.31	—	55.21
Galvanized steel pipe sleeves						
2"	P1@.130	Ea	5.59	4.84	—	10.43
2½"	P1@.150	Ea	5.67	5.58	—	11.25
3"	P1@.180	Ea	5.81	6.70	—	12.51
4"	P1@.220	Ea	6.63	8.19	—	14.82
5"	P1@.250	Ea	8.20	9.31	—	17.51
6"	P1@.270	Ea	8.85	10.00	—	18.85
8"	P1@.270	Ea	10.20	10.00	—	20.20
10"	P1@.290	Ea	11.80	10.80	—	22.60
12"	P1@.310	Ea	16.20	11.50	—	27.70
14"	P1@.330	Ea	16.20	12.30	—	28.50

Residential HVAC Assemblies

Description	Craft@Hrs	Unit	Material $	Labor $	Equipment $	Total $

Leadership in Energy & Environmental Design (LEED).

LEED offers an incentive to building owners to install energy-efficient and environmentally sensitive HVAC equipment. The LEED program awards points for application of the best available technology. The HVAC system can earn a maximum of 17 points toward LEED certification. A LEED-certified air conditioning system must comply with ASHRAE 90.1-2004 (efficiency) and ASHRAE 62.1-2004 (indoor air quality) standards. LEED certification also requires recording controls (Carrier ComfortView 3 or equal) to measure system performance. Zone sensor thermostats (Carrier Debonair or equal) are required to provide performance and efficiency data to the digital recorder. An indoor air quality CO_2 sensor, duct-mounted aspirator modules and a refrigerant such as Puron R-410A are also required.

Description	Craft@Hrs	Unit	Material $	Labor $	Equipment $	Total $
Add for a LEED-certified system with USGBC rating	—	%	20.0	—	—	—
Add for LEED registration and inspection	—	Ea	2,070.00	—	—	2,070.00
Add for LEED central digital performance monitor/recorder, ComfortView 3 or equal	—	Ea	2,480.00	—	—	2,480.00
Add for LEED-certified zone recording thermostats, Carrier DebonAir, USB remote-ready	—	Ea	367.00	—	—	367.00
Add for IAQ (Indoor Air Quality) CO_2 sensor and duct-mounted aspirator box	—	Ea	449.00	—	—	449.00

Residential A/C Cooling System.

With remote exterior condenser/compressor, field installed DX coil in existing supply air plenum. Includes electrical & control wiring, R-410A refrigerant piping & connections, programmable digital thermostat, start-up and testing. Costs based on a maximum distance between outdoor condenser and coil of 60' and include a contractor mark up of 25%. Use these figures for preliminary budget purposes. Costs are based on using 650 SF per ton as the cooling requirement with a thermal envelope of R20 or greater, i.e., a 1.5-ton A/C unit will serve a 975 SF home. Typical subcontract prices.

Description	Craft@Hrs	Unit	Material $	Labor $	Equipment $	Total $
1.5 tons	—	Ea	—	—	—	3,180.00
2 tons	—	Ea	—	—	—	3,490.00
2.5 tons	—	Ea	—	—	—	4,250.00
3 tons	—	Ea	—	—	—	5,790.00
5 tons	—	Ea	—	—	—	8,740.00
Add for high heat loads	—	%	—	—	—	12.0

Residential HVAC Assemblies

Description	Craft@Hrs	Unit	Material $	Labor $	Equipment $	Total $

Residential forced-air heating, cooling & ventilation system.
Including a remote exterior condenser/compressor, field installed R-410A refrigerant DX "A" coil in supply air plenum, a gas-fired high efficiency furnace, programmable digital thermostat, ductwork, grilles & registers, HRV (heat recovery ventilator), kitchen and bathroom exhaust fans, ventilation and exhaust ductwork and wall hoods, electrical & control wiring, gas, refrigerant & flue piping & associated connections, start-up, testing, and a contractor markup of 25%. Use these figures for preliminary budget purposes. Heating & cooling load calculations based on a thermal envelope of R20 or greater. Typical subcontract prices.

Description	Craft@Hrs	Unit	Material $	Labor $	Equipment $	Total $
900 – 1,000 SF	—	Ea	—	—	—	16,600.00
1,100 – 1,400 SF	—	Ea	—	—	—	18,800.00
1,500 – 1,750 SF	—	Ea	—	—	—	23,300.00
1,800 – 2000 SF	—	Ea	—	—	—	24,900.00
2,200 – 2,500 SF	—	Ea	—	—	—	28,000.00
2,600 – 3,000 SF	—	Ea	—	—	—	33,800.00
Deduct for mid-efficient unit	—	Ea	—	—	—	-297.00
Deduct for electric furnace	—	Ea	—	—	—	-297.00

Residential HVAC Assemblies

Description	Craft@Hrs	Unit	Material $	Labor $	Equipment $	Total $

Packaged self-contained rooftop DX air conditioning units.
EnergySmart certified, ASHRAE, U.S. Green Building Council and Underwriters Laboratories approved. Includes cooling coils, compressor, heat rejection coils, regulator valves, refrigerant tank and remote digital single-zone control package. Set in place with a 13,000 lb. truck crane. Add installation costs from the section that follows.

Description	Craft@Hrs	Unit	Material $	Labor $	Equipment $	Total $
2-T, 800 CFM	SN@5.00	Ea	4,080.00	196.00	51.00	4,327.00
2½-T, 1,000 CFM	SN@6.00	Ea	4,880.00	235.00	61.30	5,176.30
3-T, 1,200 CFM	SN@7.00	Ea	5,840.00	274.00	71.50	6,185.50
4-T, 1,600 CFM	SN@8.00	Ea	7,150.00	314.00	81.70	7,545.70
5-T, 2,000 CFM	SN@9.00	Ea	8,930.00	353.00	91.90	9,374.90
7½-T, 3,000 CFM	SN@10.0	Ea	12,400.00	392.00	102.00	12,894.00
10-T, 4,000 CFM	SN@12.0	Ea	15,500.00	470.00	122.00	16,092.00
12-T, 5,000 CFM	SN@14.0	Ea	18,400.00	549.00	143.00	19,092.00
15-T, 6,000 CFM	SN@15.0	Ea	22,600.00	588.00	153.00	23,341.00
20-T, 8,000 CFM	SN@16.0	Ea	29,900.00	627.00	163.00	30,690.00
25-T, 10,000 CFM	SN@17.0	Ea	36,400.00	666.00	174.00	37,240.00
30-T, 12,000 CFM	SN@19.0	Ea	41,600.00	745.00	194.00	42,539.00
40-T, 16,000 CFM	SN@22.0	Ea	52,000.00	862.00	225.00	53,087.00

Rooftop DX air conditioning unit, hot water coil.
EnergySmart certified, ASHRAE, U.S. Green Building Council, and Underwriters Laboratories approved. Add for the hot water coil. Single zone controls. Add installation costs from the section that follows.

Description	Craft@Hrs	Unit	Material $	Labor $	Equipment $	Total $
2-T, 800 CFM	SN@5.00	Ea	3,980.00	196.00	51.00	4,227.00
2½-T, 1,000 CFM	SN@6.00	Ea	4,780.00	235.00	61.30	5,076.30
3-T, 1,200 CFM	SN@7.00	Ea	5,720.00	274.00	71.50	6,065.50
4-T, 1,600 CFM	SN@8.00	Ea	6,970.00	314.00	81.70	7,365.70
5-T, 2,000 CFM	SN@9.00	Ea	8,730.00	353.00	91.90	9,174.90
7½-T, 3,000 CFM	SN@10.0	Ea	13,200.00	392.00	102.00	13,694.00
10-T, 4,000 CFM	SN@12.0	Ea	15,100.00	470.00	122.00	15,692.00
12-T, 5,000 CFM	SN@14.0	Ea	17,900.00	549.00	143.00	18,592.00
15-T, 6,000 CFM	SN@15.0	Ea	22,000.00	588.00	153.00	22,741.00
20-T, 8,000 CFM	SN@16.0	Ea	29,300.00	627.00	163.00	30,090.00
25-T, 10,000 CFM	SN@17.0	Ea	35,300.00	666.00	174.00	36,140.00
30-T, 12,000 CFM	SN@19.0	Ea	41,600.00	745.00	194.00	42,539.00
40-T, 16,000 CFM	SN@22.0	Ea	52,000.00	862.00	225.00	53,087.00

Air Handling Units

Description	Craft@Hrs	Unit	Material $	Labor $	Equipment $	Total $

Installation costs for packaged rooftop cooling Typical installation costs for packaged rooftop cooling. Costs will be higher for larger capacity units.

Description	Craft@Hrs	Unit	Material $	Labor $	Equipment $	Total $
Cut, frame and gasket downcomer hole in roof	S2@1.00	Ea	33.90	36.50	—	70.40
Mount duct hangers	S2@.250	Ea	4.37	9.14	—	13.51
Cut and mount sheet metal duct	S2@.350	LF	6.72	12.80	—	19.52
Apply and coat duct insulation	S2@.150	LF	2.91	5.48	—	8.39
Install piping for gas line	P1@.100	LF	4.53	3.72	—	8.25
Install piping for chilled or hot water/steam line	P1@.100	LF	3.44	3.72	—	7.16
Install electrical wiring	BE@.150	LF	2.09	6.15	—	8.24
Install HVAC controls	BE@.500	Ea	—	20.50	—	20.50
Commission and test	P1@4.00	Ea	—	149.00	—	149.00
Air balance and fine-tune	P1@4.00	Ea	—	149.00	—	149.00

Packaged air handler with chilled water and hot water/steam coil.
Costs shown based on 400 CFM per ton cooling. Unit includes insulated single wall casing, fan section, cooling coil section, heating coil section, mixing plenum section, bag filter section, fan motor, variable pitch drive, vibration isolators and drain pan. Set in place only. Make additional allowances for coil connections, controls, motor starters and power wiring. (12,000 BTUs equals 1 ton cooling.) Use these costs for preliminary estimates.

Description	Craft@Hrs	Unit	Material $	Labor $	Equipment $	Total $
3-T, 1,200 CFM	SN@4.00	Ea	3,290.00	157.00	40.80	3,487.80
4-T, 1,600 CFM	SN@5.50	Ea	4,530.00	216.00	56.10	4,802.10
5-T, 2,000 CFM	SN@7.00	Ea	5,390.00	274.00	71.50	5,735.50
7½-T, 3,000 CFM	SN@9.00	Ea	7,600.00	353.00	91.90	8,044.90
10-T, 4,000 CFM	SN@11.0	Ea	9,640.00	431.00	112.00	10,183.00
12-T, 5,000 CFM	SN@13.0	Ea	11,400.00	509.00	133.00	12,042.00
15-T, 6,000 CFM	SN@14.0	Ea	12,900.00	549.00	143.00	13,592.00
20-T, 8,000 CFM	SN@15.0	Ea	15,300.00	588.00	153.00	16,041.00
25-T, 10,000 CFM	SN@16.0	Ea	15,200.00	627.00	163.00	15,990.00
30-T, 12,000 CFM	SN@18.0	Ea	19,100.00	705.00	184.00	19,989.00
40-T, 16,000 CFM	SN@21.0	Ea	23,500.00	823.00	214.00	24,537.00

Air Handling Unit Accessories

Description	Craft@Hrs	Unit	Material $	Labor $	Equipment $	Total $

Air handling unit accessories and options. Use these costs for preliminary estimates.

Description	Craft@Hrs	Unit	Material $	Labor $	Equipment $	Total $
DDC controls per zone complete	SN@2.75	Ea	1,070.00	108.00	—	1,178.00
Electric controls per zone complete	SN@2.75	Ea	572.00	108.00	—	680.00
Pneumatic controls per zone complete	SN@2.75	Ea	538.00	108.00	—	646.00
Variable speed drive 5 HP	SN@4.00	Ea	4,310.00	157.00	—	4,467.00
Variable speed drive 7.5HP	SN@4.00	Ea	5,030.00	157.00	—	5,187.00
Variable speed drive 10 HP	SN@6.00	Ea	5,720.00	235.00	—	5,955.00
Variable speed drive 15 HP	SN@8.00	Ea	6,450.00	314.00	—	6,764.00
Variable speed drive 20 HP	SN@12.0	Ea	7,180.00	470.00	—	7,650.00
Variable speed drive 25 HP	SN@16.0	Ea	7,900.00	627.00	—	8,527.00
Variable speed drive 30 HP	SN@18.0	Ea	9,310.00	705.00	—	10,015.00
Variable speed drive 40 HP	SN@22.0	Ea	11,500.00	862.00	—	12,362.00
Variable speed drive 50 HP	SN@26.0	Ea	13,700.00	1,020.00	—	14,720.00
Variable inlet vanes 1,000-1,500 CFM	SN@4.00	Ea	215.00	157.00	2.61	374.61
Variable inlet vanes 1,600-2,500 CFM	SN@4.00	Ea	467.00	157.00	2.61	626.61
Variable inlet vanes 2,600-5,000 CFM	SN@6.00	Ea	1,310.00	235.00	3.91	1,548.91
Variable inlet vanes 6,000-10,000 CFM	SN@8.00	Ea	2,150.00	314.00	5.21	2,469.21
Variable inlet vanes 11,000-20,000 CFM	SN@10.0	Ea	3,570.00	392.00	6.52	3,968.52

Heat Recovery Ventilators — Commercial

Heat Recovery Ventilators provide a fresh air supply to tightly sealed building envelopes. An HRV extracts heat from the stale indoor air being exhausted and transfers the heat to the fresh air being drawn into the building through the HRV. Heat recovery ventilators are also excellent dehumidifiers.

HRV unit features include energy-efficient defrost cycle, cross-flow polypropylene heat exchanger, acoustically lined cabinet, outdoor air filter.

Allow 15-30 cfm per person or .04 to .05 cfm per square foot when sizing HRV unit. (ASHRAE62-19890) (cfm = cubic feet per minute)

Description	Craft@Hrs	Unit	Material $	Labor $	Equipment $	Total $

Commercial heat recovery ventilator. Set/hang in place only. Make additional allowances for duct, diffusers, controls, air balancing, electrical connections and condensate drain.

Description	Craft@Hrs	Unit	Material $	Labor $	Equipment $	Total $
HRV 700 cfm	SN@4.75	Ea	6,250.00	186.00	80.30	6,516.30
HRV 1,200 cfm	SN@5.25	Ea	6,740.00	206.00	88.70	7,034.70
HRV 2,500 cfm	SN@6.75	Ea	27,700.00	265.00	114.00	28,079.00

Swimming pool heat recovery ventilator. Set/hang in place only. Make additional allowances for duct, diffusers, controls, air balancing, electrical connections and condensate drain.

Description	Craft@Hrs	Unit	Material $	Labor $	Equipment $	Total $
HRV 700 cfm	SN@5.00	Ea	7,240.00	196.00	84.50	7,520.50
HRV 1,200 cfm	SN@5.75	Ea	8,210.00	225.00	97.20	8,532.20

Heat Recovery Ventilators — Residential

Description	Craft@Hrs	Unit	Material $	Labor $	Equipment $	Total $

Conventional heat recovery ventilator. Hang in place only. Make additional allowances for duct, diffusers, controls, air balancing, electrical connections and condensate drain.

Description	Craft@Hrs	Unit	Material $	Labor $	Equipment $	Total $
HRV 65–150 cfm	S2@2.45	Ea	1,150.00	89.50	—	1,239.50
HRV 115–200 cfm	S2@2.65	Ea	1,360.00	96.80	—	1,456.80

Compact heat recovery ventilator. Hang in place only. Make additional allowances for duct, diffusers, controls, air balancing, electrical connections and condensate drain.

Description	Craft@Hrs	Unit	Material $	Labor $	Equipment $	Total $
HRV 65–127 cfm	S2@2.45	Ea	1,230.00	89.50	—	1,319.50
HRV 115–195 cfm	S2@2.65	Ea	1,340.00	96.80	—	1,436.80

High-efficiency heat recovery ventilator. Hang in place only. Make additional allowances for duct, diffusers, controls, air balancing, electrical connections and condensate drain.

Description	Craft@Hrs	Unit	Material $	Labor $	Equipment $	Total $
HRV 65–127 cfm	S2@2.65	Ea	1,350.00	96.80	—	1,446.80
HRV 115–180cfm	S2@2.90	Ea	2,070.00	106.00	—	2,176.00
HRV 180–265cfm	S2@3.25	Ea	2,150.00	119.00	—	2,269.00

Heat recovery ventilator controls

Description	Craft@Hrs	Unit	Material $	Labor $	Equipment $	Total $
HRV basic control	S2@1.00	Ea	104.00	36.50	—	140.50
HRV std. control	S2@1.00	Ea	167.00	36.50	—	203.50
HRV auto control	S2@1.00	Ea	226.00	36.50	—	262.50
60 minute timer	S2@.600	Ea	70.00	21.90	—	91.90
Interlock relay	S2@.600	Ea	83.20	21.90	—	105.10

Heat recovery ventilator accessories

Description	Craft@Hrs	Unit	Material $	Labor $	Equipment $	Total $
6" diffusers	S2@.500	Ea	17.80	18.30	—	36.10
6" wall hoods	S2@1.50	Ea	30.70	54.80	—	85.50
6" tee fittings	S2@.350	Ea	4.11	12.80	—	16.91
6" flex duct	S2@.025	LF	.98	.91	—	1.89
6" flex insul duct	S2@.025	LF	1.87	.91	—	2.78
HRV filters	S2@.450	Ea	25.10	16.40	—	41.50

Heat Recovery Ventilators — Residential

Description	Craft@Hrs	Unit	Material $	Labor $	Equipment $	Total $

Installation costs for heat recovery ventilators

Description	Craft@Hrs	Unit	Material $	Labor $	Equipment $	Total $
Hang heat recovery ventilator, 150 CFM	S2@2.00	Ea	5.53	73.10	—	78.63
Hang heat recovery ventilator, 250 CMF	S2@2.25	LF	8.30	82.20	—	90.50
Heat recovery ventilator ducting, 150 CFM	S2@4.00	LF	369.00	146.00	—	515.00
Heat recovery ventilator ducting, 250 CFM	S2@4.75	LF	415.00	174.00	—	589.00
Heat recovery ventilator drain line, gravity	P1@1.50	LF	27.80	55.80	—	83.60
Heat recovery ventilator drain line, pumped	P1@2.00	LF	249.00	74.40	—	323.40
Heat recovery ventilator power wiring, 115 volt	BE@2.00	Ea	94.10	82.10	—	176.20
Heat recovery ventilator power wiring, 24 volt	BE@2.00	Ea	22.10	82.10	—	104.20
Commission and test	P1@1.50	Ea	—	55.80	—	55.80
Run air balance and fine tune system	P1@1.75	Ea	—	65.10	—	65.10

Water Coil Piping

Engineering drawings of coil piping details have a bad reputation among HVAC contractors and estimators. They're notorious for what they leave out. They'll rarely show more than one coil bank, no matter how big the system. Furthermore, the drawings hardly ever call out sizes for either the piping or the control valves. Don't be taken in by the apparent simplicity of the system as shown in these drawings. It's likely to be only the tip of the iceberg. For example, unless the air handling capacity of the system is less than 16,000 CFM, the single coil bank shown won't be adequate. Add one or two more coil banks and you're looking at a lot more piping — and a more complex system that takes longer to install.

You probably haven't even decided which equipment supplier to use. This is hardly the time for you to start researching heating and cooling coils. Nevertheless, you need better, more complete and realistic data to come up with a competitive estimate.

I'll tell you how and where to track down the hard data that you need, and also pass along a few tips on estimating water coil piping. They'll help you avoid leaving something out of your estimates — a real pitfall for any beginner. It's those little (but essential) items, so easily overlooked, that are so deadly to a profit margin. Finally, check the next two pages of diagrams with tables. The data given there, combined with the data you collected earlier, forms the basis for informed guesswork.

The Hard Data

There are two things you absolutely must know to estimate water coil piping. First, the size of the branch piping to the coils; second, the CFM rating of the air handling units for the system.

To find the pipe sizes, look at either the floor plans or the details for the equipment room. They'll list the sizes of the branch run-out pipes. Once you know them, you can make a good guess at the right size for the control valves. (See the diagrams and tables on pages 295 and 296.) The only information you need is the CFM ratings of the air handling units and the branch piping sizes to the coils. If the system's capacity is over 16,000 CFM, you need two or three coil banks.

A Few Tips on Estimating Water Coil Piping

1) Coil connection sizes seldom match branch pipe sizes. Be sure to include the reducing fittings you'll need in any estimate.

2) Two-way and three-way control valves are both usually one pipe size smaller than the pipe where they're installed. That means you'll need either two or three reducing fittings per control valve. Be sure you include their cost in your estimates.

Using the Diagrams

In the following diagrams, for clarity, some items aren't included. These items are: balance valves, shut-off valves, reducers, strainers, gauges and gauge taps. Any details you need about these items for your estimate are in the engineer's coil piping details.

Water Coil Piping

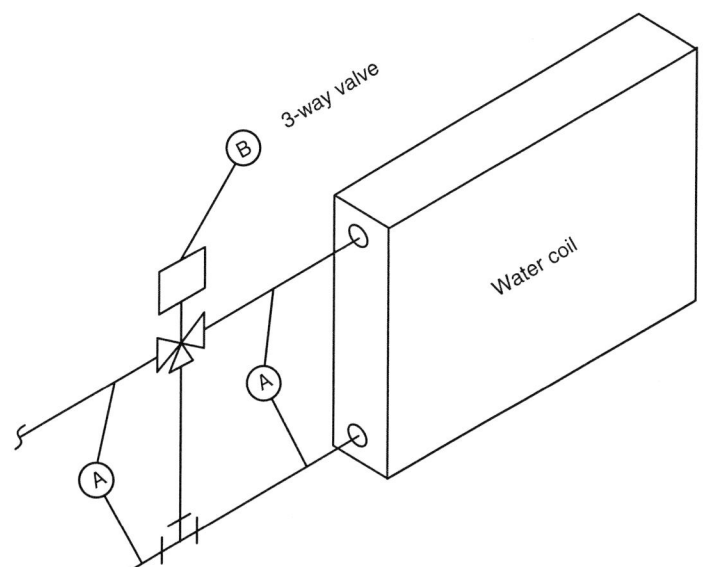

If Ⓐ is:	Then Ⓑ is:
1"	¾"
1¼"	1"
1½"	1¼"
2"	1½"
2½"	2"
3"	2½"

Typical water coil piping for A.H. units up to 16,000 CFM

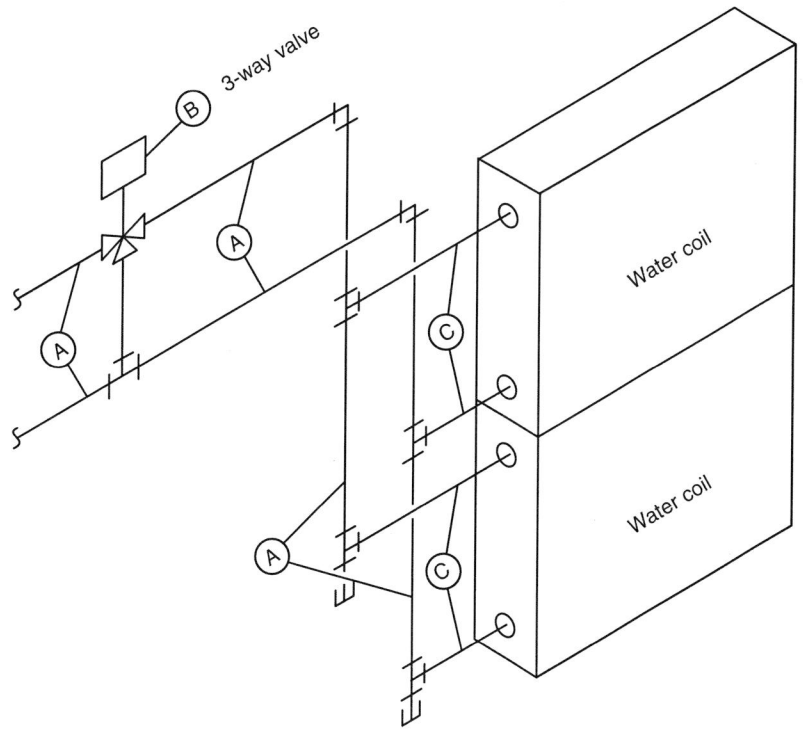

If Ⓐ is:	Then Ⓑ is:	And Ⓒ is:
2½"	2"	2"
3"	2½"	2½"
4"	3"	3"

Typical water coil piping for A.H. units from 16,000 to 26,000 CFM

Water Coil Piping

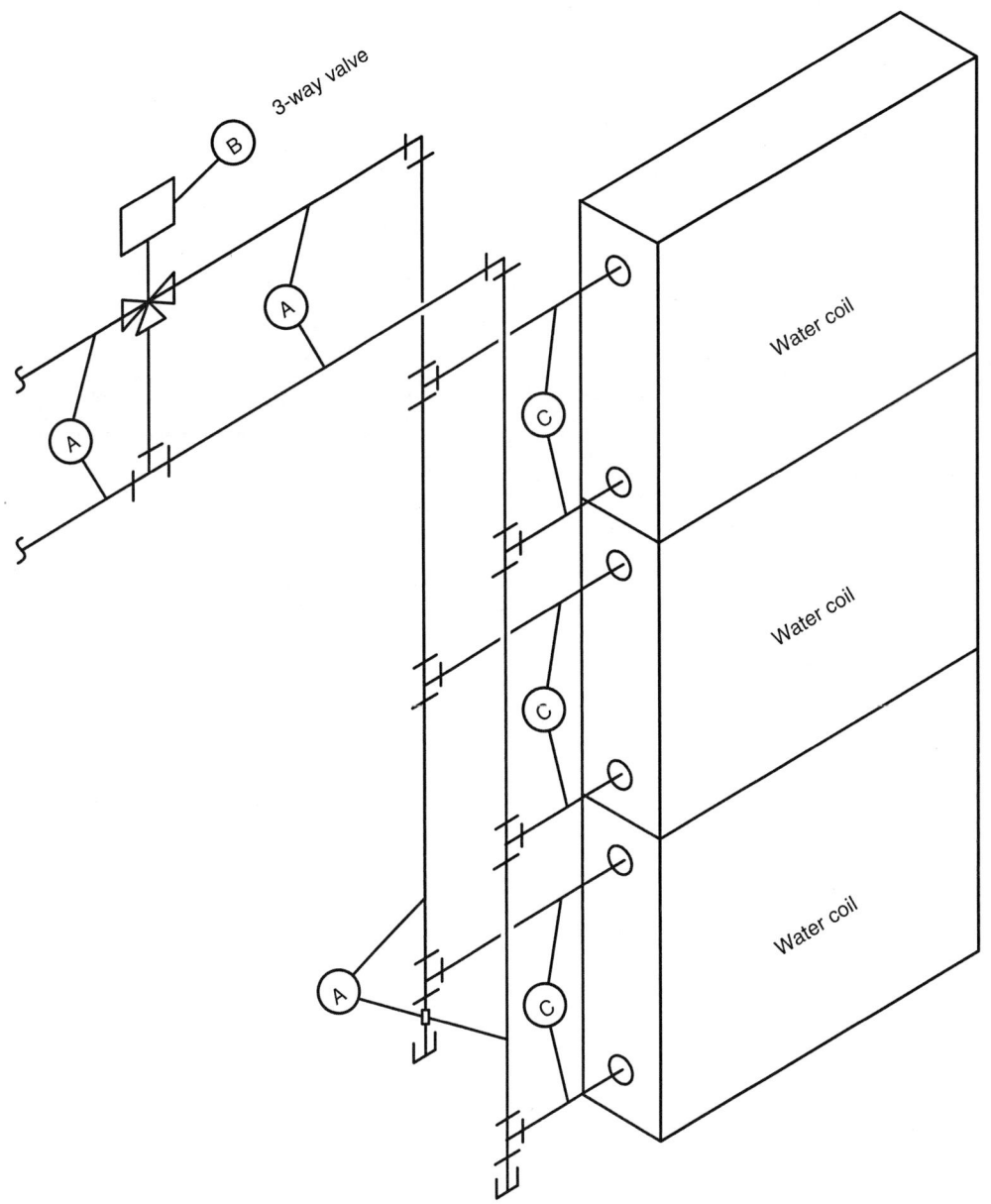

If Ⓐ is:	Then Ⓑ is:	And Ⓒ is:
4"	3"	2½"
6"	4"	3"
8"	6"	4"
10"	8"	6"

Typical water coil piping for A.H. units over 26,000 CFM

Air Handling Unit Coil Connections

Description	Craft@Hrs	Unit	Material $	Labor $	Equipment $	Total $

Air handling unit coil connection, one row coil bank, non-regulated flow. Connection assembly includes pipe, fittings, pipe insulation, valves, gauges, thermometers and vents.

Description	Craft@Hrs	Unit	Material $	Labor $	Equipment $	Total $
1½" supply	SN@14.0	Ea	1,850.00	549.00	—	2,399.00
2" supply	SN@18.0	Ea	2,280.00	705.00	—	2,985.00
2½" supply	SN@28.0	Ea	3,040.00	1,100.00	—	4,140.00
3" supply	SN@31.5	Ea	3,340.00	1,230.00	—	4,570.00
4" supply	SN@34.0	Ea	3,580.00	1,330.00	—	4,910.00
6" supply	SN@41.0	Ea	5,180.00	1,610.00	26.70	6,816.70

Air handling unit coil connection, one row coil bank, 2-way control valve design. Connection assembly includes pipe and fittings, pipe insulation, valves, gauges, thermometers and vents.

Description	Craft@Hrs	Unit	Material $	Labor $	Equipment $	Total $
1½" supply	SN@16.5	Ea	2,090.00	647.00	—	2,737.00
2" supply	SN@21.0	Ea	2,700.00	823.00	—	3,523.00
2½" supply	SN@33.6	Ea	3,280.00	1,320.00	—	4,600.00
3" supply	SN@36.0	Ea	3,520.00	1,410.00	—	4,930.00
4" supply	SN@43.8	Ea	4,280.00	1,720.00	—	6,000.00
6" supply	SN@54.0	Ea	6,850.00	2,120.00	35.20	9,005.20

Air handling unit coil connection, one row coil bank, 3-way control valve design. Connection assembly includes pipe and fittings, pipe insulation, valves, gauges, thermometers and vents.

Description	Craft@Hrs	Unit	Material $	Labor $	Equipment $	Total $
1½" supply	SN@18.0	Ea	2,410.00	705.00	—	3,115.00
2" supply	SN@26.0	Ea	3,010.00	1,020.00	—	4,030.00
2½" supply	SN@41.5	Ea	3,780.00	1,630.00	—	5,410.00
3" supply	SN@43.8	Ea	4,160.00	1,720.00	—	5,880.00
4" supply	SN@54.6	Ea	5,360.00	2,140.00	—	7,500.00
6" supply	SN@61.4	Ea	7,640.00	2,410.00	40.00	10,090.00

Air handling unit coil connection, two row coil bank, non-regulated flow. Connection assembly includes pipe and fittings, pipe insulation, valves, gauges, thermometers and vents.

Description	Craft@Hrs	Unit	Material $	Labor $	Equipment $	Total $
2½" supply	SN@48.0	Ea	4,570.00	1,880.00	—	6,450.00
3" supply	SN@51.5	Ea	5,560.00	2,020.00	—	7,580.00
4" supply	SN@63.6	Ea	7,360.00	2,490.00	—	9,850.00
6" supply	SN@79.2	Ea	10,800.00	3,100.00	51.60	13,951.60

Air handling unit coil connection, two row coil bank, 2-way control valve design. Connection assembly includes pipe and fittings, pipe insulation, valves, gauges, thermometers and vents.

Description	Craft@Hrs	Unit	Material $	Labor $	Equipment $	Total $
2½" supply	SN@54.0	Ea	5,120.00	2,120.00	—	7,240.00
3" supply	SN@58.5	Ea	6,430.00	2,290.00	—	8,720.00
4" supply	SN@67.0	Ea	7,910.00	2,630.00	—	10,540.00
6" supply	SN@83.5	Ea	11,000.00	3,270.00	54.40	14,324.40

Air Handling Unit Coil Connections

Description	Craft@Hrs	Unit	Material $	Labor $	Equipment $	Total $

Air handling unit coil connection, two row coil bank, 3-way control valve design. Connection assembly includes pipe and fittings, pipe insulation, valves, gauges, thermometers and vents.

Description	Craft@Hrs	Unit	Material $	Labor $	Equipment $	Total $
2½" supply	SN@58.8	Ea	5,360.00	2,300.00	—	7,660.00
3" supply	SN@64.0	Ea	6,760.00	2,510.00	—	9,270.00
4" supply	SN@72.0	Ea	8,560.00	2,820.00	—	11,380.00
6" supply	SN@87.0	Ea	11,400.00	3,410.00	56.70	14,866.70

Air handling unit coil connection, three-row coil bank, non-regulated flow. Connection assembly includes pipe and fittings, pipe insulation, valves, gauges, thermometers and vents.

Description	Craft@Hrs	Unit	Material $	Labor $	Equipment $	Total $
4" supply	SN@72.0	Ea	7,750.00	2,820.00	—	10,570.00
6" supply	SN@94.0	Ea	13,500.00	3,680.00	—	17,180.00
8" supply	SN@120.	Ea	18,100.00	4,700.00	78.20	22,878.20

Air handling unit coil connection, three-row coil bank, 2-way control valve design. Connection assembly includes pipe and fittings, pipe insulation, valves, gauges, thermometers and vents.

Description	Craft@Hrs	Unit	Material $	Labor $	Equipment $	Total $
4" supply	SN@78.0	Ea	8,240.00	3,060.00	—	11,300.00
6" supply	SN@105.	Ea	14,900.00	4,110.00	—	19,010.00
8" supply	SN@132.	Ea	20,300.00	5,170.00	86.00	25,556.00

Air handling unit coil connection, three-row coil bank, 3-way control valve design. Connection assembly includes pipe and fittings, pipe insulation, valves, gauges, thermometers and vents.

Description	Craft@Hrs	Unit	Material $	Labor $	Equipment $	Total $
4" supply	SN@82.0	Ea	9,540.00	3,210.00	—	12,750.00
6" supply	SN@112.	Ea	15,300.00	4,390.00	—	19,690.00
8" supply	SN@140.	Ea	22,800.00	5,490.00	91.20	28,381.20

Gas-Fired Furnaces

Description	Craft@Hrs	Unit	Material $	Labor $	Equipment $	Total $

Residential gas-fired upflow furnace. Set in place only. Make additional allowances for gas and electrical connections. Costs shown per thousand BTU per hour (MBH) Mid-efficiency, 80% AFUE.

Description	Craft@Hrs	Unit	Material $	Labor $	Equipment $	Total $
50 MBH input	SN@2.00	Ea	965.00	78.40	—	1,043.40
75 MBH input	SN@2.00	Ea	1,260.00	78.40	2.05	1,340.45
100 MBH input	SN@2.00	Ea	1,460.00	78.40	2.05	1,540.45
125 MBH input	SN@2.00	Ea	1,680.00	78.40	2.05	1,760.45

Residential combination gas-fired furnace and A/C unit. With remote exterior condenser/compressor. Set in place only. Make additional allowances for gas, electrical and DX connections. Costs shown per thousand BTU per hour (MBH) Mid-efficiency, 80% AFUE.

Description	Craft@Hrs	Unit	Material $	Labor $	Equipment $	Total $
30 MBH, 1.5 tons	SN@2.50	Ea	2,870.00	98.00	—	2,968.00
36 MBH, 2 tons	SN@2.75	Ea	3,410.00	108.00	—	3,518.00
42 MBH, 2 tons	SN@3.00	Ea	3,830.00	118.00	—	3,948.00
48 MBH, 2.5 tons	SN@3.50	Ea	4,460.00	137.00	—	4,597.00
60 MBH, 3 tons	SN@4.00	Ea	5,330.00	157.00	—	5,487.00

High-efficiency, condensing gas furnace. Direct drive, atmospheric draft, electronic ignition. Set in place only. Make additional allowances for flue, gas and electrical connections. Costs shown per thousand BTU per hour (MBH) High efficiency 94% AFUE.

Description	Craft@Hrs	Unit	Material $	Labor $	Equipment $	Total $
40 MBH input	S2@2.00	Ea	2,010.00	73.10	—	2,083.10
60 MBH input	S2@2.00	Ea	2,010.00	73.10	—	2,083.10
70 MBH input	S2@2.00	Ea	1,910.00	73.10	—	1,983.10
90 MBH input	S2@2.00	Ea	1,910.00	73.10	—	1,983.10
100 MBH input	S2@2.00	Ea	2,730.00	73.10	—	2,803.10
120 MBH input	S2@2.00	Ea	2,820.00	73.10	—	2,893.10
2 speed fan & limit control kit	S2@1.25	Ea	39.10	45.70	—	84.80
Propane conversion kit	S2@1.75	Ea	152.00	63.90	—	215.90

Gas Fired Furnaces

Description	Craft@Hrs	Unit	Material $	Labor $	Equipment $	Total $

Central dehumidification system. Upflow or counterflow, 400 CFM, 30 gallon per day baseload capacity at 0.5 water column pressure differential. With inlet filter, drain pan, drain line, heater plate, cooling draft exchanger, humidistat control sensor, control relay, actuator, cabinet, two 24" by 24" by 4' long duct transitions, gaskets, hangers and ground fault protection. Whirlpool UGD160UH or equal. Add the cost of control wiring and electrical wiring.

Description	Craft@Hrs	Unit	Material $	Labor $	Equipment $	Total $
Central dehumidification system	SW@0.50	Ea	1,790.00	21.30	—	1,811.30
Cut, frame and gasket hole in wall, ceiling or roof	SW@1.00	Ea	34.60	42.50	—	77.10
Mount duct hangers	SW@0.25	Ea	4.36	10.60	—	14.96
Cut and mount sheet metal duct	SW@0.35	LF	6.72	14.90	—	21.62
Apply and coat duct insulation	SW@0.15	LF	2.90	6.38	—	9.28
Install electrical wiring	BE@0.15	LF	2.09	6.15	—	8.24
Install dehumidifier controls	BE@0.50	Ea	—	20.50	—	20.50
Test and balance	SW@1.50	Ea	—	63.80	—	63.80

Wall furnace, mid-efficiency, pilot light ignition. Gas-fired, atmospheric draft (radiant heat, no fan). Set in place only. Make additional allowances for gas and flue connections. Costs shown per thousand BTU per hour (MBH). Mid-efficiency 80% AFUE.

Description	Craft@Hrs	Unit	Material $	Labor $	Equipment $	Total $
35 MBH input	S2@2.75	Ea	1,350.00	100.00	—	1,450.00
50 MBH input	S2@2.75	Ea	1,520.00	100.00	—	1,620.00
65 MBH input	S2@2.75	Ea	1,530.00	100.00	—	1,630.00
Fan blower kit	S2@1.75	Ea	218.00	63.90	—	281.90

Wall furnace, mid-efficiency, electronic ignition. Gas-fired, atmospheric draft (radiant heat, no fan). Set in place only. Make additional allowances for gas and flue connections. Costs shown per thousand BTU per hour (MBH). Mid-efficiency 80% AFUE.

Description	Craft@Hrs	Unit	Material $	Labor $	Equipment $	Total $
35 MBH input	S2@2.75	Ea	1,690.00	100.00	—	1,790.00
65 MBH input	S2@2.75	Ea	1,890.00	100.00	—	1,990.00
Fan blower kit	S2@1.75	Ea	218.00	63.90	—	281.90

Wall furnace, direct vent (included). Gas-fired, pilot light ignition, mid-efficiency (radiant heat, no fan). Set in place only. Make additional allowances for vent, gas and electrical connections. Costs shown per thousand BTU per hour (MBH). Mid-efficiency 80% AFUE.

Description	Craft@Hrs	Unit	Material $	Labor $	Equipment $	Total $
15 MBH input	S2@2.75	Ea	905.00	100.00	—	1,005.00
25 MBH input	S2@2.75	Ea	1,090.00	100.00	—	1,190.00
33 MBH input	S2@2.75	Ea	1,170.00	100.00	—	1,270.00
40 MBH input	S2@2.75	Ea	1,610.00	100.00	—	1,710.00
65 MBH input	S2@2.75	Ea	1,760.00	100.00	—	1,860.00
Direct vent kit	S2@2.50	Ea	153.00	91.40	—	244.40
Fan blower kit	S2@1.75	Ea	218.00	63.90	—	281.90

Energy Recovery Systems, Enthalpy

Description	Craft@Hrs	Unit	Material $	Labor $	Equipment $	Total $

Enthalpy energy recovery system. Steam-to-air or hot-water-to-air heat exchange. For boiler room downstream retrofits or new construction to qualify for EPA EnergyStar certification. Equipment cost is for a 40' rolling platform scissor lift. Add the cost of the energy recovery wheel ventilator.

Description	Craft@Hrs	Unit	Material $	Labor $	Equipment $	Total $
Reinforced attic joists	P1@3.00	Ea	72.90	112.00	—	184.90
Vibration isolators	S2@.750	LS	47.00	27.40	—	74.40
Drill and mount coil service drain	P1@.150	LF	3.51	5.58	—	9.09
Install electrical wiring	BE@.150	LF	2.09	6.15	—	8.24
Mount unit heater or enthalpy wheel	P1@.500	Ea	—	18.60	—	18.60
Connect ductwork mains to manifold	SW@4.50	Ea	179.00	191.00	—	370.00
Coil piping and condensate drain	P1@4.00	Ea	9.36	149.00	—	158.36
Zone control thermostat, manual	BE@.750	Ea	148.00	30.80	—	178.80
Master heat controller with auto change-over	BE@.750	Ea	347.00	30.80	—	377.80
Wiring of controls	BE@3.00	Ea	1.74	123.00	—	124.74
Calibration test	BE@2.00	Ea	—	82.10	—	82.10

Enthalpy energy recovery wheel. Modine Slimvent, Reznor XAWS or equal. For both heating and cooling applications in commercial and institutional buildings. Complies with the ARI Standard 1060-2000. Provides recovery of up to 75% of heat energy and 80% of cooling energy.

Description	Craft@Hrs	Unit	Material $	Labor $	Equipment $	Total $
800-1,000 CFM	SL@5.00	Ea	4,090.00	185.00	—	4,275.00
1,300-2,000 CFM	SL@5.00	Ea	5,140.00	185.00	—	5,325.00
2,400-3,000 CFM	SL@5.00	Ea	5,970.00	185.00	—	6,155.00
3,200-3,800 CFM	SL@5.00	Ea	7,850.00	185.00	—	8,035.00
4,000-4,800 CFM	SL@5.00	Ea	8,880.00	185.00	—	9,065.00
5,000-6,000 CFM	SL@5.00	Ea	9,940.00	185.00	—	10,125.00

Horizontal hot water unit heaters. Fan-driven, with integral check valves, bypass piping, drain line fitting and vacuum breaker, Modine HSB series or equal.

Description	Craft@Hrs	Unit	Material $	Labor $	Equipment $	Total $
12.5 MBtu, 200 CFM	SL@2.00	Ea	476.00	73.90	4.56	554.46
17 MBtu, 300 CFM	SL@2.00	Ea	525.00	73.90	4.56	603.46
25 MBtu, 500 CFM	SL@2.50	Ea	601.00	92.40	5.70	699.10
30 MBtu, 700 CFM	SL@3.00	Ea	633.00	111.00	6.84	750.84
50 MBtu, 1,000 CFM	SL@3.50	Ea	736.00	129.00	7.98	872.98
60 MBtu, 1,300 CFM	SL@4.00	Ea	873.00	148.00	9.12	1,030.12

Unit Heaters

Description	Craft@Hrs	Unit	Material $	Labor $	Equipment $	Total $

Vertical hot water unit heaters. Fan-driven, with integral check valves, bypass piping, drain line fitting and vacuum breaker, Modine V500 series or equal.

Description	Craft@Hrs	Unit	Material $	Labor $	Equipment $	Total $
12.5 MBtu, 200 CFM	SL@2.00	Ea	533.00	73.90	4.56	611.46
17 MBtu, 300 CFM	SL@2.00	Ea	545.00	73.90	4.56	623.46
25 MBtu, 500 CFM	SL@2.50	Ea	707.00	92.40	5.70	805.10
30 MBtu, 700 CFM	SL@3.00	Ea	738.00	111.00	6.84	855.84
50 MBtu, 1,000 CFM	SL@3.50	Ea	889.00	129.00	7.98	1,025.98
60 MBtu, 1,300 CFM	SL@4.00	Ea	976.00	148.00	9.12	1,133.12

Horizontal steam coil unit heaters. Fan-driven, with integral check valves, bypass piping, drain line fitting and vacuum breaker, Modine HSB series or equal.

Description	Craft@Hrs	Unit	Material $	Labor $	Equipment $	Total $
12.5 MBtu, 200 CFM	SL@2.00	Ea	775.00	73.90	4.56	853.46
17 MBtu, 300 CFM	SL@2.00	Ea	895.00	73.90	4.56	973.46
25 MBtu, 500 CFM	SL@2.50	Ea	982.00	92.40	5.70	1,080.10
30 MBtu, 700 CFM	SL@3.00	Ea	1,070.00	111.00	6.84	1,187.84
50 MBtu, 1,000 CFM	SL@3.50	Ea	1,300.00	129.00	7.98	1,436.98
60 MBtu, 1,300 CFM	SL@4.00	Ea	1,620.00	148.00	9.12	1,777.12

Vertical steam coil unit heaters. Fan-driven, with integral check valves, bypass piping, drain line fitting and vacuum breaker, Modine V500 series or equal.

Description	Craft@Hrs	Unit	Material $	Labor $	Equipment $	Total $
12.5 MBtu, 200 CFM	SL@2.00	Ea	836.00	73.90	4.56	914.46
17 MBtu, 300 CFM	SL@2.00	Ea	917.00	73.90	4.56	995.46
25 MBtu, 500 CFM	SL@2.50	Ea	1,070.00	92.40	5.70	1,168.10
30 MBtu, 700 CFM	SL@3.00	Ea	1,430.00	111.00	6.84	1,547.84
50 MBtu, 1,000 CFM	SL@3.50	Ea	1,680.00	129.00	7.98	1,816.98
60 MBtu, 1,300 CFM	SL@4.00	Ea	1,780.00	148.00	9.12	1,937.12

Unit Heaters

Description	Craft@Hrs	Unit	Material $	Labor $	Equipment $	Total $

Unit heater, gas, direct-fired, mid-efficiency, propeller fan, atmospheric draft, single-stage, standing pilot, aluminized burner and heat exchanger. Hang in place only. Make additional allowances for flue, gas and electrical connections. Costs shown per thousand BTU per hour (MBH). Mid-efficiency 78% AFUE.

Description	Craft@Hrs	Unit	Material $	Labor $	Equipment $	Total $
30 MBH input	S2@4.00	Ea	1,150.00	146.00	—	1,296.00
50 MBH input	S2@4.00	Ea	1,230.00	146.00	—	1,376.00
75 MBH input	S2@4.00	Ea	1,370.00	146.00	—	1,516.00
100 MBH input	S2@4.00	Ea	1,520.00	146.00	—	1,666.00
125 MBH input	S2@4.00	Ea	1,810.00	146.00	—	1,956.00
145 MBH input	S2@4.00	Ea	1,880.00	146.00	—	2,026.00
175 MBH input	S2@4.00	Ea	2,070.00	146.00	—	2,216.00
200 MBH input	S2@4.00	Ea	2,280.00	146.00	—	2,426.00
250 MBH input	S2@4.00	Ea	2,400.00	146.00	—	2,546.00
300 MBH input	S2@4.00	Ea	3,110.00	146.00	—	3,256.00
350 MBH input	S2@4.00	Ea	3,670.00	146.00	—	3,816.00
400 MBH input	S2@4.00	Ea	4,170.00	146.00	—	4,316.00

Unit heater, gas-fired, high-efficiency, propeller fan, power vented (included), direct-fired, single-stage, intermittent pilot ignition, aluminized burner and heat exchanger. Hang in place and install power vent. Make additional allowances for gas and electrical connections. Costs shown per thousand BTU per hour (MBH). High-efficiency 85% AFUE.

Description	Craft@Hrs	Unit	Material $	Labor $	Equipment $	Total $
30 MBH input	S2@5.25	Ea	1,650.00	192.00	—	1,842.00
50 MBH input	S2@5.25	Ea	1,750.00	192.00	—	1,942.00
75 MBH input	S2@5.25	Ea	2,800.00	192.00	—	2,992.00
100 MBH input	S2@5.25	Ea	2,120.00	192.00	—	2,312.00
125 MBH input	S2@5.25	Ea	2,360.00	192.00	—	2,552.00
145 MBH input	S2@5.25	Ea	2,500.00	192.00	—	2,692.00
175 MBH input	S2@5.25	Ea	2,650.00	192.00	—	2,842.00
200 MBH input	S2@5.25	Ea	2,820.00	192.00	—	3,012.00
250 MBH input	S2@5.25	Ea	3,270.00	192.00	—	3,462.00
300 MBH input	S2@5.25	Ea	3,850.00	192.00	—	4,042.00
350 MBH input	S2@5.25	Ea	4,310.00	192.00	—	4,502.00
400 MBH input	S2@5.25	Ea	4,720.00	192.00	—	4,912.00

Unit heater, gas, indirect-fired, mid-efficiency, propeller fan, power vented (included), single-stage, intermittent pilot ignition, aluminized burner and heat exchanger, Hang in place and install power vent. Make additional allowances for gas and electrical connections. Costs shown per thousand BTU per hour (MBH). Mid-efficiency 78% AFUE.

Description	Craft@Hrs	Unit	Material $	Labor $	Equipment $	Total $
130 MBH input	S2@5.25	Ea	3,010.00	192.00	—	3,202.00
150 MBH input	S2@5.25	Ea	3,270.00	192.00	—	3,462.00
170 MBH input	S2@5.25	Ea	3,380.00	192.00	—	3,572.00
225 MBH input	S2@5.25	Ea	3,790.00	192.00	—	3,982.00
280 MBH input	S2@5.25	Ea	4,500.00	192.00	—	4,692.00
340 MBH input	S2@5.25	Ea	5,090.00	192.00	—	5,282.00

Unit Heaters

Description	Craft@Hrs	Unit	Material $	Labor $	Equipment $	Total $

High-intensity infrared heater, gas-fired. Ceramic grid. Automatic spark ignition. Including reflector. Approved for unvented commercial indoor installations. Not for use in residential applications. Labor includes hang in place only. Make additional allowances for gas and 110 volt electrical connections. Refer to local gas code for proper installation requirements.

Description	Craft@Hrs	Unit	Material $	Labor $	Equipment $	Total $
30,000 BTU	P1@1.20	Ea	585.00	44.70	2.74	632.44
60,000 BTU	P1@1.80	Ea	748.00	67.00	4.11	819.11
100,000 BTU	P1@2.50	Ea	1,770.00	93.10	5.70	1,868.80
130,000 BTU	P1@2.75	Ea	2,190.00	102.00	6.27	2,298.27
160,000 BTU	P1@2.75	Ea	2,500.00	102.00	6.27	2,608.27
Add for propane	—	Ea	145.00	—	—	145.00

High-intensity infrared heater, gas-fired. Ceramic grid. Millivolt standing pilot ignition. Approved for unvented commercial indoor installations. Not for use in residential applications. Labor includes hang in place only. Make additional allowances for gas connections. Millivolt standing pilot does not require electrical source.

Description	Craft@Hrs	Unit	Material $	Labor $	Equipment $	Total $
30,000 BTU	P1@1.20	Ea	1,360.00	44.70	2.74	1,407.44
60,000 BTU	P1@1.80	Ea	1,770.00	67.00	4.11	1,841.11
100,000 BTU	P1@2.50	Ea	2,210.00	93.10	5.70	2,308.80
130,000 BTU	P1@2.75	Ea	2,640.00	102.00	6.27	2,748.27
160,000 BTU	P1@2.75	Ea	3,070.00	102.00	6.27	3,178.27
Add for propane	—	Ea	145.00	—	—	145.00

Infrared tube heater, burner unit only, gas-fired, negative pressure type. Labor includes hang in place only. Not for use in residential applications. Make additional allowances for burner tube, burner tube reflector, venting, gas piping and electrical requirements.

Description	Craft@Hrs	Unit	Material $	Labor $	Equipment $	Total $
30,000 BTU	P1@1.20	Ea	1,690.00	44.70	2.74	1,737.44
60,000 BTU	P1@1.80	Ea	1,600.00	67.00	4.11	1,671.11
100,000 BTU	P1@2.50	Ea	1,830.00	93.10	5.70	1,928.80
150,000 BTU	P1@2.75	Ea	1,940.00	102.00	6.27	2,048.27
175,000 BTU	P1@2.75	Ea	2,260.00	102.00	6.27	2,368.27
200,000 BTU	P1@2.75	Ea	2,610.00	102.00	6.27	2,718.27
Add for propane	—	Ea	145.00	—	—	145.00

Infrared tube heater, 4" diameter, 10' lengths. 16 gauge burner tube and .024" mill finish aluminum reflector. Labor includes hang in place only. Not for use in residential applications. Make additional allowances for burner unit, venting, gas piping and electrical requirements.

Description	Craft@Hrs	Unit	Material $	Labor $	Equipment $	Total $
Burner tubing	P1@.200	LF	6.93	7.44	—	14.37
Tubing U-bend	P1@.850	Ea	247.00	31.60	—	278.60
Reflector shield	P1@.200	LF	6.93	7.44	—	14.37
Vent terminal	P1@.600	Ea	109.00	22.30	—	131.30

Heat Pump Systems

Heat Pumps

In general, heat pumps are more effective in milder winter climates. In colder winter climates, secondary/supplemental electric heating elements (or equivalent heating source) are recommended.

A geothermal heat pump system's efficient operation is largely dependent upon proper design, layout, sizing and installation of the ground loop piping system for closed loop systems. In the case of well-to-well, the efficient operation of open loop systems is largely dependent upon the ability of the supply and return, or dump wells to meet the water volume demand of the heat pumps.

Rules of thumb when sizing loop & well systems:

Heating capacity, when considered or measured in tons cooling or refrigeration, is generally 3 to 4 times that of required cooling capacity for climates within 100 miles of the 49th parallel, i.e., if 2 tons of cooling is required, allow 6-8 tons of heating capacity (1 ton = 12,000 Btu).

Use 250 square feet per ton, e.g., a 2,000-square-foot home requires a 6-ton heat pump (heating capacity).

(Calculation based on R-20 wall and R-40 attic insulation, minimum dual pane argon-charged windows.)

Closed Vertical Loop:

Vertical Bore Hole Loop — ¾" or 1" diameter, socket fusioned loop = 200'-250' total bore hole depth per ton. For example, a 6-ton unit requires between 1,200'-1,500' (twelve to fifteen 5" bore holes, each 100' deep) with the associated 12-15 loop drops, spaced a minimum of 20' apart, connected to an adequately sized header, immersed in a bentonite slurry and applied with a tremie tube.

Closed Horizontal Loop:

Straight run — ¾" or 1" diameter manifolded loop = 400 horizontal feet per ton. For example, a 6-ton unit requires 2,400 horizontal feet (eight 300' runs, 150' out, 150' back, minimum 2' radius return bend, placed a minimum 10' apart, minimum 4' buried depth, or buried below potential winter frost line).

Slinky — ¾" or 1" diameter manifolded loop = 400 horizontal feet per ton. For example, a 6-ton unit requires 2,400 horizontal feet (eight 300' runs, 50' out, 50' back, coiled, placed a minimum 20' apart, minimum 4' buried depth, or buried below potential winter frost line. (Below 7' at 100 miles, or farther, north of the 49th parallel.)

Open Loop:

Well-to-Well — 2.5 gallons per minute per ton, each well (two required: supply & return). For example, a 6-ton unit requires 12-15 gallons per minute capacity per well.

Leadership in Energy & Environmental Design (LEED)

LEED offers an incentive to building owners to install energy-efficient and environmentally sensitive HVAC equipment. The LEED program awards points for application of the best available technology. The HVAC system can earn a maximum of 17 points toward LEED certification. A LEED-certified air conditioning system must comply with ASHRAE 90.1-2004 (efficiency) and ASHRAE 62.1-2004 (indoor air quality) standards. LEED certification also requires recording controls (Carrier ComfortView 3 or equal) to measure system performance. Zone sensor thermostats (Carrier Debonair or equal) are required to provide performance and efficiency data to the digital recorder. An indoor air quality CO_2 sensor, duct-mounted aspirator modules and a refrigerant such as Puron R-410A are also required.

Heat Pump Systems

Description	Craft@Hrs	Unit	Material $	Labor $	Equipment $	Total $

Geothermal ground or water source heat pump (open or closed loop), w/active cooling. Glycol/water mixture or water-to-air thermal transfer. The refrigeration circuit operates in the heating and cooling mode as required (sometimes referred to as active heating & cooling or mechanical heating & cooling). This unit is suited for either open or closed loop systems. These units provide a higher degree of cooling capacity than the open loop, passive "free" cooling units. See table below. Optional de-superheater system provides domestic hot water temperature preheat or boost. Make additional allowances for the thermal source, i.e., horizontal ground loop, vertical bore hole ground loop, supply and return water wells, pond or lake source loop, etc. Make additional allowances for connections, ancillary components, start up and testing. Based on R-410A refrigerant.

Description	Craft@Hrs	Unit	Material $	Labor $	Equipment $	Total $
Heat pump, open or closed loop, ½ ton	P1@2.25	Ea	1,890.00	83.70	—	1,973.70
Heat pump, open or closed loop, ¾ ton	P1@2.25	Ea	1,940.00	83.70	—	2,023.70
Heat pump, open or closed loop, 1 ton	P1@2.25	Ea	2,030.00	83.70	—	2,113.70
Heat pump, open or closed loop, 1½ ton	P1@2.25	Ea	2,370.00	83.70	—	2,453.70
Heat pump, open or closed loop, 2 ton	P1@2.50	Ea	2,450.00	93.10	—	2,543.10
Heat pump, open or closed loop, 2½ ton	P1@2.50	Ea	3,040.00	93.10	—	3,133.10
Heat pump, open or closed loop, 3 ton	P1@3.00	Ea	3,160.00	112.00	—	3,272.00
Heat pump, open or closed loop, 4 ton	P1@3.50	Ea	3,600.00	130.00	—	3,730.00
Heat pump, open or closed loop, 5 ton	P1@4.00	Ea	7,100.00	149.00	—	7,249.00
Heat pump, open or closed loop, 7½ ton	P1@4.75	Ea	7,610.00	177.00	—	7,787.00
Optional de-superheater coil	—	Ea	863.00	—	—	863.00
Optional supplementary electric coil 5 KW	—	Ea	187.00	—	—	187.00
Optional supplementary electric coil 10 KW	—	Ea	220.00	—	—	220.00
Optional supplementary electric coil 15 KW	—	Ea	478.00	—	—	478.00
Optional supplementary electric coil 20 KW	—	Ea	530.00	—	—	530.00
Optional supplementary electric coil 25 KW	—	Ea	651.00	—	—	651.00
Water-to-water thermal transfer, additional	—	Ea	424.00	—	—	424.00

Heat Pump Systems

Description	Craft@Hrs	Unit	Material $	Labor $	Equipment $	Total $

Geothermal water source heat pump (open loop), w/free cooling.

Water-to-air thermal transfer. The refrigeration circuit is only operated in the heating mode while passive "free" cooling is provided by pumping cold well water directly to a unit-mounted coil that bypasses the refrigeration (compressor) unit. This unit is suited to areas where the temperature of ground water supply is below 50°F. Optional de-superheater system provides domestic hot water temperature preheat or boost. Make additional allowances for the water source, i.e., supply and return water wells, pond or lake source, etc. Make additional allowances for connections, ancillary components, start up and testing.

Description	Craft@Hrs	Unit	Material $	Labor $	Equipment $	Total $
Heat pump, open loop, free cooling, 2 ton	P1@2.50	Ea	3,340.00	93.10	—	3,433.10
Heat pump, open loop, free cooling, 3 ton	P1@3.00	Ea	4,100.00	112.00	—	4,212.00
Heat pump, open loop, free cooling, 4 ton	P1@3.50	Ea	4,600.00	130.00	—	4,730.00
Heat pump, open loop, free cooling, 5 ton	P1@4.00	Ea	8,100.00	149.00	—	8,249.00
Heat pump, open loop, free cooling, 7½ ton	P1@4.75	Ea	8,780.00	177.00	—	8,957.00
Optional de-superheater coil	—	Ea	863.00	—	—	863.00
Optional supplementary electric coil 5 KW	—	Ea	187.00	—	—	187.00
Optional supplementary electric coil 10 KW	—	Ea	220.00	—	—	220.00
Optional supplementary electric coil 15 KW	—	Ea	478.00	—	—	478.00
Optional supplementary electric coil 20 KW	—	Ea	530.00	—	—	530.00
Optional supplementary electric coil 25 KW	—	Ea	651.00	—	—	651.00
Water-to-water thermal transfer additional	—	Ea	424.00	—	—	424.00

Heat Pump Systems

Description	Craft@Hrs	Unit	Material $	Labor $	Equipment $	Total $

Geothermal water source heat pump (closed refrigerant loop). Water-to-air thermal transfer. Optional de-superheater system provides domestic hot water temperature preheat or boost. Make additional allowances for the heat transfer loop, i.e., refrigerant loop. Make additional allowances for connections, ancillary components, start up and testing.

Description	Craft@Hrs	Unit	Material $	Labor $	Equipment $	Total $
Heat pump, 2 ton	P1@2.50	Ea	2,000.00	93.10	—	2,093.10
Heat pump, 3 ton	P1@3.00	Ea	2,610.00	112.00	—	2,722.00
Heat pump, 4 ton	P1@3.50	Ea	3,020.00	130.00	—	3,150.00
Heat pump, 5 ton	P1@4.00	Ea	6,360.00	149.00	—	6,509.00
Heat pump, 7½ ton	P1@4.75	Ea	6,040.00	177.00	—	6,217.00
Optional de-superheater coil	—	Ea	828.00	—	—	828.00
Optional supplementary electric coil 5 KW	—	Ea	193.00	—	—	193.00
Optional supplementary electric coil 10 KW	—	Ea	227.00	—	—	227.00
Optional supplementary electric coil 15 KW	—	Ea	497.00	—	—	497.00
Optional supplementary electric coil 20 KW	—	Ea	551.00	—	—	551.00
Optional supplementary electric coil 25 KW	—	Ea	675.00	—	—	675.00

LEED certification for geothermal heat pumps

Description	Craft@Hrs	Unit	Material $	Labor $	Equipment $	Total $
Add for a LEED-certified heat pump with USGBC rating	—	%	20.0	—	—	—
Add for LEED registration and inspection	—	Ea	2,180.00	—	—	2,180.00
Add for LEED central digital performance monitor/recorder, ComfortView 3 or equal	—	Ea	2,600.00	—	—	2,600.00
Add for LEED-certified zone recording thermostats, Carrier DebonAir, USB remote-ready	—	Ea	387.00	—	—	387.00
Add for IAQ (Indoor Air Quality) CO_2 sensor and duct-mounted aspirator box	—	Ea	468.00	—	—	468.00

Heat Pump Systems

Description	Craft@Hrs	Unit	Material $	Labor $	Equipment $	Total $

Heat pump, residential air-to-air split system. Outdoor unit installed on pad or wall bracket. Indoor coil installed in supply air plenum upstream of supply air fan (air handler). Units rated at 13SEER (Seasonal Energy Efficiency Rating). System includes outdoor condenser/compressor unit, indoor evaporative coil, programmable thermostat, indoor/outdoor sensor. Condenser/compressor unit comes pre-charged from the factory with R410A refrigerant. 208/230 Volt, single phase. Rated in nominal tonnage. (12,000 BTU's equals 1 ton cooling).

Description	Craft@Hrs	Unit	Material $	Labor $	Equipment $	Total $
Setting outdoor unit on pad or wall bracket						
2.0-ton outdoor unit	P1@1.50	Ea	2,480.00	55.80	—	2,535.80
3.0-ton outdoor unit	P1@1.75	Ea	2,720.00	65.10	—	2,785.10
3.5-ton outdoor unit	P1@2.00	Ea	3,300.00	74.40	—	3,374.40
4.0-ton outdoor unit	P1@2.25	Ea	3,530.00	83.70	—	3,613.70
5.0-ton outdoor unit	P1@2.50	Ea	4,150.00	93.10	—	4,243.10
Mounting evaporative coil in air stream						
2.0-ton evap. coil	P1@1.15	Ea	26.90	42.80	—	69.70
3.0-ton evap. coil	P1@1.35	Ea	32.40	50.20	—	82.60
3.5-ton evap. coil	P1@1.50	Ea	37.70	55.80	—	93.50
4.0-ton evap. coil	P1@2.00	Ea	43.00	74.40	—	117.40
5.0-ton evap. coil	P1@2.25	Ea	48.50	83.70	—	132.20
Install and connect (braze) refrigerant line set – 50' (Liquid x Suction). Connect refrigerant line set. Condensor/compressor unit to evaporative coil.						
2.0-ton 3/8" - 5/8"	P1@2.00	Ea	237.00	74.40	—	311.40
3.0-ton 1/2" - 7/8"	P1@2.25	Ea	274.00	83.70	—	357.70
3.5-ton 5/8" - 1"	P1@3.75	Ea	415.00	140.00	—	555.00
4.0-ton 7/8" -1-1/8"	P1@4.50	Ea	496.00	167.00	—	663.00
5.0-ton 1" - 1-3/8"	P1@6.00	Ea	668.00	223.00	—	891.00
Control wiring						
24 volt	BE@3.00	Ea	30.70	123.00	—	153.70
Power wiring compressor						
208/230 Volt	BE@2.00	Ea	184.00	82.10	—	266.10
Evaporative coil drain	P1@1.00	Ea	20.80	37.20	—	58.00
Equipment pad, concrete slab,						
4' x 4' x 2"	CF@4.00	Ea	31.40	144.00	—	175.40
Equipment pad, patio stones,						
2' x 2' x 1–1/2" (4 ea)	CF@2.00	Ea	41.70	71.80	—	113.50

Heat Pump Systems

Description	Craft@Hrs	Unit	Material $	Labor $	Equipment $	Total $

Heat pump, split system, single-speed blower, single-phase scroll compressor – 208 volt, 13 SEER (Seasonal Energy Efficiency Rating)

Air-to-air thermal transfer. Material costs include outdoor stat / low ambient, short cycle timer, high-pressure cut out and digital heat/cool thermostat with manual change-over. Labor includes setting outdoor condenser, indoor blower and coil sections in place only. Make additional allowances for refrigerant piping / connections, electrical / temperature control wiring / connections and evaporator coil condensate drain.

Description	Craft@Hrs	Unit	Material $	Labor $	Equipment $	Total $
1.5 tons cooling	S2@4.00	Ea	3,230.00	146.00	—	3,376.00
2 tons cooling	S2@4.25	Ea	3,320.00	155.00	—	3,475.00
2.5 tons cooling	S2@4.50	Ea	3,860.00	164.00	—	4,024.00
3 tons cooling	S2@5.00	Ea	4,310.00	183.00	—	4,493.00
3.5 tons cooling	S2@5.75	Ea	4,730.00	210.00	—	4,940.00
4 tons cooling	S2@6.25	Ea	5,230.00	228.00	—	5,458.00
5 tons cooling	S2@7.50	Ea	6,060.00	274.00	—	6,334.00

Heat pump, split system, single-speed blower, three-phase scroll compressor – 230 volt, 13 SEER (Seasonal Energy Efficiency Rating)

Air-to-air thermal transfer. Material costs include outdoor stat / low ambient, short cycle timer, high-pressure cut out and digital heat/cool thermostat with manual change over. Labor includes setting outdoor condenser, indoor blower and coil sections in place only. Make additional allowances for refrigerant piping / connections, electrical / temperature control wiring / connections and evaporator coil condensate drain.

Description	Craft@Hrs	Unit	Material $	Labor $	Equipment $	Total $
3.5 tons cooling	S2@5.75	Ea	5,270.00	210.00	—	5,480.00
4 tons cooling	S2@6.25	Ea	5,750.00	228.00	—	5,978.00
5 tons cooling	S2@7.50	Ea	6,590.00	274.00	—	6,864.00

Heat pump, split system, variable-speed blower, single-phase scroll compressor – 208 volt, 13 SEER (Seasonal Energy Efficiency Rating)

Air-to-air thermal transfer. Material costs include outdoor stat / low ambient, short cycle timer, high-pressure cut out and digital heat/cool thermostat with manual change over. Labor includes setting outdoor condenser, indoor blower and coil sections in place only. Make additional allowances for refrigerant piping / connections, electrical / temperature control wiring / connections and evaporator coil condensate drain.

Description	Craft@Hrs	Unit	Material $	Labor $	Equipment $	Total $
1.5 tons cooling	S2@4.00	Ea	3,870.00	146.00	—	4,016.00
2 tons cooling	S2@4.25	Ea	3,930.00	155.00	—	4,085.00
2.5 tons cooling	S2@4.50	Ea	4,510.00	164.00	—	4,674.00
3 tons cooling	S2@5.00	Ea	4,990.00	183.00	—	5,173.00
3.5 tons cooling	S2@5.75	Ea	5,360.00	210.00	—	5,570.00
4 tons cooling	S2@6.25	Ea	6,060.00	228.00	—	6,288.00
5 tons cooling	S2@7.50	Ea	6,870.00	274.00	—	7,144.00

Heat Pump Systems

Description	Craft@Hrs	Unit	Material $	Labor $	Equipment $	Total $

Heat pump, split system, variable-speed blower, three-phase scroll compressor – 230 volt, 13 SEER (Seasonal Energy Efficiency Rating)

Air-to-air thermal transfer. Material costs include outdoor stat / low ambient, short cycle timer, high-pressure cut out and digital heat/cool thermostat with manual change over. Labor includes setting outdoor condenser, indoor blower and coil sections in place only. Make additional allowances for refrigerant piping / connections, electrical / temperature control wiring / connections and evaporator coil condensate drain.

Description	Craft@Hrs	Unit	Material $	Labor $	Equipment $	Total $
3.5 tons cooling	S2@5.75	Ea	5,940.00	210.00	—	6,150.00
4 tons cooling	S2@6.25	Ea	6,570.00	228.00	—	6,798.00
5 tons cooling	S2@7.50	Ea	7,460.00	274.00	—	7,734.00

Heat pump, split system, single-speed blower, single-phase scroll compressor – 208 volt, 14 SEER (Seasonal Energy Efficiency Rating)

Air-to-air thermal transfer. Material costs include outdoor stat / low ambient, short cycle timer, high-pressure cut out and digital heat/cool thermostat with manual change over. Labor includes setting outdoor condenser, indoor blower and coil sections in place only. Make additional allowances for refrigerant piping / connections, electrical / temperature control wiring / connections and evaporator coil condensate drain.

Description	Craft@Hrs	Unit	Material $	Labor $	Equipment $	Total $
2 tons cooling	S2@4.25	Ea	4,460.00	155.00	—	4,615.00
2.5 tons cooling	S2@4.50	Ea	5,050.00	164.00	—	5,214.00
3 tons cooling	S2@5.00	Ea	5,360.00	183.00	—	5,543.00
3.5 tons cooling	S2@5.75	Ea	6,130.00	210.00	—	6,340.00
4 tons cooling	S2@6.25	Ea	6,830.00	228.00	—	7,058.00
5 tons cooling	S2@7.50	Ea	7,480.00	274.00	—	7,754.00

Heat pump, split system, variable-speed blower, single-phase scroll compressor – 208 volt, 14 SEER (Seasonal Energy Efficiency Rating)

Air-to-air thermal transfer. Material costs include outdoor stat / low ambient, short cycle timer, high-pressure cut out and digital heat/cool thermostat with manual change over. Labor includes setting outdoor condenser, indoor blower and coil sections in place only. Make additional allowances for refrigerant piping / connections, electrical / temperature control wiring / connections and evaporator coil condensate drain.

Description	Craft@Hrs	Unit	Material $	Labor $	Equipment $	Total $
2 tons cooling	S2@4.25	Ea	5,080.00	155.00	—	5,235.00
2.5 tons cooling	S2@4.50	Ea	5,560.00	164.00	—	5,724.00
3 tons cooling	S2@5.00	Ea	5,950.00	183.00	—	6,133.00
3.5 tons cooling	S2@5.75	Ea	6,710.00	210.00	—	6,920.00
4 tons cooling	S2@6.25	Ea	7,280.00	228.00	—	7,508.00
5 tons cooling	S2@7.50	Ea	10,700.00	274.00	—	10,974.00

Heat Pump System Accessories

Description	Craft@Hrs	Unit	Material $	Labor $	Equipment $	Total $

Heat pump system accessories

Description	Craft@Hrs	Unit	Material $	Labor $	Equipment $	Total $
Fossil fuel kit	S2@2.50	Ea	220.00	91.40	—	311.40
High pressure cut out kit	S2@.500	Ea	76.60	18.30	—	94.90
Outdoor stat/low ambient sensors	S2@.500	Ea	72.20	18.30	—	90.50
Short cycle protection	S2@.500	Ea	57.00	18.30	—	75.30
Digital heat/cool thermostat w/manual change-over	S2@.750	Ea	162.00	27.40	—	189.40
Digital, programmable heat/cool thermostat w/auto change-over	S2@.750	Ea	508.00	27.40	—	535.40

Heat pump, supplemental electric heating coil

Set in place only. Make additional allowances for electrical connections.

Description	Craft@Hrs	Unit	Material $	Labor $	Equipment $	Total $
5 KW, 240 Volt	S2@1.50	Ea	183.00	54.80	—	237.80
7.5 KW, 240 Volt	S2@1.75	Ea	220.00	63.90	—	283.90
10 KW, 240 Volt	S2@2.00	Ea	256.00	73.10	—	329.10
12.5 KW, 240 V	S2@2.25	Ea	425.00	82.20	—	507.20
15 KW, 240 Volt	S2@2.75	Ea	538.00	100.00	—	638.00
20 KW, 240 Volt	S2@3.50	Ea	603.00	128.00	—	731.00
25 KW, 240 Volt	S2@4.25	Ea	726.00	155.00	—	881.00

LEED certification for geothermal heat pumps

Description	Craft@Hrs	Unit	Material $	Labor $	Equipment $	Total $
Add for a LEED-certified heat pump with USGBC rating	—	%	20.0	—	—	—
Add for LEED registration and inspection	—	Ea	2,180.00	—	—	2,180.00
Add for LEED central digital performance monitor/recorder, ComfortView 3 or equal	—	Ea	2,600.00	—	—	2,600.00
Add for LEED-certified zone recording thermostats, Carrier DebonAir, USB remote-ready	—	Ea	387.00	—	—	387.00
Add for IAQ (Indoor Air Quality) CO_2 sensor and duct-mounted aspirator box	—	Ea	468.00	—	—	468.00

Water Pump Systems

Description	Craft@Hrs	Unit	Material $	Labor $	Equipment $	Total $

Water pump, submersible. Domestic water, geothermal open loop, well, lake or pond applications. Stainless steel jacket with multiple-stage impellers and integral foot valve.

Description	Craft@Hrs	Unit	Material $	Labor $	Equipment $	Total $
Submersible pump, 120/240 volt, ½ HP	P1@2.00	Ea	627.00	74.40	—	701.40
Submersible pump, 240 volt, ¾ HP	P1@2.00	Ea	781.00	74.40	—	855.40
Submersible pump, 240 volt, 1 HP	P1@2.00	Ea	1,050.00	74.40	—	1,124.40
Submersible pump, 240 volt, 1½ HP	P1@2.00	Ea	1,370.00	74.40	—	1,444.40
Submersible pump, 240 volt, 2 HP	P1@2.00	Ea	1,600.00	74.40	—	1,674.40
Submersible pump, 240 volt, 3 HP	P1@2.00	Ea	1,800.00	74.40	—	1,874.40
Submersible pump, 240 volt, 5 HP	P1@2.00	Ea	2,100.00	74.40	—	2,174.40
Pump controller – constant pressure, VSD	P1@1.50	Ea	1,060.00	55.80	—	1,115.80
Pump control box, 120 volt	P1@.600	Ea	62.70	22.30	—	85.00
Pump control box, 240 volt	P1@.600	Ea	76.90	22.30	—	99.20
Pump stand, PVC	P1@1.00	Ea	168.00	37.20	—	205.20
Pump stand, galvanized steel	P1@1.35	Ea	228.00	50.20	—	278.20

Water pump, jet. Domestic water well, lake and pond applications. 110 volt, cast iron body, single-stage impeller, including pressure switch & pressure gauge. Use for shallow depths to 25' maximum lift. Deep well applications to 120' maximum lift.

Description	Craft@Hrs	Unit	Material $	Labor $	Equipment $	Total $
Jet pump, shallow depth ½ HP	P1@2.00	Ea	266.00	74.40	—	340.40
Jet pump, shallow depth ¾ HP	P1@2.00	Ea	308.00	74.40	—	382.40
Jet pump, deep well ½ HP	P1@2.00	Ea	361.00	74.40	—	435.40
Jet pump, deep well ¾ HP	P1@2.00	Ea	419.00	74.40	—	493.40
Optional pump-mounted pressure tank 5 gallon	P1@.500	Ea	112.00	18.60	—	130.60

Water Pump System Accessories

Description	Craft@Hrs	Unit	Material $	Labor $	Equipment $	Total $

Water pump system accessories. Pressure tanks are factory pre-charged. VSD = variable speed drive.

Description	Craft@Hrs	Unit	Material $	Labor $	Equipment $	Total $
Pressure tank, bladder type, 5 gallon	P1@1.00	Ea	97.80	37.20	—	135.00
Pressure tank, bladder type, 10 gallon	P1@2.00	Ea	112.00	74.40	—	186.40
Pressure tank, bladder type, 20 gallon	P1@2.00	Ea	133.00	74.40	—	207.40
Pressure tank, bladder type, 30 gallon	P1@2.00	Ea	190.00	74.40	—	264.40
Pressure tank, bladder type, 40 gallon	P1@2.00	Ea	252.00	74.40	—	326.40
Brass tank tee, ¾" or 1" w/ union	P1@.300	Ea	58.90	11.20	—	70.10
Brass tank tee, ¾" or 1" w/o union	P1@.300	Ea	47.40	11.20	—	58.60
Constant pressure VSD pump controller	P1@1.50	Ea	1,060.00	55.80	—	1,115.80
Pressure switch, 110/220 volt	P1@.350	Ea	44.80	13.00	—	57.80
Pressure gauge, 0-100 psi, ¼"	P1@.200	Ea	11.20	7.44	—	18.64
Sediment faucet (hose bibb), ½"	P1@1.50	Ea	9.09	55.80	—	64.89
Double injector deep well venturi, 1", brass	P1@.600	Ea	238.00	22.30	—	260.30
Double injector deep well venturi, 1¼", brass	P1@.650	Ea	273.00	24.20	—	297.20
Double injector deep well venturi, 1½", brass	P1@.700	Ea	308.00	26.10	—	334.10
Double injector deep well venturi, 1", plastic	P1@.550	Ea	76.90	20.50	—	97.40
Double injector deep well venturi, 1¼", plastic	P1@.600	Ea	94.90	22.30	—	117.20
Double injector deep well venturi, 1½", plastic	P1@.650	Ea	103.00	24.20	—	127.20
Foot valve, 1", brass	P1@.400	Ea	81.00	14.90	—	95.90
Foot valve, 1¼", brass	P1@.400	Ea	83.70	14.90	—	98.60
Foot valve, 1½", brass	P1@.400	Ea	89.30	14.90	—	104.20
Foot valve, 1", plastic	P1@.400	Ea	25.20	14.90	—	40.10
Foot valve, 1¼", plastic	P1@.400	Ea	30.80	14.90	—	45.70
Foot valve, 1½", plastic	P1@.400	Ea	34.20	14.90	—	49.10
Barbed adapter, FIP/MIP, 1" brass	P1@.230	Ea	13.30	8.56	—	21.86
Barbed adapter, FIP/MIP, 1¼" brass	P1@.250	Ea	15.30	9.31	—	24.61

Water Pump System Accessories

Description	Craft@Hrs	Unit	Material $	Labor $	Equipment $	Total $
Barbed adapter, FIP/MIP, 1½" brass	P1@.265	Ea	18.70	9.86	—	28.56
Barbed adapter, FIP/MIP, 1" plastic	P1@.215	Ea	12.20	8.00	—	20.20
Barbed adapter, FIP/MIP, 1¼" plastic	P1@.220	Ea	13.40	8.19	—	21.59
Barbed adapter, FIP/MIP, 1½" plastic	P1@.235	Ea	18.00	8.75	—	26.75
Pitless adapter, 1" x 5" brass	P1@1.25	Ea	197.00	46.50	—	243.50
Pitless adapter, 1¼" x 5" brass	P1@1.25	Ea	212.00	46.50	—	258.50
Pitless adapter, 1½" x 5" brass	P1@1.25	Ea	238.00	46.50	—	284.50
Pitless adapter, 2" x 6" brass	P1@1.80	Ea	361.00	67.00	—	428.00
Poly pipe, 75 psi, 1"	P1@.020	LF	.30	.74	—	1.04
Poly pipe, 75 psi, 1¼"	P1@.021	LF	.36	.78	—	1.14
Poly pipe, 75 psi, 1½"	P1@.022	LF	.38	.82	—	1.20
Poly pipe, 100 psi, 1"	P1@.021	LF	.30	.78	—	1.08
Poly pipe, 100 psi, 1¼"	P1@.022	LF	.36	.82	—	1.18
Poly pipe, 100 psi, 1½"	P1@.023	LF	.38	.86	—	1.24
Electrical wire (110 volt), #12/4	P1@.015	LF	.38	.56	—	.94
Electrical wire (220 volt), #10/4	P1@.015	LF	.40	.56	—	.96
Heat shrink kit	—	Ea	4.87	—	—	4.87

Geothermal/Domestic Water Wells

Geothermal water wells

A geothermal water well is constructed and operates exactly like a water well used for drinking water. In many residential applications, the supply well of the geothermal system also functions as the domestic water supply well.

The cost of a drilled well depends on many variables, such as costs to get the drilling rig and crew to the site, the type of geologic formation the well drilling rig must drill through, the depth the driller must achieve to find water in that particular location and the distance from an existing available water source to provide the water required to drill (to refill the water truck if necessary).

Well drillers charge by the foot and usually have a 100' minimum charge-out fee. They can only speculate to what depth they must drill to find water. Water being found at a particular depth nearby doesn't always translate to finding water at another given location at the same depth. It's important to remember that drilling a relatively small hole in the ground and hoping to hit what is described as a vein of water as part of an underground water table is always a gamble. As such, it is, without proper research, difficult to determine a realistic budget for well drilling. It's always prudent to call the water authority or a well driller in the area for the most accurate information on the water and geological characteristics of a given location.

Use the following to begin to formulate a budget exclusive of costs associated with factors listed above. Because of the inherent uncertainties of the process, these prices are given only as typical subcontract costs.

Description	Craft@Hrs	Unit	Material $	Labor $	Equipment $	Total $
Drilled well, 5" diameter, PVC casing, set into bedrock. Based on a 2-man crew, 1 rotary drill rig and a 1,500 gallon water truck. Minimum 100' charge in most cases. Typical subcontract costs.						
Drilled through till, clay, sand	—	LF	—	—	—	37.70
Drilled through till and boulder	—	LF	—	—	—	48.70
Drilled through limestone soft bedrock	—	LF	—	—	—	54.80
Drilled through granite hard bedrock	—	LF	—	—	—	110.00
Add for galvanized steel casing	—	LF	—	—	—	15.60
Drilled well, 6" diameter, PVC casing, set into bedrock. Based on a 2-man crew, 1 rotary drill rig and a 1,500 gallon water truck. Minimum 100' charge in most cases. Typical subcontract costs.						
Drilled through till, clay, sand	—	LF	—	—	—	41.90
Drilled through till and boulder	—	LF	—	—	—	53.30
Drilled through limestone soft bedrock	—	LF	—	—	—	57.80
Drilled through granite hard bedrock	—	LF	—	—	—	118.00
Add for galvanized steel casing	—	LF	—	—	—	19.10

Geothermal Water Wells/Geothermal Closed Loops

Description	Craft@Hrs	Unit	Material $	Labor $	Equipment $	Total $

Drilled well, 8" diameter, PVC casing, set into bedrock. Based on a 2-man crew, 1-rotary drill rig and a 1,500 gallon water truck. Minimum 100' charge in most cases. Typical subcontract costs.

Description	Craft@Hrs	Unit	Material $	Labor $	Equipment $	Total $
Drilled through till, clay, sand	—	LF	—	—	—	53.30
Drilled through till and boulder	—	LF	—	—	—	60.80
Drilled through limestone soft bedrock	—	LF	—	—	—	68.50
Drilled through granite hard bedrock	—	LF	—	—	—	138.00
Add for galvanized steel casing	—	LF	—	—	—	38.10

Drilled well, 10" diameter, PVC casing, set into bedrock. Based on a 2-man crew, 1 rotary drill rig and a 1,500 gallon water truck. Minimum 100' charge in most cases. Typical subcontract costs.

Description	Craft@Hrs	Unit	Material $	Labor $	Equipment $	Total $
Drilled through till, clay, sand	—	LF	—	—	—	63.90
Drilled through till and boulder	—	LF	—	—	—	71.40
Drilled through limestone soft bedrock	—	LF	—	—	—	79.00
Drilled through granite hard bedrock	—	LF	—	—	—	183.00
Add for galvanized steel casing	—	LF	—	—	—	48.50

Geothermal bore hole (vertical loop), 3" diameter, drilled to a maximum of 30'. Includes installation of a polypropylene fluid loop using a 180-degree return bend with socket fusion joints. Bore hole filled with bentonite slurry through a tremie tube and mud pump. Includes socket fusioned header and supply & return mains to heat pump. Based on a minimum of 1,000 vertical feet (2,000' of pipe) Based on a 2-man crew, 1 skid steer-mounted rotary drill rig and a 50 gallon water tank. Typical subcontract costs.

Description	Craft@Hrs	Unit	Material $	Labor $	Equipment $	Total $
¾" vertical loop	—	LF	—	—	—	12.90
1" vertical loop	—	LF	—	—	—	14.10
Add for thermally-enhanced Bentonite	—	LF	—	—	—	3.74

Geothermal horizontal loop, directionally bored. Includes installation of a polypropylene fluid loop. Based on a minimum of 2,000 horizontal feet (2,000' of pipe). Based on a 3-man crew, 1 directional boring/horizontal drill machine. Typical subcontract costs.

Description	Craft@Hrs	Unit	Material $	Labor $	Equipment $	Total $
¾" straight horizontal loop	—	LF	—	—	—	7.82
1" straight horizontal loop	—	LF	—	—	—	8.53

Geothermal Water Wells

Description	Craft@Hrs	Unit	Material $	Labor $	Equipment $	Total $

Geothermal horizontal loop, slinky type. Includes installation of a polypropylene fluid loop. Based on a minimum of 2,000 horizontal feet (2,000' of pipe). Based on a 3-man crew, 1 excavator backhoe. Typical subcontract costs.

Description	Craft@Hrs	Unit	Material $	Labor $	Equipment $	Total $
¾" slinky loop	—	LF	—	—	—	5.93
1" slinky loop	—	LF	—	—	—	6.69
¾" slinky pond loop	—	LF	—	—	—	4.63
1" slinky pond loop	—	LF	—	—	—	5.40

Biomass-Fired Boilers

Description	Craft@Hrs	Unit	Material $	Labor $	Equipment $	Total $

Biomass-fired hot water boilers for central heating systems. For residential, commercial and light industrial applications. 15 PSI, EPA, ASME and UL rated, with wood-pellet and/or corn fuel gravity-feed hopper, ash hood cleanout, secondary natural gas or #2 oil burner, platinum stack transition emissions catalyst, forced draft fan, factory-installed boiler trim, feedwater pump and valves, and tempering valve. Add the cost of the following if required: a concrete pad, optional stainless steel 2 GPM potable hot water coil insert, expansion tank and circulating pump, stack and stack liner, CO_2 fire extinguisher system and electrical connection. Add the cost of post-installation inspection by both the fire marshal and the mechanical inspector to validate federal, state and local energy tax credits or energy tax credit offsets. See pollution control stack costs below. Equipment cost is for an appliance dolly, 3,000 lb. come-along, and ½ ton chain hoist. Listed by boiler horsepower (BHP) rating and input capacity in thousands of BTU per hour (MBH).

3 BHP, 102 MBH	SN@24.0	Ea	11,900.00	941.00	63.00	12,904.00
3 BHP, 104 MBH	SN@24.0	Ea	10,900.00	941.00	63.00	11,904.00
6 BHP, 200 MBH	SN@28.0	Ea	14,000.00	1,100.00	73.50	15,173.50
14 BHP, 500 MBH	SN@30.0	Ea	16,700.00	1,180.00	78.80	17,958.80
27 BHP, 900 MBH	SN@44.0	Ea	24,200.00	1,720.00	116.00	26,036.00

LEED certification requires 88% boiler efficiency, recording controls, compliance with SCAQMD 1146.2 emission standards and zone thermostats.

Add for LEED-certified boiler with USGBC rating	—	%	15.0	—	—	—
Add for LEED registration and inspection	—	Ea	2,150.00	—	—	2,150.00
Add for central performance monitor/ recorder	—	Ea	2,710.00	—	—	2,710.00
Add for zone recording thermostat	—	Ea	402.00	—	—	402.00

Retrofit pollution control stack for biomass hot water boilers. Stack-mounted platinum catalytic screen for biomass-fired hot water boilers for central radiant heating systems. E.P.A. and California Code standards compliant. UL rated. Includes stack mounting transition and extensions, baghouse structural steel frame, stack transition cleanout and inspection stack access bolted ports and an integral CO_2 fire extinguisher system. Add the cost of a concrete pad, electrical connection, post-installation calibration, inspection by both the fire marshal and the mechanical inspector to validate federal, state and local emission reduction tax credits and associated energy tax credit offsets. Equipment cost is for a 2-ton capacity hydraulic truck crane. Listed by boiler horsepower (BHP) rating and input capacity in thousands of BTU per hour (MBH). Use these figures to estimate the cost of adding a pollution control stack to an existing biomass boiler to meet environmental regulations.

3 BHP, 102 MBH	SN@20.0	Ea	1,430.00	784.00	52.50	2,266.50
3 BHP, 104 MBH	SN@20.0	Ea	1,470.00	784.00	52.50	2,306.50
6 BHP, 200 MBH	SN@20.0	Ea	2,200.00	784.00	52.50	3,036.50
14 BHP, 500 MBH	SN@40.0	Ea	2,700.00	1,570.00	105.00	4,375.00
27 BHP, 900 MBH	SN@40.0	Ea	3,660.00	1,570.00	105.00	5,335.00

Biomass-Fired Boilers

Description	Craft@Hrs	Unit	Material $	Labor $	Equipment $	Total $

Biomass-fired water or steam boilers for multi-dwelling, process and industrial applications. 15 PSI, EPA, ASME and UL rated. Includes wood pellet and/or corn fuel auger feed hopper, ash auger cleanout, secondary natural gas or #2 oil burner, platinum stack transition emissions catalyst, forced draft fan, factory-installed boiler trim, feedwater pump and valves, and tempering valve. Add the following costs if required: a concrete pad, optional stainless steel 10 GPM potable hot water coil insert, circulating pump, expansion tank, shell and tube heat exchanger, stack and stack liner, CO_2 fire extinguisher system, pollution control baghouse and filter, electrical connection. Add the cost of post-installation inspection by both the fire marshal and the mechanical inspector to validate federal, state and local energy tax credits or energy tax credit offsets. See pollution control stack costs below. Equipment cost is for a 5-ton capacity forklift. Listed by boiler horsepower (BHP) rating and input capacity in thousands of BTU per hour (MBH).

Description	Craft@Hrs	Unit	Material $	Labor $	Equipment $	Total $
30 BHP, 1,000 MBH	SN@168	Ea	86,600.00	6,580.00	1,220.00	94,400.00
50 BHP, 1,700 MBH	SN@200	Ea	159,000.00	7,840.00	1,450.00	168,290.00
100 BHP, 3,400 MBH	SN@400	Ea	247,000.00	15,700.00	2,910.00	265,610.00
125 BHP, 4,100 MBH	SN@400	Ea	312,000.00	15,700.00	2,910.00	330,610.00
150 BHP, 5,000 MBH	SN@500	Ea	303,000.00	19,600.00	3,630.00	326,230.00
200 BHP, 6,700 MBH	SN@600	Ea	417,000.00	23,500.00	4,360.00	444,860.00

LEED certification requires 88% boiler efficiency, recording controls, compliance with SCAQMD 1146.2 emission standards and zone thermostats.

Description	Craft@Hrs	Unit	Material $	Labor $	Equipment $	Total $
Add for LEED-certified boiler with USGBC rating	—	%	6.0	—	—	—
Add for LEED registration and inspection	—	Ea	2,150.00	—	—	2,150.00
Add for central performance monitor/recorder	—	Ea	2,710.00	—	—	2,710.00
Add for zone recording thermostat	—	Ea	402.00	—	—	402.00

Biomass boiler stack tie-in to baghouse filter. Ionically bonded silicon-lined carbon steel. Costs assume a 20-foot vertical stack height, connection via T-shaped connector with cleanout to the baghouse inlet plenum. Includes rigging, breeching, connecting stack sections from back "turnaround" plenum of boiler to baghouse inlet plenum, vendor-supplied structural steel stack support brackets and aluminized external coating. Equipment cost assumes rental of a 2-ton hydraulic truck crane. Biomass boilers, stoves and heaters smaller than 30 BHP can use a conventional boiler stack (without a baghouse) and require only a platinum catalytic screen. Costs shown assume a 20-foot stack height and 10' horizontal run from T-connector. Cost per 10' length, including a bolt and gasket set on each end. Add 15% for each 10' section (or portion thereof) beyond 20' vertically or horizontally.

Description	Craft@Hrs	Unit	Material $	Labor $	Equipment $	Total $
10" diameter	SN@12.0	Ea	3,060.00	470.00	203.00	3,733.00
12" diameter	SN@12.0	Ea	3,420.00	470.00	203.00	4,093.00
16" diameter	SN@16.0	Ea	3,660.00	627.00	270.00	4,557.00
20" diameter	SN@20.0	Ea	4,640.00	784.00	338.00	5,762.00
24" diameter	SN@24.0	Ea	5,130.00	941.00	406.00	6,477.00
30" diameter	SN@32.0	Ea	7,590.00	1,250.00	541.00	9,381.00
56" diameter	SN@40.0	Ea	12,900.00	1,570.00	676.00	15,146.00

Biomass-Fired Boilers

Description	Craft@Hrs	Unit	Material $	Labor $	Equipment $	Total $

Retrofit pollution control stack for biomass water or steam boiler.

Stack-mounted platinum screen catalytic reactor with air filtration fabric baghouse. For use with biomass-fired hot water or low-pressure steam boilers for central radiant heating or process systems. Meets EPA, NFPA, California Code and UL standards for multi-dwelling, process and industrial applications. Cost includes equipment stack transitions and fittings, transition access bolted inspection and cleanout ports, structural steel frame equipment support and ladders, bottom ash cleanout for baghouse, alarmed stack gas composition monitor and recorder, urea tank and urea pump & regulating valve and an integral CO_2 fire extinguisher system. Costs do not include the following: concrete pad, electrical tie-in, calibration and commissioning, post-installation inspection by the fire marshal and the mechanical inspector to validate federal, state and local energy tax credits or energy tax credit offsets. Equipment cost is for a 2-ton capacity hydraulic truck crane. Listed by boiler horsepower (BHP) rating and input capacity in thousands of BTU per hour (MBH). Use these figures to estimate the cost of adding a pollution control stack to an existing biomass boiler to meet environmental regulations.

Description	Craft@Hrs	Unit	Material $	Labor $	Equipment $	Total $
30 BHP, 1,000 MBH	SN@40.0	Ea	42,600.00	1,570.00	676.00	44,846.00
50 BHP, 1,700 MBH	SN@80.0	Ea	75,600.00	3,140.00	1,350.00	80,090.00
100 BHP, 3,400 MBH	SN@80.0	Ea	133,000.00	3,140.00	1,350.00	137,490.00
125 BHP, 4,100 MBH	SN@120	Ea	141,000.00	4,700.00	2,030.00	147,730.00
150 BHP, 5,000 MBH	SN@120	Ea	151,000.00	4,700.00	2,030.00	157,730.00
200 BHP, 6,700 MBH	SN@160	Ea	160,000.00	6,270.00	2,700.00	168,970.00

Biomass-fired central air space heaters for residential, multi-dwelling, and light commercial applications.

15 PSI, EPA, ASHRAE and UL rated. Includes a wood pellet and/or corn fuel gravity-feed hopper, ash hood cleanout, secondary natural gas or #2 oil burner, platinum stack transition emissions catalyst, forced draft fan, ducted hot-air circulation plenum and integral three- speed circulation duct blower. Add the cost of a concrete pad, optional stainless steel 2 GPM potable hot water coil insert, expansion tank and circulating pump, stack and stack liner, CO_2 fire extinguisher system and electrical connection. These costs do not include post-installation inspection by both the fire marshal and the mechanical inspector to validate federal, state and local energy tax credits or energy tax credit offsets. Equipment cost is for a 2-ton capacity forklift. Listed by input capacity in thousands of BTU per hour (MBH).

Description	Craft@Hrs	Unit	Material $	Labor $	Equipment $	Total $
90 MBH	SN@16.0	Ea	3,530.00	627.00	96.00	4,253.00
120 MBH	SN@16.0	Ea	4,570.00	627.00	96.00	5,293.00
170 MBH	SN@20.0	Ea	8,000.00	784.00	120.00	8,904.00
500 MBH	SN@30.0	Ea	12,400.00	1,180.00	180.00	13,760.00
950 MBH	SN@44.0	Ea	14,800.00	1,720.00	264.00	16,784.00

Biomass-Fired Boilers

Description	Craft@Hrs	Unit	Material $	Labor $	Equipment $	Total $

Water mist fire extinguishing systems for biomass boilers. 20 minute full saturation cycle, with automatic reset to enable instantaneous reactivation for re-ignition. NFPA rated. Includes high pressure spray heads, piping, pumps, valves, alarm horns, emergency halogen lighting, calibration, training, insurance inspection and safety procedure documentation. Based on remote sensor relays for the fuel hopper, fuel conveyor, combustion chamber and baghouse filter.

Description	Craft@Hrs	Unit	Material $	Labor $	Equipment $	Total $
30 BHP boiler	ST@60.0	Ea	15,300.00	2,600.00	—	17,900.00
50 BHP boiler	ST@60.0	Ea	19,200.00	2,600.00	—	21,800.00
100 BHP boiler	ST@80.0	Ea	28,100.00	3,470.00	—	31,570.00
125 BHP boiler	ST@80.0	Ea	30,400.00	3,470.00	—	33,870.00
150 BHP boiler	ST@120	Ea	35,000.00	5,200.00	—	40,200.00
200 BHP boiler	ST@120	Ea	40,700.00	5,200.00	—	45,900.00
Add for CO_2 based extinguishing system	—	%	—	—	—	20.0
Add for Halon-based extinguishing system	—	%	—	—	—	30.0

Pulse-type boilers for central radiant heating. 15 PSI, EPA, ASME and UL rated. Modular design for multiple installations. Natural gas only. With forced draft fan, factory-installed boiler trim, feedwater pump and valves, and tempering valve. For residential, commercial and light industrial radiant heating and potable water applications. Add the cost of a concrete pad, circulating pump, stack and stack liner, CO_2 fire extinguisher system, electrical connection, and post-installation inspection by both the fire marshal and the mechanical inspector to validate federal, state and local energy tax credits or energy tax credit offsets. Equipment cost is for an appliance dolly, come-along and chain hoist. Listed by boiler horsepower (BHP) rating and input capacity in thousands of BTU per hour (MBH).

Description	Craft@Hrs	Unit	Material $	Labor $	Equipment $	Total $
3 BHP, 90 MBH	P1@24.0	Ea	4,900.00	893.00	221.00	6,014.00
3 BHP, 150 MBH	P1@24.0	Ea	7,090.00	893.00	221.00	8,204.00
6 BHP, 299 MBH	P1@28.0	Ea	9,740.00	1,040.00	257.00	11,037.00

LEED certification requires 88% boiler efficiency, recording controls, compliance with SCAQMD 1146.2 emission standards and zone thermostats.

Description	Craft@Hrs	Unit	Material $	Labor $	Equipment $	Total $
Add for LEED-certified boiler with USGBC rating	—	%	15.0	—	—	—
Add for LEED registration and inspection	—	Ea	2,150.00	—	—	2,150.00
Add for central performance monitor/recorder	—	Ea	1,260.00	—	—	1,260.00
Add for zone recording thermostat	—	Ea	174.00	—	—	174.00

Solar Water Heater, Flat-Panel

Description	Craft@Hrs	Unit	Material $	Labor $	Equipment $	Total $

Passive flat-panel solar water heater. Ambient pressure, modular design for multiple installation, EPA, DOE and UL rated, with piping, tempering valve, circulating pump, controls, storage tank and heating coil. Cost does not include electrical connection, roofing modification (bracing) to accommodate the additional weight of water-filled panels, insulation or padding between roof and panels. Add the cost of post-installation inspection by the mechanical inspector to validate federal, state and local energy tax credits or energy tax credit offsets. For larger arrays (laundries, institutional facilities, food processing plants), develop an estimate based on required daily capacity and multiply these costs by the number of panels required. Equipment cost is for a 1-ton capacity hydraulic crane truck.

Description	Craft@Hrs	Unit	Material $	Labor $	Equipment $	Total $
One 32.3 SF solar panel, 80 gallons per day	P1@16.0	Ea	3,820.00	596.00	947.00	5,363.00
Two 32.3 SF solar panels, 120 gallons per day	P1@20.0	Ea	5,340.00	744.00	1,180.00	7,264.00

Fans and Blowers

Description	Craft@Hrs	Unit	Material $	Labor $	Equipment $	Total $

Centrifugal air-foil wheel blower, with motor, drive and isolation, set in place only.

1/3 HP, 12" dia.	SN@2.50	Ea	1,720.00	98.00	—	1,818.00
1/2 HP, 15" dia.	SN@4.50	Ea	2,980.00	176.00	—	3,156.00
1 HP, 18" dia.	SN@7.00	Ea	3,450.00	274.00	—	3,724.00
2 HP, 18" dia.	SN@7.50	Ea	4,080.00	294.00	—	4,374.00
3 HP, 27" dia.	SN@9.00	Ea	6,450.00	353.00	5.87	6,808.87
5 HP, 30" dia.	SN@11.0	Ea	7,610.00	431.00	7.17	8,048.17
7½ HP, 36" dia.	SN@12.0	Ea	9,220.00	470.00	7.82	9,697.82
10 HP, 40" dia.	SN@14.0	Ea	12,800.00	549.00	9.12	13,358.12
15 HP, 44" dia.	SN@17.0	Ea	17,200.00	666.00	11.10	17,877.10
20 HP, 60" dia.	SN@20.0	Ea	19,400.00	784.00	13.00	20,197.00

Utility centrifugal blower, with motor, drive and isolation, set in place only.

1/3 HP, 10" dia.	SN@2.50	Ea	1,110.00	98.00	—	1,208.00
1/2 HP, 12" dia.	SN@4.50	Ea	1,450.00	176.00	—	1,626.00
1 HP, 15" dia.	SN@6.50	Ea	2,370.00	255.00	—	2,625.00
2 HP, 18" dia.	SN@7.50	Ea	2,690.00	294.00	—	2,984.00
3 HP, 24" dia.	SN@9.00	Ea	3,880.00	353.00	5.87	4,238.87
5 HP, 27" dia.	SN@11.0	Ea	4,460.00	431.00	7.17	4,898.17
7½ HP, 36" dia.	SN@12.0	Ea	7,170.00	470.00	7.82	7,647.82
10 HP, 40" dia.	SN@14.0	Ea	8,650.00	549.00	9.12	9,208.12
15 HP, 44" dia.	SN@17.0	Ea	9,360.00	666.00	11.10	10,037.10

Vane-axial fan, set in place only.

30 HP, 36" dia.	SN@16.0	Ea	22,800.00	627.00	270.00	23,697.00
50 HP, 42" dia.	SN@24.0	Ea	25,300.00	941.00	406.00	26,647.00
100 HP, 48" dia.	SN@32.0	Ea	31,500.00	1,250.00	541.00	33,291.00
200 HP, 54" dia.	SN@40.0	Ea	44,600.00	1,570.00	676.00	46,846.00

Fans and Blowers

Description	Craft@Hrs	Unit	Material $	Labor $	Equipment $	Total $

Tube-axial fan, set in place only.

15 HP, 24" dia.	SN@10.0	Ea	6,720.00	392.00	169.00	7,281.00
25 HP, 30" dia.	SN@13.0	Ea	10,400.00	509.00	220.00	11,129.00
30 HP, 36" dia.	SN@16.0	Ea	13,900.00	627.00	270.00	14,797.00
75 HP, 40" dia.	SN@22.0	Ea	20,700.00	862.00	372.00	21,934.00
100 HP, 48" dia.	SN@32.0	Ea	31,700.00	1,250.00	541.00	33,491.00

Roof exhaust fan, ASHRAE/American National Standards Institute specification 62.2-2007 and Air Conditioning Contractors of America Standard ACCA 5QI-2007, "HVAC Indoor Air Quality Installation Specification 2007." U.S. Green Building Council, Underwriters laboratories and CSA approved. Cost includes fan and bearing, shaft mount and power transmission assembly or direct drive, enclosure, mount plates, electric motor, 110 or 408, 3-phase, 60 Hz, switch or relay assembly, motor fusing and ground fault protection, fan blade guard. Set in place only. Add installation costs from the figures below.

1/4 HP, 14" dia.	SN@2.50	Ea	1,770.00	98.00	—	1,868.00
1/3 HP, 22" dia.	SN@2.90	Ea	2,180.00	114.00	—	2,294.00
1/2 HP, 24" dia.	SN@3.50	Ea	2,490.00	137.00	—	2,627.00
1 HP, 30" dia.	SN@4.50	Ea	3,450.00	176.00	—	3,626.00
2 HP, 36" dia.	SN@5.00	Ea	4,930.00	196.00	84.50	5,210.50
3 HP, 40" dia.	SN@6.00	Ea	7,380.00	235.00	101.00	7,716.00
5 HP, 48" dia.	SN@7.00	Ea	10,200.00	274.00	118.00	10,592.00
Cut, frame and gasket downcomer hole in roof	S2@1.00	Ea	34.40	36.50	—	70.90
Mount duct hangers	S2@.250	Ea	4.42	9.14	—	13.56
Cut and mount sheet metal duct	S2@.350	LF	6.80	12.80	—	19.60
Apply and coat duct insulation	S2@.150	LF	2.95	5.48	—	8.43
Install electrical wiring	BE@.150	LF	2.13	6.15	—	8.28
Install HVAC controls	BE@.500	Ea	—	20.50	—	20.50
Commission and test	P1@2.00	Ea	—	74.40	—	74.40
Run air balance and finc-tune system	P1@1.00	Ea	—	37.20	—	37.20

Ventilators and Residential Exhaust Fans

Description	Craft@Hrs	Unit	Material $	Labor $	Equipment $	Total $

Ceiling-mounted commercial ventilator. Low-profile metal grille. Galvanized steel housing. Designed for continuous operation. Acoustically insulated. Can be installed horizontally, vertically or in-line. Labor includes layout and fasten in place only. Make additional allowances for vent duct and electrical requirements.

Description	Craft@Hrs	Unit	Material $	Labor $	Equipment $	Total $
100 CFM	P1@1.05	Ea	176.00	39.10	—	215.10
150 CFM	P1@1.15	Ea	222.00	42.80	—	264.80
200 CFM	P1@1.35	Ea	280.00	50.20	—	330.20
250 CFM	P1@1.45	Ea	352.00	54.00	—	406.00
300 CFM	P1@1.65	Ea	420.00	61.40	—	481.40
400 CFM	P1@2.15	Ea	540.00	80.00	—	620.00
500 CFM	P1@2.50	Ea	657.00	93.10	—	750.10
700 CFM	P1@2.75	Ea	914.00	102.00	—	1,016.00
900 CFM	P1@2.85	Ea	1,010.00	106.00	169.00	1,285.00
1,500 CFM	P1@3.50	Ea	1,400.00	130.00	207.00	1,737.00
2,000 CFM	P1@3.85	Ea	1,540.00	143.00	228.00	1,911.00
3,500 CFM	P1@4.25	Ea	1,760.00	158.00	251.00	2,169.00

Powered attic ventilator. Thermostatically controlled. 1/10 Hp, 120V/6-Hz motor. 8" x 24" metal housing. Labor includes layout and fasten in place only. Make additional allowances for electrical and roofing requirements.

Description	Craft@Hrs	Unit	Material $	Labor $	Equipment $	Total $
1,200 CFM	P1@2.00	Ea	222.00	74.40	—	296.40

Exterior wall/roof exhaust fan. Fully enclosed. Thermally protected. Lifetime lubricated 120V/6-Hz motor. Weather-resistant aluminum housing. Balance blower wheel. Neoprene vibration isolation. Includes backdraft damper, flashing plate and critter screen. 10" discharge outlet. Labor includes layout and fasten in place only. Make additional allowances for vent duct, electrical and roofing requirements.

Description	Craft@Hrs	Unit	Material $	Labor $	Equipment $	Total $
600 CFM	P1@2.50	Ea	657.00	93.10	—	750.10
900 CFM	P1@2.85	Ea	1,020.00	106.00	—	1,126.00

Room ventilator fan. Interior thru-wall room-to-room air transfer application. High efficiency. Manually switched operation with built-in variable speed control. 8" and 10" discharge outlets. Built in damper. Includes intake and discharge grilles. Labor includes layout and fasten in place only. Make additional allowances for electrical requirements.

Description	Craft@Hrs	Unit	Material $	Labor $	Equipment $	Total $
180 CFM	P1@1.25	Ea	136.00	46.50	—	182.50
270 CFM	P1@1.50	Ea	205.00	55.80	—	260.80
380 CFM	P1@1.75	Ea	280.00	65.10	—	345.10

Ceiling exhaust fan, with automatic motion response control. Adjustable (5-60 minutes) time interval operation. Manual override feature. 4" discharge outlet. Pre-wired outlet box. Balanced blower wheel. Permanently lubricated motor. Neoprene vibration isolators. Quiet operation. Suitable for washroom exhaust applications. Labor includes layout and fasten in place only. Make additional allowances for vent duct and electrical requirements.

Description	Craft@Hrs	Unit	Material $	Labor $	Equipment $	Total $
90 CFM	P1@1.00	Ea	210.00	37.20	—	247.20
130 CFM	P1@1.15	Ea	326.00	42.80	—	368.80

Ventilators and Residential Exhaust Fans

Description	Craft@Hrs	Unit	Material $	Labor $	Equipment $	Total $

Ceiling exhaust fan, with automatic humidity response control.
Adjustable time interval operation (5-60 minutes). Manual override feature. 4" discharge outlet. Pre-wired outlet box. Balanced blower wheel. Permanently lubricated motor. Neoprene vibration isolators. Quiet operation. Suitable for washroom exhaust applications. Labor includes layout and fasten in place only. Make additional allowances for vent duct and electrical requirements.

Description	Craft@Hrs	Unit	Material $	Labor $	Equipment $	Total $
90 CFM	P1@1.00	Ea	248.00	37.20	—	285.20
130 CFM	P1@1.15	Ea	363.00	42.80	—	405.80

Ceiling exhaust fan, manually-switched operation.
4" discharge outlet. Pre-wired outlet box. 4 pole motor. Suitable for washroom exhaust applications. Labor includes layout and fasten in place only. Make additional allowances for vent duct and electrical requirements.

Description	Craft@Hrs	Unit	Material $	Labor $	Equipment $	Total $
50 CFM	P1@1.00	Ea	47.10	37.20	—	84.30
70 CFM	P1@1.00	Ea	70.30	37.20	—	107.50
90 CFM	P1@1.00	Ea	82.00	37.20	—	119.20
110 CFM	P1@1.05	Ea	98.20	39.10	—	137.30

Ceiling exhaust fan, high efficiency.
Manually switched operation. 8" and 10" discharge outlets. Labor includes layout and fasten in place only. Suitable for kitchens, laundry rooms, rec rooms and workshops. Make additional allowances for vent duct and electrical requirements.

Description	Craft@Hrs	Unit	Material $	Labor $	Equipment $	Total $
160 CFM	P1@1.25	Ea	88.90	46.50	—	135.40
180 CFM	P1@1.25	Ea	128.00	46.50	—	174.50
270 CFM	P1@1.50	Ea	188.00	55.80	—	243.80
350 CFM	P1@1.75	Ea	236.00	65.10	—	301.10

Discharge hood ducting kit.
Including discharge hood, 4" x 10' of flexible duct, duct clamps, 3" x 4" transition coupling.

Description	Craft@Hrs	Unit	Material $	Labor $	Equipment $	Total $
Wall hood, 4"	P1@2.00	Ea	22.90	74.40	—	97.30
Roof hood, 4"	P1@2.25	Ea	45.50	83.70	—	129.20
Wall hood, 6"	P1@2.55	Ea	102.00	94.90	—	196.90

Fan controls. Switch only.

Description	Craft@Hrs	Unit	Material $	Labor $	Equipment $	Total $
1 hour timer	BE@.155	Ea	74.70	6.36	—	81.06
Humidistat	BE@.155	Ea	51.50	6.36	—	57.86
Speed controller	BE@.155	Ea	70.30	6.36	—	76.66

Ventilators and Residential Exhaust Fans

Description	Craft@Hrs	Unit	Material $	Labor $	Equipment $	Total $

Air admittance valve. For ventless sanitary plumbing. Provides a water seal in traps without roof penetration and vent piping. Complies with ASHRAE/American National Standards Institute spec. 62.2-2007 and American Society for Sanitary Engineering specs 1050 and 1051.

Description	Craft@Hrs	Unit	Material $	Labor $	Equipment $	Total $
Cut, frame and gasket for wall inset and vent valve access door	SW@1.00	Ea	36.00	42.50	—	78.50
Mount valve	P1@.500	Ea	—	18.60	—	18.60
Connect valve to sanitary S- or P-trap	P1@1.00	Ea	—	37.20	—	37.20

Bathroom fans and heaters. Comply with ASHRAE/American National Standards Institute spec 62.2-2007 and meet new indoor air quality code requirements for dehumidification, mold and mildew prevention and radon mitigation. Continental Fan, Tamarack, Broan or equal. Includes enclosure, mount plates, 110 or 408 volt electric motor, switch or relay assembly, motor fusing, ground fault protection. Add the following installation costs, if required.

Description	Craft@Hrs	Unit	Material $	Labor $	Equipment $	Total $
Cut, frame and gasket exhaust hole in sidewall	SW@1.00	Ea	36.00	42.50	—	78.50
Cut and mount sheet metal duct	SW@.150	LF	7.15	6.38	—	13.53
Apply and coat duct insulation	S2@.050	LF	3.11	1.83	—	4.94
Install electrical wiring (typical)	BE@.150	LF	2.24	6.15	—	8.39
Install controls	BE@.500	Ea	—	20.50	—	20.50
Test and adjust	P1@.250	Ea	—	9.31	—	9.31

Bathroom exhaust fan with heater, commercial grade, Continental Fan, Broan or equal.

Description	Craft@Hrs	Unit	Material $	Labor $	Equipment $	Total $
106 CFM fan, 4" duct	BE@.500	Ea	252.00	20.50	—	272.50
152 CFM fan, 4" duct	BE@.500	Ea	286.00	20.50	—	306.50
247 CFM fan, 6" duct	BE@.500	Ea	477.00	20.50	—	497.50
418 CFM fan, 8" duct	BE@.500	Ea	614.00	20.50	—	634.50

Infrared bulb heater and fan. 4-point adjustable mounting brackets span up to 24 inches. Uses 250 watt, 120 volt R-40 infrared bulbs (not included). Heater and fan units include 70 CFM, 3.5 sones ventilator fan. Damper and duct connector included. Plastic matte-white molded grille and compact housings. CFM (cubic feet of air per minute).

Description	Craft@Hrs	Unit	Material $	Labor $	Equipment $	Total $
1-bulb, 70 CFM, 3.5 sones	BE@1.00	Ea	71.40	41.00	—	112.40
2-bulb heater, 7" duct	BE@1.10	Ea	89.80	45.10	—	134.90

Ventilators and Residential Exhaust Fans

Description	Craft@Hrs	Unit	Material $	Labor $	Equipment $	Total $

Heat-A-Lamp® bulb heater and fan. NuTone. Swivel-mount. Torsion spring holds plates firmly to ceiling. 250 watt radiant heat from one R-40 infrared heat lamp. Uses 4-inch duct. Adjustable socket for ceilings up to 1-inch thick. Adjustable hanger bars. Automatic reset for thermal protection. White polymeric finish. UL listed.

Description	Craft@Hrs	Unit	Material $	Labor $	Equipment $	Total $
2.6 amp, 1 lamp, 12-1/2" x 10"	BE@1.50	Ea	73.80	61.50	—	135.30
5.0 amp, 2 lamp, 15-3/8" x 11"	BE@1.50	Ea	95.10	61.50	—	156.60

Surface-mount ceiling bath resistance heater and fan. 1,250 watt. 4,266 BTU. 120 volt. Chrome alloy wire element for instant heat. Built-in fan. Automatic overheat protection. Low-profile housing. Permanently lubricated motor. Mounts to standard 3¼-inch round or 4-inch octagonal ceiling electrical box. Satin-finish aluminum grille extends 2¾ inch from ceiling. 11" diameter x 2¾" deep.

Description	Craft@Hrs	Unit	Material $	Labor $	Equipment $	Total $
10.7 amp	BE@.750	Ea	75.20	30.80	—	106.00

Designer Series bath heater and exhaust fan. 1,500 watt fan-forced heater. 120 watt light capacity, 7 watt night-light (bulbs not included). Permanently lubricated motor. Polymeric damper. Torsion-spring grille mount. Non-glare light-diffusing glass lens. Fits single gang opening. Suitable for use with insulation. 100 CFM, 1.2 sones.

Description	Craft@Hrs	Unit	Material $	Labor $	Equipment $	Total $
4" duct, 100 watt lamp	BE@1.00	Ea	305.00	41.00	—	346.00

Ventilation exhausters

Includes all-weather enclosure, air balance control, software module and NFPA-certified fusible plug-type vertical duct riser gravity dampers, and efficiency optimization arrays. Add the cost of roof curb, control wiring, electrical wiring, permits and final inspection. For wall- or roof-mount in single- and multi-unit residences, commercial (i.e. hotels), institutional (i.e. hospitals and schools) buildings. Set in place only. Add installation costs from the section below. Costs assume rental of an appliance dolly, a 3,000 lb. come-a-long and a ½-ton chain hoist.

Downblast ventilation exhausters. Rooftop mounted, direct drive, centrifugal type. Suitable for either ducted or nonducted applications.

Description	Craft@Hrs	Unit	Material $	Labor $	Equipment $	Total $
1/4 HP, 14" dia.	SN@2.50	Ea	1,770.00	98.00	—	1,868.00
1/3 HP, 22" dia.	SN@2.90	Ea	2,180.00	114.00	—	2,294.00
1/2 HP, 24" dia.	SN@3.50	Ea	2,490.00	137.00	—	2,627.00
1 HP, 30" dia.	SN@4.50	Ea	3,450.00	176.00	—	3,626.00
2 HP, 36" dia.	SN@5.00	Ea	4,930.00	196.00	13.10	5,139.10
3 HP, 40" dia.	SN@6.00	Ea	7,380.00	235.00	15.80	7,630.80
5 HP, 48" dia.	SN@7.00	Ea	10,200.00	274.00	18.40	10,492.40

Ventilators and Residential Exhaust Fans

Description	Craft@Hrs	Unit	Material $	Labor $	Equipment $	Total $

Upblast ventilation exhausters. Direct drive, centrifugal, for roof or wall mount.

1/4 HP, 14" dia.	SN@2.50	Ea	1,910.00	98.00	—	2,008.00
1/3 HP, 22" dia.	SN@2.90	Ea	2,310.00	114.00	—	2,424.00
1/2 HP, 24" dia.	SN@3.50	Ea	2,650.00	137.00	—	2,787.00
1 HP, 30" dia.	SN@4.50	Ea	3,690.00	176.00	—	3,866.00
2 HP, 36" dia.	SN@5.00	Ea	5,180.00	196.00	13.10	5,389.10
3 HP, 40" dia.	SN@6.00	Ea	7,620.00	235.00	15.80	7,870.80
5 HP, 48" dia.	SN@7.00	Ea	10,500.00	274.00	18.40	10,792.40

Energy-efficient exhauster arrays. Upblast belt-drive centrifugal UL762 roof exhausters. Low profile metal grille. Galvanized steel housing. Designed for continuous operation. Acoustically insulated. Can be installed horizontally, vertically or in-line. Labor includes layout and fasten in place only. Make additional allowances for vent duct and electrical requirements.

100 CFM	P1@1.05	Ea	272.00	39.10	—	311.10
150 CFM	P1@1.15	Ea	300.00	42.80	—	342.80
200 CFM	P1@1.35	Ea	326.00	50.20	—	376.20
250 CFM	P1@1.45	Ea	422.00	54.00	—	476.00
300 CFM	P1@1.65	Ea	477.00	61.40	—	538.40
400 CFM	P1@2.15	Ea	598.00	80.00	—	678.00
500 CFM	P1@2.50	Ea	694.00	93.10	—	787.10
700 CFM	P1@2.75	Ea	993.00	102.00	—	1,095.00
900 CFM	P1@2.85	Ea	1,080.00	106.00	7.49	1,193.49
1,500 CFM	P1@3.50	Ea	1,650.00	130.00	9.19	1,789.19
2,000 CFM	P1@3.85	Ea	1,830.00	143.00	10.10	1,983.10
3,500 CFM	P1@4.25	Ea	2,040.00	158.00	11.20	2,209.20

Installation costs for ventilation exhausters.

Cut, frame and gasket downcomer hole in sidewall or roof	S2@1.00	Ea	34.80	36.50	—	71.30
Mount duct hangers	S2@.250	Ea	4.42	9.14	—	13.56
Cut & mount sheet metal duct	S2@.350	LF	6.80	12.80	—	19.60
Apply and coat duct insulation	S2@.150	LF	2.95	5.48	—	8.43
Install electrical wiring	BE@.150	LF	2.13	6.15	—	8.28
Install exhauster controls	BE@.500	Ea	—	20.50	—	20.50
Commission and test	P1@4.00	Ea	—	149.00	—	149.00
Run air balance and fine-tune system	P1@4.00	Ea	—	149.00	—	149.00

Apparatus Housing

Unitary or factory-packaged air conditioning units are often too small to meet the heating and cooling needs of a large single-zone space. In such cases, engineers design what's called a "built-up" system. They're custom fitted to the client's needs and the space. A typical "built-up" system includes the following equipment:

- centrifugal blowers or axial-flow fans
- damper assemblies
- coil banks
- filter banks

Once the system is assembled, it's enclosed in a shop-fabricated housing or casing. The illustration on page 332 shows a standard apparatus housing. The walls and roof are shop-built panels. These panels consist of a layer of thermal insulation covered on one or both sides with sheets of 18 gauge galvanized steel. An interior framework of steel angles or channels supports the panels. Seal all mating surfaces to minimize air leakage, using gaskets, caulking or sealant.

Apparatus Housing

There are two types of panels used in building apparatus housings: "double-skin" panels and "single-skin" panels. Double-skin panels have a layer of thermal insulation material sandwiched between two sheets of 18 gauge galvanized steel. Single-skin panels also have a layer of insulation. However, only the outside surface is clad in 18 gauge steel sheeting. On the inside the panel insulation is exposed.

The costs of constructing and installing apparatus housings vary greatly. This is due to the wide range of possible conditions and needs that you face on each job. Some of the many factors that influence costs include the following:

- internal operating pressures
- external wind loads
- seismic requirements
- physical locations

That's why I recommend that you always prepare a detailed cost estimate. Then you can take all the particulars into account. However, for budget estimating purposes only, use the following cost data:

Single-skin panels	$11.60 per SF*
Double-skin panels	$19.70 per SF*
Access doors (20" x 60")	$403.00 each

*Total area of walls and roof, less doors.

When you're estimating material costs for apparatus housings, remember to add in an extra 25 percent to cover the following costs: waste, seams, gaskets, sealants and miscellaneous assembly hardware.

Labor for Shop-Fabrication and Field Assembly of Apparatus Housings. The costs listed are manhours per square foot (SF) of panel. To find the total SF of panel, add up areas for all the panels.

Description	Shop Labor (MH per SF)	Field Labor (MH per SF)
Single-skin panels	.040	.250
Double-skin panels	.085	.300
Add for access doors	.400	.240

Apparatus Housing

Figure 1 - Typical apparatus housing

Air Devices, Registers and Grilles

Description	Craft@Hrs	Unit	Material $	Labor $	Equipment $	Total $

Supply air side-wall register, fixed pattern, with damper. Steel construction.

8" x 4"	S2@.250	Ea	10.80	9.14	—	19.94
8" x 6"	S2@.250	Ea	11.20	9.14	—	20.34
10" x 4"	S2@.250	Ea	12.20	9.14	—	21.34
10" x 6"	S2@.250	Ea	20.80	9.14	—	29.94
10" x 8"	S2@.250	Ea	19.90	9.14	—	29.04
12" x 6"	S2@.250	Ea	21.10	9.14	—	30.24
12" x 8"	S2@.250	Ea	17.10	9.14	—	26.24
14" x 6"	S2@.250	Ea	17.20	9.14	—	26.34
14" x 8"	S2@.250	Ea	19.60	9.14	—	28.74

Return air side-wall grille, fixed pattern, without damper. Steel construction.

14" x 6"	S2@.250	Ea	9.37	9.14	—	18.51
14" x 8"	S2@.250	Ea	9.06	9.14	—	18.20
16" x 16"	S2@.250	Ea	27.40	9.14	—	36.54
20" x 10"	S2@.300	Ea	40.70	11.00	—	51.70
20" x 16"	S2@.300	Ea	44.60	11.00	—	55.60
20" x 20"	S2@.300	Ea	59.00	11.00	—	70.00
24" x 24"	S2@.350	Ea	76.80	12.80	—	89.60

Supply air double-deflection side-wall register, adjustable pattern. Aluminum construction.

8" x 4"	S2@.250	Ea	35.10	9.14	—	44.24
8" x 6"	S2@.250	Ea	39.10	9.14	—	48.24
10" x 4"	S2@.250	Ea	44.90	9.14	—	54.04
10" x 6"	S2@.250	Ea	46.10	9.14	—	55.24
10" x 8"	S2@.250	Ea	49.60	9.14	—	58.74
12" x 6"	S2@.250	Ea	51.30	9.14	—	60.44
12" x 8"	S2@.250	Ea	57.40	9.14	—	66.54
12" x 10"	S2@.250	Ea	65.10	9.14	—	74.24
14" x 6"	S2@.250	Ea	65.10	9.14	—	74.24
14" x 8"	S2@.250	Ea	65.90	9.14	—	75.04
14" x 10"	S2@.250	Ea	70.10	9.14	—	79.24
16" x 8"	S2@.250	Ea	70.10	9.14	—	79.24
16" x 10"	S2@.250	Ea	76.80	9.14	—	85.94
18" x 8"	S2@.300	Ea	80.00	11.00	—	91.00
18" x 10"	S2@.300	Ea	85.80	11.00	—	96.80
18" x 12"	S2@.300	Ea	90.50	11.00	—	101.50

Air Devices, Diffusers and Grilles

Description	Craft@Hrs	Unit	Material $	Labor $	Equipment $	Total $

Return air single-deflection side-wall grille, adjustable pattern.
Aluminum construction.

Description	Craft@Hrs	Unit	Material $	Labor $	Equipment $	Total $
8" x 4"	S2@.250	Ea	31.80	9.14	—	40.94
8" x 6"	S2@.250	Ea	35.10	9.14	—	44.24
10" x 4"	S2@.250	Ea	38.50	9.14	—	47.64
10" x 6"	S2@.250	Ea	41.60	9.14	—	50.74
10" x 8"	S2@.250	Ea	44.40	9.14	—	53.54
12" x 6"	S2@.250	Ea	45.90	9.14	—	55.04
12" x 8"	S2@.250	Ea	51.60	9.14	—	60.74
12" x 10"	S2@.250	Ea	58.20	9.14	—	67.34
14" x 6"	S2@.250	Ea	58.20	9.14	—	67.34
14" x 8"	S2@.250	Ea	59.70	9.14	—	68.84
14" x 10"	S2@.250	Ea	62.60	9.14	—	71.74
16" x 8"	S2@.250	Ea	62.60	9.14	—	71.74
16" x 10"	S2@.300	Ea	69.20	11.00	—	80.20
18" x 8"	S2@.300	Ea	72.10	11.00	—	83.10
18" x 10"	S2@.300	Ea	77.40	11.00	—	88.40
18" x 12"	S2@.300	Ea	81.60	11.00	—	92.60

4-way adjustable pattern ceiling diffuser, 24" x 24" with collar and volume damper. Steel construction.

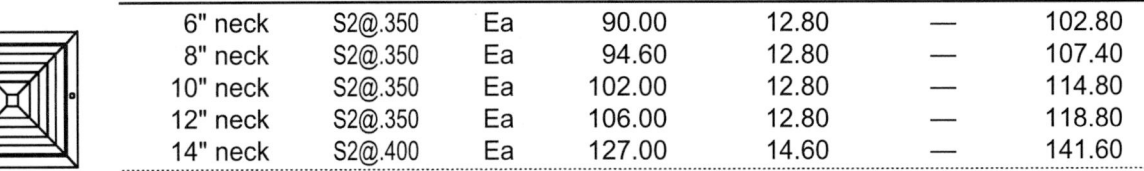

Description	Craft@Hrs	Unit	Material $	Labor $	Equipment $	Total $
6" neck	S2@.350	Ea	90.00	12.80	—	102.80
8" neck	S2@.350	Ea	94.60	12.80	—	107.40
10" neck	S2@.350	Ea	102.00	12.80	—	114.80
12" neck	S2@.350	Ea	106.00	12.80	—	118.80
14" neck	S2@.400	Ea	127.00	14.60	—	141.60

Fixed pattern ceiling diffuser, 24" x 24" with collar, without volume damper. Steel construction.

Description	Craft@Hrs	Unit	Material $	Labor $	Equipment $	Total $
6" neck	S2@.350	Ea	54.30	12.80	—	67.10
8" neck	S2@.350	Ea	58.50	12.80	—	71.30
10" neck	S2@.350	Ea	65.70	12.80	—	78.50
12" neck	S2@.350	Ea	69.60	12.80	—	82.40
14" neck	S2@.350	Ea	90.00	12.80	—	102.80

Perforated ceiling diffuser, 24" x 24" fixed pattern with collar, without volume damper. Steel construction.

Description	Craft@Hrs	Unit	Material $	Labor $	Equipment $	Total $
6" neck	S2@.350	Ea	72.20	12.80	—	85.00
8" neck	S2@.350	Ea	74.80	12.80	—	87.60
10" neck	S2@.350	Ea	80.20	12.80	—	93.00
12" neck	S2@.350	Ea	85.80	12.80	—	98.60
14" neck	S2@.400	Ea	95.90	14.60	—	110.50
16" neck	S2@.720	Ea	162.00	26.30	—	188.30

Air Devices, Diffusers and Grilles

Description	Craft@Hrs	Unit	Material $	Labor $	Equipment $	Total $

Round 3-cone, 2-position ceiling diffuser with collar, without volume damper. Steel construction.

6"	S2@.350	Ea	74.60	12.80	—	87.40
8"	S2@.350	Ea	85.80	12.80	—	98.60
10"	S2@.350	Ea	104.00	12.80	—	116.80
12"	S2@.350	Ea	122.00	12.80	—	134.80
14"	S2@.350	Ea	163.00	12.80	—	175.80

Round adjustable ceiling diffuser with collar and volume damper. Steel construction.

6"	S2@.350	Ea	112.00	12.80	—	124.80
8"	S2@.350	Ea	133.00	12.80	—	145.80
10"	S2@.350	Ea	104.00	12.80	—	116.80
12"	S2@.350	Ea	163.00	12.80	—	175.80
14"	S2@.350	Ea	303.00	12.80	—	315.80
16"	S2@.450	Ea	355.00	16.40	—	371.40
18"	S2@.450	Ea	409.00	16.40	—	425.40
20"	S2@.550	Ea	484.00	20.10	—	504.10
24"	S2@.650	Ea	665.00	23.80	—	688.80

Makeup Air Unit

Description	Craft@Hrs	Unit	Material $	Labor $	Equipment $	Total $

Makeup air unit. Makeup air unit uses 100% direct-heated outdoor air for warehouse, factory and institutional applications to provide higher thermal efficiency and balanced distribution of system air while pre-heating air for use in boiler or other HVAC applications. ANSI standard Z83.4/CSA 3.7 certified. With weatherproof metal enclosure, burner, analog electric controls for burner and fan, thermostat, actuator, natural gas burner array, and either roof-mount supports or sidewall external mounts and transition and gaskets. Set in place only. Add installation costs.

Description	Craft@Hrs	Unit	Material $	Labor $	Equipment $	Total $
200 to 399 CFM	SN@2.25	Ea	662.00	88.20	—	750.20
400 to 599 CFM	SN@2.25	Ea	731.00	88.20	—	819.20
600 to 799 CFM	SN@2.75	Ea	841.00	108.00	46.50	995.50
800 to 999 CFM	SN@3.00	Ea	1,010.00	118.00	50.70	1,178.70
1,000 to 1,499 CFM	SN@3.50	Ea	1,340.00	137.00	59.20	1,536.20
1,500 to 1,999 CFM	SN@4.00	Ea	1,690.00	157.00	67.60	1,914.60
4,000 to 5,999 CFM	SN@6.00	Ea	4,290.00	235.00	101.00	4,626.00
7,500 to 10,000 CFM	SN@8.00	Ea	7,330.00	314.00	135.00	7,779.00
15,000 to 25,000 CFM	SN@10.0	Ea	10,400.00	392.00	169.00	10,961.00
30,000 to 40,000 CFM	SN@16.0	Ea	15,400.00	627.00	270.00	16,297.00
60,000 to 80,000 CFM	SN@24.0	Ea	26,900.00	941.00	406.00	28,247.00

Installation costs for makeup air units.

Description	Craft@Hrs	Unit	Material $	Labor $	Equipment $	Total $
Cut, frame and gasket downcomer/ sidewall hole	S2@1.00	Ea	32.00	36.50	—	68.50
Mount duct hangers	S2@.250	Ea	4.11	9.14	—	13.25
Cut and mount sheet metal duct	S2@.350	LF	6.35	12.80	—	19.15
Apply and coat duct insulation	S2@.150	LF	2.77	5.48	—	8.25
Install piping for gas line	P1@.100	LF	4.29	3.72	—	8.01
Install electrical wiring	BE@.150	LF	1.99	6.15	—	8.14
Install makeup air unit controls	BE@.500	Ea	—	20.50	—	20.50
Commission and test	P1@4.00	Ea	—	149.00	—	149.00
Run air balance and fine-tune system	P1@4.00	Ea	—	149.00	—	149.00

Terminal Units (VAV)

Variable-air volume units regulate supply air to ducted air conditioning systems. These systems are certified to ANSI standard Z83.4/CSA 3.7 and can be either roof-mounted or upper sidewall externally-mounted. Costs include a weatherproof metal enclosure, heater coil or electrical heater fixture, fan and fan controls, downcomer or sidewall entry transition and gaskets, and either roof-mount supports or sidewall external mounts. Includes set in place only. Add installation costs from the table that follows.

Description	Craft@Hrs	Unit	Material $	Labor $	Equipment $	Total $

Variable-air volume terminal unit with analog electric controls, thermostat, actuator and controller.

Description	Craft@Hrs	Unit	Material $	Labor $	Equipment $	Total $
200- 400 CFM	SN@2.00	Ea	591.00	78.40	33.80	703.20
400- 600 CFM	SN@2.00	Ea	659.00	78.40	33.80	771.20
600- 800 CFM	SN@2.50	Ea	693.00	98.00	42.30	833.30
800-1,000 CFM	SN@2.75	Ea	752.00	108.00	46.50	906.50
1,000-1,500 CFM	SN@3.00	Ea	1,090.00	118.00	50.70	1,258.70
1,500-2,000 CFM	SN@3.50	Ea	1,390.00	137.00	59.20	1,586.20

Variable-air volume terminal reheat unit with analog electric controls, thermostat, actuator, controller and a two-row hot water coil.

Description	Craft@Hrs	Unit	Material $	Labor $	Equipment $	Total $
200- 400 CFM	SN@2.25	Ea	682.00	88.20	38.00	808.20
400- 600 CFM	SN@2.25	Ea	752.00	88.20	38.00	878.20
600- 800 CFM	SN@2.75	Ea	866.00	107.00	46.50	1,019.50
800-1,000 CFM	SN@3.00	Ea	1,040.00	118.00	50.70	1,208.70
1,000-1,500 CFM	SN@3.50	Ea	1,390.00	137.00	59.20	1,586.20
1,500-2,000 CFM	SN@4.00	Ea	1,750.00	157.00	67.60	1,974.60

Installation costs for variable-air volume units.

Description	Craft@Hrs	Unit	Material $	Labor $	Equipment $	Total $
Cut, frame and gasket downcomer hole in roof	S2@1.00	Ea	31.30	36.50	—	67.80
Mount duct hangers	S2@.250	Ea	4.03	9.14	—	13.17
Cut and mount sheet metal duct	S2@.350	LF	6.22	12.80	—	19.02
Apply and coat duct insulation	S2@.150	LF	2.72	5.48	—	8.20
Install piping for chilled or hot water/steam	P1@.100	LF	3.20	3.72	—	6.92
Install electrical wiring	BE@.150	LF	1.94	6.15	—	8.09
Install VAV controls	BE@.500	Ea	—	20.50	—	20.50
Commission and test	P1@4.00	Ea	—	149.00	—	149.00
Run air balance and fine-tune system	P1@4.00	Ea	—	149.00	—	149.00

Terminal Units (VAV)

Air balance software and control modules for variable-air volume ducted zone control. For commercial, institutional and industrial applications. Johnson Controls or equal. The zone and duct-mounted sensor and response unit comprises a digital controller, actuator and differential pressure transducer. These components are wired together and securely housed inside a plenum-rated enclosure which mounts directly to the VAV "serviced zone" box. Complies with Underwriters Laboratories file E107041, CCN PAZX, UL916 Energy Management Equipment. Interfaces with Windows-based commercial desktop computers per the digital control manufacturer's specifications.

Description	Craft@Hrs	Unit	Material $	Labor $	Equipment $	Total $

Fully programmable 28 point controller, 16 universal inputs, 12 universal outputs, 12 HOA switches, a battery-backed real-time clock, plus preconfigured function blocks for a real-time clock. Stores schedules and trending information.

Description	Craft@Hrs	Unit	Material $	Labor $	Equipment $	Total $
Programmable controller	BE@.500	Ea	1,010.00	20.50	—	1,030.50

Configurable variable air volume pressure independent terminal box controller with wireless transceiver (868.3 MHz)

Description	Craft@Hrs	Unit	Material $	Labor $	Equipment $	Total $
Terminal box controller	BE@.500	Ea	3,040.00	20.50	—	3,060.50

Configurable controller, preconfigured LNS plug-in, and all input devices required for typical rooftop unit application

Description	Craft@Hrs	Unit	Material $	Labor $	Equipment $	Total $
Preconfigured controller	BE@.500	Ea	2,010.00	20.50	—	2,030.50

Variable-air volume controller system software for a Windows computer. Includes a software license and support for two concurrent client browser connections with no limit on the number of concurrent users and no data tag limit.

Description	Craft@Hrs	Unit	Material $	Labor $	Equipment $	Total $
System software	BE@18.0	Ea	7,650.00	739.00	—	8,389.00

Ductwork Specialties

Description	Craft@Hrs	Unit	Material $	Labor $	Equipment $	Total $
Opposed-blade damper for rectangular ductwork						
12" x 6"	S2@.600	Ea	66.90	21.90	—	88.80
12" x 10"	S2@.650	Ea	79.10	23.80	—	102.90
12" x 12"	S2@.700	Ea	96.90	25.60	—	122.50
18" x 12"	S2@.800	Ea	111.00	29.20	—	140.20
18" x 18"	S2@.850	Ea	123.00	31.10	—	154.10
24" x 12"	S2@.850	Ea	131.00	31.10	—	162.10
24" x 18"	S2@.900	Ea	143.00	32.90	—	175.90
24" x 24"	S2@1.00	Ea	159.00	36.50	—	195.50
30" x 12"	S2@.900	Ea	149.00	32.90	—	181.90
30" x 18"	S2@1.00	Ea	166.00	36.50	—	202.50
30" x 24"	S2@1.10	Ea	189.00	40.20	—	229.20
30" x 30"	S2@1.20	Ea	202.00	43.80	—	245.80
36" x 18"	S2@1.10	Ea	195.00	40.20	—	235.20
36" x 36"	S2@1.40	Ea	267.00	51.20	—	318.20
42" x 24"	S2@1.30	Ea	272.00	47.50	—	319.50
42" x 42"	S2@1.50	Ea	372.00	54.80	—	426.80
48" x 24"	S2@1.40	Ea	283.00	51.20	—	334.20
48" x 48"	S2@1.60	Ea	392.00	58.50	—	450.50
54" x 24"	S2@1.45	Ea	358.00	53.00	—	411.00
54" x 54"	S2@1.70	Ea	447.00	62.10	—	509.10
Single-blade damper for rectangular ductwork						
12" x 6"	S2@.600	Ea	41.90	21.90	—	63.80
12" x 8"	S2@.625	Ea	45.90	22.80	—	68.70
12" x 10"	S2@.650	Ea	51.60	23.80	—	75.40
12" x 12"	S2@.700	Ea	60.50	25.60	—	86.10
18" x 12"	S2@.800	Ea	67.30	29.20	—	96.50
18" x 18"	S2@.850	Ea	76.40	31.10	—	107.50
24" x 12"	S2@.850	Ea	81.00	31.10	—	112.10
24" x 18"	S2@.900	Ea	84.30	32.90	—	117.20
24" x 24"	S2@1.00	Ea	92.50	36.50	—	129.00
Single-blade butterfly damper for round ductwork						
6"	S2@.600	Ea	41.90	21.90	—	63.80
8"	S2@.550	Ea	39.50	20.10	—	59.60
10"	S2@.600	Ea	45.80	21.90	—	67.70
12"	S2@.650	Ea	51.80	23.80	—	75.60
14"	S2@.700	Ea	59.00	25.60	—	84.60
16"	S2@.765	Ea	65.40	28.00	—	93.40
18"	S2@.835	Ea	72.40	30.50	—	102.90
20"	S2@.900	Ea	81.50	32.90	—	114.40

Ductwork Specialties

Description	Craft@Hrs	Unit	Material $	Labor $	Equipment $	Total $

Curtain-type duct fire damper, 1½-hour rating, U.L. label

Description	Craft@Hrs	Unit	Material $	Labor $	Equipment $	Total $
12" x 6"	S2@1.00	Ea	76.10	36.50	—	112.60
12" x 8"	S2@1.00	Ea	82.40	36.50	—	118.90
12" x 10"	S2@1.05	Ea	90.50	38.40	—	128.90
12" x 12"	S2@1.10	Ea	97.40	40.20	—	137.60
18" x 12"	S2@1.20	Ea	136.00	43.80	—	179.80
18" x 18"	S2@1.30	Ea	148.00	47.50	—	195.50
24" x 12"	S2@1.40	Ea	163.00	51.20	—	214.20
24" x 18"	S2@1.50	Ea	187.00	54.80	—	241.80
24" x 24"	S2@1.60	Ea	221.00	58.50	—	279.50
30" x 12"	S2@1.50	Ea	190.00	54.80	—	244.80
30" x 18"	S2@1.80	Ea	210.00	65.80	—	275.80
30" x 24"	S2@2.00	Ea	281.00	73.10	—	354.10
30" x 30"	S2@2.30	Ea	302.00	84.00	—	386.00
36" x 18"	S2@2.00	Ea	241.00	73.10	—	314.10
36" x 36"	S2@2.40	Ea	359.00	87.70	—	446.70
42" x 24"	S2@2.40	Ea	312.00	87.70	—	399.70
42" x 42"	S2@2.90	Ea	451.00	106.00	—	557.00
48" x 24"	S2@2.90	Ea	339.00	106.00	—	445.00
48" x 48"	S2@3.50	Ea	560.00	128.00	—	688.00
54" x 24"	S2@3.00	Ea	366.00	110.00	—	476.00
54" x 54"	S2@3.80	Ea	652.00	139.00	—	791.00

Correction factors:
(1) For 22 gauge break-away connection, multiply material cost by 1.30
(2) For blades out of air stream cap, multiply material cost by 1.25

Ductwork Specialties

Description	Craft@Hrs	Unit	Material $	Labor $	Equipment $	Total $

Fusible plug dampers for air duct risers. Vertical riser and header air duct fusible (melting) and heat-sensing automatic counterweight closure actuators and closure dampers. For large A/C systems in multi-story institutions such as schools, hospitals and office buildings. 1½-hour rating. The liner and insulation must meet the requirements of U.L. 181 and NFPA No. 90 A-96, as modified, "Standard for the Installation of Air Conditioning and Ventilation Systems."

Description	Craft@Hrs	Unit	Material $	Labor $	Equipment $	Total $
12" x 6"	S2@1.00	Ea	101.00	36.50	—	137.50
12" x 8"	S2@1.00	Ea	103.00	36.50	—	139.50
12" x 10"	S2@1.05	Ea	106.00	38.40	—	144.40
12" x 12"	S2@1.10	Ea	108.00	40.20	—	148.20
18" x 12"	S2@1.20	Ea	137.00	43.80	—	180.80
18" x 18"	S2@1.30	Ea	176.00	47.50	—	223.50
24" x 12"	S2@1.40	Ea	189.00	51.20	—	240.20
24" x 18"	S2@1.50	Ea	210.00	54.80	—	264.80
24" x 24"	S2@1.60	Ea	241.00	58.50	—	299.50
30" x 12"	S2@1.50	Ea	227.00	54.80	—	281.80
30" x 18"	S2@1.80	Ea	287.00	65.80	—	352.80
30" x 24"	S2@2.00	Ea	333.00	73.10	—	406.10
30" x 30"	S2@2.30	Ea	400.00	84.00	—	484.00
36" x 18"	S2@2.00	Ea	321.00	73.10	—	394.10
36" x 36"	S2@2.40	Ea	428.00	87.70	—	515.70
42" x 24"	S2@2.40	Ea	400.00	87.70	—	487.70
42" x 42"	S2@2.90	Ea	486.00	106.00	—	592.00
48" x 24"	S2@2.90	Ea	435.00	106.00	—	541.00
48" x 48"	S2@3.50	Ea	971.00	128.00	—	1,099.00
54" x 24"	S2@3.00	Ea	1,240.00	110.00	—	1,350.00
54" x 54"	S2@3.80	Ea	2,030.00	139.00	—	2,169.00
Cut, frame and gasket access hole in sidewall	S2@1.00	Ea	57.60	36.50	—	94.10
Mount duct hangers	S2@.250	Ea	4.00	9.14	—	13.14
Cut and mount sheet metal duct	S2@.350	LF	6.16	12.80	—	18.96
Apply and coat duct insulation	S2@.150	LF	2.68	5.48	—	8.16

Ductwork Specialties

Description	Craft@Hrs	Unit	Material $	Labor $	Equipment $	Total $

Single-skin 2" fixed-position turning vanes for duct

Description	Craft@Hrs	Unit	Material $	Labor $	Equipment $	Total $
12" x 6"	S2@.280	Ea	8.88	10.20	—	19.08
12" x 8"	S2@.285	Ea	11.00	10.40	—	21.40
12" x 10"	S2@.290	Ea	14.20	10.60	—	24.80
12" x 12"	S2@.300	Ea	17.50	11.00	—	28.50
18" x 12"	S2@.375	Ea	26.60	13.70	—	40.30
18" x 18"	S2@.450	Ea	32.30	16.40	—	48.70
24" x 12"	S2@.450	Ea	35.40	16.40	—	51.80
24" x 18"	S2@.500	Ea	38.50	18.30	—	56.80
24" x 24"	S2@.525	Ea	43.90	19.20	—	63.10
30" x 12"	S2@.550	Ea	44.30	20.10	—	64.40
30" x 18"	S2@.750	Ea	63.90	27.40	—	91.30
30" x 24"	S2@.800	Ea	84.90	29.20	—	114.10
30" x 30"	S2@.900	Ea	107.00	32.90	—	139.90
36" x 18"	S2@.750	Ea	75.70	27.40	—	103.10
36" x 36"	S2@.800	Ea	154.00	29.20	—	183.20
42" x 24"	S2@.900	Ea	119.00	32.90	—	151.90
42" x 42"	S2@1.00	Ea	210.00	36.50	—	246.50
48" x 24"	S2@1.10	Ea	136.00	40.20	—	176.20
48" x 48"	S2@1.50	Ea	272.00	54.80	—	326.80
54" x 24"	S2@1.15	Ea	152.00	42.00	—	194.00
54" x 54"	S2@1.25	Ea	347.00	45.70	—	392.70
60" x 24"	S2@1.30	Ea	171.00	47.50	—	218.50
60" x 36"	S2@1.40	Ea	258.00	51.20	—	309.20
60" x 60"	S2@1.50	Ea	426.00	54.80	—	480.80
72" x 36"	S2@1.65	Ea	312.00	60.30	—	372.30
72" x 54"	S2@2.40	Ea	459.00	87.70	—	546.70

Note: Labor includes cost to cut, assemble and install vanes and rails. For 4-inch air-foil vanes, multiply costs by 1.80 and labor costs by .90

Ductwork Specialties

Description	Craft@Hrs	Unit	Material $	Labor $	Equipment $	Total $
Spin-in galvanized steel collars without volume damper						
4"	S2@.200	Ea	6.62	7.31	—	13.93
5"	S2@.200	Ea	7.43	7.31	—	14.74
6"	S2@.200	Ea	8.19	7.31	—	15.50
7"	S2@.200	Ea	9.02	7.31	—	16.33
8"	S2@.250	Ea	10.40	9.14	—	19.54
9"	S2@.250	Ea	11.50	9.14	—	20.64
10"	S2@.300	Ea	12.30	11.00	—	23.30
12"	S2@.350	Ea	14.80	12.80	—	27.60
14"	S2@.400	Ea	17.60	14.60	—	32.20
16"	S2@.450	Ea	21.00	16.40	—	37.40
18"	S2@.500	Ea	26.30	18.30	—	44.60
Spin-in galvanized steel collar with volume damper						
4"	S2@.250	Ea	22.10	9.14	—	31.24
5"	S2@.250	Ea	22.50	9.14	—	31.64
6"	S2@.250	Ea	22.70	9.14	—	31.84
7"	S2@.250	Ea	23.30	9.14	—	32.44
8"	S2@.300	Ea	23.80	11.00	—	34.80
9"	S2@.300	Ea	24.10	11.00	—	35.10
10"	S2@.350	Ea	24.70	12.80	—	37.50
12"	S2@.400	Ea	26.50	14.60	—	41.10
14"	S2@.450	Ea	28.40	16.40	—	44.80
16"	S2@.500	Ea	30.10	18.30	—	48.40
18"	S2@.550	Ea	32.30	20.10	—	52.40

Description	Craft@Hrs	Unit	Material $	Labor $	Equipment $	Total $
Duct-to-equipment flexible connection						
18" x 12"	S2@.650	Ea	22.40	23.80	—	46.20
18" x 18"	S2@.850	Ea	27.10	31.10	—	58.20
24" x 12"	S2@.850	Ea	27.10	31.10	—	58.20
24" x 18"	S2@1.00	Ea	31.60	36.50	—	68.10
24" x 24"	S2@1.30	Ea	36.20	47.50	—	83.70
30" x 12"	S2@1.00	Ea	31.60	36.50	—	68.10
30" x 18"	S2@1.30	Ea	36.20	47.50	—	83.70
30" x 30"	S2@1.60	Ea	45.30	58.50	—	103.80
36" x 12"	S2@1.30	Ea	36.20	47.50	—	83.70
36" x 24"	S2@1.60	Ea	45.30	58.50	—	103.80
36" x 36"	S2@1.80	Ea	53.70	65.80	—	119.50
48" x 24"	S2@1.80	Ea	53.70	65.80	—	119.50
48" x 48"	S2@2.00	Ea	71.10	73.10	—	144.20

Galvanized Steel Ductwork

The costs for fabricating and installing galvanized steel ductwork are usually based on the total weights of the duct and fittings. Because of cost differences, fitting weights should be kept separate from straight duct weights.

Round spiral duct: Normally-used duct gauge/size relationships for low-pressure systems, up to 2 inches static pressure, are as follows:

Duct Diameter (inches)	U.S. Standard Gauge
Up to 12	26
13 through 24	24
26 through 36	22

Rectangular duct: Normally-used duct gauge/size relationships for low-pressure systems, up to 2 inches static pressure, are as follows:

Duct Size (inches)	U.S. Standard Gauge
Up to 12	26
13 through 30	24
31 through 54	22
55 through 72	20

The fitting weights in the following tables are in accordance with the above gauges.

Other material costs: All weights in the tables are net weights; they do not include bracing, cleats, scrap, hangers, end closures, sealants and miscellaneous hardware. Add 15% to calculated duct/fitting weights to cover these items if ductwork is purchased. Add 20% if ducting is manufactured in contractor's shop. The additional 5% covers the costs for scrap which is already included in an outside vendor's price.

Ductwork costs: Typical costs for purchased ductwork are as follows:

Straight duct, less than 1,000 pounds per order	$3.17/lb.
Straight duct, over 1,000 pounds per order	$2.88/lb.
Fittings, less than 1,000 pounds per order	$6.46/lb.
Fittings, over 1,000 pounds per order	$5.79/lb.
Lined ducts and fittings, add	$1.72/lb.*
Delivery costs	$0.16/lb.

*Duct sizes shown on drawings are always inside, or net, dimensions and must be increased in size when estimating lined ductwork.

If ductwork is manufactured in the contractor's fabrication shop, the above costs can be reduced by about 20%, depending on the shop's rate of productivity.

Galvanized Steel Ductwork

Installation costs: An average crew can install approximately 25 pounds of duct and fittings per manhour under normal conditions. See "Applying Correction Factors" on page 6 for situations that do not conform to the definition of a standard labor unit.

EXAMPLE:

What is the cost to furnish and install 2,585 pounds of unlined duct and 560 pounds of unlined fittings? The ductwork will be purchased from an outside vendor. A 15% allowance for miscellaneous material is included in the weights.

Material:

2,585 pounds of straight duct x $2.88/lb.	= $7,444.80
560 pounds of fittings x $6.46/lb.	= 3,617.60
Delivery cost: (2,585 + 560) x $.16/lb.	= 503.20
Total material cost	= $11,565.60*

Labor:

$$\frac{2{,}585 + 560}{25} = 125.8 \; MH; \; 125.8 \; MH \times \$36.54 = \$4{,}596.73$$

Total installed cost: $11,565.60 + $4,596.73 = $16,162.33*

Sales tax not included.

Installed Ductwork Per Pound

Description	Craft@Hrs	Unit	Material $	Labor $	Equipment $	Total $

Installed ductwork per lb. (under 1,000 lbs.). Duct purchased from an independent fabrication shop, installed by this contractor's shop. Material price includes scrap, cleats, hangers, sealant, miscellaneous hardware, delivery and fabrication shop markups.

Description	Craft@Hrs	Unit	Material $	Labor $	Equipment $	Total $
Straight duct	S2@.042	Lb	3.17	1.53	—	4.70
Duct fittings	S2@.042	Lb	6.45	1.53	—	7.98
Duct & fittings	S2@.042	Lb	4.27	1.53	—	5.80

Installed ductwork per lb. (over 1,000 lbs.). Duct purchased from an independent fabrication shop, installed by this contractor's shop. Material price includes scrap, cleats, hangers, sealant, miscellaneous hardware, delivery and subcontractor markups.

Description	Craft@Hrs	Unit	Material $	Labor $	Equipment $	Total $
Straight duct	S2@.039	Lb	2.88	1.43	—	4.31
Duct fittings	S2@.039	Lb	5.78	1.43	—	7.21
Duct & fittings	S2@.039	Lb	3.86	1.43	—	5.29

Installed lined ductwork per lb. (under 1,000 lbs.). Duct purchased from an independent fabrication shop, installed by this contractor's shop. Material price includes scrap, cleats, hangers, sealant, miscellaneous hardware, delivery and subcontractor markups.

Description	Craft@Hrs	Unit	Material $	Labor $	Equipment $	Total $
Straight duct	S2@.042	Lb	4.91	1.53	—	6.44
Duct fittings	S2@.042	Lb	8.13	1.53	—	9.66
Duct & fittings	S2@.042	Lb	5.98	1.53	—	7.51

Installed lined ductwork per lb. (over 1,000 lbs.). Duct purchased from an independent fabrication shop, installed by this contractor's shop. Material price includes scrap, cleats, hangers, sealant, miscellaneous hardware, delivery and subcontractor markups.

Description	Craft@Hrs	Unit	Material $	Labor $	Equipment $	Total $
Straight duct	S2@.039	Lb	4.61	1.43	—	6.04
Duct fittings	S2@.039	Lb	7.53	1.43	—	8.96
Duct & fittings	S2@.039	Lb	5.55	1.43	—	6.98

Installed ductwork per lb. (under 1,000 lbs.). Duct fabricated and installed by this contractor's shop. Material price includes scrap, cleats, hangers, sealant, miscellaneous hardware and delivery.

Description	Craft@Hrs	Unit	Material $	Labor $	Equipment $	Total $
Straight duct	S2@.051	Lb	1.96	1.86	—	3.82
Duct fittings	S2@.074	Lb	2.24	2.70	—	4.94
Duct & fittings	S2@.059	Lb	2.06	2.16	—	4.22

Installed ductwork per lb. (over 1,000 lbs.). Duct fabricated and installed by this contractor's shop. Material price includes scrap, cleats, hangers, sealant, miscellaneous hardware and delivery.

Description	Craft@Hrs	Unit	Material $	Labor $	Equipment $	Total $
Straight duct	S2@.046	Lb	1.81	1.68	—	3.49
Duct fittings	S2@.067	Lb	2.07	2.45	—	4.52
Duct & fittings	S2@.053	Lb	1.88	1.94	—	3.82

Installed Ductwork Per Pound

Description	Craft@Hrs	Unit	Material $	Labor $	Equipment $	Total $

Installed lined ductwork per lb. (under 1,000 lbs.). Duct fabricated and installed by this contractor's shop. Material price includes scrap, cleats, hangers, sealant, miscellaneous hardware and delivery.

Description	Craft@Hrs	Unit	Material $	Labor $	Equipment $	Total $
Straight duct	S2@.062	Lb	2.90	2.27	—	5.17
Duct fittings	S2@.085	Lb	3.20	3.11	—	6.31
Duct & fittings	S2@.070	Lb	3.01	2.56	—	5.57

Installed lined ductwork per lb. (over 1,000 lbs.). Duct fabricated and installed by this contractor's shop. Material price includes scrap, cleats, hangers, sealant, miscellaneous hardware and delivery. Use for preliminary estimates.

Description	Craft@Hrs	Unit	Material $	Labor $	Equipment $	Total $
Straight duct	S2@.057	Lb	2.75	2.08	—	4.83
Duct fittings	S2@.078	Lb	3.07	2.85	—	5.92
Duct & fittings	S2@.064	Lb	5.32	2.34	—	7.66

Galvanized Steel Spiral Ductwork

Weights of Galvanized Steel Spiral Duct (pounds per LF)					
Diameter (inches)	U.S. Standard Gauge				
	26	24	22	20	18
3	.76	1.01	1.23	1.46	—
4	1.02	1.35	1.64	1.94	—
5	1.28	1.69	2.06	2.42	—
6	1.54	2.03	2.47	2.91	3.88
7	1.79	2.37	2.88	3.40	4.52
8	2.05	2.71	3.29	3.86	5.17
9	2.31	3.05	3.71	4.37	5.82
10	2.57	3.39	4.12	4.86	6.47
12	3.08	4.07	4.95	5.83	7.75
14	3.59	4.74	5.77	6.81	9.05
16	4.11	5.42	6.60	7.78	10.34
18	4.63	6.10	7.43	8.76	11.64
20	5.15	6.78	8.25	9.73	12.93
22	5.65	7.46	9.08	10.71	14.93
24	6.16	8.14	9.91	11.68	15.52
26	6.67	8.82	10.73	12.66	16.82
28	7.18	9.50	11.56	13.63	18.11
30	7.71	10.18	12.38	14.60	19.41
32	8.22	10.84	13.21	15.58	20.71
34	—	11.52	14.05	16.55	22.00
36	—	12.20	14.90	17.53	23.29

Galvanized Steel Round Spiral Fittings

Weights of Galvanized Steel Round Spiral Fittings (pounds per piece)				
Diameter (inches)	90° elbow	45° elbow	Coupling	Reducer
3	1.3	1.0	0.5	1.0
4	2.2	1.3	0.6	1.2
5	3.3	1.9	0.7	1.4
6	4.3	2.5	0.9	1.8
7	5.8	3.3	1.0	2.0
8	7.3	4.3	1.2	2.4
9	8.8	5.3	2.6	3.3
10	11.8	7.5	2.9	4.4
12	16.3	10.0	3.5	5.3
14	22.0	13.0	4.1	6.2
16	28.3	15.8	4.6	6.9
18	34.5	19.0	5.2	7.8
20	41.5	23.5	5.8	8.7
22	48.3	27.5	6.4	9.6
24	57.5	32.0	6.9	10.3
26	68.8	37.5	7.5	11.2
28	76.8	42.8	8.1	12.1
30	87.0	48.0	8.7	13.0
32	99.5	55.0	9.2	13.8
34	112.0	61.0	9.8	14.7
36	161.0	89.5	10.4	15.6

Largest run diam. (inches)	Weights of Galvanized Steel Round Spiral Fittings (pounds per piece) Tee with reducing run and branch Branch diameter (inches)												
	3	4	5	6	7	8	9	10	12	14	16	18	20	22
4	2.3													
5	2.8	3.0												
6	3.2	3.5	3.7											
7	3.7	3.9	4.1	4.5										
8	4.1	4.4	4.7	5.0	5.3									
9	4.5	4.8	5.1	5.5	5.8	6.2								
10	6.4	6.7	7.1	7.5	7.8	8.2	8.6							
12	7.7	8.1	8.5	8.9	9.4	9.9	10.3	10.7						
14	8.9	9.4	10.0	10.5	11.0	11.5	12.1	12.6	13.6					
16	10.2	10.8	11.4	12.0	12.5	13.1	13.7	14.3	15.5	16.7				
18	11.5	12.1	12.7	13.5	14.1	14.8	15.5	16.1	17.5	18.8	19.2			
20	12.7	13.4	14.2	14.9	17.7	16.4	17.2	17.9	19.4	20.9	22.4	23.9		
22	14.0	14.8	15.6	16.4	17.2	18.1	18.9	19.7	21.3	23.0	24.6	26.2	27.9	
24	15.3	16.1	17.0	17.8	18.7	19.6	20.5	21.4	23.2	25.0	26.8	28.5	30.3	32.1
26	16.5	17.5	18.5	19.5	20.4	21.3	22.3	23.3	25.2	26.2	29.1	31.1	33.0	34.9
28	17.7	18.7	19.8	20.8	21.9	22.9	23.9	25.0	27.0	29.2	31.2	33.2	35.4	37.4
30	18.9	20.0	21.1	22.2	23.3	24.5	25.5	26.7	28.9	31.1	33.3	35.5	37.7	40.0
32	20.4	21.6	22.6	24.0	25.2	26.4	27.6	28.8	31.2	33.6	36.0	38.4	40.8	42.2
34	21.6	22.9	24.2	25.4	26.7	28.0	29.3	30.5	33.1	35.6	38.1	40.7	42.2	45.6
36	29.6	31.4	33.1	34.8	36.6	38.6	40.1	41.8	45.3	48.8	52.2	55.7	59.2	62.7

Galvanized Steel Round Spiral Fittings

Run diam. (inches)	Weights of Galvanized Steel Round Spiral Fittings (pounds per piece) Cross with reducing run and branch Branch diameter (inches)												
	3	4	5	6	7	8	9	10	12	14	16	18	20	22
4	2.6													
5	3.2	3.6												
6	3.8	4.2	4.6											
7	4.4	4.9	5.4	5.8										
8	4.9	5.4	6.0	6.5	7.0									
9	5.4	6.0	6.6	7.2	7.7	8.3								
10	7.6	8.2	8.9	9.5	10.2	10.8	11.4							
12	8.8	9.6	10.3	11.0	11.7	12.5	13.2	13.8						
14	10.1	10.9	11.7	12.6	13.4	14.2	15.1	15.9	17.5					
16	11.3	12.2	13.1	14.0	15.0	15.9	16.8	17.7	19.5	21.4				
18	12.5	13.5	14.5	15.5	16.5	17.5	18.5	19.5	21.5	23.5	25.5			
20	13.8	15.0	16.1	17.2	18.2	19.4	20.6	21.7	23.9	26.2	28.4	30.6		
22	15.2	16.4	17.7	18.9	20.1	21.4	22.6	23.8	26.2	28.7	31.2	33.6	36.1	
24	16.6	17.9	19.3	20.6	22.0	23.3	24.6	26.0	28.7	31.5	34.1	36.7	39.4	42.1
26	18.0	19.4	20.9	22.3	23.8	25.2	26.7	28.2	31.1	33.9	36.9	39.8	42.7	45.6
28	19.3	20.8	22.4	24.0	25.5	27.1	28.6	30.2	32.3	36.4	39.6	42.7	45.8	48.9
30	20.6	22.3	23.9	25.6	27.3	28.9	30.6	32.3	35.6	38.9	42.3	45.6	49.9	52.2
32	22.2	24.0	25.8	27.6	29.4	31.2	33.0	34.8	38.4	42.0	45.6	49.1	52.8	56.5
34	23.5	25.4	27.3	29.2	31.1	33.1	34.9	36.9	40.7	44.5	48.3	52.1	55.9	59.8
36	32.2	34.8	37.4	40.0	42.8	45.3	48.0	50.5	55.7	61.0	66.2	71.5	76.7	82.0

Size (inches)	Weights of 26 Gauge Galvanized Steel Rectangular Duct (pounds per LF)								
	4	5	6	7	8	9	10	11	12
4	1.21								
5	1.36	1.51							
6	1.51	1.66	1.81						
7	1.66	1.81	1.96	2.11					
8	1.81	1.96	2.11	2.27	2.42				
9	1.96	2.11	2.27	2.42	2.57	2.72			
10	2.11	2.27	2.42	2.57	2.72	2.87	3.02		
11	2.27	2.42	2.57	2.72	2.87	3.02	3.17	3.32	
12	2.42	2.57	2.72	2.87	3.02	3.17	3.32	3.47	3.62

Galvanized Steel Rectangular Ductwork

Weights of 24 Gauge Galvanized Steel Rectangular Duct (pounds per LF)

Size (inches)	14	16	18	20	22	24	26	28	30
6	3.85	4.24	4.62	5.01	5.40	5.78	6.17	6.55	6.94
8	4.24	4.62	5.01	5.40	5.78	6.17	6.55	6.94	7.32
10	4.62	5.01	5.40	5.78	6.17	6.55	6.94	7.32	7.71
12	5.01	5.40	5.78	6.17	6.55	6.94	7.32	7.71	8.09
14	5.40	5.78	6.17	6.55	6.94	7.32	7.71	8.09	8.48
16	5.78	6.17	6.55	6.94	7.32	7.71	8.09	8.48	8.86
18	6.17	6.55	6.94	7.32	7.71	8.09	8.48	8.86	9.25
20	6.55	6.94	7.32	7.71	8.09	8.48	8.86	9.25	9.64
22	6.94	7.32	7.71	8.09	8.48	8.86	9.25	9.64	10.00
24	7.32	7.71	8.09	8.48	8.86	9.25	9.64	10.00	10.40
26	7.71	8.09	8.48	8.86	9.25	9.64	10.00	10.40	10.80
28	8.09	8.48	8.86	9.25	9.64	10.00	10.40	10.80	11.20
30	8.48	8.86	9.25	9.64	10.00	10.40	10.80	11.20	11.60

Weights of 22 Gauge Galvanized Steel Rectangular Duct (pounds per LF)

Size (inches)	32	34	36	38	40	42	44	46	48	50	52	54
8	9.37	9.84	10.3	10.8	11.3	11.7	12.2	12.7	13.1	13.6	14.1	14.5
10	9.84	10.3	10.8	11.3	11.7	12.2	12.7	13.1	13.6	14.1	14.5	15.0
12	10.3	10.8	11.3	11.7	12.2	12.7	13.1	13.6	14.1	14.5	15.0	15.5
14	10.8	11.3	11.7	12.2	12.7	13.1	13.6	14.1	14.5	15.0	15.5	15.9
16	11.3	11.7	12.2	12.7	13.1	13.6	14.1	14.5	15.0	15.5	15.9	16.4
18	11.7	12.2	12.7	13.1	13.6	14.1	14.5	15.0	15.5	15.9	16.4	16.9
20	12.2	12.7	13.1	13.6	14.1	14.5	15.0	15.5	15.9	16.4	16.9	17.3
22	12.7	13.1	13.6	14.1	14.5	15.0	15.5	15.9	16.4	16.9	17.3	17.8
24	13.1	13.6	14.1	14.5	15.0	15.5	15.9	16.4	16.9	17.3	17.8	18.3
26	13.6	14.1	14.5	15.0	15.5	15.9	16.4	16.9	17.3	17.8	18.3	18.7
28	14.1	14.5	15.0	15.5	15.9	16.4	16.9	17.3	17.8	18.3	18.7	19.2
30	14.5	15.0	15.5	15.9	16.4	16.9	17.3	17.8	18.3	18.7	19.2	19.7
32	15.0	15.5	15.9	16.4	16.9	17.3	17.8	18.3	18.7	19.2	19.7	20.2
34	15.5	15.9	16.4	16.9	17.3	17.8	18.3	18.7	19.2	19.7	20.2	20.6
36	15.9	16.4	16.9	17.3	17.8	18.3	18.7	19.2	19.7	20.2	20.6	21.1
38	16.4	16.9	17.3	17.8	18.3	18.7	19.2	19.7	20.2	20.6	21.1	21.6
40	16.9	17.3	17.8	18.3	18.7	19.2	19.7	20.2	20.6	21.1	21.6	22.0
42	17.3	17.8	18.3	18.7	19.2	19.7	20.2	20.6	21.1	21.6	22.0	22.5
44	17.8	18.3	18.7	19.2	19.7	20.2	20.6	21.1	21.6	22.0	22.5	23.0
46	18.3	18.7	19.2	19.7	20.2	20.6	21.1	21.6	22.0	22.5	23.0	23.4
48	18.7	19.2	19.7	20.2	20.6	21.1	21.6	22.0	22.5	23.0	23.4	23.9
50	19.2	19.7	20.2	20.6	21.1	21.6	22.0	22.5	23.0	23.4	23.9	24.4
52	19.7	20.2	20.6	21.1	21.6	22.0	22.5	23.0	23.4	23.9	24.4	24.8
54	20.2	20.6	21.1	21.6	22.0	22.5	23.0	23.4	23.9	24.4	24.8	25.3

Galvanized Steel Rectangular Ductwork

Weights of 20 Gauge Galvanized Steel Rectangular Duct (pounds per LF)								
Size (inches)	56	58	60	62	64	66	68	70	72
18	20.4	21.0	21.5	22.1	22.6	23.2	23.7	24.3	24.8
20	21.0	21.5	22.1	22.6	23.2	23.7	24.3	24.8	25.4
22	21.5	22.1	22.6	23.2	23.7	24.3	24.8	25.4	25.9
24	22.1	22.6	23.2	23.7	24.3	24.8	25.4	25.9	26.5
26	22.6	23.2	23.7	24.3	24.8	25.4	25.9	26.5	27.0
28	23.2	23.7	24.3	24.8	25.4	25.9	26.5	27.0	27.6
30	23.7	24.3	24.8	25.4	25.9	26.5	27.0	27.6	28.2
32	24.3	24.8	25.4	25.9	26.5	27.0	27.6	28.2	28.7
34	24.8	25.4	25.9	26.5	27.0	27.6	28.2	28.7	29.3
36	25.4	25.9	26.5	27.0	27.6	28.2	28.7	29.3	29.8
38	25.9	26.5	27.0	27.6	28.2	28.7	29.3	29.8	30.4
40	26.5	27.0	27.6	28.2	28.7	29.3	29.8	30.4	30.9
42	27.0	27.6	28.2	28.7	29.3	29.8	30.4	30.9	31.5
44	27.6	28.2	28.7	29.3	29.8	30.4	30.9	31.5	32.0
46	28.2	28.7	29.3	29.8	30.4	30.9	31.5	32.0	32.6
48	28.7	29.3	29.8	30.4	30.9	31.5	32.0	32.6	33.1
50	29.3	29.8	30.4	30.9	31.5	32.0	32.6	33.1	33.7
52	29.8	30.4	30.9	31.5	32.0	32.6	33.1	33.7	34.2
54	30.4	30.9	31.5	32.0	32.6	33.1	33.7	34.2	34.8
56	30.9	31.5	32.0	32.6	33.1	33.7	34.2	34.8	35.3
58	31.5	32.0	32.6	33.1	33.7	34.2	34.8	35.3	35.9
60	32.0	32.6	33.1	33.7	34.2	34.8	35.3	35.9	36.4
62	32.6	33.1	33.7	34.2	34.8	35.3	35.9	36.4	37.0
64	33.1	33.7	34.2	34.8	35.3	35.9	36.4	37.0	37.5
66	33.7	34.2	34.8	35.3	35.9	36.4	37.0	37.5	38.1
68	34.2	34.8	35.3	35.9	36.4	37.0	37.5	38.1	38.6
70	34.8	35.3	35.9	36.4	37.0	37.5	38.1	38.6	39.2
72	35.3	35.9	36.4	37.0	37.5	38.1	38.6	39.2	39.7

Galvanized Steel Rectangular 90-Degree Elbows

Weights of Galvanized Steel Rectangular 90 Degree Elbow
(pounds per piece)

Size (inches)	4	6	8	10	12	14	16	18	20
4	1.05	1.45	1.95	2.60	3.30	5.30	6.50	7.80	9.25
6	1.45	1.95	2.60	3.30	3.70	5.80	7.75	8.75	9.75
8	1.95	2.60	3.30	3.70	4.25	6.50	8.50	9.75	10.4
10	2.60	3.30	3.70	4.25	4.85	7.30	9.50	10.7	11.2
12	3.30	3.70	4.25	4.85	5.60	8.20	10.5	11.7	12.2
14	5.30	5.80	6.50	7.30	8.20	9.25	11.5	12.6	13.2
16	6.50	7.50	8.50	9.50	10.5	11.5	12.5	13.4	14.4
18	7.80	8.75	9.75	10.7	11.7	12.6	13.4	14.5	15.7
20	9.25	9.75	10.4	11.2	12.2	13.8	14.4	15.7	17.2
22	10.8	11.8	12.8	13.8	14.8	25.8	16.8	17.8	18.8
24	12.6	13.6	14.6	15.6	16.6	17.6	18.6	19.6	20.5
26	14.4	15.4	16.4	17.4	18.4	19.4	20.4	21.4	22.3
28	16.4	17.4	18.4	19.4	20.4	21.4	22.4	23.4	24.3
30	18.5	19.5	20.5	21.5	22.5	23.5	24.5	25.5	26.4
32	24.3	25.7	27.0	28.4	29.7	31.1	32.4	33.7	35.0
34	28.2	29.4	30.6	31.8	33.0	34.3	35.5	36.7	37.9
36	31.3	32.5	33.7	34.9	36.1	37.3	38.5	39.7	40.9
38	34.5	35.7	36.9	38.1	39.3	40.5	41.7	42.9	44.1
40	37.9	39.1	40.3	41.5	42.7	43.9	45.1	46.3	47.5
42	41.4	42.6	43.8	45.0	46.2	47.4	48.6	49.8	51.0
44	45.1	46.4	47.6	48.8	50.0	51.2	52.4	53.6	54.8
46	49.0	50.2	51.4	52.6	53.8	55.0	56.2	57.4	58.6
48	53.0	54.2	55.4	56.6	57.8	59.0	60.2	61.4	62.6
50	57.2	58.5	59.7	60.9	62.1	63.3	64.5	65.7	66.9

Note: Elbow weights do not include turning vanes.

Galvanized Steel Rectangular 90-Degree Elbows

Weights of Galvanized Steel Rectangular 90 Degree Elbows (pounds per piece)									
Size (inches)	22	24	26	28	30	32	34	36	38
22	19.6	21.9	24.1	26.3	28.5	37.0	40.8	44.5	48.3
24	21.9	24.2	26.4	28.6	30.9	39.5	43.6	47.2	50.8
26	24.1	26.4	28.7	30.9	33.2	42.0	46.4	49.8	53.3
28	26.3	28.6	30.9	33.3	35.6	44.5	49.2	52.5	55.8
30	28.5	30.9	33.2	35.6	37.9	47.0	52.0	55.1	58.3
32	37.0	39.5	42.0	44.5	47.0	49.5	52.0	57.8	60.8
34	40.8	43.6	46.4	49.2	52.0	54.8	57.6	60.5	63.3
36	44.5	47.2	49.8	52.5	55.1	57.8	60.5	63.1	65.8
38	48.3	50.8	53.3	55.8	58.3	60.8	63.3	65.8	68.3
40	52.1	54.6	57.1	59.6	62.1	64.6	67.1	69.6	72.1
42	55.9	58.4	60.9	63.4	65.9	68.4	70.9	73.4	75.9
44	59.6	62.1	64.6	67.1	69.6	72.1	74.6	77.1	79.6
46	63.4	65.9	68.4	70.9	73.4	75.9	78.4	80.9	83.4
48	67.2	69.7	72.2	74.7	77.2	79.7	82.2	84.7	87.2
50	70.9	73.4	75.9	78.4	80.9	83.4	85.9	88.4	90.9
52	74.8	77.3	79.8	82.3	84.8	87.3	89.8	92.3	94.7
54	78.5	81.0	83.5	86.0	88.5	91.0	93.5	96.0	98.5
56	97.2	100	103	106	109	112	115	118	121
58	103	106	109	112	115	118	121	124	127
60	110	113	116	119	122	125	128	131	134
62	116	119	122	125	128	131	134	137	140
64	122	125	129	132	135	138	141	144	147
66	128	131	135	138	141	144	148	151	154
68	135	138	142	145	148	151	154	157	160
70	141	144	148	151	154	157	160	163	166
72	147	150	154	157	160	163	166	169	172

Note: Elbow weights do not include turning vanes.

Weights of Galvanized Steel Rectangular 90 Degree Elbows (pounds per piece)									
Size (inches)	40	42	44	46	48	50	52	54	56
40	74.6	78.4	82.1	85.9	89.7	93.4	97.2	102	125
42	78.4	82.3	86.2	90.1	94.0	97.9	102	106	130
44	82.1	86.2	90.3	94.4	98.5	102	106	110	135
46	85.9	90.1	94.4	98.7	102	106	110	113	140
48	89.7	94.0	98.5	102	106	110	113	117	144
50	93.4	97.9	102	106	110	113	117	122	149
52	97.2	102	106	110	113	117	122	126	154
54	102	106	110	113	117	122	126	131	159
56	126	130	134	139	145	150	155	160	165
58	132	137	142	147	151	156	161	166	170
60	137	142	147	152	157	162	167	172	176
62	143	148	153	158	163	168	173	178	182
64	150	155	160	165	169	174	179	184	188
66	156	161	166	171	176	181	186	191	195
68	163	168	173	178	182	187	192	197	201
70	170	175	180	185	189	194	199	204	208
72	176	181	185	190	194	199	204	210	215

Note: Elbow weights do not include turning vanes.

Galvanized Steel Rectangular 90-Degree Elbows

Weights of Galvanized Steel Rectangular 90 Degree Elbows (pounds per piece)								
Size (inches)	58	60	62	64	66	68	70	72
58	176	182	187	194	200	207	214	221
60	182	187	194	200	207	214	221	227
62	187	194	200	207	214	221	227	233
64	194	200	207	214	221	227	233	239
66	200	207	214	221	227	233	239	246
68	207	214	221	227	233	239	246	252
70	214	221	227	233	239	246	252	259
72	221	227	233	239	246	252	259	266

Note: Elbow weights do not include turning vanes.

Weights of Galvanized Steel Rectangular Drops and Tap-In Tees (pounds per piece)								
Size (inches)	6	8	10	12	16	20	24	28	32
6	1.81	2.11	2.42	2.72	4.24	5.01	5.78	6.55	8.90
8	2.11	2.42	2.72	3.02	4.62	5.40	6.17	6.94	9.37
10	2.42	2.72	3.02	3.32	5.01	5.78	6.55	7.32	9.84
12	2.72	3.02	3.32	3.62	5.40	6.17	6.94	7.71	10.3
16	4.24	4.62	5.01	5.40	6.17	6.94	7.71	8.48	11.3
20	5.01	5.40	5.78	6.17	6.94	7.71	8.48	9.25	12.2
24	5.78	6.17	6.55	6.94	7.71	8.48	9.25	10.0	13.1
28	6.55	6.94	7.32	7.71	8.48	9.25	10.0	10.8	14.1
32	8.90	9.37	9.84	10.3	11.3	12.2	13.1	14.1	15.0
36	9.84	10.3	10.8	11.3	12.2	13.1	14.1	15.0	15.9
40	10.8	11.3	11.7	12.2	13.1	14.1	15.0	15.9	16.9
44	11.7	12.2	12.7	13.1	14.1	15.0	15.9	16.9	17.8
48	12.7	13.1	13.6	14.1	15.0	15.9	16.9	17.8	18.7
52	13.6	14.1	14.5	15.0	15.9	16.9	17.8	18.7	19.7
56	16.0	17.6	18.2	18.8	19.6	21.0	22.1	23.2	24.3
60	18.7	19.0	19.3	19.9	21.0	22.1	23.2	24.3	25.4

Note: Weights of drops and tap-in tees do not include splitter dampers.

Galvanized Steel Spiral Duct

Costs for Steel Spiral Duct. Costs per linear foot of duct, based on quantities less than 1,000 pounds. Add 50% to the material cost for lined duct. These costs include delivery, typical bracing, cleats, scrap, hangers, end closures, sealants and miscellaneous hardware.

Description	Craft@Hrs	Unit	Material $	Labor $	Equipment $	Total $
Galvanized steel spiral duct 26 gauge						
3"	S2@.030	LF	2.67	1.10	—	3.77
4"	S2@.041	LF	3.61	1.50	—	5.11
5"	S2@.051	LF	4.54	1.86	—	6.40
6"	S2@.061	LF	5.41	2.23	—	7.64
7"	S2@.072	LF	6.28	2.63	—	8.91
8"	S2@.082	LF	7.20	3.00	—	10.20
9"	S2@.092	LF	8.13	3.36	—	11.49
10"	S2@.103	LF	9.01	3.76	—	12.77
12"	S2@.123	LF	10.70	4.49	—	15.19
14" - 16"	S2@.174	LF	13.50	6.36	—	19.86
18" - 20"	S2@.195	LF	17.10	7.13	—	24.23
22" - 24"	S2@.236	LF	20.80	8.62	—	29.42
26" - 28"	S2@.277	LF	24.50	10.10	—	34.60
30" - 32"	S2@.318	LF	28.30	11.60	—	39.90
Galvanized steel spiral duct 24 gauge						
3"	S2@.041	LF	3.60	1.50	—	5.10
4"	S2@.054	LF	4.71	1.97	—	6.68
5"	S2@.067	LF	5.92	2.45	—	8.37
6"	S2@.081	LF	7.12	2.96	—	10.08
7"	S2@.095	LF	8.28	3.47	—	11.75
8"	S2@.108	LF	9.53	3.95	—	13.48
9"	S2@.122	LF	10.70	4.46	—	15.16
10"	S2@.135	LF	11.80	4.93	—	16.73
12"	S2@.163	LF	14.20	5.96	—	20.16
14" - 16"	S2@.203	LF	17.70	7.42	—	25.12
18" - 20"	S2@.257	LF	22.50	9.39	—	31.89
22" - 24"	S2@.312	LF	27.50	11.40	—	38.90
26" - 28"	S2@.366	LF	32.30	13.40	—	45.70
30" - 32"	S2@.419	LF	36.90	15.30	—	52.20
34" - 36"	S2@.474	LF	42.00	17.30	—	59.30

Galvanized Steel Spiral Duct

Description	Craft@Hrs	Unit	Material $	Labor $	Equipment $	Total $
Galvanized steel spiral duct 22 gauge						
3"	S2@.049	LF	4.42	1.79	—	6.21
4"	S2@.065	LF	5.83	2.38	—	8.21
5"	S2@.082	LF	7.34	3.00	—	10.34
6"	S2@.099	LF	8.73	3.62	—	12.35
7"	S2@.115	LF	10.20	4.20	—	14.40
8"	S2@.132	LF	11.50	4.82	—	16.32
9"	S2@.148	LF	13.10	5.41	—	18.51
10"	S2@.165	LF	14.80	6.03	—	20.83
12"	S2@.198	LF	17.60	7.23	—	24.83
14" - 16"	S2@.247	LF	21.80	9.03	—	30.83
18" - 20"	S2@.313	LF	28.10	11.40	—	39.50
22" - 24"	S2@.380	LF	33.70	13.90	—	47.60
26" - 28"	S2@.444	LF	39.80	16.20	—	56.00
30" - 32"	S2@.491	LF	45.80	17.90	—	63.70
34" - 36"	S2@.576	LF	51.80	21.00	—	72.80
Galvanized steel spiral duct 20 gauge						
3"	S2@.058	LF	5.21	2.12	—	7.33
4"	S2@.078	LF	6.93	2.85	—	9.78
5"	S2@.097	LF	8.61	3.54	—	12.15
6"	S2@.116	LF	10.30	4.24	—	14.54
7"	S2@.136	LF	11.90	4.97	—	16.87
8"	S2@.154	LF	13.70	5.63	—	19.33
9"	S2@.175	LF	15.40	6.39	—	21.79
10"	S2@.194	LF	17.20	7.09	—	24.29
12"	S2@.233	LF	20.90	8.51	—	29.41
14" - 16"	S2@.291	LF	25.60	10.60	—	36.20
18" - 20"	S2@.369	LF	32.80	13.50	—	46.30
22" - 24"	S2@.446	LF	39.80	16.30	—	56.10
26" - 28"	S2@.525	LF	46.80	19.20	—	66.00
30" - 32"	S2@.604	LF	53.40	22.10	—	75.50
34" - 36"	S2@.681	LF	60.40	24.90	—	85.30
Galvanized steel spiral duct 18 gauge						
6"	S2@.155	LF	13.70	5.66	—	19.36
7"	S2@.181	LF	16.00	6.61	—	22.61
8"	S2@.207	LF	18.20	7.56	—	25.76
9"	S2@.233	LF	20.60	8.51	—	29.11
10"	S2@.259	LF	22.70	9.46	—	32.16
12"	S2@.310	LF	27.80	11.30	—	39.10
14" - 16"	S2@.387	LF	34.60	14.10	—	48.70
18" - 20"	S2@.491	LF	44.10	17.90	—	62.00
22" - 24"	S2@.608	LF	53.90	22.20	—	76.10
26" - 28"	S2@.696	LF	61.90	25.40	—	87.30
30" - 32"	S2@.803	LF	71.10	29.30	—	100.40
34" - 36"	S2@.905	LF	80.50	33.10	—	113.60

Galvanized Steel Spiral Duct Fittings

Costs for Steel Spiral Duct Fittings. Costs per fitting, based on quantities less than 1,000 pounds. For quantities over 1,000 pounds, deduct 15% from material costs. For lined duct add 50%. These costs include delivery, typical bracing, cleats, scrap, hangers, end closures, sealants and miscellaneous hardware.

Description	Craft@Hrs	Unit	Material $	Labor $	Equipment $	Total $
Galvanized steel spiral duct 90-degree elbows						
3"	S2@.052	Ea	3.34	1.90	—	5.24
4"	S2@.088	Ea	5.63	3.22	—	8.85
5"	S2@.132	Ea	8.44	4.82	—	13.26
6"	S2@.172	Ea	11.00	6.28	—	17.28
7"	S2@.232	Ea	15.00	8.48	—	23.48
8"	S2@.292	Ea	18.90	10.70	—	29.60
9"	S2@.352	Ea	22.60	12.90	—	35.50
10"	S2@.470	Ea	30.20	17.20	—	47.40
12"	S2@.653	Ea	41.80	23.90	—	65.70
14" - 16"	S2@1.01	Ea	64.30	36.90	—	101.20
18" - 20"	S2@1.52	Ea	97.20	55.50	—	152.70
22" - 24"	S2@2.11	Ea	135.00	77.10	—	212.10
26" - 28"	S2@2.91	Ea	189.00	106.00	—	295.00
30" - 32"	S2@3.73	Ea	240.00	136.00	—	376.00
34" - 36"	S2@5.47	Ea	351.00	200.00	—	551.00
Galvanized steel spiral duct 45-degree elbows						
3"	S2@.040	Ea	2.56	1.46	—	4.02
4"	S2@.052	Ea	3.34	1.90	—	5.24
5"	S2@.076	Ea	4.86	2.78	—	7.64
6"	S2@.100	Ea	6.43	3.65	—	10.08
7"	S2@.132	Ea	8.44	4.82	—	13.26
8"	S2@.172	Ea	11.00	6.28	—	17.28
9"	S2@.212	Ea	13.50	7.75	—	21.25
10"	S2@.300	Ea	19.10	11.00	—	30.10
12"	S2@.400	Ea	25.60	14.60	—	40.20
14" - 16"	S2@.576	Ea	36.60	21.00	—	57.60
18" - 20"	S2@.850	Ea	54.60	31.10	—	85.70
22" - 24"	S2@1.19	Ea	76.00	43.50	—	119.50
26" - 28"	S2@1.61	Ea	103.00	58.80	—	161.80
30" - 32"	S2@2.06	Ea	133.00	75.30	—	208.30
34" - 36"	S2@3.01	Ea	192.00	110.00	—	302.00

Galvanized Steel Spiral Duct Fittings

Description	Craft@Hrs	Unit	Material $	Labor $	Equipment $	Total $

Galvanized steel spiral duct coupling

Description	Craft@Hrs	Unit	Material $	Labor $	Equipment $	Total $
3"	S2@.020	Ea	2.54	.73	—	3.27
4"	S2@.024	Ea	3.13	.88	—	4.01
5"	S2@.028	Ea	3.56	1.02	—	4.58
6"	S2@.036	Ea	4.59	1.32	—	5.91
7"	S2@.040	Ea	5.13	1.46	—	6.59
8"	S2@.048	Ea	6.16	1.75	—	7.91
9"	S2@.104	Ea	13.30	3.80	—	17.10
10"	S2@.116	Ea	15.00	4.24	—	19.24
12"	S2@.140	Ea	18.00	5.12	—	23.12
14" - 16"	S2@.174	Ea	22.40	6.36	—	28.76
18" - 20"	S2@.220	Ea	27.90	8.04	—	35.94
22" - 24"	S2@.266	Ea	34.20	9.72	—	43.92
26" - 28"	S2@.312	Ea	39.90	11.40	—	51.30
30" - 32"	S2@.357	Ea	45.90	13.00	—	58.90
34" - 36"	S2@.403	Ea	51.70	14.70	—	66.40

Galvanized steel spiral duct with reducer

Description	Craft@Hrs	Unit	Material $	Labor $	Equipment $	Total $
3"	S2@.040	Ea	5.13	1.46	—	6.59
4"	S2@.048	Ea	6.16	1.75	—	7.91
5"	S2@.056	Ea	7.18	2.05	—	9.23
6"	S2@.072	Ea	9.27	2.63	—	11.90
7"	S2@.080	Ea	10.20	2.92	—	13.12
8"	S2@.096	Ea	12.10	3.51	—	15.61
9"	S2@.132	Ea	16.80	4.82	—	21.62
10"	S2@.176	Ea	22.40	6.43	—	28.83
12"	S2@.212	Ea	27.30	7.75	—	35.05
14" - 16"	S2@.262	Ea	33.70	9.57	—	43.27
18" - 20"	S2@.330	Ea	42.40	12.10	—	54.50
22" - 24"	S2@.398	Ea	50.90	14.50	—	65.40
26" - 28"	S2@.465	Ea	59.60	17.00	—	76.60
30" - 32"	S2@.536	Ea	68.20	19.60	—	87.80
34" - 36"	S2@.606	Ea	77.40	22.10	—	99.50

Galvanized Steel Spiral Tees

Costs for Steel Spiral Tees with Reducing Branch. Costs per tee, based on quantities less than 1,000 pounds. Deduct 15% from the material cost for quantities over 1,000 pounds. Add 50% to the material cost for lined duct. These costs include delivery, typical bracing, cleats, scrap, hangers, end closures, sealants and miscellaneous hardware.

Description	Craft@Hrs	Unit	Material $	Labor $	Equipment $	Total $

Galvanized steel spiral duct tees with 3" reducing branch

Description	Craft@Hrs	Unit	Material $	Labor $	Equipment $	Total $
4"	S2@.109	Ea	11.40	3.98	—	15.38
5"	S2@.112	Ea	13.40	4.09	—	17.49
6"	S2@.128	Ea	15.40	4.68	—	20.08
7"	S2@.148	Ea	17.30	5.41	—	22.71
8"	S2@.164	Ea	19.50	5.99	—	25.49
9"	S2@.180	Ea	20.90	6.58	—	27.48
10"	S2@.256	Ea	33.00	9.35	—	42.35
12"	S2@.308	Ea	39.10	11.30	—	50.40
14" - 16"	S2@.382	Ea	49.50	14.00	—	63.50
18" - 20"	S2@.485	Ea	62.10	17.70	—	79.80
22" - 24"	S2@.585	Ea	74.90	21.40	—	96.30
26" - 28"	S2@.683	Ea	87.30	25.00	—	112.30
30" - 32"	S2@.786	Ea	101.00	28.70	—	129.70
34" - 36"	S2@1.02	Ea	131.00	37.30	—	168.30

Galvanized steel spiral duct tees with 4" reducing branch

Description	Craft@Hrs	Unit	Material $	Labor $	Equipment $	Total $
5"	S2@.120	Ea	14.90	4.38	—	19.28
6"	S2@.140	Ea	17.30	5.12	—	22.42
7"	S2@.156	Ea	18.90	5.70	—	24.60
8"	S2@.176	Ea	21.40	6.43	—	27.83
9"	S2@.192	Ea	22.70	7.02	—	29.72
10"	S2@.268	Ea	35.00	9.79	—	44.79
12"	S2@.324	Ea	42.40	11.80	—	54.20
14" - 16"	S2@.403	Ea	52.60	14.70	—	67.30
18" - 20"	S2@.508	Ea	66.50	18.60	—	85.10
22" - 24"	S2@.619	Ea	80.70	22.60	—	103.30
26" - 28"	S2@.724	Ea	94.50	26.50	—	121.00
30" - 32"	S2@.830	Ea	108.00	30.30	—	138.30
34" - 36"	S2@1.08	Ea	140.00	39.50	—	179.50

Galvanized steel spiral duct tees with 5" reducing branch

Description	Craft@Hrs	Unit	Material $	Labor $	Equipment $	Total $
6"	S2@.148	Ea	18.90	5.41	—	24.31
7"	S2@.164	Ea	20.90	5.99	—	26.89
8"	S2@.188	Ea	23.40	6.87	—	30.27
9"	S2@.204	Ea	25.40	7.45	—	32.85
10"	S2@.284	Ea	37.80	10.40	—	48.20
12"	S2@.340	Ea	45.10	12.40	—	57.50
14" - 16"	S2@.428	Ea	56.80	15.60	—	72.40
18" - 20"	S2@.538	Ea	72.00	19.70	—	91.70
22" - 24"	S2@.651	Ea	86.10	23.80	—	109.90
26" - 28"	S2@.764	Ea	101.00	27.90	—	128.90
30" - 32"	S2@.873	Ea	116.00	31.90	—	147.90
34" - 36"	S2@1.15	Ea	152.00	42.00	—	194.00

Galvanized Steel Spiral Tees

Description	Craft@Hrs	Unit	Material $	Labor $	Equipment $	Total $

Galvanized steel spiral duct tees with 6" reducing branch

Description	Craft@Hrs	Unit	Material $	Labor $	Equipment $	Total $
7"	S2@.180	Ea	23.40	6.58	—	29.98
8"	S2@.200	Ea	25.40	7.31	—	32.71
9"	S2@.220	Ea	27.70	8.04	—	35.74
10"	S2@.300	Ea	40.50	11.00	—	51.50
12"	S2@.356	Ea	48.30	13.00	—	61.30
14" - 16"	S2@.449	Ea	60.50	16.40	—	76.90
18" - 20"	S2@.566	Ea	76.40	20.70	—	97.10
22" - 24"	S2@.683	Ea	92.00	25.00	—	117.00
26" - 28"	S2@.807	Ea	107.00	29.50	—	136.50
30" - 32"	S2@.924	Ea	123.00	33.80	—	156.80
34" - 36"	S2@1.20	Ea	161.00	43.80	—	204.80

Galvanized steel spiral duct tees with 7" reducing branch

Description	Craft@Hrs	Unit	Material $	Labor $	Equipment $	Total $
8"	S2@.164	Ea	27.70	5.99	—	33.69
9"	S2@.180	Ea	30.00	6.58	—	36.58
10"	S2@.252	Ea	42.40	9.21	—	51.61
12"	S2@.304	Ea	51.30	11.10	—	62.40
14" - 16"	S2@.381	Ea	64.40	13.90	—	78.30
18" - 20"	S2@.482	Ea	80.70	17.60	—	98.30
22" - 24"	S2@.579	Ea	97.20	21.20	—	118.40
26" - 28"	S2@.683	Ea	115.00	25.00	—	140.00
30" - 32"	S2@.784	Ea	131.00	28.60	—	159.60
34" - 36"	S2@1.02	Ea	171.00	37.30	—	208.30

Galvanized steel spiral duct tees with 8" reducing branch

Description	Craft@Hrs	Unit	Material $	Labor $	Equipment $	Total $
9"	S2@.228	Ea	38.60	8.33	—	46.93
10"	S2@.268	Ea	45.10	9.79	—	54.89
12"	S2@.324	Ea	54.30	11.80	—	66.10
14" - 16"	S2@.401	Ea	67.60	14.70	—	82.30
18" - 20"	S2@.508	Ea	85.70	18.60	—	104.30
22" - 24"	S2@.617	Ea	103.00	22.50	—	125.50
26" - 28"	S2@.722	Ea	120.00	26.40	—	146.40
30" - 32"	S2@.830	Ea	138.00	30.30	—	168.30
34" - 36"	S2@1.08	Ea	181.00	39.50	—	220.50

Galvanized steel spiral duct tees with 9" reducing branch

Description	Craft@Hrs	Unit	Material $	Labor $	Equipment $	Total $
10"	S2@.284	Ea	48.00	10.40	—	58.40
12"	S2@.340	Ea	57.10	12.40	—	69.50
14" - 16"	S2@.426	Ea	72.00	15.60	—	87.60
18" - 20"	S2@.540	Ea	90.40	19.70	—	110.10
22" - 24"	S2@.649	Ea	108.00	23.70	—	131.70
26" - 28"	S2@.760	Ea	126.00	27.80	—	153.80
30" - 32"	S2@.875	Ea	149.00	32.00	—	181.00
34" - 36"	S2@1.15	Ea	191.00	42.00	—	233.00

Galvanized Steel Spiral Tees

Description	Craft@Hrs	Unit	Material $	Labor $	Equipment $	Total $
Galvanized steel spiral duct tees with 10" reducing branch						
12"	S2@.360	Ea	60.50	13.20	—	73.70
14" - 16"	S2@.447	Ea	74.90	16.30	—	91.20
18" - 20"	S2@.564	Ea	95.20	20.60	—	115.80
22" - 24"	S2@.683	Ea	115.00	25.00	—	140.00
26" - 28"	S2@.805	Ea	135.00	29.40	—	164.40
30" - 32"	S2@.924	Ea	154.00	33.80	—	187.80
34" - 36"	S2@1.20	Ea	203.00	43.80	—	246.80
Galvanized steel spiral duct tees with 12" reducing branch						
14" - 16"	S2@.572	Ea	82.70	20.90	—	103.60
18" - 20"	S2@.681	Ea	104.00	24.90	—	128.90
22" - 24"	S2@.820	Ea	124.00	30.00	—	154.00
26" - 28"	S2@.965	Ea	149.00	35.30	—	184.30
30" - 32"	S2@1.11	Ea	170.00	40.60	—	210.60
34" - 36"	S2@1.44	Ea	224.00	52.60	—	276.60
Galvanized steel spiral duct tees with 14" reducing branch						
14" - 16"	S2@.572	Ea	95.90	20.90	—	116.80
18" - 20"	S2@.681	Ea	114.00	24.90	—	138.90
22" - 24"	S2@.820	Ea	137.00	30.00	—	167.00
26" - 28"	S2@.965	Ea	161.00	35.30	—	196.30
30" - 32"	S2@1.11	Ea	187.00	40.60	—	227.60
34" - 36"	S2@1.44	Ea	245.00	52.60	—	297.60
Galvanized steel spiral duct tees with 16" reducing branch						
18" - 20"	S2@.739	Ea	123.00	27.00	—	150.00
22" - 24"	S2@.888	Ea	151.00	32.40	—	183.40
26" - 28"	S2@1.04	Ea	174.00	38.00	—	212.00
30" - 32"	S2@1.20	Ea	201.00	43.80	—	244.80
34" - 36"	S2@1.56	Ea	265.00	57.00	—	322.00
Galvanized steel spiral duct tees with 18" reducing branch						
18" - 20"	S2@.837	Ea	139.00	30.60	—	169.60
22" - 24"	S2@.954	Ea	160.00	34.90	—	194.90
26" - 28"	S2@1.12	Ea	189.00	40.90	—	229.90
30" - 32"	S2@1.29	Ea	216.00	47.10	—	263.10
34" - 36"	S2@1.68	Ea	287.00	61.40	—	348.40
Galvanized steel spiral duct tees with 20" reducing branch						
22" - 24"	S2@1.02	Ea	172.00	37.30	—	209.30
26" - 28"	S2@1.20	Ea	203.00	43.80	—	246.80
30" - 32"	S2@1.38	Ea	231.00	50.40	—	281.40
34" - 36"	S2@1.81	Ea	303.00	66.10	—	369.10
Galvanized steel spiral duct tees with 22" reducing branch						
22" - 24"	S2@1.57	Ea	191.00	57.40	—	248.40
26" - 28"	S2@1.32	Ea	224.00	48.20	—	272.20
30" - 32"	S2@1.48	Ea	249.00	54.10	—	303.10
34" - 36"	S2@1.92	Ea	327.00	70.20	—	397.20

Galvanized Steel Spiral Tees

Costs for Steel Spiral Tees with Reducing Run and Branch. Costs per tee, based on quantities less than 1,000 pounds. Deduct 15% from the material cost for quantities over 1,000 pounds. Add 50% to the material cost for lined duct. These costs include delivery, typical bracing, cleats, scrap, hangers, end closures, sealants and miscellaneous hardware.

Description	Craft@Hrs	Unit	Material $	Labor $	Equipment $	Total $

Galvanized steel spiral duct tees with 3" run and branch

4"	S2@.109	Ea	18.60	3.98	—	22.58
5"	S2@.112	Ea	18.90	4.09	—	22.99
6"	S2@.128	Ea	21.40	4.68	—	26.08
7"	S2@.148	Ea	24.90	5.41	—	30.31
8"	S2@.164	Ea	27.70	5.99	—	33.69
9"	S2@.180	Ea	30.00	6.58	—	36.58
10"	S2@.256	Ea	43.20	9.35	—	52.55
12"	S2@.308	Ea	51.50	11.30	—	62.80
14" - 16"	S2@.382	Ea	64.40	14.00	—	78.40
18" - 20"	S2@.485	Ea	81.20	17.70	—	98.90
22" - 24"	S2@.585	Ea	98.50	21.40	—	119.90
26" - 28"	S2@.683	Ea	115.00	25.00	—	140.00
30" - 32"	S2@.786	Ea	133.00	28.70	—	161.70
34" - 36"	S2@1.02	Ea	171.00	37.30	—	208.30

Galvanized steel spiral duct tees with 4" run and branch

5"	S2@.120	Ea	20.10	4.38	—	24.48
6"	S2@.140	Ea	23.40	5.12	—	28.52
7"	S2@.156	Ea	26.50	5.70	—	32.20
8"	S2@.176	Ea	29.40	6.43	—	35.83
9"	S2@.192	Ea	32.10	7.02	—	39.12
10"	S2@.268	Ea	45.10	9.79	—	54.89
12"	S2@.324	Ea	54.30	11.80	—	66.10
14" - 16"	S2@.403	Ea	67.90	14.70	—	82.60
18" - 20"	S2@.508	Ea	85.70	18.60	—	104.30
22" - 24"	S2@.619	Ea	103.00	22.60	—	125.60
26" - 28"	S2@.724	Ea	120.00	26.50	—	146.50
30" - 32"	S2@.830	Ea	138.00	30.30	—	168.30
34" - 36"	S2@1.08	Ea	182.00	39.50	—	221.50

Galvanized steel spiral duct tees with 5" run and branch

6"	S2@.148	Ea	24.90	5.41	—	30.31
7"	S2@.164	Ea	27.70	5.99	—	33.69
8"	S2@.188	Ea	31.70	6.87	—	38.57
9"	S2@.204	Ea	34.40	7.45	—	41.85
10"	S2@.284	Ea	48.00	10.40	—	58.40
12"	S2@.340	Ea	57.10	12.40	—	69.50
14" - 16"	S2@.428	Ea	72.00	15.60	—	87.60
18" - 20"	S2@.538	Ea	90.20	19.70	—	109.90
22" - 24"	S2@.651	Ea	109.00	23.80	—	132.80
26" - 28"	S2@.764	Ea	128.00	27.90	—	155.90
30" - 32"	S2@.873	Ea	149.00	31.90	—	180.90
34" - 36"	S2@1.15	Ea	191.00	42.00	—	233.00

Galvanized Steel Spiral Tees

Description	Craft@Hrs	Unit	Material $	Labor $	Equipment $	Total $
Galvanized steel spiral duct tees with 6" run and branch						
7"	S2@.180	Ea	30.00	6.58	—	36.58
8"	S2@.200	Ea	33.60	7.31	—	40.91
9"	S2@.220	Ea	36.50	8.04	—	44.54
10"	S2@.300	Ea	50.50	11.00	—	61.50
12"	S2@.356	Ea	60.10	13.00	—	73.10
14" - 16"	S2@.449	Ea	75.30	16.40	—	91.70
18" - 20"	S2@.566	Ea	95.20	20.70	—	115.90
22" - 24"	S2@.683	Ea	115.00	25.00	—	140.00
26" - 28"	S2@.807	Ea	135.00	29.50	—	164.50
30" - 32"	S2@.924	Ea	154.00	33.80	—	187.80
34" - 36"	S2@1.20	Ea	203.00	43.80	—	246.80
Galvanized steel spiral duct tees with 7" run and branch						
8"	S2@.212	Ea	35.60	7.75	—	43.35
9"	S2@.232	Ea	39.10	8.48	—	47.58
10"	S2@.312	Ea	52.10	11.40	—	63.50
12"	S2@.376	Ea	63.20	13.70	—	76.90
14" - 16"	S2@.470	Ea	79.30	17.20	—	96.50
18" - 20"	S2@.636	Ea	106.00	23.20	—	129.20
22" - 24"	S2@.717	Ea	120.00	26.20	—	146.20
26" - 28"	S2@.845	Ea	140.00	30.90	—	170.90
30" - 32"	S2@.969	Ea	165.00	35.40	—	200.40
34" - 36"	S2@1.27	Ea	213.00	46.40	—	259.40
Galvanized steel spiral duct tees with 8" run and branch						
9"	S2@.248	Ea	41.80	9.06	—	50.86
10"	S2@.328	Ea	55.50	12.00	—	67.50
12"	S2@.396	Ea	66.40	14.50	—	80.90
14" - 16"	S2@.493	Ea	82.30	18.00	—	100.30
18" - 20"	S2@.623	Ea	104.00	22.80	—	126.80
22" - 24"	S2@.751	Ea	124.00	27.40	—	151.40
26" - 28"	S2@.882	Ea	150.00	32.20	—	182.20
30" - 32"	S2@1.02	Ea	170.00	37.30	—	207.30
34" - 36"	S2@1.33	Ea	224.00	48.60	—	272.60
Galvanized steel spiral duct tees with 9" run and branch						
10"	S2@.344	Ea	57.60	12.60	—	70.20
12"	S2@.412	Ea	69.00	15.10	—	84.10
14" - 16"	S2@.515	Ea	86.70	18.80	—	105.50
18" - 20"	S2@.653	Ea	109.00	23.90	—	132.90
22" - 24"	S2@.779	Ea	133.00	28.50	—	161.50
26" - 28"	S2@.924	Ea	154.00	33.80	—	187.80
30" - 32"	S2@1.06	Ea	179.00	38.70	—	217.70
34" - 36"	S2@1.39	Ea	231.00	50.80	—	281.80

Galvanized Steel Spiral Tees

Description	Craft@Hrs	Unit	Material $	Labor $	Equipment $	Total $

Galvanized steel spiral duct tees with 10" run and branch

Description	Craft@Hrs	Unit	Material $	Labor $	Equipment $	Total $
12"	S2@.427	Ea	36.10	15.60	—	51.70
14" - 16"	S2@.538	Ea	45.50	19.70	—	65.20
18" - 20"	S2@.681	Ea	113.00	24.90	—	137.90
22" - 24"	S2@.820	Ea	137.00	30.00	—	167.00
26" - 28"	S2@.965	Ea	161.00	35.30	—	196.30
30" - 32"	S2@1.11	Ea	187.00	40.60	—	227.60
34" - 36"	S2@1.44	Ea	243.00	52.60	—	295.60

Galvanized steel spiral duct tees with 12" run and branch

Description	Craft@Hrs	Unit	Material $	Labor $	Equipment $	Total $
14" - 16"	S2@.581	Ea	97.30	21.20	—	118.50
18" - 20"	S2@.739	Ea	123.00	27.00	—	150.00
22" - 24"	S2@.888	Ea	151.00	32.40	—	183.40
26" - 28"	S2@1.04	Ea	174.00	38.00	—	212.00
30" - 32"	S2@1.20	Ea	201.00	43.80	—	244.80
34" - 36"	S2@1.57	Ea	265.00	57.40	—	322.40

Galvanized steel spiral duct tees with 14" run and branch

Description	Craft@Hrs	Unit	Material $	Labor $	Equipment $	Total $
14" - 16"	S2@.666	Ea	111.00	24.30	—	135.30
18" - 20"	S2@.794	Ea	134.00	29.00	—	163.00
22" - 24"	S2@.959	Ea	160.00	35.00	—	195.00
26" - 28"	S2@1.11	Ea	187.00	40.60	—	227.60
30" - 32"	S2@1.29	Ea	216.00	47.10	—	263.10
34" - 36"	S2@1.69	Ea	287.00	61.80	—	348.80

Galvanized steel spiral duct tees with 16" run and branch

Description	Craft@Hrs	Unit	Material $	Labor $	Equipment $	Total $
18" - 20"	S2@.833	Ea	138.00	30.40	—	168.40
22" - 24"	S2@1.03	Ea	172.00	37.60	—	209.60
26" - 28"	S2@1.20	Ea	203.00	43.80	—	246.80
30" - 32"	S2@1.39	Ea	231.00	50.80	—	281.80
34" - 36"	S2@1.81	Ea	303.00	66.10	—	369.10

Galvanized steel spiral duct tees with 18" run and branch

Description	Craft@Hrs	Unit	Material $	Labor $	Equipment $	Total $
18" - 20"	S2@.956	Ea	160.00	34.90	—	194.90
22" - 24"	S2@1.09	Ea	182.00	39.80	—	221.80
26" - 28"	S2@1.29	Ea	215.00	47.10	—	262.10
30" - 32"	S2@1.48	Ea	249.00	54.10	—	303.10
34" - 36"	S2@1.93	Ea	327.00	70.50	—	397.50

Galvanized steel spiral duct tees with 20" run and branch

Description	Craft@Hrs	Unit	Material $	Labor $	Equipment $	Total $
22" - 24"	S2@1.16	Ea	195.00	42.40	—	237.40
26" - 28"	S2@1.37	Ea	230.00	50.10	—	280.10
30" - 32"	S2@1.57	Ea	265.00	57.40	—	322.40
34" - 36"	S2@2.03	Ea	344.00	74.20	—	418.20

Galvanized steel spiral duct tees with 22" run and branch

Description	Craft@Hrs	Unit	Material $	Labor $	Equipment $	Total $
22" - 24"	S2@1.28	Ea	215.00	46.80	—	261.80
26" - 28"	S2@1.45	Ea	245.00	53.00	—	298.00
30" - 32"	S2@1.64	Ea	278.00	59.90	—	337.90
34" - 36"	S2@2.16	Ea	362.00	78.90	—	440.90

Galvanized Steel Spiral Crosses

Costs for Steel Spiral Crosses with Reducing Branches. Costs per cross, based on quantities less than 1,000 pounds. Deduct 15% from the material cost for quantities over 1,000 pounds. Add 50% to the material cost for lined duct. These costs include delivery, typical bracing, cleats, scrap, hangers, end closures, sealants and miscellaneous hardware.

Description	Craft@Hrs	Unit	Material $	Labor $	Equipment $	Total $
Galvanized steel spiral duct crosses with 3" branch						
4"	S2@.104	Ea	17.30	3.80	—	21.10
5"	S2@.128	Ea	21.40	4.68	—	26.08
6"	S2@.152	Ea	25.40	5.55	—	30.95
7"	S2@.176	Ea	29.40	6.43	—	35.83
8"	S2@.196	Ea	33.00	7.16	—	40.16
9"	S2@.216	Ea	36.10	7.89	—	43.99
10"	S2@.304	Ea	51.30	11.10	—	62.40
12"	S2@.352	Ea	59.60	12.90	—	72.50
14" - 16"	S2@.428	Ea	72.40	15.60	—	88.00
18" - 20"	S2@.525	Ea	88.20	19.20	—	107.40
22" - 24"	S2@.634	Ea	106.00	23.20	—	129.20
26" - 28"	S2@.745	Ea	123.00	27.20	—	150.20
30" - 32"	S2@.856	Ea	142.00	31.30	—	173.30
34" - 36"	S2@1.11	Ea	188.00	40.60	—	228.60

Description	Craft@Hrs	Unit	Material $	Labor $	Equipment $	Total $
Galvanized steel spiral duct crosses with 4" branch						
5"	S2@.144	Ea	24.30	5.26	—	29.56
6"	S2@.168	Ea	28.70	6.14	—	34.84
7"	S2@.196	Ea	33.00	7.16	—	40.16
8"	S2@.216	Ea	36.10	7.89	—	43.99
9"	S2@.240	Ea	40.50	8.77	—	49.27
10"	S2@.328	Ea	55.50	12.00	—	67.50
12"	S2@.384	Ea	65.10	14.00	—	79.10
14" - 16"	S2@.461	Ea	77.60	16.80	—	94.40
18" - 20"	S2@.568	Ea	95.70	20.80	—	116.50
22" - 24"	S2@.685	Ea	115.00	25.00	—	140.00
26" - 28"	S2@.805	Ea	135.00	29.40	—	164.40
30" - 32"	S2@.927	Ea	156.00	33.90	—	189.90
34" - 36"	S2@1.20	Ea	203.00	43.80	—	246.80

Description	Craft@Hrs	Unit	Material $	Labor $	Equipment $	Total $
Galvanized steel spiral duct crosses with 5" branch						
6"	S2@.184	Ea	31.20	6.72	—	37.92
7"	S2@.216	Ea	36.10	7.89	—	43.99
8"	S2@.240	Ea	40.50	8.77	—	49.27
9"	S2@.264	Ea	44.70	9.65	—	54.35
10"	S2@.356	Ea	60.10	13.00	—	73.10
12"	S2@.412	Ea	69.00	15.10	—	84.10
14" - 16"	S2@.497	Ea	83.00	18.20	—	101.20
18" - 20"	S2@.613	Ea	102.00	22.40	—	124.40
22" - 24"	S2@.741	Ea	123.00	27.10	—	150.10
26" - 28"	S2@.867	Ea	148.00	31.70	—	179.70
30" - 32"	S2@.993	Ea	167.00	36.30	—	203.30
34" - 36"	S2@1.29	Ea	216.00	47.10	—	263.10

Galvanized Steel Spiral Crosses

Description	Craft@Hrs	Unit	Material $	Labor $	Equipment $	Total $

Galvanized steel spiral duct crosses with 6" branch

7"	S2@.232	Ea	39.10	8.48	—	47.58
8"	S2@.260	Ea	44.00	9.50	—	53.50
9"	S2@.288	Ea	48.30	10.50	—	58.80
10"	S2@.380	Ea	64.30	13.90	—	78.20
12"	S2@.440	Ea	74.00	16.10	—	90.10
14" - 16"	S2@.532	Ea	89.40	19.40	—	108.80
18" - 20"	S2@.653	Ea	109.00	23.90	—	132.90
22" - 24"	S2@.790	Ea	134.00	28.90	—	162.90
26" - 28"	S2@.927	Ea	156.00	33.90	—	189.90
30" - 32"	S2@1.06	Ea	179.00	38.70	—	217.70
34" - 36"	S2@1.38	Ea	231.00	50.40	—	281.40

Galvanized steel spiral duct crosses with 7" branch

8"	S2@.280	Ea	46.80	10.20	—	57.00
9"	S2@.308	Ea	51.50	11.30	—	62.80
10"	S2@.408	Ea	68.50	14.90	—	83.40
12"	S2@.470	Ea	78.50	17.20	—	95.70
14" - 16"	S2@.566	Ea	95.20	20.70	—	115.90
18" - 20"	S2@.692	Ea	116.00	25.30	—	141.30
22" - 24"	S2@.841	Ea	139.00	30.70	—	169.70
26" - 28"	S2@.986	Ea	166.00	36.00	—	202.00
30" - 32"	S2@1.13	Ea	190.00	41.30	—	231.30
34" - 36"	S2@1.48	Ea	249.00	54.10	—	303.10

Galvanized steel spiral duct crosses with 8" branch

9"	S2@.310	Ea	55.80	11.30	—	67.10
10"	S2@.403	Ea	73.10	14.70	—	87.80
12"	S2@.470	Ea	83.50	17.20	—	100.70
14" - 16"	S2@.561	Ea	101.00	20.50	—	121.50
18" - 20"	S2@.690	Ea	123.00	25.20	—	148.20
22" - 24"	S2@.835	Ea	151.00	30.50	—	181.50
26" - 28"	S2@.980	Ea	174.00	35.80	—	209.80
30" - 32"	S2@1.12	Ea	203.00	40.90	—	243.90
34" - 36"	S2@1.47	Ea	265.00	53.70	—	318.70

Galvanized steel spiral duct crosses with 9" branch

10"	S2@.427	Ea	76.40	15.60	—	92.00
12"	S2@.495	Ea	88.40	18.10	—	106.50
14" - 16"	S2@.596	Ea	107.00	21.80	—	128.80
18" - 20"	S2@.732	Ea	131.00	26.70	—	157.70
22" - 24"	S2@.882	Ea	159.00	32.20	—	191.20
26" - 28"	S2@1.04	Ea	187.00	38.00	—	225.00
30" - 32"	S2@1.19	Ea	213.00	43.50	—	256.50
34" - 36"	S2@1.55	Ea	278.00	56.60	—	334.60

Galvanized Steel Spiral Crosses

Description	Craft@Hrs	Unit	Material $	Labor $	Equipment $	Total $

Galvanized steel spiral duct crosses with 12" branch

14" - 16"	S2@.692	Ea	123.00	25.30	—	148.30
18" - 20"	S2@.850	Ea	153.00	31.10	—	184.10
22" - 24"	S2@1.03	Ea	186.00	37.60	—	223.60
26" - 28"	S2@1.18	Ea	213.00	43.10	—	256.10
30" - 32"	S2@1.38	Ea	249.00	50.40	—	299.40
34" - 36"	S2@1.80	Ea	327.00	65.80	—	392.80

Galvanized steel spiral duct crosses with 14" branch

14" - 16"	S2@1.81	Ea	63.20	66.10	—	129.30
18" - 20"	S2@2.11	Ea	73.40	77.10	—	150.50
22" - 24"	S2@2.55	Ea	89.40	93.20	—	182.60
26" - 28"	S2@2.98	Ea	104.00	109.00	—	213.00
30" - 32"	S2@3.43	Ea	118.00	125.00	—	243.00
34" - 36"	S2@4.47	Ea	156.00	163.00	—	319.00

Galvanized steel spiral duct crosses with 16" branch

18" - 20"	S2@1.08	Ea	181.00	39.50	—	220.50
22" - 24"	S2@1.30	Ea	218.00	47.50	—	265.50
26" - 28"	S2@1.53	Ea	257.00	55.90	—	312.90
30" - 32"	S2@1.76	Ea	294.00	64.30	—	358.30
34" - 36"	S2@2.31	Ea	389.00	84.40	—	473.40

Galvanized steel spiral duct crosses with 18" branch

18" - 20"	S2@1.22	Ea	205.00	44.60	—	249.60
22" - 24"	S2@1.40	Ea	237.00	51.20	—	288.20
26" - 28"	S2@1.65	Ea	278.00	60.30	—	338.30
30" - 32"	S2@1.89	Ea	318.00	69.10	—	387.10
34" - 36"	S2@2.47	Ea	418.00	90.30	—	508.30

Galvanized steel spiral duct crosses with 20" branch

22" - 24"	S2@1.51	Ea	253.00	55.20	—	308.20
26" - 28"	S2@1.77	Ea	298.00	64.70	—	362.70
30" - 32"	S2@2.05	Ea	349.00	74.90	—	423.90
34" - 36"	S2@2.65	Ea	447.00	96.80	—	543.80

Galvanized steel spiral duct crosses with 22" branch

22" - 24"	S2@1.68	Ea	287.00	61.40	—	348.40
26" - 28"	S2@1.89	Ea	318.00	69.10	—	387.10
30" - 32"	S2@2.17	Ea	362.00	79.30	—	441.30
34" - 36"	S2@2.84	Ea	480.00	104.00	—	584.00

Galvanized Steel Rectangular Ductwork

Costs for Steel 26 Gauge Rectangular Duct. Costs per linear foot of duct, based on quantities less than 1,000 pounds of prefabricated duct. Deduct 15% from the material cost for quantities over 1,000 pounds. Add 50% to the material cost for lined duct. These costs include delivery, typical bracing, cleats, scrap, hangers, end closures, sealants and miscellaneous hardware.

Description	Craft@Hrs	Unit	Material $	Labor $	Equipment $	Total $

Galvanized steel 4" rectangular duct 26 gauge

4"	S2@.048	LF	4.15	1.75	—	5.90
5"	S2@.054	LF	4.59	1.97	—	6.56
6"	S2@.060	LF	5.13	2.19	—	7.32
7"	S2@.066	LF	5.67	2.41	—	8.08
8"	S2@.072	LF	6.18	2.63	—	8.81
9"	S2@.078	LF	6.64	2.85	—	9.49
10"	S2@.084	LF	7.17	3.07	—	10.24
11"	S2@.091	LF	7.70	3.33	—	11.03
12"	S2@.097	LF	8.22	3.54	—	11.76

Galvanized steel 5" rectangular duct 26 gauge

5"	S2@.060	LF	5.13	2.19	—	7.32
6"	S2@.066	LF	5.67	2.41	—	8.08
7"	S2@.072	LF	6.18	2.63	—	8.81
8"	S2@.078	LF	6.64	2.85	—	9.49
9"	S2@.084	LF	7.17	3.07	—	10.24
10"	S2@.091	LF	7.70	3.33	—	11.03
11"	S2@.097	LF	8.22	3.54	—	11.76
12"	S2@.103	LF	8.70	3.76	—	12.46

Galvanized steel 6" rectangular duct 26 gauge

6"	S2@.072	LF	6.18	2.63	—	8.81
7"	S2@.078	LF	6.64	2.85	—	9.49
8"	S2@.084	LF	7.17	3.07	—	10.24
9"	S2@.091	LF	7.70	3.33	—	11.03
10"	S2@.097	LF	8.22	3.54	—	11.76
11"	S2@.103	LF	8.70	3.76	—	12.46
12"	S2@.109	LF	9.21	3.98	—	13.19

Galvanized steel 7" rectangular duct 26 gauge

7"	S2@.084	LF	7.17	3.07	—	10.24
8"	S2@.091	LF	7.70	3.33	—	11.03
9"	S2@.097	LF	8.22	3.54	—	11.76
10"	S2@.103	LF	8.70	3.76	—	12.46
11"	S2@.109	LF	9.21	3.98	—	13.19
12"	S2@.115	LF	9.72	4.20	—	13.92

Galvanized Steel Rectangular Ductwork

Description	Craft@Hrs	Unit	Material $	Labor $	Equipment $	Total $
Galvanized steel 8" rectangular duct 26 gauge						
8"	S2@.097	LF	8.22	3.54	—	11.76
9"	S2@.103	LF	8.70	3.76	—	12.46
10"	S2@.109	LF	9.21	3.98	—	13.19
11"	S2@.115	LF	9.72	4.20	—	13.92
12"	S2@.121	LF	10.20	4.42	—	14.62
Galvanized steel 9" rectangular duct 26 gauge						
9"	S2@.109	LF	9.21	3.98	—	13.19
10"	S2@.115	LF	9.72	4.20	—	13.92
11"	S2@.121	LF	10.20	4.42	—	14.62
12"	S2@.127	LF	10.70	4.64	—	15.34
Galvanized steel 10" rectangular duct 26 gauge						
10"	S2@.129	LF	10.20	4.71	—	14.91
11"	S2@.127	LF	10.70	4.64	—	15.34
12"	S2@.133	LF	11.20	4.86	—	16.06
Galvanized steel 11" rectangular duct 26 gauge						
11"	S2@.133	LF	11.20	4.86	—	16.06
12"	S2@.139	LF	11.60	5.08	—	16.68
Galvanized steel 12" rectangular duct 26 gauge						
12"	S2@.145	LF	12.10	5.30	—	17.40

Galvanized Steel Rectangular Ductwork

Costs for Galvanized 24 Gauge Rectangular Duct. Costs per linear foot of duct, based on quantities less than 1,000 pounds of prefabricated duct. Deduct 15% from the material cost for over 1,000 pounds. Add 50% to the material cost for lined duct. These costs include delivery, typical bracing, cleats, scrap, hangers, end closures.

Description	Craft@Hrs	Unit	Material $	Labor $	Equipment $	Total $

Galvanized steel 14" rectangular duct 24 gauge

6"	S2@.154	LF	13.10	5.63	—	18.73
8"	S2@.170	LF	14.40	6.21	—	20.61
10"	S2@.184	LF	15.80	6.72	—	22.52
12"	S2@.200	LF	17.10	7.31	—	24.41
14" - 16"	S2@.223	LF	19.00	8.15	—	27.15
18" - 20"	S2@.254	LF	21.60	9.28	—	30.88
22" - 24"	S2@.285	LF	24.50	10.40	—	34.90
26" - 28"	S2@.316	LF	27.30	11.50	—	38.80
30"	S2@.339	LF	29.10	12.40	—	41.50

Galvanized steel 16" rectangular duct 24 gauge

6"	S2@.170	LF	14.40	6.21	—	20.61
8"	S2@.184	LF	15.80	6.72	—	22.52
10"	S2@.200	LF	17.10	7.31	—	24.41
12"	S2@.216	LF	18.60	7.89	—	26.49
14" - 16"	S2@.239	LF	20.40	8.73	—	29.13
18" - 20"	S2@.269	LF	23.00	9.83	—	32.83
22" - 24"	S2@.300	LF	25.40	11.00	—	36.40
26" - 28"	S2@.331	LF	28.70	12.10	—	40.80
30"	S2@.354	LF	30.70	12.90	—	43.60

Galvanized steel 18" rectangular duct 24 gauge

6"	S2@.184	LF	15.80	6.72	—	22.52
8"	S2@.200	LF	17.10	7.31	—	24.41
10"	S2@.216	LF	18.60	7.89	—	26.49
12"	S2@.231	LF	19.70	8.44	—	28.14
14" - 16"	S2@.254	LF	21.60	9.28	—	30.88
18" - 20"	S2@.285	LF	24.50	10.40	—	34.90
22" - 24"	S2@.316	LF	27.30	11.50	—	38.80
26" - 28"	S2@.346	LF	29.40	12.60	—	42.00
30"	S2@.370	LF	31.80	13.50	—	45.30

Galvanized Steel Rectangular Ductwork

Description	Craft@Hrs	Unit	Material $	Labor $	Equipment $	Total $

Galvanized steel 20" rectangular duct 24 gauge

Description	Craft@Hrs	Unit	Material $	Labor $	Equipment $	Total $
6"	S2@.200	LF	17.10	7.31	—	24.41
8"	S2@.216	LF	18.60	7.89	—	26.49
10"	S2@.231	LF	19.70	8.44	—	28.14
12"	S2@.247	LF	21.10	9.03	—	30.13
14" - 16"	S2@.269	LF	23.00	9.83	—	32.83
18" - 20"	S2@.300	LF	25.40	11.00	—	36.40
22" - 24"	S2@.331	LF	28.70	12.10	—	40.80
26" - 28"	S2@.362	LF	31.20	13.20	—	44.40
30"	S2@.385	LF	33.00	14.10	—	47.10

Galvanized steel 22" rectangular duct 24 gauge

Description	Craft@Hrs	Unit	Material $	Labor $	Equipment $	Total $
6"	S2@.216	LF	18.60	7.89	—	26.49
8"	S2@.231	LF	19.70	8.44	—	28.14
10"	S2@.247	LF	21.10	9.03	—	30.13
12"	S2@.262	LF	22.40	9.57	—	31.97
14" - 16"	S2@.285	LF	24.50	10.40	—	34.90
18" - 20"	S2@.316	LF	27.30	11.50	—	38.80
22" - 24"	S2@.346	LF	29.40	12.60	—	42.00
26" - 28"	S2@.377	LF	32.70	13.80	—	46.50
30"	S2@.400	LF	34.40	14.60	—	49.00

Galvanized steel 24" rectangular duct 24 gauge

Description	Craft@Hrs	Unit	Material $	Labor $	Equipment $	Total $
6"	S2@.231	LF	19.70	8.44	—	28.14
8"	S2@.247	LF	21.10	9.03	—	30.13
10"	S2@.262	LF	22.40	9.57	—	31.97
12"	S2@.278	LF	23.70	10.20	—	33.90
14" - 16"	S2@.300	LF	25.40	11.00	—	36.40
18" - 20"	S2@.331	LF	28.70	12.10	—	40.80
22" - 24"	S2@.362	LF	31.20	13.20	—	44.40
26" - 28"	S2@.392	LF	33.60	14.30	—	47.90
30"	S2@.415	LF	35.60	15.20	—	50.80

Galvanized steel 26" rectangular duct 24 gauge

Description	Craft@Hrs	Unit	Material $	Labor $	Equipment $	Total $
6"	S2@.247	LF	21.10	9.03	—	30.13
8"	S2@.262	LF	22.40	9.57	—	31.97
10"	S2@.278	LF	23.70	10.20	—	33.90
12"	S2@.292	LF	25.00	10.70	—	35.70
14" - 16"	S2@.316	LF	27.30	11.50	—	38.80
18" - 20"	S2@.346	LF	29.40	12.60	—	42.00
22" - 24"	S2@.377	LF	32.70	13.80	—	46.50
26" - 28"	S2@.407	LF	35.00	14.90	—	49.90
30"	S2@.431	LF	36.50	15.70	—	52.20

Galvanized Steel Rectangular Ductwork

Description	Craft@Hrs	Unit	Material $	Labor $	Equipment $	Total $

Galvanized steel 28" rectangular duct 24 gauge

Description	Craft@Hrs	Unit	Material $	Labor $	Equipment $	Total $
6"	S2@.262	LF	22.40	9.57	—	31.97
8"	S2@.278	LF	23.70	10.20	—	33.90
10"	S2@.292	LF	25.00	10.70	—	35.70
12"	S2@.308	LF	26.50	11.30	—	37.80
14" - 16"	S2@.331	LF	28.70	12.10	—	40.80
18" - 20"	S2@.362	LF	31.20	13.20	—	44.40
22" - 24"	S2@.392	LF	33.60	14.30	—	47.90
26" - 28"	S2@.423	LF	36.20	15.50	—	51.70
30"	S2@.448	LF	38.60	16.40	—	55.00

Galvanized steel 30" rectangular duct 24 gauge

Description	Craft@Hrs	Unit	Material $	Labor $	Equipment $	Total $
6"	S2@.278	LF	23.70	10.20	—	33.90
8"	S2@.292	LF	25.00	10.70	—	35.70
10"	S2@.308	LF	26.50	11.30	—	37.80
12"	S2@.323	LF	27.80	11.80	—	39.60
14" - 16"	S2@.346	LF	29.40	12.60	—	42.00
18" - 20"	S2@.377	LF	32.70	13.80	—	46.50
22" - 24"	S2@.407	LF	35.00	14.90	—	49.90
26" - 28"	S2@.440	LF	37.50	16.10	—	53.60
30"	S2@.465	LF	39.60	17.00	—	56.60

Costs for Galvanized 22 Gauge Rectangular Duct. Costs per linear foot of duct, based on quantities less than 1,000 pounds of prefabricated duct. Deduct 15% from the material cost for over 1,000 pounds. Add 50% to the material cost for lined duct. These costs include delivery, typical bracing, cleats, scrap, hangers, end closures.

Galvanized steel 32" rectangular duct 22 gauge

Description	Craft@Hrs	Unit	Material $	Labor $	Equipment $	Total $
8"	S2@.374	LF	32.10	13.70	—	45.80
10"	S2@.393	LF	33.60	14.40	—	48.00
12"	S2@.412	LF	35.20	15.10	—	50.30
14" - 16"	S2@.442	LF	37.80	16.20	—	54.00
18" - 20"	S2@.478	LF	41.10	17.50	—	58.60
22" - 24"	S2@.517	LF	44.30	18.90	—	63.20
26" - 28"	S2@.553	LF	47.30	20.20	—	67.50
30" - 32"	S2@.589	LF	50.70	21.50	—	72.20
34" - 36"	S2@.628	LF	53.80	22.90	—	76.70
38" - 40"	S2@.664	LF	57.10	24.30	—	81.40
42" - 44"	S2@.702	LF	59.80	25.70	—	85.50
46" - 48"	S2@.739	LF	63.20	27.00	—	90.20
50" - 52"	S2@.777	LF	66.40	28.40	—	94.80
54"	S2@.807	LF	68.70	29.50	—	98.20

Galvanized Steel Rectangular Ductwork

Description	Craft@Hrs	Unit	Material $	Labor $	Equipment $	Total $
Galvanized steel 34" rectangular duct 22 gauge						
8"	S2@.393	LF	33.60	14.40	—	48.00
10"	S2@.412	LF	35.20	15.10	—	50.30
12"	S2@.431	LF	36.50	15.70	—	52.20
14" - 16"	S2@.461	LF	39.30	16.80	—	56.10
18" - 20"	S2@.497	LF	43.10	18.20	—	61.30
22" - 24"	S2@.534	LF	45.50	19.50	—	65.00
26" - 28"	S2@.572	LF	49.20	20.90	—	70.10
30" - 32"	S2@.608	LF	52.00	22.20	—	74.20
34" - 36"	S2@.645	LF	55.50	23.60	—	79.10
38" - 40"	S2@.683	LF	58.40	25.00	—	83.40
42" - 44"	S2@.722	LF	62.00	26.40	—	88.40
46" - 48"	S2@.758	LF	65.10	27.70	—	92.80
50" - 52"	S2@.796	LF	67.90	29.10	—	97.00
54"	S2@.824	LF	70.70	30.10	—	100.80
Galvanized steel 36" rectangular duct 22 gauge						
8"	S2@.412	LF	35.20	15.10	—	50.30
10"	S2@.431	LF	36.50	15.70	—	52.20
12"	S2@.453	LF	38.90	16.60	—	55.50
14" - 16"	S2@.478	LF	41.10	17.50	—	58.60
18" - 20"	S2@.517	LF	44.30	18.90	—	63.20
22" - 24"	S2@.553	LF	47.30	20.20	—	67.50
26" - 28"	S2@.589	LF	50.70	21.50	—	72.20
30" - 32"	S2@.628	LF	53.80	22.90	—	76.70
34" - 36"	S2@.664	LF	57.10	24.30	—	81.40
38" - 40"	S2@.702	LF	59.80	25.70	—	85.50
42" - 44"	S2@.739	LF	63.20	27.00	—	90.20
46" - 48"	S2@.777	LF	66.40	28.40	—	94.80
50" - 52"	S2@.816	LF	69.20	29.80	—	99.00
54"	S2@.841	LF	72.40	30.70	—	103.10
Galvanized steel 38" rectangular duct 22 gauge						
8"	S2@.431	LF	36.50	15.70	—	52.20
10"	S2@.453	LF	38.90	16.60	—	55.50
12"	S2@.470	LF	40.00	17.20	—	57.20
14" - 16"	S2@.497	LF	43.10	18.20	—	61.30
18" - 20"	S2@.534	LF	45.50	19.50	—	65.00
22" - 24"	S2@.572	LF	49.20	20.90	—	70.10
26" - 28"	S2@.608	LF	52.00	22.20	—	74.20
30" - 32"	S2@.645	LF	55.50	23.60	—	79.10
34" - 36"	S2@.683	LF	58.40	25.00	—	83.40
38" - 40"	S2@.722	LF	62.00	26.40	—	88.40
42" - 44"	S2@.758	LF	65.10	27.70	—	92.80
46" - 48"	S2@.796	LF	67.90	29.10	—	97.00
50" - 52"	S2@.833	LF	71.40	30.40	—	101.80
54"	S2@.863	LF	73.50	31.50	—	105.00

Galvanized Steel Rectangular Ductwork

Description	Craft@Hrs	Unit	Material $	Labor $	Equipment $	Total $

Galvanized steel 40" rectangular duct 22 gauge

Description	Craft@Hrs	Unit	Material $	Labor $	Equipment $	Total $
8"	S2@.453	LF	38.90	16.60	—	55.50
10"	S2@.470	LF	40.00	17.20	—	57.20
12"	S2@.487	LF	41.90	17.80	—	59.70
14" - 16"	S2@.517	LF	44.30	18.90	—	63.20
18" - 20"	S2@.553	LF	47.30	20.20	—	67.50
22" - 24"	S2@.589	LF	50.70	21.50	—	72.20
26" - 28"	S2@.628	LF	53.80	22.90	—	76.70
30" - 32"	S2@.664	LF	57.10	24.30	—	81.40
34" - 36"	S2@.702	LF	59.80	25.70	—	85.50
38" - 40"	S2@.739	LF	63.20	27.00	—	90.20
42" - 44"	S2@.777	LF	66.40	28.40	—	94.80
46" - 48"	S2@.816	LF	69.20	29.80	—	99.00
50" - 52"	S2@.852	LF	73.10	31.10	—	104.20
54"	S2@.880	LF	74.90	32.20	—	107.10

Galvanized steel 42" rectangular duct 22 gauge

Description	Craft@Hrs	Unit	Material $	Labor $	Equipment $	Total $
8"	S2@.470	LF	40.00	17.20	—	57.20
10"	S2@.487	LF	41.90	17.80	—	59.70
12"	S2@.508	LF	43.80	18.60	—	62.40
14" - 16"	S2@.534	LF	45.50	19.50	—	65.00
18" - 20"	S2@.572	LF	49.20	20.90	—	70.10
22" - 24"	S2@.611	LF	52.00	22.30	—	74.30
26" - 28"	S2@.645	LF	55.50	23.60	—	79.10
30" - 32"	S2@.683	LF	58.40	25.00	—	83.40
34" - 36"	S2@.722	LF	62.00	26.40	—	88.40
38" - 40"	S2@.758	LF	65.10	27.70	—	92.80
42" - 44"	S2@.798	LF	67.90	29.20	—	97.10
46" - 48"	S2@.833	LF	71.40	30.40	—	101.80
50" - 52"	S2@.871	LF	74.60	31.80	—	106.40
54"	S2@.901	LF	76.70	32.90	—	109.60

Galvanized steel 44" rectangular duct 22 gauge

Description	Craft@Hrs	Unit	Material $	Labor $	Equipment $	Total $
8"	S2@.487	LF	41.80	17.80	—	59.60
10"	S2@.508	LF	43.20	18.60	—	61.80
12"	S2@.525	LF	44.70	19.20	—	63.90
14" - 16"	S2@.555	LF	46.80	20.30	—	67.10
18" - 20"	S2@.589	LF	50.50	21.50	—	72.00
22" - 24"	S2@.628	LF	53.20	22.90	—	76.10
26" - 28"	S2@.666	LF	56.80	24.30	—	81.10
30" - 32"	S2@.705	LF	59.60	25.80	—	85.40
34" - 36"	S2@.739	LF	63.00	27.00	—	90.00
38" - 40"	S2@.777	LF	66.10	28.40	—	94.50
42" - 44"	S2@.816	LF	69.00	29.80	—	98.80
46" - 48"	S2@.852	LF	72.70	31.10	—	103.80
50" - 52"	S2@.890	LF	75.40	32.50	—	107.90
54"	S2@.918	LF	78.00	33.50	—	111.50

Galvanized Steel Rectangular Ductwork

Description	Craft@Hrs	Unit	Material $	Labor $	Equipment $	Total $

Galvanized steel 46" rectangular duct 22 gauge

Description	Craft@Hrs	Unit	Material $	Labor $	Equipment $	Total $
8"	S2@.508	LF	43.80	18.60	—	62.40
10"	S2@.525	LF	44.70	19.20	—	63.90
12"	S2@.542	LF	46.40	19.80	—	66.20
14" - 16"	S2@.572	LF	49.20	20.90	—	70.10
18" - 20"	S2@.611	LF	52.00	22.30	—	74.30
22" - 24"	S2@.649	LF	55.50	23.70	—	79.20
26" - 28"	S2@.683	LF	58.40	25.00	—	83.40
30" - 32"	S2@.722	LF	62.00	26.40	—	88.40
34" - 36"	S2@.758	LF	65.10	27.70	—	92.80
38" - 40"	S2@.796	LF	67.90	29.10	—	97.00
42" - 44"	S2@.833	LF	71.40	30.40	—	101.80
46" - 48"	S2@.871	LF	74.60	31.80	—	106.40
50" - 52"	S2@.909	LF	77.60	33.20	—	110.80
54"	S2@.935	LF	80.00	34.20	—	114.20

Galvanized steel 48" rectangular duct 22 gauge

Description	Craft@Hrs	Unit	Material $	Labor $	Equipment $	Total $
8"	S2@.525	LF	44.70	19.20	—	63.90
10"	S2@.542	LF	46.40	19.80	—	66.20
12"	S2@.564	LF	48.10	20.60	—	68.70
14" - 16"	S2@.589	LF	50.70	21.50	—	72.20
18" - 20"	S2@.628	LF	53.80	22.90	—	76.70
22" - 24"	S2@.664	LF	57.10	24.30	—	81.40
26" - 28"	S2@.702	LF	59.80	25.70	—	85.50
30" - 32"	S2@.739	LF	63.20	27.00	—	90.20
34" - 36"	S2@.777	LF	66.40	28.40	—	94.80
38" - 40"	S2@.816	LF	69.20	29.80	—	99.00
42" - 44"	S2@.852	LF	73.10	31.10	—	104.20
46" - 48"	S2@.890	LF	75.90	32.50	—	108.40
50" - 52"	S2@.927	LF	79.60	33.90	—	113.50
54"	S2@.956	LF	81.20	34.90	—	116.10

Galvanized steel 50" rectangular duct 22 gauge

Description	Craft@Hrs	Unit	Material $	Labor $	Equipment $	Total $
8"	S2@.542	LF	46.40	19.80	—	66.20
10"	S2@.564	LF	48.10	20.60	—	68.70
12"	S2@.581	LF	49.70	21.20	—	70.90
14" - 16"	S2@.611	LF	52.00	22.30	—	74.30
18" - 20"	S2@.645	LF	55.50	23.60	—	79.10
22" - 24"	S2@.683	LF	58.40	25.00	—	83.40
26" - 28"	S2@.722	LF	62.00	26.40	—	88.40
30" - 32"	S2@.758	LF	65.10	27.70	—	92.80
34" - 36"	S2@.798	LF	67.90	29.20	—	97.10
38" - 40"	S2@.833	LF	71.40	30.40	—	101.80
42" - 44"	S2@.871	LF	74.60	31.80	—	106.40
46" - 48"	S2@.909	LF	77.60	33.20	—	110.80
50" - 52"	S2@.946	LF	80.70	34.60	—	115.30
54"	S2@.974	LF	83.00	35.60	—	118.60

Galvanized Steel Rectangular Ductwork

Description	Craft@Hrs	Unit	Material $	Labor $	Equipment $	Total $

Galvanized steel 52" rectangular duct 22 gauge

8"	S2@.564	LF	48.10	20.60	—	68.70
10"	S2@.581	LF	49.70	21.20	—	70.90
12"	S2@.598	LF	51.30	21.90	—	73.20
14" - 16"	S2@.628	LF	53.80	22.90	—	76.70
18" - 20"	S2@.662	LF	57.10	24.20	—	81.30
22" - 24"	S2@.700	LF	59.80	25.60	—	85.40
26" - 28"	S2@.739	LF	63.20	27.00	—	90.20
30" - 32"	S2@.777	LF	66.40	28.40	—	94.80
34" - 36"	S2@.816	LF	69.20	29.80	—	99.00
38" - 40"	S2@.852	LF	73.10	31.10	—	104.20
42" - 44"	S2@.890	LF	75.90	32.50	—	108.40
46" - 48"	S2@.927	LF	79.60	33.90	—	113.50
50" - 52"	S2@.976	LF	82.30	35.70	—	118.00
54"	S2@.991	LF	84.40	36.20	—	120.60

Galvanized steel 54" rectangular duct 22 gauge

8"	S2@.581	LF	49.70	21.20	—	70.90
10"	S2@.598	LF	51.30	21.90	—	73.20
12"	S2@.619	LF	52.80	22.60	—	75.40
14" - 16"	S2@.645	LF	55.50	23.60	—	79.10
18" - 20"	S2@.683	LF	58.40	25.00	—	83.40
22" - 24"	S2@.722	LF	62.00	26.40	—	88.40
26" - 28"	S2@.758	LF	65.10	27.70	—	92.80
30" - 32"	S2@.796	LF	67.90	29.10	—	97.00
34" - 36"	S2@.833	LF	71.40	30.40	—	101.80
38" - 40"	S2@.871	LF	74.60	31.80	—	106.40
42" - 44"	S2@.909	LF	77.60	33.20	—	110.80
46" - 48"	S2@.946	LF	80.70	34.60	—	115.30
50" - 52"	S2@.982	LF	83.50	35.90	—	119.40
54"	S2@1.01	LF	86.70	36.90	—	123.60

Galvanized Steel Rectangular Ductwork

Costs for Galvanized 20 Gauge Rectangular Duct. Costs per linear foot of duct, based on quantities less than 1,000 pounds of prefabricated duct. Deduct 15% from the material cost for over 1,000 pounds. Add 50% to the material cost for lined duct. These costs include delivery, typical bracing, cleats, scrap, hangers, end closures, sealants and miscellaneous hardware.

Description	Craft@Hrs	Unit	Material $	Labor $	Equipment $	Total $
Galvanized steel 56" rectangular duct 20 gauge						
18" - 20"	S2@.828	LF	71.40	30.30	—	101.70
22" - 24"	S2@.871	LF	74.90	31.80	—	106.70
26" - 28"	S2@.916	LF	78.50	33.50	—	112.00
30" - 32"	S2@.959	LF	82.70	35.00	—	117.70
34" - 36"	S2@1.00	LF	86.70	36.50	—	123.20
38" - 40"	S2@1.05	LF	90.40	38.40	—	128.80
42" - 44"	S2@1.09	LF	94.20	39.80	—	134.00
46" - 48"	S2@1.14	LF	97.90	41.70	—	139.60
50" - 52"	S2@1.18	LF	102.00	43.10	—	145.10
54" - 56"	S2@1.23	LF	105.00	44.90	—	149.90
58" - 60"	S2@1.27	LF	109.00	46.40	—	155.40
62" - 64"	S2@1.31	LF	113.00	47.90	—	160.90
66" - 68"	S2@1.36	LF	116.00	49.70	—	165.70
70" - 72"	S2@1.40	LF	120.00	51.20	—	171.20
Galvanized steel 58" rectangular duct 20 gauge						
18" - 20"	S2@.850	LF	73.20	31.10	—	104.30
22" - 24"	S2@.895	LF	77.00	32.70	—	109.70
26" - 28"	S2@.937	LF	80.70	34.20	—	114.90
30" - 32"	S2@.980	LF	84.70	35.80	—	120.50
34" - 36"	S2@1.02	LF	88.20	37.30	—	125.50
38" - 40"	S2@1.07	LF	92.10	39.10	—	131.20
42" - 44"	S2@1.11	LF	96.10	40.60	—	136.70
46" - 48"	S2@1.16	LF	99.60	42.40	—	142.00
50" - 52"	S2@1.02	LF	104.00	37.30	—	141.30
54" - 56"	S2@1.25	LF	107.00	45.70	—	152.70
58" - 60"	S2@1.29	LF	110.00	47.10	—	157.10
62" - 64"	S2@1.33	LF	115.00	48.60	—	163.60
66" - 68"	S2@1.38	LF	118.00	50.40	—	168.40
70" - 72"	S2@1.42	LF	121.00	51.90	—	172.90
Galvanized steel 60" rectangular duct 20 gauge						
18" - 20"	S2@.871	LF	74.90	31.80	—	106.70
22" - 24"	S2@.916	LF	78.50	33.50	—	112.00
26" - 28"	S2@.959	LF	82.70	35.00	—	117.70
30" - 32"	S2@1.00	LF	86.70	36.50	—	123.20
34" - 36"	S2@1.05	LF	90.40	38.40	—	128.80
38" - 40"	S2@1.09	LF	94.20	39.80	—	134.00
42" - 44"	S2@1.14	LF	97.90	41.70	—	139.60
46" - 48"	S2@1.18	LF	102.00	43.10	—	145.10
50" - 52"	S2@1.23	LF	105.00	44.90	—	149.90
54" - 56"	S2@1.27	LF	109.00	46.40	—	155.40
58" - 60"	S2@1.31	LF	113.00	47.90	—	160.90
62" - 64"	S2@1.36	LF	116.00	49.70	—	165.70
66" - 68"	S2@1.40	LF	120.00	51.20	—	171.20
70" - 72"	S2@1.45	LF	123.00	53.00	—	176.00

Galvanized Steel Rectangular Ductwork

Description	Craft@Hrs	Unit	Material $	Labor $	Equipment $	Total $

Galvanized steel 62" rectangular duct 20 gauge

Description	Craft@Hrs	Unit	Material $	Labor $	Equipment $	Total $
18" - 20"	S2@.895	LF	77.00	32.70	—	109.70
22" - 24"	S2@.937	LF	80.70	34.20	—	114.90
26" - 28"	S2@.980	LF	84.70	35.80	—	120.50
30" - 32"	S2@1.02	LF	88.20	37.30	—	125.50
34" - 36"	S2@1.07	LF	92.10	39.10	—	131.20
38" - 40"	S2@1.11	LF	96.10	40.60	—	136.70
42" - 44"	S2@1.16	LF	99.60	42.40	—	142.00
46" - 48"	S2@1.20	LF	104.00	43.80	—	147.80
50" - 52"	S2@1.25	LF	107.00	45.70	—	152.70
54" - 56"	S2@1.29	LF	110.00	47.10	—	157.10
58" - 60"	S2@1.33	LF	115.00	48.60	—	163.60
62" - 64"	S2@1.38	LF	118.00	50.40	—	168.40
66" - 68"	S2@1.42	LF	121.00	51.90	—	172.90
70" - 72"	S2@1.47	LF	124.00	53.70	—	177.70

Galvanized steel 64" rectangular duct 20 gauge

Description	Craft@Hrs	Unit	Material $	Labor $	Equipment $	Total $
18" - 20"	S2@.916	LF	78.50	33.50	—	112.00
22" - 24"	S2@.959	LF	82.70	35.00	—	117.70
26" - 28"	S2@1.00	LF	86.70	36.50	—	123.20
30" - 32"	S2@1.05	LF	90.40	38.40	—	128.80
34" - 36"	S2@1.09	LF	94.20	39.80	—	134.00
38" - 40"	S2@1.14	LF	97.90	41.70	—	139.60
42" - 44"	S2@1.18	LF	102.00	43.10	—	145.10
46" - 48"	S2@1.23	LF	105.00	44.90	—	149.90
50" - 52"	S2@1.27	LF	109.00	46.40	—	155.40
54" - 56"	S2@1.31	LF	113.00	47.90	—	160.90
58" - 60"	S2@1.36	LF	116.00	49.70	—	165.70
62" - 64"	S2@1.40	LF	120.00	51.20	—	171.20
66" - 68"	S2@1.45	LF	123.00	53.00	—	176.00
70" - 72"	S2@1.49	LF	128.00	54.40	—	182.40

Galvanized steel 66" rectangular duct 20 gauge

Description	Craft@Hrs	Unit	Material $	Labor $	Equipment $	Total $
18" - 20"	S2@.937	LF	80.70	34.20	—	114.90
22" - 24"	S2@.980	LF	84.70	35.80	—	120.50
26" - 28"	S2@1.02	LF	88.20	37.30	—	125.50
30" - 32"	S2@1.07	LF	92.10	39.10	—	131.20
34" - 36"	S2@1.11	LF	96.10	40.60	—	136.70
38" - 40"	S2@1.16	LF	99.60	42.40	—	142.00
42" - 44"	S2@1.20	LF	104.00	43.80	—	147.80
46" - 48"	S2@1.25	LF	107.00	45.70	—	152.70
50" - 52"	S2@1.29	LF	110.00	47.10	—	157.10
54" - 56"	S2@1.33	LF	115.00	48.60	—	163.60
58" - 60"	S2@1.38	LF	118.00	50.40	—	168.40
62" - 64"	S2@1.42	LF	121.00	51.90	—	172.90
66" - 68"	S2@1.47	LF	124.00	53.70	—	177.70
70" - 72"	S2@1.51	LF	129.00	55.20	—	184.20

Galvanized Steel Rectangular Ductwork

Description	Craft@Hrs	Unit	Material $	Labor $	Equipment $	Total $

Galvanized steel 68" rectangular duct 20 gauge

Description	Craft@Hrs	Unit	Material $	Labor $	Equipment $	Total $
18" - 20"	S2@.959	LF	82.70	35.00	—	117.70
22" - 24"	S2@1.00	LF	86.70	36.50	—	123.20
26" - 28"	S2@1.05	LF	90.40	38.40	—	128.80
30" - 32"	S2@1.09	LF	94.20	39.80	—	134.00
34" - 36"	S2@1.14	LF	97.90	41.70	—	139.60
38" - 40"	S2@1.18	LF	102.00	43.10	—	145.10
42" - 44"	S2@1.23	LF	105.00	44.90	—	149.90
46" - 48"	S2@1.27	LF	109.00	46.40	—	155.40
50" - 52"	S2@1.31	LF	113.00	47.90	—	160.90
54" - 56"	S2@1.36	LF	116.00	49.70	—	165.70
58" - 60"	S2@1.40	LF	120.00	51.20	—	171.20
62" - 64"	S2@1.45	LF	123.00	53.00	—	176.00
66" - 68"	S2@1.49	LF	128.00	54.40	—	182.40
70" - 72"	S2@1.53	LF	133.00	55.90	—	188.90

Galvanized steel 70" rectangular duct 20 gauge

Description	Craft@Hrs	Unit	Material $	Labor $	Equipment $	Total $
18" - 20"	S2@.980	LF	84.70	35.80	—	120.50
22" - 24"	S2@1.02	LF	88.20	37.30	—	125.50
26" - 28"	S2@1.07	LF	92.10	39.10	—	131.20
30" - 32"	S2@1.11	LF	96.10	40.60	—	136.70
34" - 36"	S2@1.16	LF	99.60	42.40	—	142.00
38" - 40"	S2@1.20	LF	104.00	43.80	—	147.80
42" - 44"	S2@1.25	LF	107.00	45.70	—	152.70
46" - 48"	S2@1.29	LF	110.00	47.10	—	157.10
50" - 52"	S2@1.33	LF	115.00	48.60	—	163.60
54" - 56"	S2@1.38	LF	118.00	50.40	—	168.40
58" - 60"	S2@1.42	LF	121.00	51.90	—	172.90
62" - 64"	S2@1.47	LF	124.00	53.70	—	177.70
66" - 68"	S2@1.51	LF	129.00	55.20	—	184.20
70" - 72"	S2@1.55	LF	134.00	56.60	—	190.60

Galvanized steel 72" rectangular duct 20 gauge

Description	Craft@Hrs	Unit	Material $	Labor $	Equipment $	Total $
18" - 20"	S2@1.00	LF	86.70	36.50	—	123.20
22" - 24"	S2@1.05	LF	90.40	38.40	—	128.80
26" - 28"	S2@1.09	LF	94.20	39.80	—	134.00
30" - 32"	S2@1.14	LF	97.90	41.70	—	139.60
34" - 36"	S2@1.18	LF	102.00	43.10	—	145.10
38" - 40"	S2@1.23	LF	105.00	44.90	—	149.90
42" - 44"	S2@1.27	LF	109.00	46.40	—	155.40
46" - 48"	S2@1.31	LF	113.00	47.90	—	160.90
50" - 52"	S2@1.36	LF	116.00	49.70	—	165.70
54" - 56"	S2@1.40	LF	120.00	51.20	—	171.20
58" - 60"	S2@1.45	LF	123.00	53.00	—	176.00
62" - 64"	S2@1.49	LF	128.00	54.40	—	182.40
66" - 68"	S2@1.53	LF	133.00	55.90	—	188.90
70" - 72"	S2@1.58	LF	135.00	57.70	—	192.70

Galvanized Steel Rectangular Elbows

Costs for Galvanized Rectangular 90-Degree Elbows. Costs per elbow, based on quantities less than 1,000 pounds. Deduct 15% from the material cost for over 1,000 pounds. Add 50% to the material cost for lined duct. These costs include delivery, typical bracing, cleats, scrap, hangers, end closures, sealants and miscellaneous hardware. For turning vanes, please see page 342.

Description	Craft@Hrs	Unit	Material $	Labor $	Equipment $	Total $

Galvanized steel 4" rectangular duct 90-degree elbow

Description	Craft@Hrs	Unit	Material $	Labor $	Equipment $	Total $
4"	S2@.042	Ea	7.24	1.53	—	8.77
6"	S2@.058	Ea	9.95	2.12	—	12.07
8"	S2@.078	Ea	13.40	2.85	—	16.25
10"	S2@.104	Ea	17.90	3.80	—	21.70
12"	S2@.132	Ea	22.60	4.82	—	27.42
14" - 16"	S2@.236	Ea	40.50	8.62	—	49.12
18" - 20"	S2@.339	Ea	58.40	12.40	—	70.80
22" - 24"	S2@.468	Ea	80.00	17.10	—	97.10
26" - 28"	S2@.615	Ea	104.00	22.50	—	126.50
30" - 32"	S2@.854	Ea	149.00	31.20	—	180.20
34" - 36"	S2@1.19	Ea	204.00	43.50	—	247.50
38" - 40"	S2@1.45	Ea	249.00	53.00	—	302.00
42" - 44"	S2@1.73	Ea	294.00	63.20	—	357.20
46" - 48"	S2@2.04	Ea	350.00	74.50	—	424.50
50"	S2@2.28	Ea	393.00	83.30	—	476.30

Galvanized steel 6" rectangular duct 90-degree elbow

Description	Craft@Hrs	Unit	Material $	Labor $	Equipment $	Total $
4"	S2@.058	Ea	9.95	2.12	—	12.07
6"	S2@.078	Ea	13.40	2.85	—	16.25
8"	S2@.104	Ea	17.90	3.80	—	21.70
10"	S2@.132	Ea	22.60	4.82	—	27.42
12"	S2@.148	Ea	25.30	5.41	—	30.71
14" - 16"	S2@.266	Ea	45.90	9.72	—	55.62
18" - 20"	S2@.370	Ea	63.20	13.50	—	76.70
22" - 24"	S2@.506	Ea	86.90	18.50	—	105.40
26" - 28"	S2@.655	Ea	111.00	23.90	—	134.90
30" - 32"	S2@.905	Ea	154.00	33.10	—	187.10
34" - 36"	S2@1.24	Ea	213.00	45.30	—	258.30
38" - 40"	S2@1.49	Ea	254.00	54.40	—	308.40
42" - 44"	S2@1.78	Ea	303.00	65.00	—	368.00
46" - 48"	S2@2.09	Ea	358.00	76.40	—	434.40
50"	S2@2.34	Ea	403.00	85.50	—	488.50

Galvanized Steel Rectangular Elbows

Description	Craft@Hrs	Unit	Material $	Labor $	Equipment $	Total $

Galvanized steel 8" rectangular duct 90-degree elbow

Description	Craft@Hrs	Unit	Material $	Labor $	Equipment $	Total $
4"	S2@.078	Ea	13.40	2.85	—	16.25
6"	S2@.104	Ea	17.90	3.80	—	21.70
8"	S2@.132	Ea	22.60	4.82	—	27.42
10"	S2@.148	Ea	25.30	5.41	—	30.71
12"	S2@.170	Ea	29.30	6.21	—	35.51
14" - 16"	S2@.300	Ea	51.30	11.00	—	62.30
18" - 20"	S2@.403	Ea	69.00	14.70	—	83.70
22" - 24"	S2@.549	Ea	94.10	20.10	—	114.20
26" - 28"	S2@.694	Ea	118.00	25.40	—	143.40
30" - 32"	S2@.950	Ea	165.00	34.70	—	199.70
34" - 36"	S2@1.28	Ea	219.00	46.80	—	265.80
38" - 40"	S2@1.54	Ea	267.00	56.30	—	323.30
42" - 44"	S2@1.83	Ea	314.00	66.90	—	380.90
46" - 48"	S2@2.13	Ea	365.00	77.80	—	442.80
50"	S2@2.39	Ea	415.00	87.30	—	502.30

Galvanized steel 10" rectangular duct 90-degree elbow

Description	Craft@Hrs	Unit	Material $	Labor $	Equipment $	Total $
4"	S2@.104	Ea	17.90	3.80	—	21.70
6"	S2@.132	Ea	22.60	4.82	—	27.42
8"	S2@.148	Ea	25.30	5.41	—	30.71
10"	S2@.170	Ea	29.30	6.21	—	35.51
12"	S2@.194	Ea	33.20	7.09	—	40.29
14" - 16"	S2@.336	Ea	57.20	12.30	—	69.50
18" - 20"	S2@.438	Ea	74.90	16.00	—	90.90
22" - 24"	S2@.587	Ea	101.00	21.40	—	122.40
26" - 28"	S2@.737	Ea	124.00	26.90	—	150.90
30" - 32"	S2@.997	Ea	171.00	36.40	—	207.40
34" - 36"	S2@1.33	Ea	227.00	48.60	—	275.60
38" - 40"	S2@1.59	Ea	275.00	58.10	—	333.10
42" - 44"	S2@1.87	Ea	321.00	68.30	—	389.30
46" - 48"	S2@2.18	Ea	375.00	79.70	—	454.70
50"	S2@2.43	Ea	419.00	88.80	—	507.80

Galvanized steel 12" rectangular duct 90-degree elbow

Description	Craft@Hrs	Unit	Material $	Labor $	Equipment $	Total $
4"	S2@.132	Ea	22.60	4.82	—	27.42
6"	S2@.148	Ea	25.30	5.41	—	30.71
8"	S2@.170	Ea	29.30	6.21	—	35.51
10"	S2@.194	Ea	33.20	7.09	—	40.29
12"	S2@.224	Ea	38.60	8.18	—	46.78
18" - 20"	S2@.478	Ea	81.20	17.50	—	98.70
22" - 24"	S2@.628	Ea	107.00	22.90	—	129.90
26" - 28"	S2@.775	Ea	134.00	28.30	—	162.30
30" - 32"	S2@1.04	Ea	179.00	38.00	—	217.00
34" - 36"	S2@1.38	Ea	237.00	50.40	—	287.40
38" - 40"	S2@1.64	Ea	281.00	59.90	—	340.90
42" - 44"	S2@1.92	Ea	330.00	70.20	—	400.20
46" - 48"	S2@2.23	Ea	386.00	81.50	—	467.50
50"	S2@2.48	Ea	424.00	90.60	—	514.60

Galvanized Steel Rectangular Elbows

Description	Craft@Hrs	Unit	Material $	Labor $	Equipment $	Total $

Galvanized steel 14" rectangular duct 90-degree elbow

4"	S2@.212	Ea	36.20	7.75	—	43.95
6"	S2@.232	Ea	40.00	8.48	—	48.48
8"	S2@.260	Ea	44.60	9.50	—	54.10
10"	S2@.292	Ea	50.10	10.70	—	60.80
12"	S2@.328	Ea	56.30	12.00	—	68.30
14" - 16"	S2@.415	Ea	71.40	15.20	—	86.60
18" - 20"	S2@.527	Ea	90.40	19.30	—	109.70
22" - 24"	S2@.867	Ea	114.00	31.70	—	145.70
26" - 28"	S2@.816	Ea	138.00	29.80	—	167.80
30" - 32"	S2@1.09	Ea	188.00	39.80	—	227.80
34" - 36"	S2@1.43	Ea	247.00	52.30	—	299.30
38" - 40"	S2@1.69	Ea	290.00	61.80	—	351.80
42" - 44"	S2@1.97	Ea	341.00	72.00	—	413.00
46" - 48"	S2@2.28	Ea	391.00	83.30	—	474.30
50"	S2@2.53	Ea	438.00	92.40	—	530.40

Galvanized steel 16" rectangular duct 90-degree elbow

4"	S2@.260	Ea	44.60	9.50	—	54.10
6"	S2@.310	Ea	53.20	11.30	—	64.50
8"	S2@.340	Ea	58.30	12.40	—	70.70
10"	S2@.380	Ea	65.50	13.90	—	79.40
12"	S2@.420	Ea	72.00	15.30	—	87.30
14" - 16"	S2@.480	Ea	82.20	17.50	—	99.70
18" - 20"	S2@.555	Ea	95.20	20.30	—	115.50
22" - 24"	S2@.707	Ea	120.00	25.80	—	145.80
26" - 28"	S2@.856	Ea	148.00	31.30	—	179.30
30" - 32"	S2@1.14	Ea	195.00	41.70	—	236.70
34" - 36"	S2@1.48	Ea	253.00	54.10	—	307.10
38" - 40"	S2@1.73	Ea	298.00	63.20	—	361.20
42" - 44"	S2@2.02	Ea	349.00	73.80	—	422.80
46" - 48"	S2@2.32	Ea	403.00	84.80	—	487.80
50"	S2@2.58	Ea	443.00	94.30	—	537.30

Galvanized steel 18" rectangular duct 90-degree elbow

4"	S2@.312	Ea	53.80	11.40	—	65.20
6"	S2@.350	Ea	60.30	12.80	—	73.10
8"	S2@.390	Ea	67.20	14.30	—	81.50
10"	S2@.427	Ea	73.40	15.60	—	89.00
12"	S2@.470	Ea	80.00	17.20	—	97.20
14" - 16"	S2@.519	Ea	89.40	19.00	—	108.40
18" - 20"	S2@.604	Ea	103.00	22.10	—	125.10
22" - 24"	S2@.747	Ea	128.00	27.30	—	155.30
26" - 28"	S2@.895	Ea	153.00	32.70	—	185.70
30" - 32"	S2@1.18	Ea	203.00	43.10	—	246.10
34" - 36"	S2@1.53	Ea	265.00	55.90	—	320.90
38" - 40"	S2@1.78	Ea	303.00	65.00	—	368.00
42" - 44"	S2@2.07	Ea	356.00	75.60	—	431.60
46" - 48"	S2@2.37	Ea	411.00	86.60	—	497.60
50"	S2@2.63	Ea	451.00	96.10	—	547.10

Galvanized Steel Rectangular Elbows

Description	Craft@Hrs	Unit	Material $	Labor $	Equipment $	Total $

Galvanized steel 20" rectangular duct 90-degree elbow

Description	Craft@Hrs	Unit	Material $	Labor $	Equipment $	Total $
4"	S2@.370	Ea	63.20	13.50	—	76.70
6"	S2@.390	Ea	67.20	14.30	—	81.50
8"	S2@.415	Ea	71.40	15.20	—	86.60
10"	S2@.448	Ea	76.80	16.40	—	93.20
12"	S2@.487	Ea	83.10	17.80	—	100.90
14" - 16"	S2@.553	Ea	94.70	20.20	—	114.90
18" - 20"	S2@.658	Ea	112.00	24.00	—	136.00
22" - 24"	S2@.786	Ea	135.00	28.70	—	163.70
26" - 28"	S2@.931	Ea	159.00	34.00	—	193.00
30" - 32"	S2@1.23	Ea	210.00	44.90	—	254.90
34" - 36"	S2@1.58	Ea	273.00	57.70	—	330.70
38" - 40"	S2@1.83	Ea	317.00	66.90	—	383.90
42" - 44"	S2@2.11	Ea	362.00	77.10	—	439.10
46" - 48"	S2@2.42	Ea	418.00	88.40	—	506.40
50"	S2@2.67	Ea	463.00	97.60	—	560.60

Galvanized steel rectangular 22" duct 90-degree elbow

Description	Craft@Hrs	Unit	Material $	Labor $	Equipment $	Total $
22" - 24"	S2@.828	Ea	140.00	30.30	—	170.30
26" - 28"	S2@1.01	Ea	172.00	36.90	—	208.90
30" - 32"	S2@1.31	Ea	224.00	47.90	—	271.90
34" - 36"	S2@1.71	Ea	293.00	62.50	—	355.50
38" - 40"	S2@2.01	Ea	349.00	73.40	—	422.40
42" - 44"	S2@2.31	Ea	393.00	84.40	—	477.40
46" - 48"	S2@2.61	Ea	447.00	95.40	—	542.40
50" - 52"	S2@2.91	Ea	501.00	106.00	—	607.00
54" - 56"	S2@3.51	Ea	603.00	128.00	—	731.00
58" - 60"	S2@4.26	Ea	732.00	156.00	—	888.00
62" - 64"	S2@4.76	Ea	811.00	174.00	—	985.00
66" - 68"	S2@5.25	Ea	902.00	192.00	—	1,094.00
70" - 72"	S2@5.76	Ea	985.00	210.00	—	1,195.00

Galvanized steel 24" rectangular duct 90-degree elbow

Description	Craft@Hrs	Unit	Material $	Labor $	Equipment $	Total $
22" - 24"	S2@.922	Ea	158.00	33.70	—	191.70
26" - 28"	S2@1.10	Ea	189.00	40.20	—	229.20
30" - 32"	S2@1.41	Ea	243.00	51.50	—	294.50
34" - 36"	S2@1.81	Ea	312.00	66.10	—	378.10
38" - 40"	S2@2.10	Ea	362.00	76.70	—	438.70
42" - 44"	S2@2.41	Ea	415.00	88.10	—	503.10
46" - 48"	S2@2.71	Ea	464.00	99.00	—	563.00
50" - 52"	S2@3.01	Ea	513.00	110.00	—	623.00
54" - 56"	S2@3.62	Ea	621.00	132.00	—	753.00
58" - 60"	S2@4.38	Ea	749.00	160.00	—	909.00
62" - 64"	S2@4.87	Ea	831.00	178.00	—	1,009.00
66" - 68"	S2@5.38	Ea	920.00	197.00	—	1,117.00
70" - 72"	S2@5.87	Ea	1,010.00	214.00	—	1,224.00

Galvanized Steel Rectangular Elbows

Description	Craft@Hrs	Unit	Material $	Labor $	Equipment $	Total $

Galvanized steel 26" rectangular duct 90-degree elbow

22" - 24"	S2@1.01	Ea	172.00	36.90	—	208.90
26" - 28"	S2@1.19	Ea	204.00	43.50	—	247.50
30" - 32"	S2@1.50	Ea	257.00	54.80	—	311.80
34" - 36"	S2@1.92	Ea	330.00	70.20	—	400.20
38" - 40"	S2@2.21	Ea	378.00	80.80	—	458.80
42" - 44"	S2@2.51	Ea	432.00	91.70	—	523.70
46" - 48"	S2@2.81	Ea	480.00	103.00	—	583.00
50" - 52"	S2@3.11	Ea	538.00	114.00	—	652.00
54" - 56"	S2@3.73	Ea	644.00	136.00	—	780.00
58" - 60"	S2@4.50	Ea	770.00	164.00	—	934.00
62" - 64"	S2@5.02	Ea	857.00	183.00	—	1,040.00
66" - 68"	S2@5.53	Ea	952.00	202.00	—	1,154.00
70" - 72"	S2@6.04	Ea	1,030.00	221.00	—	1,251.00

Galvanized steel 28" rectangular duct 90-degree elbow

22" - 24"	S2@1.10	Ea	188.00	40.20	—	228.20
26" - 28"	S2@1.28	Ea	219.00	46.80	—	265.80
30" - 32"	S2@1.60	Ea	277.00	58.50	—	335.50
34" - 36"	S2@2.03	Ea	350.00	74.20	—	424.20
38" - 40"	S2@2.31	Ea	393.00	84.40	—	477.40
42" - 44"	S2@2.61	Ea	447.00	95.40	—	542.40
46" - 48"	S2@2.91	Ea	501.00	106.00	—	607.00
50" - 52"	S2@3.21	Ea	555.00	117.00	—	672.00
54" - 56"	S2@3.84	Ea	661.00	140.00	—	801.00
58" - 60"	S2@4.61	Ea	795.00	168.00	—	963.00
62" - 64"	S2@5.15	Ea	873.00	188.00	—	1,061.00
66" - 68"	S2@5.66	Ea	965.00	207.00	—	1,172.00
70" - 72"	S2@6.15	Ea	1,040.00	225.00	—	1,265.00

Galvanized steel 30" rectangular duct 90-degree elbow

22" - 24"	S2@1.19	Ea	203.00	43.50	—	246.50
26" - 28"	S2@1.37	Ea	237.00	50.10	—	287.10
30" - 32"	S2@1.70	Ea	291.00	62.10	—	353.10
34" - 36"	S2@2.14	Ea	365.00	78.20	—	443.20
38" - 40"	S2@2.41	Ea	415.00	88.10	—	503.10
42" - 44"	S2@2.71	Ea	464.00	99.00	—	563.00
46" - 48"	S2@3.01	Ea	513.00	110.00	—	623.00
50" - 52"	S2@3.31	Ea	568.00	121.00	—	689.00
54" - 56"	S2@3.95	Ea	676.00	144.00	—	820.00
58" - 60"	S2@4.74	Ea	809.00	173.00	—	982.00
62" - 64"	S2@5.25	Ea	902.00	192.00	—	1,094.00
66" - 68"	S2@5.79	Ea	987.00	212.00	—	1,199.00
70" - 72"	S2@6.28	Ea	1,070.00	229.00	—	1,299.00

Galvanized Steel Rectangular Elbows

Description	Craft@Hrs	Unit	Material $	Labor $	Equipment $	Total $

Galvanized steel 32" rectangular duct 90-degree elbow

Description	Craft@Hrs	Unit	Material $	Labor $	Equipment $	Total $
22" - 24"	S2@1.53	Ea	265.00	55.90	—	320.90
26" - 28"	S2@1.73	Ea	294.00	63.20	—	357.20
30" - 32"	S2@1.93	Ea	332.00	70.50	—	402.50
34" - 36"	S2@2.25	Ea	389.00	82.20	—	471.20
38" - 40"	S2@2.51	Ea	432.00	91.70	—	523.70
42" - 44"	S2@2.81	Ea	480.00	103.00	—	583.00
46" - 48"	S2@3.11	Ea	532.00	114.00	—	646.00
50" - 52"	S2@3.41	Ea	584.00	125.00	—	709.00
54" - 56"	S2@4.06	Ea	692.00	148.00	—	840.00
58" - 60"	S2@4.85	Ea	830.00	177.00	—	1,007.00
62" - 64"	S2@5.38	Ea	920.00	197.00	—	1,117.00
66" - 68"	S2@5.89	Ea	1,010.00	215.00	—	1,225.00
70" - 72"	S2@6.40	Ea	1,090.00	234.00	—	1,324.00

Galvanized steel 34" rectangular duct 90-degree elbow

Description	Craft@Hrs	Unit	Material $	Labor $	Equipment $	Total $
22" - 24"	S2@1.67	Ea	288.00	61.00	—	349.00
26" - 28"	S2@1.91	Ea	330.00	69.80	—	399.80
30" - 32"	S2@2.12	Ea	362.00	77.50	—	439.50
34" - 36"	S2@2.36	Ea	405.00	86.20	—	491.20
38" - 40"	S2@2.60	Ea	447.00	95.00	—	542.00
42" - 44"	S2@2.91	Ea	501.00	106.00	—	607.00
46" - 48"	S2@3.21	Ea	544.00	117.00	—	661.00
50" - 52"	S2@3.51	Ea	603.00	128.00	—	731.00
54" - 56"	S2@4.17	Ea	717.00	152.00	—	869.00
58" - 60"	S2@4.97	Ea	855.00	182.00	—	1,037.00
62" - 64"	S2@5.49	Ea	942.00	201.00	—	1,143.00
66" - 68"	S2@6.04	Ea	1,030.00	221.00	—	1,251.00
70" - 72"	S2@6.51	Ea	1,100.00	238.00	—	1,338.00

Galvanized steel 36" rectangular duct 90-degree elbow

Description	Craft@Hrs	Unit	Material $	Labor $	Equipment $	Total $
22" - 24"	S2@1.83	Ea	317.00	66.90	—	383.90
26" - 28"	S2@2.04	Ea	352.00	74.50	—	426.50
30" - 32"	S2@2.26	Ea	389.00	82.60	—	471.60
34" - 36"	S2@2.47	Ea	424.00	90.30	—	514.30
38" - 40"	S2@2.70	Ea	464.00	98.70	—	562.70
42" - 44"	S2@3.01	Ea	513.00	110.00	—	623.00
46" - 48"	S2@3.31	Ea	568.00	121.00	—	689.00
50" - 52"	S2@3.61	Ea	620.00	132.00	—	752.00
54" - 56"	S2@4.27	Ea	732.00	156.00	—	888.00
58" - 60"	S2@5.10	Ea	870.00	186.00	—	1,056.00
62" - 64"	S2@5.61	Ea	961.00	205.00	—	1,166.00
66" - 68"	S2@6.15	Ea	1,040.00	225.00	—	1,265.00
70" - 72"	S2@6.64	Ea	1,130.00	243.00	—	1,373.00

Galvanized Steel Rectangular Elbows

Description	Craft@Hrs	Unit	Material $	Labor $	Equipment $	Total $

Galvanized steel 38" rectangular duct 90-degree elbow

Description	Craft@Hrs	Unit	Material $	Labor $	Equipment $	Total $
22" - 24"	S2@1.98	Ea	341.00	72.30	—	413.30
26" - 28"	S2@2.18	Ea	375.00	79.70	—	454.70
30" - 32"	S2@2.38	Ea	411.00	87.00	—	498.00
34" - 36"	S2@2.58	Ea	446.00	94.30	—	540.30
38" - 40"	S2@2.81	Ea	480.00	103.00	—	583.00
42" - 44"	S2@3.11	Ea	532.00	114.00	—	646.00
46" - 48"	S2@3.41	Ea	584.00	125.00	—	709.00
50" - 52"	S2@3.71	Ea	633.00	136.00	—	769.00
54" - 56"	S2@4.38	Ea	749.00	160.00	—	909.00
58" - 60"	S2@5.21	Ea	898.00	190.00	—	1,088.00
62" - 64"	S2@5.74	Ea	979.00	210.00	—	1,189.00
66" - 68"	S2@6.28	Ea	1,070.00	229.00	—	1,299.00
70" - 72"	S2@6.75	Ea	1,150.00	247.00	—	1,397.00

Galvanized steel 40" rectangular duct 90-degree elbow

Description	Craft@Hrs	Unit	Material $	Labor $	Equipment $	Total $
40" - 42"	S2@3.06	Ea	526.00	112.00	—	638.00
44" - 46"	S2@3.36	Ea	572.00	123.00	—	695.00
48" - 50"	S2@3.66	Ea	629.00	134.00	—	763.00
52" - 54"	S2@3.98	Ea	681.00	145.00	—	826.00
56" - 58"	S2@5.17	Ea	882.00	189.00	—	1,071.00
60" - 62"	S2@5.59	Ea	959.00	204.00	—	1,163.00
64" - 66"	S2@6.11	Ea	1,040.00	223.00	—	1,263.00
68" - 70"	S2@6.66	Ea	1,140.00	243.00	—	1,383.00
72"	S2@7.05	Ea	1,190.00	258.00	—	1,448.00

Galvanized steel 42" rectangular duct 90-degree elbow

Description	Craft@Hrs	Unit	Material $	Labor $	Equipment $	Total $
40" - 42"	S2@3.21	Ea	555.00	117.00	—	672.00
44" - 46"	S2@3.52	Ea	605.00	129.00	—	734.00
48" - 50"	S2@3.83	Ea	661.00	140.00	—	801.00
52" - 54"	S2@4.16	Ea	714.00	152.00	—	866.00
56" - 58"	S2@5.34	Ea	913.00	195.00	—	1,108.00
60" - 62"	S2@5.81	Ea	991.00	212.00	—	1,203.00
64" - 66"	S2@6.32	Ea	1,080.00	231.00	—	1,311.00
68" - 70"	S2@6.85	Ea	1,160.00	250.00	—	1,410.00
72"	S2@7.22	Ea	1,230.00	264.00	—	1,494.00

Galvanized steel 44" rectangular duct 90-degree elbow

Description	Craft@Hrs	Unit	Material $	Labor $	Equipment $	Total $
40" - 42"	S2@3.36	Ea	572.00	123.00	—	695.00
44" - 46"	S2@3.69	Ea	632.00	135.00	—	767.00
48" - 50"	S2@4.01	Ea	685.00	147.00	—	832.00
52" - 54"	S2@4.32	Ea	740.00	158.00	—	898.00
56" - 58"	S2@5.51	Ea	947.00	201.00	—	1,148.00
60" - 62"	S2@6.00	Ea	1,020.00	219.00	—	1,239.00
64" - 66"	S2@6.51	Ea	1,100.00	238.00	—	1,338.00
68" - 70"	S2@7.05	Ea	1,200.00	258.00	—	1,458.00
72"	S2@7.39	Ea	1,240.00	270.00	—	1,510.00

Galvanized Steel Rectangular Elbows

Description	Craft@Hrs	Unit	Material $	Labor $	Equipment $	Total $

Galvanized steel 46" rectangular duct 90-degree elbow

Description	Craft@Hrs	Unit	Material $	Labor $	Equipment $	Total $
40" - 42"	S2@3.52	Ea	605.00	129.00	—	734.00
44" - 46"	S2@3.86	Ea	664.00	141.00	—	805.00
48" - 50"	S2@4.16	Ea	714.00	152.00	—	866.00
52" - 54"	S2@4.32	Ea	767.00	158.00	—	925.00
56" - 58"	S2@5.72	Ea	976.00	209.00	—	1,185.00
60" - 62"	S2@6.19	Ea	1,050.00	226.00	—	1,276.00
64" - 66"	S2@6.70	Ea	1,150.00	245.00	—	1,395.00
68" - 70"	S2@7.26	Ea	1,230.00	265.00	—	1,495.00
72"	S2@7.60	Ea	1,290.00	278.00	—	1,568.00

Galvanized steel 48" rectangular duct 90-degree elbow

Description	Craft@Hrs	Unit	Material $	Labor $	Equipment $	Total $
40" - 42"	S2@3.67	Ea	630.00	134.00	—	764.00
44" - 46"	S2@4.01	Ea	685.00	147.00	—	832.00
48" - 50"	S2@4.32	Ea	740.00	158.00	—	898.00
52" - 54"	S2@4.61	Ea	785.00	168.00	—	953.00
56" - 58"	S2@5.91	Ea	1,020.00	216.00	—	1,236.00
60" - 62"	S2@6.40	Ea	1,090.00	234.00	—	1,324.00
64" - 66"	S2@6.90	Ea	1,170.00	252.00	—	1,422.00
68" - 70"	S2@7.41	Ea	1,260.00	271.00	—	1,531.00
72"	S2@7.77	Ea	1,340.00	284.00	—	1,624.00

Galvanized steel 50" rectangular duct 90-degree elbow

Description	Craft@Hrs	Unit	Material $	Labor $	Equipment $	Total $
40" - 42"	S2@3.82	Ea	661.00	140.00	—	801.00
44" - 46"	S2@4.16	Ea	714.00	152.00	—	866.00
48" - 50"	S2@4.46	Ea	767.00	163.00	—	930.00
52" - 54"	S2@4.78	Ea	812.00	175.00	—	987.00
56" - 58"	S2@6.11	Ea	104.00	223.00	—	327.00
60" - 62"	S2@6.60	Ea	1,130.00	241.00	—	1,371.00
64" - 66"	S2@7.09	Ea	1,200.00	259.00	—	1,459.00
68" - 70"	S2@7.62	Ea	1,310.00	278.00	—	1,588.00
72"	S2@7.94	Ea	1,360.00	290.00	—	1,650.00

Galvanized steel 52" rectangular duct 90-degree elbow

Description	Craft@Hrs	Unit	Material $	Labor $	Equipment $	Total $
40" - 42"	S2@3.98	Ea	681.00	145.00	—	826.00
44" - 46"	S2@4.32	Ea	740.00	158.00	—	898.00
48" - 50"	S2@4.61	Ea	785.00	168.00	—	953.00
52" - 54"	S2@4.95	Ea	847.00	181.00	—	1,028.00
56" - 58"	S2@6.32	Ea	1,080.00	231.00	—	1,311.00
60" - 62"	S2@6.79	Ea	1,160.00	248.00	—	1,408.00
64" - 66"	S2@7.30	Ea	1,230.00	267.00	—	1,497.00
68" - 70"	S2@7.81	Ea	1,340.00	285.00	—	1,625.00
72"	S2@8.16	Ea	1,380.00	298.00	—	1,678.00

Galvanized Steel Rectangular Elbows

Description	Craft@Hrs	Unit	Material $	Labor $	Equipment $	Total $

Galvanized steel 54" rectangular duct 90-degree elbow

40" - 42"	S2@4.16	Ea	714.00	152.00	—	866.00
44" - 46"	S2@4.46	Ea	767.00	163.00	—	930.00
48" - 50"	S2@4.78	Ea	812.00	175.00	—	987.00
52" - 54"	S2@5.15	Ea	873.00	188.00	—	1,061.00
56" - 58"	S2@6.51	Ea	1,100.00	238.00	—	1,338.00
60" - 62"	S2@7.00	Ea	1,190.00	256.00	—	1,446.00
64" - 66"	S2@7.49	Ea	1,280.00	274.00	—	1,554.00
68" - 70"	S2@8.01	Ea	1,370.00	293.00	—	1,663.00
72"	S2@8.41	Ea	1,420.00	307.00	—	1,727.00

Galvanized steel 56" rectangular duct 90-degree elbow

40" - 42"	S2@5.10	Ea	870.00	186.00	—	1,056.00
44" - 46"	S2@5.49	Ea	942.00	201.00	—	1,143.00
48" - 50"	S2@5.85	Ea	1,010.00	214.00	—	1,224.00
52" - 54"	S2@6.26	Ea	1,070.00	229.00	—	1,299.00
56" - 58"	S2@6.68	Ea	1,140.00	244.00	—	1,384.00
60" - 62"	S2@7.15	Ea	1,210.00	261.00	—	1,471.00
64" - 66"	S2@7.66	Ea	1,310.00	280.00	—	1,590.00
68" - 70"	S2@8.18	Ea	1,380.00	299.00	—	1,679.00
72"	S2@8.58	Ea	1,490.00	314.00	—	1,804.00

Galvanized steel 58" rectangular duct 90-degree elbow

58" - 60"	S2@7.15	Ea	1,230.00	261.00	—	1,491.00
62" - 64"	S2@7.62	Ea	1,310.00	278.00	—	1,588.00
66" - 68"	S2@8.13	Ea	1,390.00	297.00	—	1,687.00
70" - 72"	S2@8.69	Ea	1,510.00	318.00	—	1,828.00

Galvanized steel 60" rectangular duct 90-degree elbow

58" - 60"	S2@7.37	Ea	1,260.00	269.00	—	1,529.00
62" - 64"	S2@7.88	Ea	1,360.00	288.00	—	1,648.00
66" - 68"	S2@8.41	Ea	1,480.00	307.00	—	1,787.00
70" - 72"	S2@8.95	Ea	1,540.00	327.00	—	1,867.00

Galvanized steel 62" rectangular duct 90-degree elbow

58" - 60"	S2@7.62	Ea	1,310.00	278.00	—	1,588.00
62" - 64"	S2@8.13	Ea	1,380.00	297.00	—	1,677.00
66" - 68"	S2@8.69	Ea	1,500.00	318.00	—	1,818.00
70" - 72"	S2@9.18	Ea	1,580.00	335.00	—	1,915.00

Galvanized steel 64" rectangular duct 90-degree elbow

58" - 60"	S2@7.88	Ea	1,350.00	288.00	—	1,638.00
62" - 64"	S2@8.41	Ea	1,440.00	307.00	—	1,747.00
66" - 68"	S2@8.95	Ea	1,530.00	327.00	—	1,857.00
70" - 72"	S2@9.44	Ea	1,610.00	345.00	—	1,955.00

Galvanized Steel Rectangular Elbows

Description	Craft@Hrs	Unit	Material $	Labor $	Equipment $	Total $
Galvanized steel 66" rectangular duct 90-degree elbow						
58" - 60"	S2@8.13	Ea	1,380.00	297.00	—	1,677.00
62" - 64"	S2@8.69	Ea	1,500.00	318.00	—	1,818.00
66" - 68"	S2@9.18	Ea	1,580.00	335.00	—	1,915.00
70" - 72"	S2@9.69	Ea	1,670.00	354.00	—	2,024.00
Galvanized steel 68" rectangular duct 90-degree elbow						
58" - 60"	S2@8.41	Ea	1,440.00	307.00	—	1,747.00
62" - 64"	S2@8.95	Ea	1,530.00	327.00	—	1,857.00
66" - 68"	S2@9.44	Ea	1,610.00	345.00	—	1,955.00
70" - 72"	S2@9.95	Ea	1,710.00	364.00	—	2,074.00
Galvanized steel 70" rectangular duct 90-degree elbow						
58" - 60"	S2@8.69	Ea	1,500.00	318.00	—	1,818.00
62" - 64"	S2@9.18	Ea	1,580.00	335.00	—	1,915.00
66" - 68"	S2@9.69	Ea	1,670.00	354.00	—	2,024.00
70" - 72"	S2@10.2	Ea	1,740.00	373.00	—	2,113.00
Galvanized steel 72" rectangular duct 90-degree elbow						
58" - 60"	S2@8.95	Ea	1,530.00	327.00	—	1,857.00
62" - 64"	S2@9.44	Ea	1,610.00	345.00	—	1,955.00
66" - 68"	S2@9.95	Ea	1,710.00	364.00	—	2,074.00
70" - 72"	S2@10.5	Ea	1,800.00	384.00	—	2,184.00

Galvanized Steel Rectangular Drops and Tees

Costs for Galvanized Rectangular Drops and Tap-in Tees. Costs per drop or tap-in tee, based on quantities less than 1,000 pounds. Deduct 15% from the material cost for over 1,000 pounds. Add 50% to the material cost for the lined duct. These costs include delivery, typical bracing, cleats, scrap, hangers, end closures, sealants and miscellaneous hardware. For turning vanes, please see page 342.

Description	Craft@Hrs	Unit	Material $	Labor $	Equipment $	Total $

Galvanized steel 6" rectangular duct drops and tap-in tees

Description	Craft@Hrs	Unit	Material $	Labor $	Equipment $	Total $
6"	S2@.072	Ea	13.50	2.63	—	16.13
8"	S2@.084	Ea	15.80	3.07	—	18.87
10"	S2@.097	Ea	18.00	3.54	—	21.54
12" - 16"	S2@.139	Ea	26.00	5.08	—	31.08
20" - 24"	S2@.216	Ea	40.30	7.89	—	48.19
28" - 32"	S2@.309	Ea	57.20	11.30	—	68.50
36" - 40"	S2@.412	Ea	76.80	15.10	—	91.90
44" - 48"	S2@.489	Ea	91.10	17.90	—	109.00
52" - 56"	S2@.591	Ea	109.00	21.60	—	130.60
60"	S2@.747	Ea	138.00	27.30	—	165.30

Galvanized steel 8" rectangular duct drops and tap-in tees

Description	Craft@Hrs	Unit	Material $	Labor $	Equipment $	Total $
6"	S2@.084	Ea	15.80	3.07	—	18.87
8"	S2@.097	Ea	18.00	3.54	—	21.54
10"	S2@.109	Ea	20.30	3.98	—	24.28
12" - 16"	S2@.152	Ea	28.80	5.55	—	34.35
20" - 24"	S2@.231	Ea	43.20	8.44	—	51.64
28" - 32"	S2@.326	Ea	61.30	11.90	—	73.20
36" - 40"	S2@.432	Ea	80.70	15.80	—	96.50
44" - 48"	S2@.506	Ea	94.20	18.50	—	112.70
52" - 56"	S2@.634	Ea	117.00	23.20	—	140.20
60"	S2@.760	Ea	140.00	27.80	—	167.80

Galvanized steel 10" rectangular duct drops and tap-in tees

Description	Craft@Hrs	Unit	Material $	Labor $	Equipment $	Total $
6"	S2@.097	Ea	18.00	3.54	—	21.54
8"	S2@.109	Ea	20.30	3.98	—	24.28
10"	S2@.121	Ea	22.60	4.42	—	27.02
12" - 16"	S2@.167	Ea	31.20	6.10	—	37.30
20" - 24"	S2@.246	Ea	46.30	8.99	—	55.29
28" - 32"	S2@.343	Ea	64.40	12.50	—	76.90
36" - 40"	S2@.450	Ea	83.50	16.40	—	99.90
44" - 48"	S2@.525	Ea	97.60	19.20	—	116.80
52" - 56"	S2@.653	Ea	120.00	23.90	—	143.90
60"	S2@.773	Ea	142.00	28.20	—	170.20

Galvanized Steel Rectangular Drops and Tees

Description	Craft@Hrs	Unit	Material $	Labor $	Equipment $	Total $
Galvanized steel 12" rectangular duct drops and tap-in tees						
6"	S2@.109	Ea	20.30	3.98	—	24.28
8"	S2@.121	Ea	22.60	4.42	—	27.02
10"	S2@.133	Ea	24.90	4.86	—	29.76
12" - 16"	S2@.180	Ea	33.60	6.58	—	40.18
20" - 24"	S2@.262	Ea	49.20	9.57	—	58.77
28" - 32"	S2@.360	Ea	67.20	13.20	—	80.40
36" - 40"	S2@.470	Ea	87.00	17.20	—	104.20
44" - 48"	S2@.544	Ea	101.00	19.90	—	120.90
52" - 56"	S2@.675	Ea	124.00	24.70	—	148.70
60"	S2@.794	Ea	149.00	29.00	—	178.00
Galvanized steel 16" rectangular duct drops and tap-in tees						
6"	S2@.170	Ea	31.70	6.21	—	37.91
8"	S2@.184	Ea	34.90	6.72	—	41.62
10"	S2@.200	Ea	37.50	7.31	—	44.81
12" - 16"	S2@.231	Ea	43.20	8.44	—	51.64
20" - 24"	S2@.293	Ea	54.30	10.70	—	65.00
28" - 32"	S2@.396	Ea	73.40	14.50	—	87.90
36" - 40"	S2@.506	Ea	94.20	18.50	—	112.70
44" - 48"	S2@.581	Ea	108.00	21.20	—	129.20
52" - 56"	S2@.709	Ea	133.00	25.90	—	158.90
60"	S2@.841	Ea	156.00	30.70	—	186.70
Galvanized steel 20" rectangular duct drops and tap-in tees						
6"	S2@.200	Ea	37.50	7.31	—	44.81
8"	S2@.216	Ea	40.30	7.89	—	48.19
10"	S2@.231	Ea	43.20	8.44	—	51.64
12" - 16"	S2@.262	Ea	49.20	9.57	—	58.77
20" - 24"	S2@.324	Ea	60.50	11.80	—	72.30
28" - 32"	S2@.428	Ea	79.80	15.60	—	95.40
36" - 40"	S2@.544	Ea	101.00	19.90	—	120.90
44" - 48"	S2@.617	Ea	114.00	22.50	—	136.50
52" - 56"	S2@.758	Ea	139.00	27.70	—	166.70
60"	S2@.884	Ea	165.00	32.30	—	197.30

Galvanized Steel Rectangular Drops and Tees

Description	Craft@Hrs	Unit	Material $	Labor $	Equipment $	Total $

Galvanized steel 24" rectangular duct drops and tap-in tees

Description	Craft@Hrs	Unit	Material $	Labor $	Equipment $	Total $
6"	S2@.231	Ea	43.20	8.44	—	51.64
8"	S2@.247	Ea	46.30	9.03	—	55.33
10"	S2@.262	Ea	49.20	9.57	—	58.77
12" - 16"	S2@.293	Ea	54.30	10.70	—	65.00
20" - 24"	S2@.354	Ea	66.10	12.90	—	79.00
28" - 32"	S2@.462	Ea	85.50	16.90	—	102.40
36" - 40"	S2@.581	Ea	108.00	21.20	—	129.20
44" - 48"	S2@.655	Ea	120.00	23.90	—	143.90
52" - 56"	S2@.798	Ea	150.00	29.20	—	179.20
60"	S2@.927	Ea	172.00	33.90	—	205.90

Galvanized steel 28" rectangular duct drops and tap-in tees

Description	Craft@Hrs	Unit	Material $	Labor $	Equipment $	Total $
6"	S2@.262	Ea	49.20	9.57	—	58.77
8"	S2@.278	Ea	51.30	10.20	—	61.50
10"	S2@.292	Ea	54.30	10.70	—	65.00
12" - 16"	S2@.324	Ea	60.50	11.80	—	72.30
20" - 24"	S2@.385	Ea	71.70	14.10	—	85.80
28" - 32"	S2@.497	Ea	92.60	18.20	—	110.80
36" - 40"	S2@.617	Ea	114.00	22.50	—	136.50
44" - 48"	S2@.694	Ea	128.00	25.40	—	153.40
52" - 56"	S2@.837	Ea	156.00	30.60	—	186.60
60"	S2@.969	Ea	181.00	35.40	—	216.40

Galvanized steel 32" rectangular duct drops and tap-in tees

Description	Craft@Hrs	Unit	Material $	Labor $	Equipment $	Total $
6"	S2@.356	Ea	66.40	13.00	—	79.40
8"	S2@.374	Ea	70.50	13.70	—	84.20
10"	S2@.393	Ea	73.20	14.40	—	87.60
12" - 16"	S2@.432	Ea	80.70	15.80	—	96.50
20" - 24"	S2@.506	Ea	94.20	18.50	—	112.70
28" - 32"	S2@.581	Ea	108.00	21.20	—	129.20
36" - 40"	S2@.655	Ea	120.00	23.90	—	143.90
44" - 48"	S2@.730	Ea	135.00	26.70	—	161.70
52" - 56"	S2@.877	Ea	165.00	32.00	—	197.00
60"	S2@1.02	Ea	189.00	37.30	—	226.30

Galvanized Steel Round Ductwork

Description	Craft@Hrs	Unit	Material $	Labor $	Equipment $	Total $

Galvanized steel round duct snap-lock with slip joints

3"	S2@.080	LF	3.07	2.92	—	5.99
4"	S2@.090	LF	3.17	3.29	—	6.46
5"	S2@.095	LF	3.40	3.47	—	6.87
6"	S2@.100	LF	3.96	3.65	—	7.61
7"	S2@.110	LF	6.08	4.02	—	10.10
8"	S2@.120	LF	6.28	4.38	—	10.66
9"	S2@.140	LF	6.57	5.12	—	11.69
10"	S2@.160	LF	6.84	5.85	—	12.69

Galvanized steel round duct 90-degree adjustable elbow

3"	S2@.250	Ea	5.93	9.14	—	15.07
4"	S2@.300	Ea	8.18	11.00	—	19.18
5"	S2@.350	Ea	9.88	12.80	—	22.68
6"	S2@.400	Ea	12.80	14.60	—	27.40
7"	S2@.450	Ea	44.00	16.40	—	60.40
8"	S2@.500	Ea	49.20	18.30	—	67.50
9"	S2@.600	Ea	53.20	21.90	—	75.10
10"	S2@.700	Ea	58.10	25.60	—	83.70

Galvanized steel round duct 45-degree adjustable elbow

3"	S2@.250	Ea	5.93	9.14	—	15.07
4"	S2@.300	Ea	8.18	11.00	—	19.18
5"	S2@.350	Ea	9.88	12.80	—	22.68
6"	S2@.400	Ea	12.80	14.60	—	27.40
7"	S2@.450	Ea	44.00	16.40	—	60.40
8"	S2@.500	Ea	49.20	18.30	—	67.50
9"	S2@.600	Ea	53.20	21.90	—	75.10
10"	S2@.700	Ea	58.10	25.60	—	83.70

Fiberglass Ductwork

Standard fiberglass ductwork, or duct board, should not be used in systems where air velocities exceed 2,500 feet per minute, or where static pressures are higher than 2 inches of water gauge.

Additionally, it should not be used for the following applications:

    1) Equipment rooms

    2) Underground or outdoors

    3) Final connections to air handling equipment

    4) Plenums or casings

    5) Fume, heat or moisture exhaust systems

The costs for manufacturing and installing fiberglass ductwork are based on the net areas of the duct and fittings.

Here are approximate current material costs, per square foot:

Fiberglass Type	Up to 2,500 SF Material $/SF	Over 2,500 SF Material $/SF
475	4.64	4.24
800	5.09	4.62
1400	6.24	6.00

Additional costs: Approximately 25 cents per square foot should be added for the cost of tie rods, hangers, staples, tape and waste.

Type 475 fiberglass board is used almost exclusively for residential and small commercial air systems, while Types 800 and 1400 are used in larger commercial and industrial systems.

	Fiberglass Ductwork Fabrication Labor*					
	Size ranges (inches)					
	Up through 16		17 through 32		33 through 48	
Activity	SF/Hour Output	Labor $/SF	SF/Hour Output	Labor $/SF	SF/Hour Output	Labor $/SF
Duct only	60	.61	58	.63	56	.65
Fittings only	40	.91	38	.96	35	1.04
Duct & 15% fittings	55	.66	53	.69	50	.73
Duct & 25% fittings	53	.69	51	.72	48	.76
Duct & 35% fittings	51	.72	49	.75	46	.79

Labor rates are based on using a standard grooving machine and include all tie-rod reinforcing.

Labor correction factors:

    1) Hand grooving in lieu of machine grooving: 1.58

    2) Use of an auto-closer: .82

    3) Tie-rods enclosed in conduit: 1.36

    4) T-bar or channel reinforcing in lieu of tie-rods: 1.25

In the following installation table the measurements refer to the "semi-perimeter", the width plus depth of the ductwork. Example – 12" duct (8" wide plus 4" high) has 2.00 SF of area for each linear foot.

Fiberglass Ductwork

Description	Craft@Hrs	Unit	Material $	Labor $	Equipment $	Total $

Fiberglass ductwork installation

Description	Craft@Hrs	Unit	Material $	Labor $	Equipment $	Total $
12" (2.00 SF/LF)	S2@.070	LF	—	2.56	—	2.56
14" (2.35 SF/LF)	S2@.082	LF	—	3.00	—	3.00
16" (2.65 SF/LF)	S2@.093	LF	—	3.40	—	3.40
18" (3.00 SF/LF)	S2@.105	LF	—	3.84	—	3.84
20" (3.35 SF/LF)	S2@.117	LF	—	4.28	—	4.28
22" (3.65 SF/LF)	S2@.128	LF	—	4.68	—	4.68
24" (4.00 SF/LF)	S2@.140	LF	—	5.12	—	5.12
26" (4.35 SF/LF)	S2@.152	LF	—	5.55	—	5.55
28" (4.65 SF/LF)	S2@.163	LF	—	5.96	—	5.96
30" (5.00 SF/LF)	S2@.175	LF	—	6.39	—	6.39
32" (5.35 SF/LF)	S2@.187	LF	—	6.83	—	6.83
34" (5.65 SF/LF)	S2@.198	LF	—	7.23	—	7.23
36" (6.00 SF/LF)	S2@.210	LF	—	7.67	—	7.67
38" (6.35 SF/LF)	S2@.222	LF	—	8.11	—	8.11
40" (6.65 SF/LF)	S2@.232	LF	—	8.48	—	8.48
42" (7.00 SF/LF)	S2@.245	LF	—	8.95	—	8.95
44" (7.35 SF/LF)	S2@.257	LF	—	9.39	—	9.39
46" (7.65 SF/LF)	S2@.268	LF	—	9.79	—	9.79
48" (8.00 SF/LF)	S2@.280	LF	—	10.20	—	10.20
50" (8.35 SF/LF)	S2@.292	LF	—	10.70	—	10.70
52" (8.65 SF/LF)	S2@.303	LF	—	11.10	—	11.10
54" (9.00 SF/LF)	S2@.315	LF	—	11.50	—	11.50
56" (9.35 SF/LF)	S2@.327	LF	—	11.90	—	11.90
58" (9.65 SF/LF)	S2@.338	LF	—	12.40	—	12.40
60" (10.0 SF/LF)	S2@.350	LF	—	12.80	—	12.80
64" (10.6 SF/LF)	S2@.373	LF	—	13.60	—	13.60
68" (11.4 SF/LF)	S2@.397	LF	—	14.50	—	14.50
72" (12.0 SF/LF)	S2@.420	LF	—	15.30	—	15.30
76" (12.6 SF/LF)	S2@.443	LF	—	16.20	—	16.20
80" (13.4 SF/LF)	S2@.467	LF	—	17.10	—	17.10
84" (14.0 SF/LF)	S2@.490	LF	—	17.90	—	17.90
88" (14.6 SF/LF)	S2@.513	LF	—	18.70	—	18.70
92" (15.4 SF/LF)	S2@.537	LF	—	19.60	—	19.60
96" (16.0 SF/LF)	S2@.560	LF	—	20.50	—	20.50

For estimating purposes, when measuring the length of ductwork, include the fitting. Then apply the labor per run foot to the total.

1¼" thick wire reinforced flexible fiberglass duct with vinyl cover

Description	Craft@Hrs	Unit	Material $	Labor $	Equipment $	Total $
4"	S2@.060	LF	8.38	2.19	—	10.57
5"	S2@.070	LF	8.67	2.56	—	11.23
6"	S2@.080	LF	9.49	2.92	—	12.41
8"	S2@.090	LF	12.10	3.29	—	15.39
9"	S2@.100	LF	12.90	3.65	—	16.55
10"	S2@.120	LF	13.60	4.38	—	17.98
12"	S2@.140	LF	17.40	5.12	—	22.52
14"	S2@.160	LF	20.80	5.85	—	26.65
16"	S2@.180	LF	24.20	6.58	—	30.78
18"	S2@.200	LF	27.50	7.31	—	34.81

Fiberglass Pipe Insulation

Fiberglass Pipe Insulation with AP-T Plus (all-purpose self-sealing) Jacket. By nominal pipe diameter. R factor equals 2.56 at 300 degrees F. These costs do not include scaffolding; add for same, if required. Also see fittings, flanges and valves at the end of this section.

Description	Craft@Hrs	Unit	Material $	Labor $	Equipment $	Total $

Jacketed fiberglass insulation for ½" pipe

Description	Craft@Hrs	Unit	Material $	Labor $	Equipment $	Total $
½" thick	P1@.038	LF	2.57	1.41	—	3.98
1" thick	P1@.038	LF	3.02	1.41	—	4.43
1½" thick	P1@.040	LF	6.15	1.49	—	7.64

Jacketed fiberglass insulation for ¾" pipe

Description	Craft@Hrs	Unit	Material $	Labor $	Equipment $	Total $
½" thick	P1@.038	LF	2.78	1.41	—	4.19
1" thick	P1@.038	LF	3.43	1.41	—	4.84
1½" thick	P1@.040	LF	6.38	1.49	—	7.87

Jacketed fiberglass insulation for 1" pipe

Description	Craft@Hrs	Unit	Material $	Labor $	Equipment $	Total $
½" thick	P1@.040	LF	2.96	1.49	—	4.45
1" thick	P1@.040	LF	3.93	1.49	—	5.42
1½" thick	P1@.042	LF	6.75	1.56	—	8.31

Jacketed fiberglass insulation for 1¼" pipe

Description	Craft@Hrs	Unit	Material $	Labor $	Equipment $	Total $
½" thick	P1@.040	LF	3.28	1.49	—	4.77
1" thick	P1@.040	LF	3.97	1.49	—	5.46
1½" thick	P1@.042	LF	7.19	1.56	—	8.75

Jacketed fiberglass insulation for 1½" pipe

Description	Craft@Hrs	Unit	Material $	Labor $	Equipment $	Total $
½" thick	P1@.042	LF	3.52	1.56	—	5.08
1" thick	P1@.042	LF	4.47	1.56	—	6.03
1½" thick	P1@.044	LF	7.73	1.64	—	9.37

Jacketed fiberglass insulation for 2" pipe

Description	Craft@Hrs	Unit	Material $	Labor $	Equipment $	Total $
½" thick	P1@.044	LF	3.68	1.64	—	5.32
1" thick	P1@.044	LF	5.13	1.64	—	6.77
1½" thick	P1@.046	LF	8.30	1.71	—	10.01

Jacketed fiberglass insulation for 2½" pipe

Description	Craft@Hrs	Unit	Material $	Labor $	Equipment $	Total $
1" thick	P1@.047	LF	5.29	1.75	—	7.04
1½" thick	P1@.048	LF	7.51	1.79	—	9.30
2" thick	P1@.050	LF	11.90	1.86	—	13.76

Jacketed fiberglass insulation for 3" pipe

Description	Craft@Hrs	Unit	Material $	Labor $	Equipment $	Total $
1" thick	P1@.051	LF	5.83	1.90	—	7.73
1½" thick	P1@.054	LF	8.21	2.01	—	10.22
2" thick	P1@.056	LF	13.20	2.08	—	15.28

Fiberglass Pipe Insulation

Description	Craft@Hrs	Unit	Material $	Labor $	Equipment $	Total $
Jacketed fiberglass insulation for 4" pipe						
1" thick	P1@.063	LF	6.86	2.34	—	9.20
1½" thick	P1@.066	LF	9.15	2.46	—	11.61
2" thick	P1@.069	LF	14.70	2.57	—	17.27
Jacketed fiberglass insulation for 6" pipe						
1" thick	P1@.077	LF	7.98	2.87	—	10.85
1½" thick	P1@.080	LF	11.00	2.98	—	13.98
2" thick	P1@.085	LF	18.60	3.16	—	21.76
Jacketed fiberglass insulation for 8" pipe						
1" thick	P1@.117	LF	11.00	4.35	—	15.35
1½" thick	P1@.123	LF	13.40	4.58	—	17.98
2" thick	P1@.129	LF	21.90	4.80	—	26.70
Jacketed fiberglass insulation for 10" pipe						
1" thick	P1@.140	LF	13.10	5.21	—	18.31
1½" thick	P1@.147	LF	16.50	5.47	—	21.97
2" thick	P1@.154	LF	26.10	5.73	—	31.83
Jacketed fiberglass insulation for 12" pipe						
1" thick	P1@.155	LF	15.10	5.77	—	20.87
1½" thick	P1@.163	LF	18.60	6.07	—	24.67
2" thick	P1@.171	LF	28.50	6.36	—	34.86
Add for .016" aluminum jacket						
Per SF of surface	P1@.035	SF	2.16	1.30	—	3.46

Pipe fittings and flanges
For each fitting or flange use the cost for 3 LF of pipe of the same pipe size

Valves
Body only: for each valve body use the cost for 5 LF of pipe of the same pipe size

Body and bonnet or yoke: for each valve use the cost for 10 LF of pipe of the same pipe size

Flanged valves: add the cost for flanges per above

Calcium Silicate Pipe Insulation with Aluminum Jacket

Calcium Silicate Pipe Insulation with .016" Aluminum Jacket. By nominal pipe diameter. R factor equals 2.20 at 300 degrees F. Manufactured by Pabco. These costs do not include scaffolding; add for same, if required. Also see fittings, flanges and valves below.

Description	Craft@Hrs	Unit	Material $	Labor $	Equipment $	Total $

Jacketed calcium silicate insulation for 6" pipe

2" thick	P1@.183	LF	35.50	6.81	—	42.31
4" thick	P1@.218	LF	76.90	8.11	—	85.01
6" thick	P1@.288	LF	108.00	10.70	—	118.70

Jacketed calcium silicate insulation for 8" pipe

2" thick	P1@.194	LF	42.60	7.22	—	49.82
4" thick	P1@.228	LF	90.10	8.49	—	98.59
6" thick	P1@.308	LF	137.00	11.50	—	148.50

Jacketed calcium silicate insulation for 10" pipe

2" thick	P1@.202	LF	49.70	7.52	—	57.22
4" thick	P1@.239	LF	105.00	8.90	—	113.90
6" thick	P1@.337	LF	165.00	12.50	—	177.50

Jacketed calcium silicate insulation for 12" pipe

2" thick	P1@.208	LF	57.50	7.74	—	65.24
4" thick	P1@.261	LF	118.00	9.71	—	127.71
6" thick	P1@.388	LF	187.00	14.40	—	201.40

Pipe fittings and flanges
For each fitting or flange use the cost for 3 LF of pipe of the same pipe size

Valves
Body only: for each valve body use the cost for 5 LF of pipe of the same pipe size

Body and bonnet or yoke: for each valve use the cost for 10 LF of pipe of the same pipe size

Flanged valves: add the cost for flanges per above

Closed Cell Elastomeric Pipe Insulation

Description	Craft@Hrs	Unit	Material $	Labor $	Equipment $	Total $

Closed Cell Elastomeric Pipe and Tubing Insulation. Semi split, by nominal pipe or tube diameter with insulation wall thickness as shown, no cover. R factor equals 3.58 at 220 degrees F. Manufactured by Rubatex. These costs do not include scaffolding; add for same, if required. Also see fittings, flanges and valves below.

Description	Craft@Hrs	Unit	Material $	Labor $	Equipment $	Total $
¼", ½" thick	P1@.039	LF	1.07	1.45	—	2.52
3/8", ½" thick	P1@.039	LF	1.10	1.45	—	2.55
½", ½" thick	P1@.039	LF	1.22	1.45	—	2.67
¾", ½" thick	P1@.039	LF	1.40	1.45	—	2.85
1", ½" thick	P1@.042	LF	1.56	1.56	—	3.12
1¼", ½" thick	P1@.042	LF	1.74	1.56	—	3.30
1½", ½" thick	P1@.042	LF	2.04	1.56	—	3.60
2", ½" thick	P1@.046	LF	2.55	1.71	—	4.26
2½", 3/8" thick	P1@.056	LF	3.27	2.08	—	5.35
2½", ½" thick	P1@.056	LF	4.39	2.08	—	6.47
2½", ¾" thick	P1@.056	LF	5.66	2.08	—	7.74
3", 3/8" thick	P1@.062	LF	3.73	2.31	—	6.04
3", ½" thick	P1@.062	LF	4.93	2.31	—	7.24
3", ¾" thick	P1@.062	LF	6.42	2.31	—	8.73
4", 3/8" thick	P1@.068	LF	4.39	2.53	—	6.92
4", ½" thick	P1@.068	LF	5.78	2.53	—	8.31
4", ¾" thick	P1@.068	LF	7.53	2.53	—	10.06

Pipe fittings and flanges
For each fitting or flange use the cost for 3 LF of pipe of the same pipe size

Valves
Body only: for each valve body use the cost for 5 LF of pipe of the same pipe size

Body and bonnet or yoke: for each valve use the cost for 10 LF of pipe of the same pipe size

Flanged valves: add the cost for flanges per above

Thermal Duct Insulation

Description	Craft@Hrs	Unit	Material $	Labor $	Equipment $	Total $

Fiberglass flexible wrap duct insulation with foil facing vapor barrier to 250 degrees F

Description	Craft@Hrs	Unit	Material $	Labor $	Equipment $	Total $
1"	S2@.023	SF	1.33	.84	—	2.17
1½"	S2@.023	SF	1.61	.84	—	2.45
2"	S2@.025	SF	2.18	.91	—	3.09
Deduct foil facing	—	%	−30.0	—	—	—

Fiberglass rigid board duct insulation plain, *without* foil facing vapor barrier to 450 degrees F

Description	Craft@Hrs	Unit	Material $	Labor $	Equipment $	Total $
1"	S2@.067	SF	1.64	2.45	—	4.09
1½"	S2@.067	SF	2.78	2.45	—	5.23
2"	S2@.075	SF	3.28	2.74	—	6.02
3"	S2@.085	SF	4.90	3.11	—	8.01

Fiberglass rigid board duct insulation with FSK (foil) facing vapor barrier to 250 degrees F

Description	Craft@Hrs	Unit	Material $	Labor $	Equipment $	Total $
1"	S2@.067	SF	2.18	2.45	—	4.63
1½"	S2@.067	SF	2.66	2.45	—	5.11
2"	S2@.075	SF	3.16	2.74	—	5.90
3"	S2@.085	SF	4.50	3.11	—	7.61

Fiberglass flexible blanket duct lining to 250 degrees F

Description	Craft@Hrs	Unit	Material $	Labor $	Equipment $	Total $
½"	S2@.032	SF	1.00	1.17	—	2.17
1"	S2@.032	SF	1.30	1.17	—	2.47
1½"	S2@.035	SF	1.86	1.28	—	3.14
2"	S2@.035	SF	2.42	1.28	—	3.70

Fiberglass rigid duct lining to 250 degrees F

Description	Craft@Hrs	Unit	Material $	Labor $	Equipment $	Total $
5/8"	S2@.032	SF	2.17	1.17	—	3.34
1"	S2@.032	SF	2.81	1.17	—	3.98
1½"	S2@.035	SF	3.70	1.28	—	4.98
2"	S2@.035	SF	4.70	1.28	—	5.98

Note: Material prices include all necessary tape, pins, mastic and 15% waste.

Balancing of HVAC Systems

Description	Craft@Hrs	Unit	Material $	Labor $	Equipment $	Total $
Air balancing – air handling units						
Central station	S2@6.00	Ea	—	219.00	—	219.00
Multi-zone A/C	S2@4.00	Ea	—	146.00	—	146.00
Single-zone A/C	S2@3.00	Ea	—	110.00	—	110.00
Packaged A/C	S2@2.50	Ea	—	91.40	—	91.40
Built-up low pressure	S2@4.50	Ea	—	164.00	—	164.00
Built-up high pres.	S2@6.00	Ea	—	219.00	—	219.00
Built-up VAV	S2@6.00	Ea	—	219.00	—	219.00
Built-up dual duct	S2@6.00	Ea	—	219.00	—	219.00
Air balancing – terminal boxes						
Variable air volume	S2@0.40	Ea	—	14.60	—	14.60
Constant volume	S2@0.40	Ea	—	14.60	—	14.60
Dual duct box	S2@0.70	Ea	—	25.60	—	25.60
Terminal box and all downstream diffusers cost per zone	S2@1.75	Ea	—	63.90	—	63.90
Air balancing – diffusers						
Ceiling diffusers to 500 CFM	S2@.250	Ea	—	9.14	—	9.14
Ceiling diffusers over 500 CFM	S2@.350	Ea	—	12.80	—	12.80
Ceiling diffusers perforated	S2@.370	Ea	—	13.50	—	13.50
Sidewall diffusers double deflection	S2@.250	Ea	—	9.14	—	9.14
Linear slot diffusers 2 slot	S2@.060	LF	—	2.19	—	2.19
Linear slot diffusers 3 slot	S2@.070	LF	—	2.56	—	2.56
Linear slot diffusers 4 slot	S2@.080	LF	—	2.92	—	2.92
Plug-in diffuser	S2@.250	Ea	—	9.14	—	9.14
Light troffers 1 slot	S2@.195	Ea	—	7.13	—	7.13
Light troffers 2 slot	S2@.220	Ea	—	8.04	—	8.04
Laminar flow diff.	S2@.750	Ea	—	27.40	—	27.40
Laminar flow diffusers with HEPA filter	S2@3.00	Ea	—	110.00	—	110.00

Balancing of HVAC Systems

Description	Craft@Hrs	Unit	Material $	Labor $	Equipment $	Total $

Air balancing – grilles

Description	Craft@Hrs	Unit	Material $	Labor $	Equipment $	Total $
Return air grilles to 800 CFM	S2@.250	Ea	—	9.14	—	9.14
Return air grilles over 800 CFM	S2@.350	Ea	—	12.80	—	12.80
Bar type return air grilles 2 slot	S2@.060	LF	—	2.19	—	2.19
Bar type return air grilles 4 slot	S2@.070	LF	—	2.56	—	2.56
Bar type return air grilles 6 slot	S2@.080	LF	—	2.92	—	2.92

Air balancing – centrifugal fans

Description	Craft@Hrs	Unit	Material $	Labor $	Equipment $	Total $
Supply and return air to 3,000 CFM	S2@1.00	Ea	—	36.50	—	36.50
Supply and return air to 5,000 CFM	S2@1.50	Ea	—	54.80	—	54.80
Supply and return air to 10,000 CFM	S2@2.00	Ea	—	73.10	—	73.10
Supply and return air to 20,000 CFM	S2@2.75	Ea	—	100.00	—	100.00
Roof exhaust to 6,000 CFM	S2@.725	Ea	—	26.50	—	26.50
Roof exhaust to 10,000 CFM	S2@.925	Ea	—	33.80	—	33.80
Roof exhaust to 20,000 CFM	S2@1.50	Ea	—	54.80	—	54.80
Range hood exhaust to 10,000 CFM	S2@1.75	Ea	—	63.90	—	63.90
Range hood exhaust to 20,000 CFM	S2@2.50	Ea	—	91.40	—	91.40

Air balancing – fan coil units

Description	Craft@Hrs	Unit	Material $	Labor $	Equipment $	Total $
Vertical floor model	S2@.500	Ea	—	18.30	—	18.30
Horiz. ceiling mtd	S2@.650	Ea	—	23.80	—	23.80

Air balancing – fume hoods

Description	Craft@Hrs	Unit	Material $	Labor $	Equipment $	Total $
Lab fume hood	S2@2.00	Ea	—	73.10	—	73.10
General purpose	S2@.450	Ea	—	16.40	—	16.40
Dust collectors	S2@1.50	Ea	—	54.80	—	54.80

Balancing of HVAC Systems

Description	Craft@Hrs	Unit	Material $	Labor $	Equipment $	Total $

Wet balancing

Description	Craft@Hrs	Unit	Material $	Labor $	Equipment $	Total $
Boiler	S2@1.65	Ea	—	60.30	—	60.30
Chiller	S2@3.50	Ea	—	128.00	—	128.00
Cooling tower	S2@2.00	Ea	—	73.10	—	73.10
Circulating pump	S2@1.65	Ea	—	60.30	—	60.30
Heat exchanger	S2@1.75	Ea	—	63.90	—	63.90
AHU water coil	S2@2.50	Ea	—	91.40	—	91.40
Reheat coil	S2@.450	Ea	—	16.40	—	16.40
Wall fin radiation	S2@.350	Ea	—	12.80	—	12.80
Radiant heat panel	S2@.450	Ea	—	16.40	—	16.40
Fan coil unit, 2 pipe	S2@.350	Ea	—	12.80	—	12.80
Fan coil unit, 4 pipe	S2@.450	Ea	—	16.40	—	16.40
Cabinet heater	S2@.350	Ea	—	12.80	—	12.80
Unit ventilator	S2@.450	Ea	—	16.40	—	16.40
Unit heater	S2@.350	Ea	—	12.80	—	12.80

Commercial HVAC controls. Temperature, pressure and humidity control for any device. 0-10 Volts DC or 4-20 milliamp input. Honeywell T775U or equal. Complies with ANSI/ACCA standard 5QI-2007. With two modulating and two relay outputs, each with their own setpoint.

Description	Craft@Hrs	Unit	Material $	Labor $	Equipment $	Total $
Install control module enclosure	BE@1.00	Ea	—	41.00	—	41.00
Twisted pair Category 6 PVC cable	BE@.008	LF	.57	.33	—	.90
Panel wiring	BE@.250	Ea	3.00	10.30	—	13.30
Relay and damper actuator wiring	BE@.250	Ea	2.69	10.30	—	12.99
Sensor wiring	BE@.300	Ea	2.69	12.30	—	14.99
Zone control module wiring	BE@.400	Ea	2.52	16.40	—	18.92
Control module and wiring	BE@2.00	Ea	480.00	82.10	—	562.10
Ground fault protection wiring	BE@1.25	Ea	150.00	51.30	—	201.30
Programming and system test	BE@3.00	Ea	—	123.00	—	123.00
Commissioning and final test	BE@8.00	Ea	—	328.00	—	328.00

Sensor and zone control modules. Honeywell Webstat TR23 or equal. 20K non-linear sensor with setpoint adjust 55-85, and override LED. With LonWorks interface card and power supply. Add the cost of control wiring.

Description	Craft@Hrs	Unit	Material $	Labor $	Equipment $	Total $
Room sensor	BE@.500	Ea	89.50	20.50	—	110.00

Damper actuators. Modutrol IV or equal. For rectangular volume control dampers. Add the cost of wiring and the damper from the section Ductwork Specialties.

Description	Craft@Hrs	Unit	Material $	Labor $	Equipment $	Total $
Lever arm actuator	BE@.500	Ea	557.00	20.50	—	577.50

Temperature Controls

Description	Craft@Hrs	Unit	Material $	Labor $	Equipment $	Total $

Remote CO$_2$ sensors. Honeywell C72321 or equal, with LonWorks interface card and power supply. Add the cost of wiring.

Panel mount	BE@.500	Ea	95.30	20.50	—	115.80

Monitoring and recording equipment. Honeywell DR4500A digital circular chart recorder or equal, with LonWorks card and power supply. Add the cost of wiring.

Panel mount	BE@.500	Ea	4,530.00	20.50	—	4,550.50

Thermostat, electric. Install and connect. Make additional allowances for sub-subcontractor markups and control wiring as required.

Description	Craft@Hrs	Unit	Material $	Labor $	Equipment $	Total $
Heat/cool low voltage	BE@.575	Ea	47.60	23.60	—	71.20
Heat only low voltage	BE@.500	Ea	39.70	20.50	—	60.20
Heat/cool line voltage	BE@.650	Ea	144.00	26.70	—	170.70
Programmable – residential low voltage	BE@.850	LF	218.00	34.90	—	252.90
Programmable – commercial w/subbase low voltage	BE@1.10	LF	450.00	45.10	—	495.10

Thermostat, pneumatic. Install and connect. Make additional allowances for sub-subcontractor markups and pneumatic tubing as required.

Description	Craft@Hrs	Unit	Material $	Labor $	Equipment $	Total $
Direct acting 55 to 85 Deg. F.	BE@.675	Ea	123.00	27.70	—	150.70
Reverse acting 55 to 85 Deg. F.	BE@.675	Ea	102.00	27.70	—	129.70
Floating	BE@.675	Ea	219.00	27.70	—	246.70

Control valve, 2-way, electric. 6.9 Volt Input. Proportional Spring Return. Connect only. N.O. means: "normally open." Make additional allowances for sub-subcontractor markups, control wiring and valve installation by piping trades as required. (See Plumbing and Piping Specialties)

Description	Craft@Hrs	Unit	Material $	Labor $	Equipment $	Total $
½", 2-way, N.O.	BE@.400	Ea	558.00	16.40	—	574.40
¾", 2-way, N.O.	BE@.400	Ea	593.00	16.40	—	609.40
1", 2-way, N.O.	BE@.400	Ea	639.00	16.40	—	655.40
1¼" 2-way, N.O.	BE@.400	Ea	698.00	16.40	—	714.40
1½" 2-way, N.O.	BE@.400	Ea	1,500.00	16.40	—	1,516.40
2", 2-way, N.O.	BE@.400	Ea	1,640.00	16.40	—	1,656.40

Control valve, 3-way, electric. 6.9 Volt Input, Proportional Spring Return. Connect only. Make additional allowances for sub-subcontractor markups, control wiring and valve installation by piping trades as required. (See Plumbing and Piping Specialties)

Description	Craft@Hrs	Unit	Material $	Labor $	Equipment $	Total $
½", 3-way	BE@.450	Ea	582.00	18.50	—	600.50
¾", 3-way	BE@.450	Ea	633.00	18.50	—	651.50
1", 3-way	BE@.450	Ea	758.00	18.50	—	776.50

Temperature Controls

Description	Craft@Hrs	Unit	Material $	Labor $	Equipment $	Total $

Control valve, 2-way, pneumatic. Proportional Spring Return, 3 to 7 psi Input, Connect only. N.O. means "normally open." Make additional allowances for sub-subcontractor markups, pneumatic tubing and valve installation by piping trades as required. (See Plumbing and Piping Specialties)

Description	Craft@Hrs	Unit	Material $	Labor $	Equipment $	Total $
½", 2-way, N.O.	BE@.400	Ea	229.00	16.40	—	245.40
¾", 2-way, N.O.	BE@.400	Ea	268.00	16.40	—	284.40
1", 2-way, N.O.	BE@.400	Ea	316.00	16.40	—	332.40
1¼" 2-way, N.O.	BE@.400	Ea	374.00	16.40	—	390.40
1½" 2-way, N.O.	BE@.400	Ea	666.00	16.40	—	682.40
2", 2-way, N.O.	BE@.400	Ea	795.00	16.40	—	811.40

Control valve, 3-way, pneumatic. Proportional Spring Return, 3 to 7 psi Input, Connect only. Make additional allowances for sub-subcontractor markups, pneumatic tubing and valve installation by piping trades as required. (See Plumbing and Piping Specialties)

Description	Craft@Hrs	Unit	Material $	Labor $	Equipment $	Total $
½", 3-way	BE@.450	Ea	257.00	18.50	—	275.50
¾", 3-way	BE@.450	Ea	309.00	18.50	—	327.50
1", 3-way	BE@.450	Ea	428.00	18.50	—	446.50
1¼", 3-way	BE@.450	Ea	496.00	18.50	—	514.50
1½", 3-way	BE@.450	Ea	712.00	18.50	—	730.50
2", 3-way	BE@.450	Ea	832.00	18.50	—	850.50

Air compressor. Reciprocating piston-type. Larger systems include HVAC actuator controls. Underwriters Laboratories and CSA rated. With ASME U-stamped pressure tank, pressure safety relief valve, silencer, lubrication system with lube cooling radiator and fan. Single-phase 110 volt electric motor, drive belt and OSHA-rated safety belt guard. Add the cost of compressor hose and fittings and hose reel, if required. Set in place only. Add installation costs from the table below.

Description	Craft@Hrs	Unit	Material $	Labor $	Equipment $	Total $
½ hp	SL@3.00	Ea	432.00	111.00	—	543.00
¾ hp	SL@3.00	Ea	473.00	111.00	—	584.00
1 hp	SL@3.50	Ea	550.00	129.00	—	679.00
1½ hp	SL@3.50	Ea	609.00	129.00	—	738.00
2 hp	SL@3.50	Ea	647.00	129.00	—	776.00

Installation of air compressors.

Description	Craft@Hrs	Unit	Material $	Labor $	Equipment $	Total $
Cut, frame support and gasket hole in sidewall	S2@1.00	Ea	32.10	36.50	—	68.60
Install piping for compressor line	P1@.100	LF	4.31	3.72	—	8.03
Install electrical wiring	BE@.150	LF	2.00	6.15	—	8.15
Install compressor controls	BE@.500	Ea	—	20.50	—	20.50
Commission and test	P1@2.00	Ea	—	74.40	—	74.40
Fine-tune system	P1@1.00	Ea	—	37.20	—	37.20

Ductile Iron, Class 153, Cement-Lined with Mechanical Joints

Description	Craft@Hrs	Unit	Material $	Labor $	Equipment $	Total $

Cement-lined Class 153 ductile iron pipe with mechanical joints

4"	P1@.140	LF	13.20	5.21	—	18.41
6"	ER@.180	LF	19.10	7.19	3.04	29.33
8"	ER@.200	LF	26.10	7.99	3.38	37.47
10"	ER@.230	LF	38.70	9.19	3.89	51.78
12"	ER@.270	LF	52.80	10.80	4.56	68.16
14"	ER@.310	LF	80.20	12.40	5.24	97.84
16"	ER@.350	LF	103.00	14.00	5.92	122.92

Cement-lined Class 153 ductile iron 1/16 bend with mechanical joints

4"	P1@.900	Ea	52.30	33.50	—	85.80
6"	ER@1.30	Ea	80.70	51.90	22.00	154.60
8"	ER@1.70	Ea	127.00	67.90	28.70	223.60
10"	ER@2.15	Ea	173.00	85.90	36.30	295.20
12"	ER@2.40	Ea	240.00	95.90	40.60	376.50
14"	ER@2.70	Ea	526.00	108.00	45.60	679.60
16"	ER@3.00	Ea	594.00	120.00	50.70	764.70

Cement-lined Class 153 ductile iron 1/8 bend with mechanical joints

4"	P1@.900	Ea	54.00	33.50	—	87.50
6"	ER@1.30	Ea	80.20	51.90	22.00	154.10
8"	ER@1.70	Ea	129.00	67.90	28.70	225.60
10"	ER@2.15	Ea	176.00	85.90	36.30	298.20
12"	ER@2.40	Ea	244.00	95.90	40.60	380.50
14"	ER@2.70	Ea	523.00	108.00	45.60	676.60
16"	ER@3.00	Ea	612.00	120.00	50.70	782.70

Cement-lined Class 153 ductile iron 1/4 bend with mechanical joints

4"	P1@.900	Ea	57.80	33.50	—	91.30
6"	ER@1.30	Ea	94.00	51.90	22.00	167.90
8"	ER@1.70	Ea	197.00	67.90	28.70	293.60
10"	ER@2.15	Ea	296.00	85.90	36.30	418.20
12"	ER@2.40	Ea	405.00	95.90	40.60	541.50
14"	ER@2.70	Ea	1,040.00	108.00	45.60	1,193.60
16"	ER@3.00	Ea	1,090.00	120.00	50.70	1,260.70

Cement-lined Class 153 ductile iron tee with mechanical joints

4"	P1@1.65	Ea	87.40	61.40	—	148.80
6"	ER@2.35	Ea	129.00	93.90	39.70	262.60
8"	ER@3.15	Ea	197.00	126.00	53.20	376.20
10"	ER@4.00	Ea	296.00	160.00	67.60	523.60
12"	ER@4.45	Ea	405.00	178.00	75.20	658.20
14"	ER@4.90	Ea	1,040.00	196.00	82.80	1,318.80
16"	ER@5.30	Ea	1,090.00	212.00	89.60	1,391.60

Ductile Iron, Class 153, Cement-Lined with Mechanical Joints

Description	Craft@Hrs	Unit	Material $	Labor $	Equipment $	Total $
Cement-lined Class 153 ductile iron wye with mechanical joints						
4"	P1@1.65	Ea	197.00	61.40	—	258.40
6"	ER@2.35	Ea	244.00	93.90	39.70	377.60
8"	ER@3.15	Ea	369.00	126.00	53.20	548.20
10"	ER@4.00	Ea	686.00	160.00	67.60	913.60
12"	ER@4.45	Ea	1,030.00	178.00	75.20	1,283.20
14"	ER@4.90	Ea	2,120.00	196.00	82.80	2,398.80
16"	ER@5.30	Ea	2,650.00	212.00	89.60	2,951.60
Cement-lined Class 153 ductile iron reducer with mechanical joints						
6"	ER@1.45	Ea	64.20	57.90	24.50	146.60
8"	ER@2.00	Ea	101.00	79.90	33.80	214.70
10"	ER@2.55	Ea	135.00	102.00	43.10	280.10
12"	ER@3.05	Ea	173.00	122.00	51.60	346.60
14"	ER@3.60	Ea	387.00	144.00	60.90	591.90
16"	ER@4.20	Ea	471.00	168.00	71.00	710.00

Ductile Iron, 153, Double Cement-Lined with Mechanical Joints

Description	Craft@Hrs	Unit	Material $	Labor $	Equipment $	Total $

Double cement-lined Class 153 ductile iron pipe with mechanical joints

4"	P1@.140	LF	14.00	5.21	—	19.21
6"	ER@.180	LF	20.20	7.19	3.04	30.43
8"	ER@.200	LF	27.60	7.99	3.38	38.97
10"	ER@.230	LF	41.20	9.19	3.89	54.28
12"	ER@.270	LF	56.20	10.80	4.56	71.56
14"	ER@.310	LF	85.40	12.40	5.24	103.04
16"	ER@.350	LF	110.00	14.00	5.92	129.92

Double cement-lined Class 153 ductile iron 1/16 bend with mechanical joints

4"	P1@.900	Ea	61.60	33.50	—	95.10
6"	ER@1.30	Ea	95.30	51.90	22.00	169.20
8"	ER@1.70	Ea	152.00	67.90	28.70	248.60
10"	ER@2.15	Ea	204.00	85.90	36.30	326.20
12"	ER@2.40	Ea	279.00	95.90	40.60	415.50
14"	ER@2.70	Ea	625.00	108.00	45.60	778.60
16"	ER@3.00	Ea	708.00	120.00	50.70	878.70

Double cement-lined Class 153 ductile iron 1/8 bend with mechanical joints

4"	P1@.900	Ea	61.60	33.50	—	95.10
6"	ER@1.30	Ea	92.70	51.90	22.00	166.60
8"	ER@1.70	Ea	148.00	67.90	28.70	244.60
10"	ER@2.15	Ea	203.00	85.90	36.30	325.20
12"	ER@2.40	Ea	276.00	95.90	40.60	412.50
14"	ER@2.70	Ea	608.00	108.00	45.60	761.60
16"	ER@3.00	Ea	708.00	120.00	50.70	878.70

Double cement-lined Class 153 ductile iron 1/4 bend with mechanical joints

4"	P1@.900	Ea	66.60	33.50	—	100.10
6"	ER@1.30	Ea	108.00	51.90	22.00	181.90
8"	ER@1.70	Ea	168.00	67.90	28.70	264.60
10"	ER@2.15	Ea	261.00	85.90	36.30	383.20
12"	ER@2.40	Ea	339.00	95.90	40.60	475.50
14"	ER@2.70	Ea	807.00	108.00	45.60	960.60
16"	ER@3.00	Ea	916.00	120.00	50.70	1,086.70

Ductile Iron, Class 153, Double Cement-Lined with Mechanical Joints

Description	Craft@Hrs	Unit	Material $	Labor $	Equipment $	Total $
Double cement-lined Class 153 ductile iron tee with mechanical joints						
4"	P1@1.65	Ea	102.00	61.40	—	163.40
6"	ER@2.35	Ea	148.00	93.90	39.70	281.60
8"	ER@3.15	Ea	226.00	126.00	53.20	405.20
10"	ER@4.00	Ea	339.00	160.00	67.60	566.60
12"	ER@4.45	Ea	464.00	178.00	75.20	717.20
14"	ER@4.90	Ea	1,200.00	196.00	82.80	1,478.80
16"	ER@5.30	Ea	1,250.00	212.00	89.60	1,551.60
Double cement-lined Class 153 ductile iron wye with mechanical joints						
4"	P1@1.65	Ea	226.00	61.40	—	287.40
6"	ER@2.35	Ea	288.00	93.90	39.70	421.60
8"	ER@3.15	Ea	424.00	126.00	53.20	603.20
10"	ER@4.00	Ea	800.00	160.00	67.60	1,027.60
12"	ER@4.45	Ea	1,190.00	178.00	75.20	1,443.20
14"	ER@4.90	Ea	2,440.00	196.00	82.80	2,718.80
16"	ER@5.30	Ea	3,010.00	212.00	89.60	3,311.60
Double cement-lined Class 153 ductile iron reducer with mechanical joints						
6"	ER@1.45	Ea	74.00	57.90	24.50	156.40
8"	ER@2.00	Ea	105.00	79.90	33.80	218.70
10"	ER@2.55	Ea	137.00	102.00	43.10	282.10
12"	ER@3.05	Ea	201.00	122.00	51.60	374.60
14"	ER@3.60	Ea	430.00	144.00	60.90	634.90
16"	ER@4.20	Ea	541.00	168.00	71.00	780.00

Ductile Iron, Class 110, Cement-Lined with Mechanical Joints

Description	Craft@Hrs	Unit	Material $	Labor $	Equipment $	Total $

Cement-lined Class 110 ductile iron pipe with mechanical joints

Description	Craft@Hrs	Unit	Material $	Labor $	Equipment $	Total $
4"	P1@.180	LF	15.30	6.70	—	22.00
6"	ER@.220	LF	21.70	8.79	3.72	34.21
8"	ER@.240	LF	30.00	9.59	4.06	43.65
10"	ER@.270	LF	44.80	10.80	4.56	60.16
12"	ER@.320	LF	60.90	12.80	5.41	79.11
16"	ER@.400	LF	119.00	16.00	6.76	141.76

Cement-lined Class 110 ductile iron 1/16 bend with mechanical joints

Description	Craft@Hrs	Unit	Material $	Labor $	Equipment $	Total $
4"	P1@1.40	Ea	153.00	52.10	—	205.10
6"	ER@2.00	Ea	204.00	79.90	33.80	317.70
8"	ER@2.50	Ea	280.00	99.90	42.30	422.20
10"	ER@3.20	Ea	427.00	128.00	54.10	609.10
12"	ER@3.60	Ea	543.00	144.00	60.90	747.90
16"	ER@4.50	Ea	1,410.00	180.00	76.10	1,666.10

Cement-lined Class 110 ductile iron 1/8 bend with mechanical joints

Description	Craft@Hrs	Unit	Material $	Labor $	Equipment $	Total $
4"	P1@1.40	Ea	151.00	52.10	—	203.10
6"	ER@2.00	Ea	202.00	79.90	33.80	315.70
8"	ER@2.50	Ea	280.00	99.90	42.30	422.20
10"	ER@3.20	Ea	411.00	128.00	54.10	593.10
12"	ER@3.60	Ea	542.00	144.00	60.90	746.90
16"	ER@4.50	Ea	1,390.00	180.00	76.10	1,646.10

Cement-lined Class 110 ductile iron 1/4 bend with mechanical joints

Description	Craft@Hrs	Unit	Material $	Labor $	Equipment $	Total $
4"	P1@1.40	Ea	166.00	52.10	—	218.10
6"	ER@2.00	Ea	223.00	79.90	33.80	336.70
8"	ER@2.50	Ea	320.00	99.90	42.30	462.20
10"	ER@3.20	Ea	496.00	128.00	54.10	678.10
12"	ER@3.60	Ea	743.00	144.00	60.90	947.90
16"	ER@4.50	Ea	1,810.00	180.00	76.10	2,066.10

Cement-lined Class 110 ductile iron tee with mechanical joints

Description	Craft@Hrs	Unit	Material $	Labor $	Equipment $	Total $
4"	P1@2.50	Ea	251.00	93.10	—	344.10
6"	ER@3.50	Ea	332.00	140.00	59.20	531.20
8"	ER@4.70	Ea	475.00	188.00	79.40	742.40
10"	ER@6.00	Ea	820.00	240.00	101.00	1,161.00
12"	ER@6.70	Ea	1,040.00	268.00	113.00	1,421.00
16"	ER@8.00	Ea	2,740.00	320.00	135.00	3,195.00

Cement-lined Class 110 ductile iron reducer with mechanical joints

Description	Craft@Hrs	Unit	Material $	Labor $	Equipment $	Total $
6"	ER@2.20	Ea	166.00	87.90	37.20	291.10
8"	ER@3.00	Ea	229.00	120.00	50.70	399.70
10"	ER@3.80	Ea	375.00	152.00	64.20	591.20
12"	ER@4.60	Ea	443.00	184.00	77.80	704.80
16"	ER@6.30	Ea	1,210.00	252.00	106.00	1,568.00

Cast Iron, Class 150 with Mechanical Joints

Description	Craft@Hrs	Unit	Material $	Labor $	Equipment $	Total $

Class 150 cast iron pipe with mechanical joints

1½"	P1@.080	LF	7.44	2.98	—	10.42
2"	P1@.100	LF	7.28	3.72	—	11.00
3"	P1@.120	LF	10.00	4.47	—	14.47
4"	P1@.140	LF	13.10	5.21	—	18.31
6"	ER@.180	LF	22.50	7.19	3.04	32.73
8"	ER@.200	LF	35.90	7.99	3.38	47.27

Class 150 cast iron 1/8 bend with mechanical joints

1½"	P1@.750	Ea	8.20	27.90	—	36.10
2"	P1@.800	Ea	9.13	29.80	—	38.93
3"	P1@1.00	Ea	12.30	37.20	—	49.50
4"	P1@1.20	Ea	15.60	44.70	—	60.30
6"	ER@1.70	Ea	36.70	67.90	28.70	133.30
8"	ER@2.25	Ea	105.00	89.90	38.00	232.90

Class 150 cast iron 1/4 bend with mechanical joints

2"	P1@.800	Ea	13.20	29.80	—	43.00
3"	P1@1.00	Ea	19.00	37.20	—	56.20
4"	P1@1.20	Ea	29.70	44.70	—	74.40
6"	ER@1.70	Ea	61.60	67.90	28.70	158.20
8"	ER@2.25	Ea	167.00	89.90	38.00	294.90

Class 150 cast iron sanitary tee with mechanical joints

1½"	P1@.750	Ea	13.70	27.90	—	41.60
2"	P1@1.00	Ea	15.10	37.20	—	52.30
3"	P1@1.20	Ea	18.20	44.70	—	62.90
4"	P1@1.65	Ea	27.90	61.40	—	89.30
6"	ER@2.35	Ea	79.70	93.90	39.70	213.30
8"	ER@3.15	Ea	255.00	126.00	53.20	434.20

Class 150 cast iron wye with mechanical joints

1½"	P1@.750	Ea	14.00	27.90	—	41.90
2"	P1@1.00	Ea	14.00	37.20	—	51.20
3"	P1@1.20	Ea	19.70	44.70	—	64.40
4"	P1@1.65	Ea	32.00	61.40	—	93.40
6"	ER@2.35	Ea	85.70	93.90	39.70	219.30
8"	ER@3.15	Ea	202.00	126.00	53.20	381.20

Class 150 cast iron reducer

2" x 1½"	P1@.600	Ea	7.49	22.30	—	29.79
3" x 2"	P1@.800	Ea	7.43	29.80	—	37.23
4" x 2"	P1@1.00	Ea	11.70	37.20	—	48.90
6" x 4"	ER@1.45	Ea	30.80	57.90	24.50	113.20
8" x 6"	ER@2.00	Ea	52.80	79.90	33.80	166.50

Asbestos-Cement, Class 2400 or 3000 with Mechanical Joints

Description	Craft@Hrs	Unit	Material $	Labor $	Equipment $	Total $
Asbestos-cement pipe with mechanical joints						
4"	ER@.080	LF	12.00	3.20	1.35	16.55
6"	ER@.100	LF	7.82	4.00	1.69	13.51
8"	ER@.140	LF	12.20	5.59	2.37	20.16
10"	ER@.200	LF	19.20	7.99	3.38	30.57
12"	ER@.260	LF	25.50	10.40	4.39	40.29
14"	ER@.300	LF	34.30	12.00	5.07	51.37
16"	ER@.350	LF	36.10	14.00	5.92	56.02
Asbestos-cement 1/16 bend with mechanical joints						
4"	ER@1.00	Ea	176.00	40.00	16.90	232.90
6"	ER@1.25	Ea	231.00	50.00	21.10	302.10
8"	ER@1.70	Ea	317.00	67.90	28.70	413.60
10"	ER@2.20	Ea	487.00	87.90	37.20	612.10
12"	ER@2.70	Ea	557.00	108.00	45.60	710.60
14"	ER@3.40	Ea	1,160.00	136.00	57.50	1,353.50
16"	ER@3.90	Ea	1,610.00	156.00	65.90	1,831.90
Asbestos-cement 1/8 bend with mechanical joints						
4"	ER@1.00	Ea	168.00	40.00	16.90	224.90
6"	ER@1.25	Ea	230.00	50.00	21.10	301.10
8"	ER@1.70	Ea	317.00	67.90	28.70	413.60
10"	ER@2.20	Ea	467.00	87.90	37.20	592.10
12"	ER@2.70	Ea	622.00	108.00	45.60	775.60
14"	ER@3.40	Ea	1,110.00	136.00	57.50	1,303.50
16"	ER@3.90	Ea	1,600.00	156.00	65.90	1,821.90
Asbestos-cement 1/4 bend with mechanical joints						
4"	ER@1.00	Ea	189.00	40.00	16.90	245.90
6"	ER@1.25	Ea	255.00	50.00	21.10	326.10
8"	ER@1.70	Ea	365.00	67.90	28.70	461.60
10"	ER@2.20	Ea	556.00	87.90	37.20	681.10
12"	ER@2.70	Ea	737.00	108.00	45.60	890.60
14"	ER@3.40	Ea	1,360.00	136.00	57.50	1,553.50
16"	ER@3.90	Ea	2,020.00	156.00	65.90	2,241.90

Asbestos-Cement, Class 2400 or 3000 with Mechanical Joints

Description	Craft@Hrs	Unit	Material $	Labor $	Equipment $	Total $
Asbestos-cement tee or wye with mechanical joints						
4"	ER@1.40	Ea	284.00	55.90	23.70	363.60
6"	ER@1.75	Ea	376.00	69.90	29.60	475.50
8"	ER@2.40	Ea	538.00	95.90	40.60	674.50
10"	ER@3.10	Ea	929.00	124.00	52.40	1,105.40
12"	ER@3.80	Ea	1,170.00	152.00	64.20	1,386.20
14"	ER@4.75	Ea	2,130.00	190.00	80.30	2,400.30
16"	ER@5.45	Ea	3,070.00	218.00	92.10	3,380.10
Asbestos-cement reducer with mechanical joints						
6" x 4"	ER@1.20	Ea	199.00	48.00	20.30	267.30
8" x 6"	ER@1.50	Ea	294.00	59.90	25.40	379.30
10" x 8"	ER@1.95	Ea	427.00	77.90	33.00	537.90
12" x 10"	ER@2.45	Ea	594.00	97.90	41.40	733.30

Fiberglass Tanks

Description	Craft@Hrs	Unit	Material $	Labor $	Equipment $	Total $

Fiberglass septic tank. 2 compartments. Shallow burial (2 feet of earth cover). 32" manway entrance. One-piece construction. Including excavation, backfill and equipment costs.

750 gallon	P1@8.00	Ea	1,560.00	298.00	473.00	2,331.00
1,000 gallon	P1@9.00	Ea	2,070.00	335.00	532.00	2,937.00
1,250 gallon	P1@10.0	Ea	2,650.00	372.00	592.00	3,614.00
1,500 gallon	P1@11.5	Ea	3,000.00	428.00	680.00	4,108.00
1,750 gallon	P1@12.0	Ea	3,350.00	447.00	710.00	4,507.00
2,000 gallon	P1@12.5	Ea	3,690.00	465.00	739.00	4,894.00

Fiberglass septic tank. 2 compartments. Deep burial (7 feet of earth cover). 32" x 40" manway extension. One-piece construction. Including excavation, backfill and equipment costs.

750 gallon	P1@8.00	Ea	1,760.00	298.00	473.00	2,531.00
1,000 gallon	P1@9.00	Ea	2,310.00	335.00	532.00	3,177.00
1,250 gallon	P1@10.0	Ea	3,030.00	372.00	592.00	3,994.00
1,500 gallon	P1@11.5	Ea	3,400.00	428.00	680.00	4,508.00
1,750 gallon	P1@12.0	Ea	3,770.00	447.00	710.00	4,927.00
2,000 gallon	P1@12.5	Ea	4,150.00	465.00	739.00	5,354.00
Manway extension	P1@.350	LF	52.20	13.00	—	65.20

Fiberglass holding tank. Single compartment. Shallow burial (2 feet of earth cover). 32" manway entrance. One-piece construction. Including excavation, backfill and equipment costs. Can be used for drink water storage or sewage holding.

750 gallon	P1@8.00	Ea	1,360.00	298.00	473.00	2,131.00
1,000 gallon	P1@9.00	Ea	1,770.00	335.00	532.00	2,637.00
1,250 gallon	P1@10.0	Ea	2,310.00	372.00	592.00	3,274.00
1,500 gallon	P1@11.5	Ea	2,650.00	428.00	680.00	3,758.00
1,750 gallon	P1@12.0	Ea	3,000.00	447.00	710.00	4,157.00
2,000 gallon	P1@12.5	Ea	3,350.00	465.00	739.00	4,554.00

Fiberglass holding tank. Single compartment. Deep burial (7 feet of earth cover). 32" x 40" manway extension. One-piece construction. Including excavation, backfill and equipment costs. Can be used for drink water storage or sewage holding.

750 gallon	P1@8.00	Ea	1,590.00	298.00	473.00	2,361.00
1,000 gallon	P1@9.00	Ea	2,010.00	335.00	532.00	2,877.00
1,250 gallon	P1@10.0	Ea	2,690.00	372.00	592.00	3,654.00
1,500 gallon	P1@11.5	Ea	3,060.00	428.00	680.00	4,168.00
1,750 gallon	P1@12.0	Ea	3,420.00	447.00	710.00	4,577.00
2,000 gallon	P1@12.5	Ea	3,810.00	465.00	739.00	5,014.00
Manway extension	P1@.350	LF	52.20	13.00	—	65.20

Plastic Tanks

Description	Craft@Hrs	Unit	Material $	Labor $	Equipment $	Total $

Plastic water tank. Suitable for potable water. Labor includes setting in place only. Freight, hoisting & rigging excluded.

Description	Craft@Hrs	Unit	Material $	Labor $	Equipment $	Total $
Vertical tank, D31" x H55"						
165 gallon	P1@2.00	Ea	640.00	74.40	—	714.40
Vertical tank, D48" x H102"						
500 gallon	P1@4.00	Ea	1,190.00	149.00	—	1,339.00
Vertical tank, D64" x H79"						
1,000 gallon	P1@4.00	Ea	1,690.00	149.00	—	1,839.00
Vertical tank, D87" x H65"						
1,500 gallon	P1@5.00	Ea	2,220.00	186.00	—	2,406.00
Vertical tank, D95" x H89"						
2,500 gallon	P1@5.00	Ea	2,300.00	186.00	—	2,486.00
Vertical tank, D95" x H107"						
3,000 gallon	P1@6.00	Ea	3,620.00	223.00	—	3,843.00
Vertical tank, D102" x H125"						
4,000 gallon	P1@6.00	Ea	3,850.00	223.00	—	4,073.00
Vertical tank, D102" x H150"						
5,000 gallon	P1@8.00	Ea	7,230.00	298.00	—	7,528.00
Vertical tank, D141" x H192"						
12,000 gallon	P1@9.00	Ea	11,200.00	335.00	—	11,535.00

Plastic holding tank. Single compartment. One-piece polyethylene construction. Shallow burial (2 feet of earth cover). 28" manway entrance. Including excavation, backfill and equipment costs. Used for drink water storage.

Description	Craft@Hrs	Unit	Material $	Labor $	Equipment $	Total $
30 gallon	P1@3.00	Ea	214.00	112.00	—	326.00
40 gallon	P1@3.00	Ea	244.00	112.00	—	356.00
75 gallon	P1@3.00	Ea	271.00	112.00	177.00	560.00
300 gallon	P1@5.00	Ea	1,200.00	186.00	296.00	1,682.00
525 gallon	P1@6.00	Ea	1,720.00	223.00	355.00	2,298.00
1,150 gallon	P1@8.00	Ea	3,010.00	298.00	473.00	3,781.00
1,700 gallon	P1@11.0	Ea	3,880.00	409.00	651.00	4,940.00
1,950 gallon	P1@12.0	Ea	4,130.00	447.00	710.00	5,287.00

Plastic holding tank. Single compartment. One-piece polyethylene construction. Above ground applications only. Used for drink water storage.

Description	Craft@Hrs	Unit	Material $	Labor $	Equipment $	Total $
180 gallon	P1@3.00	Ea	558.00	112.00	177.00	847.00
200 gallon	P1@3.75	Ea	827.00	140.00	222.00	1,189.00
300 gallon	P1@4.25	Ea	1,070.00	158.00	252.00	1,480.00
525 gallon	P1@5.00	Ea	1,590.00	186.00	296.00	2,072.00
1,000 gallon	P1@5.75	Ea	2,140.00	214.00	340.00	2,694.00
1,150 gallon	P1@6.25	Ea	2,630.00	233.00	370.00	3,233.00
1,700 gallon	P1@7.00	Ea	3,480.00	261.00	414.00	4,155.00

Plastic septic tank. 750 gallon. Single compartment. One-piece polyethylene construction. Including excavation, backfill and equipment costs.

Description	Craft@Hrs	Unit	Material $	Labor $	Equipment $	Total $
Trickle tank	P1@8.00	Ea	2,440.00	298.00	473.00	3,211.00
Shallow buried	P1@8.00	Ea	2,440.00	298.00	473.00	3,211.00
Deep buried	P1@8.00	Ea	2,570.00	298.00	473.00	3,341.00
Manway extension						
	P1@.350	LF	97.20	13.00	—	110.20

Plastic Tanks

Description	Craft@Hrs	Unit	Material $	Labor $	Equipment $	Total $

Plastic sump pit. One-piece polyethylene construction. Complete with radon gasket electrical and discharge grommets. Including excavation, backfill and equipment costs.

Description	Craft@Hrs	Unit	Material $	Labor $	Equipment $	Total $
30 gallon	P1@3.00	Ea	200.00	112.00	—	312.00
40 gallon	P1@3.00	Ea	244.00	112.00	—	356.00
75 gallon	P1@3.00	Ea	288.00	112.00	177.00	577.00
110 gallon	P1@3.25	Ea	456.00	121.00	192.00	769.00
150 gallon	P1@3.75	Ea	615.00	140.00	222.00	977.00
250 gallon	P1@4.75	Ea	972.00	177.00	281.00	1,430.00
380 gallon	P1@5.50	Ea	1,520.00	205.00	325.00	2,050.00

Plastic sewage lift tank. One-piece polyethylene construction. Complete with radon gasket, inlet, vent, electrical and discharge grommets. Including excavation, backfill and equipment costs.

Description	Craft@Hrs	Unit	Material $	Labor $	Equipment $	Total $
30 gallon	P1@3.00	Ea	251.00	112.00	—	363.00
40 gallon	P1@3.00	Ea	277.00	112.00	—	389.00
75 gallon	P1@3.00	Ea	311.00	112.00	177.00	600.00
110 gallon	P1@3.25	Ea	528.00	121.00	192.00	841.00
150 gallon	P1@3.75	Ea	641.00	140.00	222.00	1,003.00
250 gallon	P1@4.75	Ea	1,050.00	177.00	281.00	1,508.00
380 gallon	P1@5.50	Ea	1,570.00	205.00	325.00	2,100.00

Plastic grey water treatment tank. 75 gallon. One-piece polyethylene construction. Including excavation, backfill and equipment costs.

Description	Craft@Hrs	Unit	Material $	Labor $	Equipment $	Total $
Buried	P1@3.00	Ea	500.00	112.00	177.00	789.00
Non-buried	P1@1.50	Ea	456.00	55.80	—	511.80

Trenching

Because of varying soil compositions and densities, trenching costs can sometimes be difficult to accurately estimate. The tables in this section are based on average soil conditions, which are loosely defined as dirt, soft clay or gravel mixed with rocks. The costs for trenching through limestone, sandstone or shale can increase by as much as 1,000 percent. For these conditions, it is best to consult with experts or to subcontract the work to specialists.

The primary consideration in excavations is to ensure the safety of the pipe layers. Current OSHA statutes require all excavations over 5 feet deep to be sloped, shored, sheeted or braced to prevent cave-ins. The use of shoring minimizes trenching costs by eliminating the need to slope the trench walls, but increases pipe-laying costs because of the difficulty in working between cross-bracing.

The estimator should carefully compare the costs for installing and removing shoring, and reduced pipe installation productivity, against the additional excavation costs for sloping the walls of the trench. For average soil conditions, not exceeding 6 to 8 feet in depth, it's usually more economical to slope the trench walls.

The current average cost for a backhoe with operator is $105.00 per hour. Some trenching contractors additionally charge $25.00 move-on and $25.00 move-off costs, particularly for small projects.

Total trenching, backfill and compaction costs are calculated by multiplying the backhoe hours from the following charts by $105.00 per hour, and adding move-on and move-off charges, if applicable.

Backhoe Trenching						
Trench walls at a 90-degree angle, based on average soil, including trenching, backfill and compaction. Hours per 100 linear feet of trench						
Trench depth (feet)	Trench width (feet)					
	1	2	3	4	5	6
2	2.6	5.1	7.7	10.2	12.8	15.3
3	3.8	7.7	11.5	15.3	19.2	23.0
4	5.1	10.2	15.3	20.4	25.6	30.7
5	6.4	12.8	19.2	25.6	32.0	38.3
6	7.7	15.3	23.0	30.7	38.3	46.0
7	8.9	17.9	26.8	35.8	44.7	53.7
8	10.2	20.4	30.7	40.9	51.1	61.3
9	11.5	23.0	34.5	46.0	57.5	69.0
10	12.8	25.5	38.3	51.1	63.9	76.6

Notes:
1) Compaction is based on the use of a compactor wheel.
2) Additional labor costs must be added if watering-in, stone removal or excess spoil removal is required.

Trenching

	Backhoe Trenching					
	Trench walls at a 45-degree angle, based on average soil including trenching, backfill and compaction. Hours per 100 linear feet of trench					
Trench depth (feet)	Trench width (feet)					
	1	2	3	4	5	6
2	7.6	10.3	12.6	15.5	17.9	20.3
3	15.2	22.6	22.6	26.6	30.7	34.5
4	25.5	30.7	35.5	40.7	45.9	51.0
5	38.3	44.5	51.0	57.6	63.8	70.0
6	53.4	61.0	69.0	76.6	84.1	91.7
7	71.4	80.3	89.3	98.3	107.2	115.9
8	92.1	102.1	112.4	122.4	132.8	143.1
9	114.8	126.6	137.9	149.3	161.0	172.4
10	140.3	153.1	165.9	178.6	191.4	204.5

Notes:
1) Compaction is based on the use of a compactor wheel.
2) Additional labor costs must be added if watering-in, stone removal or excess spoil removal is required.

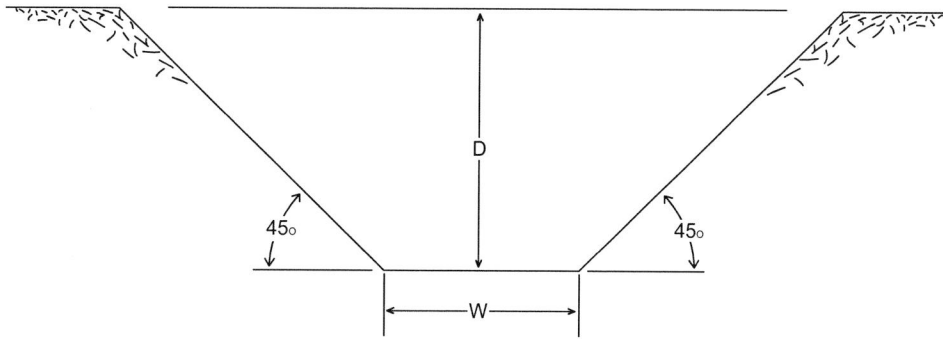

Equipment Rental

Equipment Rental			
Description	Per day ($)	Per week ($)	Per month ($)
Air compressor, gas or diesel, 100-125 CFM with 100' hose, air tool and 4 bits	266.00	787.00	1,970.00
Air compressor, gas or diesel, 165-185 CFM with 100' hose, air tool and 4 bits	307.00	913.00	2,340.00
Air compressor, gas or diesel, 100-125 CFM, no accessories	124.00	372.00	1,120.00
Air compressor, gas or diesel, 165-185 CFM, no accessories	165.00	498.00	1,490.00
Backhoe with bucket, 55 HP	380.00	1,380.00	3,210.00
Backhoe with bucket, 100 HP	423.00	1,330.00	3,970.00
Backhoe hydraulic compactor, 5-10 ton	194.00	499.00	1,240.00
Bender, hydraulic	91.00	335.00	946.00
Bevel machine with torch	30.00	97.00	240.00
Boom lift, 31' - 40'	490.00	1,210.00	2,490.00
Boom lift, 51' - 60'	590.00	1,270.00	2,630.00
Chain hoist, to 2 tons	33.00	105.00	230.00
Chain hoist, 10 tons	172.00	345.00	574.00
Come-along, ½ to 1 ton	28.00	38.00	216.00
Come-along, 1½ to 2 tons	48.00	124.00	378.00
Come-along, 4-ton	68.00	196.00	607.00
Compactor, soil, gas or diesel, 125 lb	115.00	351.00	745.00
Crane, truck-mounted, 13,000 lb capacity	420.00	1,220.00	3,820.00
Crane, truck-mounted, 30,000 lb capacity	526.00	1,620.00	4,590.00
Forklift, 2,000 pound capacity	216.00	573.00	1,550.00
Forklift, 4,000 pound capacity	207.00	563.00	1,700.00
Forklift, 6,000 pound capacity	271.00	1,210.00	1,920.00
Front-end loader, 1 yard capacity	358.00	1,040.00	2,920.00
Front-end loader, 2 yard capacity	583.00	1,650.00	4,360.00
Front-end loader, 3¼ yard capacity	867.00	1,470.00	6,750.00
Pipe machine, up to 2" capacity	68.00	203.00	656.00
Pipe machine, 2½" to 4" capacity	75.00	236.00	806.00
Ramset (nail gun)	29.00	116.00	422.00
Scissor-lift, to 20', diesel, self-propelled	110.00	274.00	714.00
Scissor-lift, 21' – 30'	305.00	650.00	1,320.00
Skid steer loader 1,750 lb capacity	310.00	902.00	2,040.00
Trailer, office, 25'	65.00	207.00	574.00
Truck, ¾-ton pickup	99.00	384.00	1,550.00
Truck, 1-ton pickup	149.00	470.00	1,490.00
Truck, 1-ton stake bed	202.00	608.00	1,790.00
Van, storage, 26'	—	—	376.00
Van, storage, 40'	—	—	468.00
Welding machine, gas, 200 amp, trailer-mounted with 50' bare lead	81.00	207.00	453.00
Welding machine, diesel, 400 amp, trailer-mounted with 50' bare lead	139.00	319.00	741.00

Close-Out Items

Description	Craft@Hrs	Unit	Material $	Labor $	Equipment $	Total $

1½" brass valve tags with black filled numbers
Typical costs	P1@.150	Ea	3.83	5.58	—	9.41

1½" blank brass valve tags
Typical costs	P1@.150	Ea	1.89	5.58	—	7.47

1½" plastic valve tags with engraved numbers
Typical costs	P1@.150	Ea	3.17	5.58	—	8.75

Engraved plastic equipment nameplates
Typical costs	P1@.400	Ea	8.22	14.90	—	23.12

Pipe and duct markers
Typical costs	P1@.250	Ea	4.70	9.31	—	14.01

As-built drawings
Typical costs	P1@.500	Ea	—	18.60	—	18.60

Instructing owner's operating personnel
Typical costs	P1@8.00	Ea	—	298.00	—	298.00

HVAC & Plumbing Demolition

Description	Craft@Hrs	Unit	Material $	Labor $	Equipment $	Total $

Remove rectangular galvanized ductwork. 22 to 30 gauge galvanized sheet metal (dimension/gauge per SMACNA standards). Per linear foot of sheet metal ductwork. The cost of removing ductwork includes the cost of removing the support hangers. Add the cost of disconnecting and removing heating coils, air volume zone boxes, air handing units, condensate drains, motorized dampers or other associated accessories, piping or wiring. These figures assume demolished materials are piled on site. Disposal costs or salvage values excluded. (SMACNA = Sheetmetal & Air Conditioning National Association.) Equipment may be needed for some sizes and locations. Refer to the equipment rental costs at the front of this section.

Description	Craft@Hrs	Unit	Material $	Labor $	Equipment $	Total $
4" to 12" rectangular/square duct, remove	S2@.015	LF	—	.55	—	.55
14" to 24" rectangular/square duct, remove	S2@.020	LF	—	.73	—	.73
26" to 36" rectangular/square duct, remove	S2@.065	LF	—	2.38	—	2.38
38" to 60" rectangular/square duct, remove	S2@.090	LF	—	3.29	—	3.29

Remove round galvanized ductwork. 22 to 30 gauge galvanized sheet metal (dimension/gauge per SMACNA standards). Per linear foot of sheet metal ductwork. The cost of removing ductwork includes the cost of removing the support hangers. Add the cost of disconnecting and removing heating coils, air volume zone boxes, air handing units, condensate drains, motorized dampers or other associated accessories, piping or wiring. These figures assume demolished materials are piled on site. Disposal costs or salvage values excluded. (SMACNA = Sheetmetal & Air Conditioning National Association.) Equipment may be needed for some sizes and locations. Refer to the equipment rental costs at the front of this section.

Description	Craft@Hrs	Unit	Material $	Labor $	Equipment $	Total $
4" to 12" round duct, remove	S2@.012	LF	—	.44	—	.44
14" to 24" round duct, remove	S2@.016	LF	—	.58	—	.58
26" to 36" round duct, remove	S2@.052	LF	—	1.90	—	1.90
38" to 60" round duct, remove	S2@.072	LF	—	2.63	—	2.63

HVAC & Plumbing Demolition

Description	Craft@Hrs	Unit	Material $	Labor $	Equipment $	Total $

Remove welded steel ductwork. 10 to 16 gauge mild steel. Per linear foot of welded milled steel ductwork. The cost of removing ductwork includes the cost of removing the support hangers. Add the cost of disconnecting and removing exhaust fans, condensate drains, sprinkler heads or other associated accessories, piping or wiring. These figures assume demolished materials are piled on site. Disposal costs or salvage values excluded. Equipment may be needed for some sizes and locations. Refer to the equipment rental costs at the front of this section.

Description	Craft@Hrs	Unit	Material $	Labor $	Equipment $	Total $
4" to 12", square or round duct, remove	S2@.045	LF	—	1.64	—	1.64
14" to 24", square or round duct, remove	S2@.060	LF	—	2.19	—	2.19
26" to 36", square or round duct, remove	S2@.095	LF	—	3.47	—	3.47
38" to 60", square or round duct, remove	S2@.150	LF	—	5.48	—	5.48

Remove diffuser or grille. These figures assume demolished materials are piled on site. Disposal costs or salvage values excluded. Equipment may be needed for some sizes and locations. Refer to the equipment rental costs at the front of this section.

Description	Craft@Hrs	Unit	Material $	Labor $	Equipment $	Total $
6" to 12", square or round diffuser/grille, remove	S2@.150	Ea	—	5.48	—	5.48
14" to 24", square or round diffuser/grille, remove	S2@.180	Ea	—	6.58	—	6.58
26" to 36", square or round diffuser/grille, remove	S2@.400	Ea	—	14.60	—	14.60
38" to 60" square or round diffuser/grille, remove	S2@.800	Ea	—	29.20	—	29.20

Remove air mixing box. These figures include isolating and disconnecting hot water, chilled water, steam, refrigerant and/or pneumatic control tube or control wiring. These figures assume demolished materials are piled on site. Disposal costs or salvage values excluded. Equipment may be needed for some sizes and locations. Refer to the equipment rental costs at the front of this section.

Description	Craft@Hrs	Unit	Material $	Labor $	Equipment $	Total $
Air box, 50 to 150 CFM, remove	S2@.500	Ea	—	18.30	—	18.30
Air box, 160 to 500 CFM, remove	S2@.650	Ea	—	23.80	—	23.80
Air box, 550 to 750 CFM, remove	S2@.900	Ea	—	32.90	—	32.90

HVAC & Plumbing Demolition

Description	Craft@Hrs	Unit	Material $	Labor $	Equipment $	Total $

Remove air handling unit. These figures include isolating and disconnecting hot water, chilled water, steam or gas and/or ductwork, electrical connections and unit disassembly when required for removal. Disposal costs or salvage values excluded. Equipment may be needed for some sizes and locations. Refer to the equipment rental costs at the front of this section.

Description	Craft@Hrs	Unit	Material $	Labor $	Equipment $	Total $
AHU, 2,000-5,000 CFM, remove	P1@8.00	Ea	—	298.00	—	298.00
AHU, 5,000-10,000 CFM, remove	P1@12.0	Ea	—	447.00	—	447.00
AHU, 10,000-20,000 CFM, remove	P1@18.0	Ea	—	670.00	—	670.00
AHU, over 20,000 CFM, remove	P1@28.0	Ea	—	1,040.00	—	1,040.00

Remove rooftop unit. These figures include isolating and disconnecting hot water, chilled water, steam or gas and/or ductwork, electrical connections and unit disassembly when required for removal. Disposal costs or salvage values excluded. Make additional hoisting allowances for roof heights above 40'. Equipment may be needed for some sizes and locations. Refer to the equipment rental costs at the front of this section.

Description	Craft@Hrs	Unit	Material $	Labor $	Equipment $	Total $
RTU, 2,000-5,000 CFM, remove	P1@6.00	Ea	—	223.00	—	223.00
RTU, 5,000-10,000 CFM, remove	P1@10.0	Ea	—	372.00	—	372.00
RTU. 10,000-20,000 CFM, remove	P1@16.0	Ea	—	596.00	—	596.00
RTU, over 20,000 CFM, remove	P1@26.0	Ea	—	968.00	—	968.00

HVAC & Plumbing Demolition

Description	Craft@Hrs	Unit	Material $	Labor $	Equipment $	Total $

Remove boiler. These figures include isolating and disconnecting hot water, steam or gas and/or electrical connections and unit disassembly when required for removal. Disposal costs or salvage values excluded. Equipment may be needed for some sizes and locations. Refer to the equipment rental costs at the front of this section.

Description	Craft@Hrs	Unit	Material $	Labor $	Equipment $	Total $
10-50 KW electric boiler, remove	P1@8.00	Ea	—	298.00	—	298.00
50-100 KW electric boiler, remove	P1@12.0	Ea	—	447.00	—	447.00
100-200 KW electric boiler, remove	P1@26.0	Ea	—	968.00	—	968.00
200-500 KW electric boiler, remove	P1@36.0	Ea	—	1,340.00	—	1,340.00
over 500 KW electric boiler, remove	P1@44.0	Ea	—	1,640.00	—	1,640.00
to 150 Mbh gas-fired boiler, remove	P1@9.00	Ea	—	335.00	—	335.00
150-300 Mbh gas-fired boiler, remove	P1@14.0	Ea	—	521.00	—	521.00
300-750 Mbh gas-fired boiler, remove	P1@30.0	Ea	—	1,120.00	—	1,120.00
750-1,500 Mbh gas-fired boiler, remove	P1@46.0	Ea	—	1,710.00	—	1,710.00
1,500-2,000 Mbh gas-fired boiler, remove	P1@65.0	Ea	—	2,420.00	—	2,420.00
over 2,000 Mbh gas-fired boiler, remove	P1@80.0	Ea	—	2,980.00	—	2,980.00

Remove furnace. These figures include isolating and disconnecting plenums, fuel and/or electrical connections. Disposal costs or salvage values excluded. Equipment may be needed for some sizes and locations. Refer to the equipment rental costs at the front of this section.

Description	Craft@Hrs	Unit	Material $	Labor $	Equipment $	Total $
5-30 KW electric furnace, remove	P1@2.50	Ea	—	93.10	—	93.10
30-60 KW electric furnace, remove	P1@3.75	Ea	—	140.00	—	140.00
30-150 Mbh gas-fired furnace, remove	P1@5.00	Ea	—	186.00	—	186.00
150-225 Mbh gas-fired furnace, remove	P1@7.00	Ea	—	261.00	—	261.00

HVAC & Plumbing Demolition

Description	Craft@Hrs	Unit	Material $	Labor $	Equipment $	Total $

Remove chiller. These figures include isolating and disconnecting chilled water and electrical connections, evacuating refrigerant to storage vessels and unit disassembly when required for removal. Disposal costs or salvage values excluded. Equipment may be needed for some sizes and locations. Refer to the equipment rental costs at the front of this section.

Description	Craft@Hrs	Unit	Material $	Labor $	Equipment $	Total $
Chiller, 20-50 tons, remove	P1@34.0	Ea	—	1,270.00	—	1,270.00
Chiller, 50-100 tons, remove	P1@48.0	Ea	—	1,790.00	—	1,790.00
Chiller, 100-150 tons, remove	P1@56.0	Ea	—	2,080.00	—	2,080.00
Chiller, 150-200 tons, remove	P1@68.0	Ea	—	2,530.00	—	2,530.00
Chiller, 200-250 tons, remove	P1@80.0	Ea	—	2,980.00	—	2,980.00
Chiller, over 250 tons, remove	P1@100	Ea	—	3,720.00	—	3,720.00

Remove air-cooled condenser. These figures include isolating and disconnecting refrigerant lines and electrical connections, evacuating refrigerant to storage vessels and unit disassembly when required for removal. Disposal costs or salvage values excluded. Make additional hoisting allowances for roof heights above 40'. Equipment may be needed for some sizes and locations. Refer to the equipment rental costs at the front of this section.

Description	Craft@Hrs	Unit	Material $	Labor $	Equipment $	Total $
Air-cooled condenser, 2-4 tons, remove	P1@4.00	Ea	—	149.00	—	149.00
Air-cooled condenser, 5-10 tons, remove	P1@8.00	Ea	—	298.00	—	298.00
Air-cooled condenser, 10-20 tons, remove	P1@20.0	Ea	—	744.00	—	744.00
Air-cooled condenser, 20-30 tons, remove	P1@28.0	Ea	—	1,040.00	—	1,040.00
Air-cooled condenser, 30-50 tons, remove	P1@36.0	Ea	—	1,340.00	—	1,340.00
Air-cooled condenser, 50-150 tons, remove	P1@44.0	Ea	—	1,640.00	—	1,640.00
Air-cooled condenser, 150-250 tons, remove	P1@52.0	Ea	—	1,940.00	—	1,940.00

HVAC & Plumbing Demolition

Description	Craft@Hrs	Unit	Material $	Labor $	Equipment $	Total $

Remove cooling tower. These figures include isolating and disconnecting chilled water lines and electrical connections, and unit disassembly when required for removal. Disposal costs or salvage values excluded. Make additional hoisting allowances for roof heights above 40'. Equipment may be needed for some sizes and locations. Refer to the equipment rental costs at the front of this section.

Description	Craft@Hrs	Unit	Material $	Labor $	Equipment $	Total $
Cooling tower, 5-10 tons, remove	P1@8.00	Ea	—	298.00	—	298.00
Cooling tower, 10-20 tons, remove	P1@16.0	Ea	—	596.00	—	596.00
Cooling tower, 20-30 tons, remove	P1@24.0	Ea	—	893.00	—	893.00
Cooling tower, 30-50 tons, remove	P1@32.0	Ea	—	1,190.00	—	1,190.00
Cooling tower, 50-150 tons, remove	P1@40.0	Ea	—	1,490.00	—	1,490.00
Cooling tower, 150-250 tons, remove	P1@48.0	Ea	—	1,790.00	—	1,790.00
Cooling tower, 250-500 tons, remove	P1@70.0	Ea	—	2,610.00	—	2,610.00

Remove fan coil. These figures include isolating and disconnecting chilled water and/or hot water lines and/or duct and electrical connections, and unit disassembly when required for removal. Disposal costs or salvage values excluded. Equipment may be needed for some sizes and locations. Refer to the equipment rental costs at the front of this section.

Description	Craft@Hrs	Unit	Material $	Labor $	Equipment $	Total $
2-pipe fan coil, 300-500 CFM, remove	P1@4.00	Ea	—	149.00	—	149.00
2-pipe fan coil, 550-750 CFM, remove	P1@4.50	Ea	—	167.00	—	167.00
2-pipe fan coil, 800-1,200 CFM, remove	P1@6.00	Ea	—	223.00	—	223.00
4-pipe fan coil, 300-500 CFM, remove	P1@4.50	Ea	—	167.00	—	167.00
4-pipe fan coil, 550-750 CFM, remove	P1@5.00	Ea	—	186.00	—	186.00
4-pipe fan coil, 800-1,200 CFM, remove	P1@6.50	Ea	—	242.00	—	242.00

HVAC & Plumbing Demolition

Description	Craft@Hrs	Unit	Material $	Labor $	Equipment $	Total $

Remove unit heater. These figures include isolating and disconnecting hot water, steam lines or gas and/or electrical connections. Disposal costs or salvage values excluded. Equipment may be needed for some sizes and locations. Refer to the equipment rental costs at the front of this section.

Description	Craft@Hrs	Unit	Material $	Labor $	Equipment $	Total $
Gas-fired unit heater, 50-150 Mbh, remove	P1@2.25	Ea	—	83.70	—	83.70
Gas-fired unit heater, 200-350 Mbh, remove	P1@3.50	Ea	—	130.00	—	130.00
Steam unit heater, 75-200 Mbh, remove	P1@3.25	Ea	—	121.00	—	121.00
Steam unit heater, 250-350 Mbh, remove	P1@4.75	Ea	—	177.00	—	177.00
Electric unit heater, 5-25 KW, remove	P1@2.00	Ea	—	74.40	—	74.40
Electric unit heater, 30-50 KW, remove	P1@2.75	Ea	—	102.00	—	102.00

Remove heat pump. These figures include isolating and disconnecting pipe, duct and electrical connections and evacuating refrigerant to storage vessels when required for removal. Disposal costs or salvage values excluded. Equipment may be needed for some sizes and locations. Refer to the equipment rental costs at the front of this section.

Description	Craft@Hrs	Unit	Material $	Labor $	Equipment $	Total $
Heat pump, split system, 2-5 ton, remove	P1@4.75	Ea	—	177.00	—	177.00
Heat pump, split system, 7.5-10 ton, remove	P1@8.50	Ea	—	316.00	—	316.00
Heat pump, packaged system, 2-5 ton, remove	P1@3.25	Ea	—	121.00	—	121.00
Heat pump, pkg. system, 7.5-10 ton, remove	P1@5.50	Ea	—	205.00	—	205.00

HVAC & Plumbing Demolition

Description	Craft@Hrs	Unit	Material $	Labor $	Equipment $	Total $

Remove heat exchanger. These figures include isolating and disconnecting chilled water, hot water and/or steam line connections. Disposal costs or salvage values excluded. Equipment may be needed for some sizes and locations. Refer to the equipment rental costs at the front of this section.

Description	Craft@Hrs	Unit	Material $	Labor $	Equipment $	Total $
Heat exchanger, 6" x 30", remove	P1@2.75	Ea	—	102.00	—	102.00
Heat exchanger, 10" x 48", remove	P1@4.00	Ea	—	149.00	—	149.00
Heat exchanger, 12" x 72", remove	P1@5.25	Ea	—	195.00	—	195.00
Heat exchanger, 16" x 90", remove	P1@6.00	Ea	—	223.00	—	223.00

Remove duct-mounted coil. These figures include isolating and disconnecting steam, chilled water, hot water, condensate, refrigerant or electrical and duct connections. Disposal costs or salvage values excluded. Refrigerant coil removal costs include system refrigerant evacuation and storage. Equipment may be needed for some sizes and locations. Refer to the equipment rental costs at the front of this section.

Description	Craft@Hrs	Unit	Material $	Labor $	Equipment $	Total $
Single-row water coil, 12" to 18", remove	P1@2.00	Ea	—	74.40	—	74.40
Single-row water coil, 20" to 30", remove	P1@3.50	Ea	—	130.00	—	130.00
Single-row water coil, 32" to 60", remove	P1@8.00	Ea	—	298.00	—	298.00
Single-row steam coil, 12" to 18", remove	P1@2.25	Ea	—	83.70	—	83.70
Single-row steam coil, 20" to 30", remove	P1@3.75	Ea	—	140.00	—	140.00
Single-row steam coil, 32" to 60", remove	P1@9.00	Ea	—	335.00	—	335.00
Electric resistance, 12" to 18", remove	P1@1.25	Ea	—	46.50	—	46.50
Electric resistance, 20" to 30", remove	P1@1.75	Ea	—	65.10	—	65.10
Electric resistance, 32" to 60", remove	P1@4.00	Ea	—	149.00	—	149.00
Refrigerant coil, 2-3 ton, remove	P1@4.00	Ea	—	149.00	—	149.00
Refrigerant coil, 4-5 ton, remove	P1@6.50	Ea	—	242.00	—	242.00
Refrigerant coil, 6-10 ton, remove	P1@12.0	Ea	—	447.00	—	447.00
Refrigerant coil, 12-15 ton, remove	P1@16.0	Ea	—	596.00	—	596.00

HVAC & Plumbing Demolition

Description	Craft@Hrs	Unit	Material $	Labor $	Equipment $	Total $

Remove steel piping. These figures include cutting out and removing threaded or welded steel piping. These figures assume demolished materials are piled on site. Equipment, fixture or apparatus disconnection costs and salvage values excluded. Equipment may be needed for some sizes and locations. Refer to the equipment rental costs at the front of this section.

Description	Craft@Hrs	Unit	Material $	Labor $	Equipment $	Total $
Steel pipe, Sch. 40, ½"-1" remove	P1@.038	LF	—	1.41	—	1.41
Steel pipe, Sch. 40, 1½"-2" remove	P1@.068	LF	—	2.53	—	2.53
Steel pipe, Sch. 40, 2½"-4" remove	P1@.108	LF	—	4.02	—	4.02
Steel pipe, Sch. 40, 6"-8" remove	P1@.230	LF	—	8.56	—	8.56
Steel pipe, Sch. 40, 10"-12" remove	P1@.330	LF	—	12.30	—	12.30
Steel pipe, Sch. 80, ½"-1" remove	P1@.042	LF	—	1.56	—	1.56
Steel pipe, Sch. 80, 1½"-2" remove	P1@.075	LF	—	2.79	—	2.79
Steel pipe, Sch. 80, 2½"-4" remove	P1@.118	LF	—	4.39	—	4.39
Steel pipe, Sch. 80, 6"-8" remove	P1@.290	LF	—	10.80	—	10.80
Steel pipe, Sch. 80, 10"-12" remove	P1@.360	LF	—	13.40	—	13.40
Remove steel pipe, Sch. 160, add	—	%	—	30.00	—	—

Remove copper piping. These figures include cutting out and removing copper piping. These figures assume demolished materials are piled on site. Equipment, fixture or apparatus disconnection costs and salvage values excluded.

Description	Craft@Hrs	Unit	Material $	Labor $	Equipment $	Total $
Copper pipe, ½"-1" remove	P1@.022	LF	—	.82	.08	.90
Copper pipe, 1½"-2" remove	P1@.028	LF	—	1.04	.09	1.13
Copper pipe, 2½"-4" remove	P1@.044	LF	—	1.64	.12	1.76

Remove plastic piping. These figures include cutting out and removing plastic piping. These figures assume demolished materials are piled on site. Equipment, fixture or apparatus disconnection costs and salvage values excluded. Equipment may be needed for some sizes and locations. Refer to the equipment rental costs at the front of this section.

Description	Craft@Hrs	Unit	Material $	Labor $	Equipment $	Total $
Plastic pipe, ½"-1" remove	P1@.015	LF	—	.56	—	.56
Plastic pipe, 1½"-2" remove	P1@.030	LF	—	1.12	—	1.12
Plastic pipe, 2½"-4" remove	P1@.038	LF	—	1.41	—	1.41
Plastic pipe, 6"-8" remove	P1@.080	LF	—	2.91	—	2.91

HVAC & Plumbing Demolition

Description	Craft@Hrs	Unit	Material $	Labor $	Equipment $	Total $

Remove pump. These figures include isolating and disconnecting pipe and electrical connections. Disposal costs or salvage values excluded. Equipment may be needed for some sizes and locations. Refer to the equipment rental costs at the front of this section.

Description	Craft@Hrs	Unit	Material $	Labor $	Equipment $	Total $
Base-mounted pump, ½-1 HP, remove	P1@1.00	Ea	—	37.20	—	37.20
Base-mounted pump, 1½-3 HP, remove	P1@1.75	Ea	—	65.10	—	65.10
Base-mounted pump, 5-10 HP, remove	P1@2.75	Ea	—	102.00	—	102.00
Base-mounted pump, 15-25 HP, remove	P1@3.75	Ea	—	140.00	—	140.00
Base-mounted pump, 30-50 HP, remove	P1@4.55	Ea	—	169.00	—	169.00
In-line mounted pump, 1/12-1/3 HP, remove	P1@.650	Ea	—	24.20	—	24.20
In-line mounted pump, ½-1 HP, remove	P1@.750	Ea	—	27.90	—	27.90
In-line mounted pump, 3-5 HP, remove	P1@1.85	Ea	—	68.90	—	68.90
In-line mounted pump, 7.5-10 HP, remove	P1@2.55	Ea	—	94.90	—	94.90
In-line mounted pump, 15-25 HP, remove	P1@3.90	Ea	—	145.00	—	145.00
In-line mounted pump, 30-50 HP, remove	P1@4.85	Ea	—	181.00	—	181.00

Remove valve. These figures include isolation and disconnection. Disposal costs or salvage values excluded. Equipment may be needed for some sizes and locations. Refer to the equipment rental costs at the front of this section.

Description	Craft@Hrs	Unit	Material $	Labor $	Equipment $	Total $
Soldered valve, ½"-1", remove	P1@.165	Ea	—	6.14	—	6.14
Soldered valve, 1½"-2", remove	P1@.225	Ea	—	8.37	—	8.37
Soldered valve, 2½"-4", remove	P1@.650	Ea	—	24.20	—	24.20
Threaded valve, ½"-1", remove	P1@.150	Ea	—	5.58	—	5.58
Threaded valve, 1½"-2", remove	P1@.235	Ea	—	8.75	—	8.75
Threaded valve, 2½"-4", remove	P1@.680	Ea	—	25.30	—	25.30
Flanged valve, ½"-1", remove	P1@.200	Ea	—	7.44	—	7.44
Flanged valve, 1½"-2", remove	P1@.280	Ea	—	10.40	—	10.40
Flanged valve, 2½"-4", remove	P1@.625	Ea	—	23.30	—	23.30
Flanged valve, 6"-8", remove	P1@1.70	Ea	—	63.30	—	63.30
Flanged valve, 10"-12", remove	P1@2.65	Ea	—	98.60	—	98.60

HVAC & Plumbing Demolition

Description	Craft@Hrs	Unit	Material $	Labor $	Equipment $	Total $

Remove hot water tank. These figures include isolation and disconnection. Disposal costs or salvage values excluded. Equipment may be needed for some sizes and locations. Refer to the equipment rental costs at the front of this section.

Description	Craft@Hrs	Unit	Material $	Labor $	Equipment $	Total $
Gas-fired hot water tank, 40 gallon, remove	P1@2.00	Ea	—	74.40	—	74.40
Gas-fired hot water tank, 60 gallon, remove	P1@2.50	Ea	—	93.10	—	93.10
Gas-fired hot water tank, 100 gallon, remove	P1@3.00	Ea	—	112.00	—	112.00
Electric hot water tank, 40 gallon, remove	P1@1.60	Ea	—	59.60	—	59.60
Electric hot water tank, 60 gallon, remove	P1@2.00	Ea	—	74.40	—	74.40
Electric hot water tank, 100 gallon, remove	P1@2.60	Ea	—	96.80	—	96.80

Remove plumbing fixture. These figures include isolation and disconnection. Disposal costs or salvage values excluded. Equipment may be needed for some sizes and locations. Refer to the equipment rental costs at the front of this section.

Description	Craft@Hrs	Unit	Material $	Labor $	Equipment $	Total $
Floor-mount, tank-type water closet, remove	P1@.850	ea	—	31.60	—	31.60
Wall-mount, tank-type water closet, remove	P1@1.35	Ea	—	50.20	—	50.20
Floor-mount, flush valve water closet, remove	P1@.950	Ea	—	35.40	—	35.40
Wal- mount, flush valve water closet, remove	P1@1.55	Ea	—	57.70	—	57.70
Wall-mount, flush valve urinal, remove	P1@1.15	Ea	—	42.80	—	42.80
Wall-mount, tank-type urinal, remove	P1@1.65	Ea	—	61.40	—	61.40
Kitchen sink, remove	P1@.850	Ea	—	31.60	—	31.60
Bathroom sink, remove	P1@.850	Ea	—	31.60	—	31.60
Bathtub, remove	P1@2.25	Ea	—	83.70	—	83.70
Shower stall, remove	P1@2.00	Ea	—	74.40	—	74.40
Tub and shower combination, remove	P1@2.85	Ea	—	106.00	—	106.00
Laundry sink, remove	P1@.950	Ea	—	35.40	—	35.40
Mop sink, remove	P1@1.50	Ea	—	55.80	—	55.80
Slop sink, remove	P1@1.75	Ea	—	65.10	—	65.10
Drinking fountain, remove	P1@1.40	Ea	—	52.10	—	52.10

HVAC & Plumbing Demolition

Description	Craft@Hrs	Unit	Material $	Labor $	Equipment $	Total $

Disconnect plumbing fixture. These figures include isolation and disconnection only. Fixtures removed by others. Disposal costs or salvage values excluded. Equipment may be needed for some sizes and locations. Refer to the equipment rental costs at the front of this section.

Description	Craft@Hrs	Unit	Material $	Labor $	Equipment $	Total $
Floor-mount, tank-type water closet, disconnect	P1@.450	Ea	—	16.70	—	16.70
Wall-mount, tank-type water closet, disconnect	P1@.650	Ea	—	24.20	—	24.20
Floor-mount, flush valve water closet, disconnect	P1@.550	Ea	—	20.50	—	20.50
Wall-mount, flush valve water closet, disconnect	P1@.650	Ea	—	24.20	—	24.20
Wall-mount, flush valve urinal, disconnect	P1@.650	Ea	—	24.20	—	24.20
Wall-mount, tank-type urinal, disconnect	P1@.950	Ea	—	35.40	—	35.40
Kitchen sink, disconnect	P1@.450	Ea	—	16.70	—	16.70
Bathroom sink, disconnect	P1@.450	Ea	—	16.70	—	16.70
Bathtub, disconnect	P1@.650	Ea	—	24.20	—	24.20
Shower stall, disconnect	P1@.650	Ea	—	24.20	—	24.20
Tub and shower combination, disconnect	P1@.650	Ea	—	24.20	—	24.20
Laundry sink, disconnect	P1@.450	Ea	—	16.70	—	16.70
Mop sink, disconnect	P1@.450	Ea	—	16.70	—	16.70
Slop sink, disconnect	P1@.750	Ea	—	27.90	—	27.90
Drinking fountain, disconnect	P1@.650	Ea	—	24.20	—	24.20
Gas-fired hot water tank, 40 gallon, disconnect	P1@.900	Ea	—	33.50	—	33.50
Gas-fired hot water tank, 60 gallon, disconnect	P1@1.15	Ea	—	42.80	—	42.80
Gas-fired hot water tank, 100 gallon, disconnect	P1@1.35	Ea	—	50.20	—	50.20
Electric hot water tank, 40 gallon, disconnect	P1@.600	Ea	—	22.30	—	22.30
Electric hot water tank, 60 gallon, disconnect	P1@.900	Ea	—	33.50	—	33.50
Electric hot water tank, 100 gallon, disconnect	P1@1.00	Ea	—	37.20	—	37.20

HVAC & Plumbing Demolition

Description	Craft@Hrs	Unit	Material $	Labor $	Equipment $	Total $

Remove pipe insulation. These figures assume demolished materials are piled on site. Disposal costs or salvage values excluded.

Description	Craft@Hrs	Unit	Material $	Labor $	Equipment $	Total $
Fiberglass pipe insulation, ½"-1", remove	P1@.020	LF	—	.74	—	.74
Fiberglass pipe insulation, 1½"-2", remove	P1@.025	LF	—	.93	—	.93
Fiberglass pipe insulation, 2½"-4", remove	P1@.035	LF	—	1.30	—	1.30
Fiberglass pipe insulation, 6"-8", remove	P1@.060	LF	—	2.23	—	2.23
Fiberglass pipe insulation, 10"-12", remove	P1@.090	LF	—	3.35	—	3.35
Pipe insul. with canvas jacket – add	—	%	—	10.00	—	—
Pipe insul. with alum. jacket – add	—	%	—	20.00	—	—
Calcium silicate pipe insul. – add	—	%	—	15.00	—	—

Remove duct insulation. These figures assume demolished materials are piled on site. Disposal costs or salvage values excluded.

Description	Craft@Hrs	Unit	Material $	Labor $	Equipment $	Total $
Fiberglass blanket insulation, remove	S2@.020	SF	—	.73	—	.73
Fiberglass rigid board insulation, remove	S2@.025	SF	—	.91	—	.91
Insul. with canvas jacket – add	—	%	—	8.00	—	—
Insul. with alum jacket – add	—	%	—	10.00	—	—

Budget Estimating

Budget Estimates for Plumbing and HVAC Work

General contractors and owners often ask mechanical subcontractors to prepare budget cost estimates for use in planning projects. A budget estimate isn't a bid for the contract and you probably won't be paid for preparing the estimate. That's why you won't want to spend too much time on these estimates. Accuracy within 8 to 10 percent of the final bid price will usually be acceptable. Normally you'll have only preliminary architectural drawings and possibly a guide specification to work from. The tables that follow will help you prepare budget estimates for nearly any plumbing or HVAC job.

The costs in these tables are averages and will vary considerably with design requirements and by geographic area. Make it clear to the owner or architect requesting the figures that these are budget estimates and will be revised when final plans and specifications are available.

Budget Estimates for Plumbing: First, count the number of plumbing fixtures. Multiply the number of fixtures of each type by the price per fixture shown in the table below. These prices include all the usual labor and material for a complete average plumbing installation. Prices are based on branch piping lengths averaging 35 feet between the fixture and the main. Typical main piping is also included in the prices.

Description	Craft@Hrs	Unit	Material $	Labor $	Equipment $	Total $
Budget plumbing estimate, cast iron DWV, copper supply pipe						
Bathtubs	P1@15.0	Ea	2,470.00	558.00	—	3,028.00
Showers	P1@9.00	Ea	1,340.00	335.00	—	1,675.00
Lavatories	P1@11.0	Ea	2,310.00	409.00	—	2,719.00
Kitchen sinks	P1@12.0	Ea	2,420.00	447.00	—	2,867.00
Service sinks	P1@10.0	Ea	2,080.00	372.00	—	2,452.00
Bar sinks	P1@9.00	Ea	1,750.00	335.00	—	2,085.00
Floor sinks	P1@7.00	Ea	1,070.00	261.00	—	1,331.00
Water closets						
Tank-type	P1@11.0	Ea	1,750.00	409.00	—	2,159.00
Flush valve	P1@12.0	Ea	1,970.00	447.00	—	2,417.00
Urinals	P1@8.00	Ea	2,530.00	298.00	—	2,828.00
Drinking fountains						
(refrigerated)	P1@12.0	Ea	2,920.00	447.00	—	3,367.00
Wash fountains	P1@25.0	Ea	6,280.00	931.00	—	7,211.00
Can washers	P1@18.0	Ea	3,200.00	670.00	—	3,870.00
Floor drains	P1@8.00	Ea	988.00	298.00	—	1,286.00
Area drains	P1@6.00	Ea	849.00	223.00	—	1,072.00
Roof drains	P1@18.0	Ea	2,250.00	670.00	—	2,920.00
Overflow drains	P1@18.0	Ea	2,250.00	670.00	—	2,920.00
Deck drains	P1@4.00	Ea	699.00	149.00	—	848.00
Cleanouts	P1@5.00	Ea	755.00	186.00	—	941.00
Trap primers	P1@2.00	Ea	383.00	74.40	—	457.40
Sump pumps	P1@6.00	Ea	3,730.00	223.00	—	3,953.00
Water heaters						
Gas, 40 gal.	P1@16.0	Ea	4,400.00	596.00	—	4,996.00
Gas, 80 gal.	P1@17.0	Ea	4,880.00	633.00	—	5,513.00
Gas, 120 gal.	P1@18.0	Ea	5,710.00	670.00	—	6,380.00

Budget Estimating

Description	Craft@Hrs	Unit	Material $	Labor $	Equipment $	Total $

Budget plumbing estimate, plastic DWV and plastic supply pipe

Description	Craft@Hrs	Unit	Material $	Labor $	Equipment $	Total $
Bathtubs	P1@12.0	Ea	1,970.00	447.00	—	2,417.00
Showers	P1@7.00	Ea	953.00	261.00	—	1,214.00
Lavatories	P1@9.00	Ea	1,820.00	335.00	—	2,155.00
Kitchen sinks	P1@10.0	Ea	2,030.00	372.00	—	2,402.00
Service sinks	P1@8.00	Ea	1,630.00	298.00	—	1,928.00
Bar sinks	P1@7.00	Ea	1,380.00	261.00	—	1,641.00
Floor sinks	P1@6.00	Ea	828.00	223.00	—	1,051.00
Water closets						
Tank-type	P1@8.00	Ea	1,270.00	298.00	—	1,568.00
Flush valve	P1@9.00	Ea	1,530.00	335.00	—	1,865.00
Urinals	P1@6.00	Ea	2,070.00	223.00	—	2,293.00
Drinking fountains						
(refrigerated)	P1@10.0	Ea	2,410.00	372.00	—	2,782.00
Wash fountains	P1@22.0	Ea	5,480.00	819.00	—	6,299.00
Can washers	P1@15.0	Ea	2,690.00	558.00	—	3,248.00
Floor drains	P1@7.00	Ea	685.00	261.00	—	946.00
Area drains	P1@5.00	Ea	561.00	186.00	—	747.00
Roof drains	P1@16.0	Ea	1,920.00	596.00	—	2,516.00
Overflow drains	P1@16.0	Ea	1,920.00	596.00	—	2,516.00
Deck drains	P1@3.00	Ea	476.00	112.00	—	588.00
Cleanouts	P1@4.00	Ea	582.00	149.00	—	731.00
Trap primers	P1@2.00	Ea	289.00	74.40	—	363.40
Sump pumps	P1@5.00	Ea	3,360.00	186.00	—	3,546.00
Water heaters						
Gas, 40 gal.	P1@15.0	Ea	4,060.00	558.00	—	4,618.00
Gas, 80 gal.	P1@16.0	Ea	4,440.00	596.00	—	5,036.00
Gas, 120 gal.	P1@17.0	Ea	5,160.00	633.00	—	5,793.00

Budget Estimating

Budget Estimates for HVAC Work: Budget estimates for heating, ventilating and air conditioning work are figured by the square foot of floor. Multiply the area of conditioned space by the cost per square foot in the table that follows. The result is the estimated cost including a complete heating, ventilating and air conditioning system and 20 percent overhead and profit. The table also shows average heating and cooling requirements for various types of buildings.

Description	Cooling (Btu/SF)	Heating (Btu/SF)	Total Selling Price ($) Per SF	Total Selling Price ($) Per Ton
Apartments and condominiums	25	25	14.10	5,200.00
Auditoriums	40	40	36.10	8,310.00
Banks	45	30	37.40	8,290.00
Barber shops	40	40	36.90	8,460.00
Beauty shops	60	30	43.80	6,700.00
Bowling alleys	40	40	28.80	6,610.00
Churches	35	35	31.30	8,150.00
Classrooms	45	40	56.90	11,500.00
Cocktail lounges	70	30	39.50	5,200.00
Computer rooms	145	20	80.50	5,100.00
Department stores	35	30	25.30	6,660.00
Laboratories	60	45	70.00	10,700.00
Libraries	45	35	41.80	8,530.00
Manufacturing plants	40	35	16.30	3,730.00
Medical buildings	35	35	27.60	7,280.00
Motels	30	30	19.40	5,910.00
Museums	35	40	30.80	8,090.00
Nursing homes	45	35	35.50	7,310.00
Office buildings, low-rise	35	35	28.10	7,390.00
Office buildings, high-rise	40	30	38.30	8,760.00
Residences	20	30	8.64	3,990.00
Retail shops	45	30	28.80	5,870.00
Supermarkets	30	30	15.10	4,580.00
Theaters	40	40	28.10	6,500.00

Note that these are budget estimates and can vary widely with conditions which may be unique to the job site. Many of these conditions are identified below. Avoid submitting even a budget estimate without either a site walk-through, a conference with the consulting engineer or architect or a careful review of the plans.

Be alert to any of the following conditions. Any of these can have a major impact on cost:

1. Do HVAC unit dimensions exceed the size of available openings?
2. If a crane is required, is there ample clearance for crane operations?
3. Will any doors, permanent ladders, scaffolding or load-bearing supports obstruct installation?
4. Do the plans show all vents, drains and cleanouts required by the code? Many items like these are not specified by the engineer but will be required by the inspector.
5. For retrofit work, consider carefully problems associated with demolition, including removal of the existing HVAC system, structural modifications and any encased asbestos that may be present.
6. In retrofit jobs, the type and size of HVAC equipment is limited by the window area and installed insulation.
7. How will available power, gas and water lines be extended to the new HVAC equipment?

Plumbing & HVAC Forms and Letters

The following pages contain business forms and letters most plumbing and HVAC contractors can use.

Change Orders	Purchase Order
Change Estimate Takeoff	Construction Schedule
Change Estimate Worksheet	
Change Estimate Summary	Letter of Intent
Change Order Log	Submittal Data Cover Sheet
Subcontract Forms	Submittal Index
Standard Form Subcontract	Billing Breakdown Worksheet
Subcontract Change Order	

An explanation is provided on how the forms function in the contractor's office, how to fill them out, and a sample provided. Where appropriate, a blank copy of the form is provided for your use.

Permission is hereby given by the publisher for the duplication of any of the blank forms on a copy machine or at a printer, provided they are for your own use and not to be sold at a cost exceeding the actual printing and paper cost.

Space has been provided at the top of the forms so you can add your own letterhead or logo.

Change Estimates

Very few projects are completed without some changes to the original plans. Most changes in the scope of work are a result of either errors by the designer or requests by the owner.

Here's the usual sequence in handling changes.

1) The project architect usually initiates a change by sending a Request for Quotation (R.F.Q.) to the general contractor with a detailed description of the work to be modified.

2) The general contractor prepares a Change Estimate (C.E.) if the work is to be done by the general.

3) The subcontractor prepares a change estimate which shows the work to be added to or deleted from the contract and the cost of changes, including markup. The general contractor reviews the change estimate and forwards a summary to the architect for decision.

4) If the decision is to go ahead with the change, the architect issues a Change Order (C.O.) to the general contractor. This change order is authorization to proceed with the change at the price quoted.

Your costs for making changes will normally be higher than your costs for similar items on the original estimate. Here's why:

Equipment and materials for changes are almost always purchased in smaller quantities. Volume discounts are seldom available.

Special handling and delivery of small orders will further increase equipment and material costs.

Scheduled work will be disrupted.

Overhead costs will be higher because special purchase orders or subcontracts will be required.

Vendor restocking charges for deleted items will range from 20 to 50 percent of the original cost.

Large change orders may require additional manpower, which usually results in decreased efficiency.

The cumulative effect of change orders will be to increase overhead costs if the project completion date is extended.

To compensate for these higher costs, most estimators use a higher markup on change estimates:

Add an extra 10 percent markup to equipment and material costs to cover increased overhead and handling costs, prior to adding final Overhead and Profit markup.

Increase competitive labor units (manhours required to perform a specific task) by 15 percent to compensate for inefficiencies such as disruptions in the work schedule, extra handling of equipment and material, and additional manpower.

Mark up subcontractor prices by 10 percent.

Figure overhead and profit markup at 15/10 or 10/10.

By following these suggestions, the mathematical gross profit will appear to be 40 to 45 percent, but the true gross will be closer to 25 percent because your costs for change order work are higher, as listed above.

Change Estimate Example

Let's assume that the general contractor on your job has sent a R.F.Q. to your attention. The extra work is a run of hot water piping that will connect to equipment being furnished and set by others. The first step is to list all material and labor required to do the work. Notice the items entered on the Change Estimate Take-off form on page 440.

When the take-off is complete, transfer the manhour total to the Change Estimate Worksheet on page 441. To complete this form:

1) Multiply manhours by the cost per hour.

2) Add the cost of supervision.

3) Total direct labor cost.

4) Add the cost of preparing the change order estimate (C.E.).

5) Add equipment and tool costs.

6) Add subcontract costs.

Then bring totals forward to the Change Estimate Summary on page 442:

1) Enter the material cost total from the Change Estimate Take-off form on page 444.

2) Enter direct and indirect labor cost totals from the Change Estimate Worksheet.

3) Enter equipment, tool and subcontract cost totals from the Change Estimate Worksheet.

Notice the additional bond charge on line M of the Change Estimate Summary. This is the extra premium charged by the bonding company to cover the increase in contract value. The next line, Service Reserve, is the allowance for service work (if any is required) during the warranty period. A sample Change Order Log, used to record the status of changes, is also included on page 443.

ACME
Mechanical Contractors

7600 Oak Avenue - Smallville, U.S.A. 12345-6789
Voice 1-234-5678 / FAX 1-234-5680

Change Estimate Take-off

Project CBC Bank Building **C. E. Number** One

Take-off by DCO **Job No** 7777

Date 2-2-2022

Item	Equipment and Material Description	Quantity	Material $ Unit	Material $ Total	Labor Manhours Unit	Labor Manhours Total
1	1" Type L hard copper pipe	40 LF	3.87	154.80	.038	1.52
2	1" 90 degree copper ell	6 Ea	4.21	25.26	.23	1.38
3	1" copper coupling	4 Ea	2.54	10.16	.24	.96
4	1" bronze ball valve	2 Ea	17.30	34.60	.36	.72
5	2" x 1" reducing copper tee	2 Ea	20.60	41.20	.40	.80
6	1" copper union	2 Ea	19.10	38.20	.26	.52
7	1" pipe hanger	8 Ea	4.33	34.64	.25	2.00
8	Miscellaneous material	1 Ea	LS	50.00	1.78	1.78
		Totals		388.86		9.68

ACME
Mechanical Contractors

7600 Oak Avenue - Smallville, U.S.A. 12345-6789
Voice 1-234-5678 / FAX 1-234-5680

Change Estimate Worksheet

Estimated by DCO **C. E. Number** One

Date 2-3-2022 **Job No** 7777

Direct Labor Expense (including benefits)

Plumber or pipefitter journeyman	9.68 manhours at $	41.85	per hour =	$ 405.11
Plumber or pipefitter supervisor	1.00 manhours at $	45.00	per hour =	$ 45.00
Sheet metal journeyman	manhours at $		per hour =	$
Sheet metal supervisor	manhours at $		per hour =	$
Laborer journeyman	manhours at $		per hour =	$
	manhours at $		per hour =	$

Total Direct Labor Cost $ 450.11

Indirect Costs

Travel		$
Subsistence		$
Labor	Cost Estimate Preparation (1 Hour)	$ 40.00

Total Indirect Cost $ 40.00

Equipment and Tool Costs

	Ramset rental	$ 35.00
	Propane	$ 4.50

Total Equipment and Tool Cost $ 39.50

Subcontract Costs

	Quikrap Insulation Co.	$ 100.00
		$

Total Subcontract Cost $ 100.00

ACME
Mechanical Contractors

7600 Oak Avenue - Smallville, U.S.A. 12345-6789
Voice 1-234-5678 / FAX 1-234-5680

Change Estimate Summary

Quotation to ABC General Contractors, Inc.　　**Date** 2-3-2022

Address 380 First Street　　**Change Order No.** One

Address 　　**Job Number** 7777

City / State / Zip Smallville, CA 63876

Attention: Mr. J.H. Smith

Job Name CBC Bank Building

Reference: Your R.F.O. No. 7 Labor and Material to furnish and install additional hot water piping.

A.	Materials and equipment:	$ 388.86		
B.	Sales tax: 7% .	$ 27.22		
C.	Direct labor:	$ 450.11		
D.	Indirect costs:	$ 40.00		
E.	Equipment and tools:	$ 39.50		
F.			**Subtotal** $	945.69
G.	Overhead at : 15% of line F:	$ 141.85		
H.	Subcontracts:	$ 100.00		
I.	Overhead at : 10% of line H:	$ 10.00		
J.			**Subtotal** $	1,197.54
K.	Profit at 10% of line J:	$ 119.75		
L.			**Subtotal** $	1,317.29
M.	Bond premium at 1% of line L:	$ 13.17		
N.	Service reserve at 0.5% of line L:	$ 6.59		
O.	**Total cost estimate, lines L thru N:**	☐ Add ☐ Deduct	$	1,337.05

P. Exclusions from this estimate: Per Base
Q. This quotation is valid for 30 days ☐
R. We require 1 day(s) extension of the contract time. ☐
S. We are proceeding with this work per your authorization. ☐
T. Please forward your confirming change order. ☐

Signed by: _____
　　　　　　　　　　　Project Manager

Change Order Log

Job Name _____ **Job Number** _____

Change Number	Customer Number	Date	Amount	Date Approved	Amount Approved	Remarks

Use this space to insert your company logo and letterhead.

Change Estimate Take-off

Project _____   **C. E. Number** _____

Take-off by _____   **Job No** _____

Date _____

Item	Equipment and Material		Material $		Labor Manhours	
	Description	Quantity	Unit	Total	Unit	Total
		Totals				

Use this space to insert your company logo and letterhead.

Change Estimate Worksheet

Estimated by _____ **C. E. Number** _____

Date _____ **Job No** _____

Direct Labor Expense (including benefits)

Plumber or pipefitter journeyman	_____ manhours at $ _____	per hour =	$ _____
Plumber or pipefitter supervisor	_____ manhours at $ _____	per hour =	$ _____
Sheet metal journeyman	_____ manhours at $ _____	per hour =	$ _____
Sheet metal supervisor	_____ manhours at $ _____	per hour =	$ _____
Laborer journeyman	_____ manhours at $ _____	per hour =	$ _____
_____	_____ manhours at $ _____	per hour =	$ _____
		Total Direct Labor Cost	$ _____

Indirect Costs

Travel	_____	$ _____
Subsistence	_____	$ _____
Labor	_____	$ _____
	Total Indirect Cost	$ _____

Equipment and Tool Costs

_____ $ _____

_____ $ _____

Total Equipment and Tool Cost $ _____

Subcontract Costs

_____ $ _____

_____ $ _____

Total Subcontract Cost $ _____

Use this space to insert your company logo and letterhead.

Change Estimate Summary

Quotation to _____ **Date** _____
Address _____ **Change Order No.** _____
Address _____ **Job Number** _____
City / State / Zip _____
Attention: _____
Job Name _____
Reference: _____

A. Materials and equipment: $ _____
B. Sales tax: . $ _____
C. Direct labor: $ _____
D. Indirect costs: $ _____
E. Equipment and tools: $ _____
F.   **Subtotal** $ _____
G. Overhead at _____ of line F: $ _____
H. Subcontracts: $ _____
I. Overhead at _____ of line H: $ _____
J.   **Subtotal** $ _____
K. Profit at _____ of line J: $ _____
L.   **Subtotal** $ _____
M. Bond premium at 1% of line L: $ _____
N. Service reserve at 0.5% of line L: $ _____
O. **Total cost estimate, lines L thru N:** ☐ Add ☐ Deduct $ _____
P. Exclusions from this estimate: _____
Q. This quotation is valid for _____ days
R. We require _____ day(s) extension of the contract time.
S. We are proceeding with this work per your authorization.
T. Please forward your confirming change order.

Signed by: _____
                            Project Manager

Subcontract Forms

Another term for the general contractor is *prime contractor*. Subcontractors working directly for the prime contractor are called *second-tier subcontractors*. Contractors working for these second-tier subs are usually known as *third-tier subcontractors*. On a large job, many second-tier subs may hire third-tier subs to perform specialized work such as insulation, fireproofing, test and balance, water treatment, instrumentation and trenching.

If a portion of your work will be subcontracted to others, I recommend drawing up a subcontract agreement for that work. Any subcontractors who furnish both material and labor for a job should have a written contract.

Most larger plumbing, heating and A/C subcontractors use a two-part *Standard Form Subcontract* published by the Associated General Contractors of America. This two-part form is used to write the base subcontract. The A.G.C. also publishes a *Subcontract Change Order* form which is used when adding or deleting items after the base subcontract is signed. These forms are sold at modest cost at most A.G.C. offices.

The first section of page one of the Standard Form Subcontract identifies the two contractors entering into the agreement. The party issuing the subcontract is the *contractor*. Recipient of the agreement is the *subcontractor*.

Under *Recitals,* the contractor must affirm the right to issue a subcontract by declaring that a legal contract exists with the *owner* (general contractor).

Section 1 - Entire Contract describes the contract documents: the project plans, specifications and addenda. This list of documents should always be the same as the documents identified in the second-tier subcontractor's contract with the general contractor.

Section 2 - Scope describes the specific scope of work to be performed.

Section 3 - Contract Price. Complete this section by inserting the agreed-upon price for doing the work described.

Section 4 - Payment Schedule on page two should show the same payment terms as your contract with the general contractor.

Section 7 - Special Provisions is used to list any special agreements or exclusions.

When the contract is typed, do not sign it. Instead, forward it to the third-tier subcontractor with instructions to sign it and return it intact. When the signed contract is received, sign it and send a copy to the third-tier subcontractor.

For contracts that comply with your state's regulations, please visit www.constructioncontractwriter.com

Standard Form Subcontract

Subcontract No. 7777-1

THIS AGREEMENT, made and entered into at Smallville, CA. this 3rd day of May, 2021, by and between Acme Mechanical Contractors hereinafter called **CONTRACTOR,** with principal office at 7600 Oak Avenue, Smallville, California, and Quikrap Insulation Co., 320 Maple Blvd., Smallville, CA hereinafter called a **SUBCONTRACTOR.**

RECITALS

On or about the 2nd day of January , 20 22, CONTRACTOR entered into a prime contract with

ABC General Contractors, Inc. hereinafter called OWNER, whose address is 380 First Street, Smallville, California to perform the following construction work: The installation of all mechanical and plumbing systems for the proposed Mile-Hi office building to be located at 2600 Second, North, Smallville, California.

Said work is to be performed in accordance with the prime contract and the plans and specifications. Said plans and specifications have been prepared by or on behalf of Quik-Draw and Associates, Smallville, California , **ARCHITECT**

SECTION 1 - ENTIRE CONTRACT

SUBCONTRACTOR certifies and agrees that he is fully familiar with all of the terms, conditions and obligations of the Contract Documents, as hereinafter defined, the location of the job site, and the conditions under which the work is to be performed, and that he enters into this Agreement based upon his investigation of all of such matters and is in no way relying upon any opinions or representations of CONTRACTOR. It is agreed that this Agreement represents the entire agreement. It is further agreed that the Contract Documents are incorporated in this Agreement by this reference, with the same force and effect as if the same were set forth at length herein, and that SUBCONTRACTOR and his subcontractors will be and are bound by any and all of said Contract Documents insofar as they relate in any part or in any way, directly or indirectly to the work covered by this Agreement. SUBCONTRACTOR agrees to be bound to CONTRACTOR in the same manner and to the same extent as CONTRACTOR is bound to OWNER under the Contract Documents, to the extent of the work provided for in this Agreement, and that where, in the Contract Documents, reference is made to CONTRACTOR and the work or specification therein pertains to SUBCONTRACTOR'S trade, craft, or type of work, then such work or specification shall be interpreted to apply to SUBCONTRACTOR instead of CONTRACTOR. The phrase "Contract Documents" is defined to mean and include:

Drawing Nos. A-1 thru A-17, C-1 thru C-4, S-1 thru S-7, E-1 thru E-6 and M-1 thru M-7, all dated 10-4-2021, Specifications dated 10-2-2021 and Addendum No. 1 dated 10-18-2021.

SECTION 2 - SCOPE

SUBCONTRACTOR agrees to furnish all labor, services, materials, installation, cartage, hoisting, supplies, insurance, equipment, scaffolding, tools and other facilities of every kind and description required for the prompt and efficient execution of the work described herein and to perform the work necessary or incidental to complete thermal insulation for all plumbing, HHW, CHW and condensate piping systems, including pumps P-1, P-2 and P-3 for the project in strict accordance with the Contract Documents and as more particularly, though not exclusively, specified in:

Division 15, Section 400 entitled "Thermal Insulation systems" and General Conditions, as applicable.

SECTION 3 - CONTRACT PRICE

CONTRACTOR agrees to pay SUBCONTRACTOR for the strict performance of his work, the sum of:

Forty-two thousand, three hundred dollars ($42,300.00), subject to additions and deductions for changes in the work as may be agreed upon, and to make payment in accordance with the Payment Schedule, Section 4.

SECTION 4 - PAYMENT SCHEDULE

CONTRACTOR agrees to pay SUBCONTRACTOR in monthly payments of 90 % of labor and materials which have been placed in position and for which payment has been made by OWNER to CONTRACTOR. The remaining 10 % shall be retained by CONTRACTOR until he receives final payment from OWNER, but not less than thirty-five days after the entire work required by the prime contract has been fully completed in conformity with the Contract Documents and has been delivered and accepted by OWNER, ARCHITECT, and CONTRACTOR. Subject to the provisions of the next sentence, the retained percentage shall be paid SUBCONTRACTOR promptly after CONTRACTOR receives his final payment from OWNER. SUBCONTRACTOR agrees to furnish, if and when required by CONTRACTOR, payroll affidavits, receipts, vouchers, release of claims for labor, material and subcontractors performing work or furnishing materials under this Agreement, all in form satisfactory to CONTRACTOR, and it is agreed that no payment hereunder shall be made, except at CONTRACTOR'S option, until and unless such payroll affidavits, receipts, vouchers or release; or any or all of them, have been furnished. And payment made hereunder prior to completion and acceptance of the work, as referred to above, shall not be construed as evidence of acceptance of any part of SUBCONTRACTOR'S work.

SECTION 5 - GENERAL SUBCONTRACT PROVISIONS

General Subcontract Provisions on back of Pages 1 and 2 are an integral part of this Agreement.

SECTION 6 - GENERAL PROVISIONS

1. *SUBCONTRACTOR* agrees to begin work as soon as instructed by the CONTRACTOR, and shall carry on said work promptly, efficiently and at a speed that will not cause delay in the progress of the CONTRACTOR'S work or work of other subcontractors. If, in the opinion of the CONTRACTOR, the SUBCONTRACTOR falls behind in the progress of the work, the CONTRACTOR may direct the SUBCONTRACTOR to take such steps as the CONTRACTOR deems necessary to improve the rate of progress, including, without limitation, requiring the SUBCONTRACTOR to increase the number of shifts, personnel, overtime operations, days of work, equipment, amount of plant, or other remedies and to submit to CONTRACTOR for CONTRACTOR'S approval an outline schedule demonstrating the manner in which the required rate of progress will be regained, without additional cost to the CONTRACTOR. CONTRACTOR may require SUBCONTRACTOR to prosecute, in preference to other parts of the work, such part or parts of the work as CONTRACTOR may specify.

The SUBCONTRACTOR shall complete the work as required by the progress schedule prepared by the CONTRACTOR, which may be amended from time to time. The progress schedule may be reviewed in the office of the CONTRACTOR and sequence of construction will be as directed by the CONTRACTOR.

The SUBCONTRACTOR agrees to have an acceptable representative (an officer of SUBCONTRACTOR if requested by the CONTRACTOR) present at all job meetings and to submit weekly progress reports in writing if requested by the CONTRACTOR. Any job progress schedules are hereby made a part of and incorporated herein by reference.

2. **Reserved Gate Usage**

SUBCONTRACTOR shall notify in writing, and assign its employees, material men and suppliers, to such gates or entrances as may be established for their use by CONTRACTOR and in accordance with such conditions and at such times as may be imposed by CONTRACTOR. Strict compliance with CONTRACTOR'S gate usage procedures shall be required by the SUBCONTRACTOR who shall be responsible for such gate usage by its employees, material men, suppliers, subcontractors, and their material men and suppliers.

3. **Staggered Days and Hours of Work and for Deliveries**

SUBCONTRACTOR shall schedule the work and the presence of its employees at the jobsite and any deliveries of supplies or materials by its material men and suppliers to the jobsite on such days, and at such times and during such hours, as may be directed by CONTRACTOR. SUBCONTRACTOR shall assume responsibility for such schedule compliance not only for its employees but for all its material men, suppliers and subcontractors, and their material men and suppliers.

SECTION 7 - SPECIAL PROVISIONS

1. Pipe hanger inserts will be furnished and installed by Contractor.
2. All trash pickup by Subcontractor; haul-away will be by General Contractor.

Contractors are required by law to be licensed and regulated by the Contractors' State License Board. Any questions concerning a contractor may be referred to the registrar of the board whose address is:

Contractors' State License Board -- P. O. Box 26000, Sacramento, California 95826

IN WITNESS WHEREOF: The parties hereto have executed this Agreement for themselves, their heirs, executors, successors, administrators, and assignees on the day and year first above written.

SUBCONTRACTOR	CONTRACTOR
Quikrap Insulation Company	Acme Mechanical Contractors
By _____	By _____
Name Title	Name Title
☐ Corporation ☐ Partnership ☐ Proprietorship	
(Seal)	
Contractor's State License No. _____	Contractor's State License No. _____

Revised 5-09   Published by AGC San Diego

For contracts that comply with your state's regulations, please visit www.constructioncontractwriter.com

Subcontract Change Order

Number One

Date 6-11-2022

To Quickrap Insulation Co.

Subcontract Number 7777-1

Our Job Number

Our Proposal Number

Architect's C.O. Number

Effective Date of Charge 6-17-2022

Subject to all the provisions of this Change Order, you are hereby directed to make the following change(s)

Delete HHW and CHW piping insulation located above Room No. 322 per ABC General Contractor's Change Order No. 18, dated 6-5-22.

The following change(s) will alter the price provided in your subcontract by the __deduction__ of $ 465.00

*Surety Consent		
This Change is Approved Date	Adjusted Subcontract Price Through C.O. Number 0	$ 42,300.00
Name:	Amount this C.O. Number:	$ (465.00)
By:		
Title:	Current Adjusted Subcontract Price:	$ 41,835.00

When this Change Order is signed by both parties (and by Subcontractor's surety is subcontract is bonded), it constitutes their agreement:

(A) That the subcontract price is adjusted as shown above and that no further adjustment in that price by reason of the change(s) provided herein shall be made; and (B) That all the terms and conditions of the subcontract, except as modified by this and any previous changes, shall remain in full force and effect and apply to the work as so changed.

Accepted and Agreed:	Date:
By _____ Authorized Signature – Subcontractor	By _____ Authorized Signature – Contractor
Title	Title

* Subcontractor: Sign pink copy and return immediately. If subcontract is bonded, obtain consent of surety endorsed thereon. No payments on account of this Change Order will be made until you have complied with the foregoing.

Published by San Diego Chapter A.G.C.

Purchase Order

A purchase order is a legal document used to order equipment and materials from a vendor, with the implied promise to pay for these goods within a specified time after receipt of the items. Purchase orders are sometimes used to write subcontracts whose value does not exceed one or two thousand dollars. Refer to the sample purchase order on page 452.

While most of the purchase order is self-explanatory, certain items need to be emphasized:

"Description" should include all pertinent data such as model numbers, arrangements, performance requirements, applicable standards, electrical characteristics and so on.

"Tag" is used to identify a particular piece of equipment to facilitate unloading it at the proper location at the jobsite. Tag numbers for all major equipment can usually be found on the contract drawings.

"48 hour delivery notice": This can be very important if the material or equipment being delivered is too heavy to be unloaded by hand. Time will be required to arrange for a crane or forklift to unload the truck.

"Sales tax" is the current state sales tax which must be included unless the goods are to be delivered and installed at a project located out-of-state.

"Terms": The usual payment terms are "Net 30" which means the entire amount of the purchase order must be paid within thirty days after receipt of all goods, undamaged. If items are received damaged, payment should be withheld until satisfactory repair or replacement is made.

ACME
Mechanical Contractors

7600 Oak Avenue - Smallville, U.S.A. 12345-6789
Voice 1-234-5678 / FAX 1-234-5680

Purchase Order

P.O. Number: 7777-36 **Date:** 1-6-2022
Job: CBC Bank Building
Job No: 7777

Seller:	Smallville Industrial Sales Co. 3657 Third Avenue Northwest Smallville, CA 63876	Ship to:	Acme Mechanical Contractors c/o CBC Bank Building 465 Commercial Street Smallville, CA 63876

Attn: Mr. L.H. Seller   **By:** RJS

Delivery date: 2-28-22   ☒ **24 Hr. Delivery Notice Required**   **Payment terms:** Net 30

Quantity	Description	Unit	Total
Two	Big-Blo Model No. 1JN 86-13 centrifugal	$443.00	$886.00
	fans with 480/3/60 one-half H.P., O.D.P.		
	motor. 3250 CFM @ 0.50 T.S.P.		
	Tag one EF-13 and one EF-23		

Number of copies of operation and maintenance manuals required   2.

All shipments are to be F.O.B. jobsite or shop unless otherwise noted.

Total before tax	$886.00
Sales tax	$62.02
Freight	Included
Total Order	$948.02

Buyer: RJS

Acceptance by seller _____   Date _____

Construction Schedules

The general contractor's construction schedule may range from a simple bar chart to a complicated computer-generated CPM (Critical Path Method) diagram. More complex jobs require more complete and detailed schedules. On larger jobs, lower-tier subcontractors may be required to prepare a bar chart showing their construction schedule.

If you have to prepare a construction schedule, make conservative estimates of the time required to complete each part of the job. A slower schedule makes it possible to use smaller crews which are usually more efficient.

You can't schedule your own work until the general contractor has supplied a schedule for the balance of the project. When you have that schedule, prepare a list of tasks your crews will perform. Show this list to the general contractor to be certain there are no conflicts with the master schedule.

When the list of tasks to be performed has been approved, begin recording the manhours required for each task. Take these manhours from your estimate. When the duration for each task has been determined, figure the crew sizes required.

For example, suppose the estimate shows that it will take 1,040 manhours to install the underground plumbing for a project. The work must be completed in two months. Assuming one worker averages 173 hours of work in one month, how many workers will be required to meet the schedule? Here's the solution:

$$\frac{1{,}040 \text{ manhours}}{173 \text{ MH} / \text{Mo.} \times 2 \text{ months}} = 3 \text{ workers}$$

The first of the two bar charts on the following pages shows the schedule a general contractor might expect to receive from a plumbing and HVAC subcontractor. The chart on page 454 is intended for use by the subcontractor. It includes information a mechanical subcontractor needs to determine proper crew sizes for each task. The *Cumulative Manhours* line shows the budgeted manhours for each month of the project. If labor costs are to be kept within budget, actual hours should not exceed estimated hours. Monitor labor costs each month by comparing the line *Actual MH Used* (actual manhours used) with the line *Cumulative MH* (estimated cumulative manhours).

Construction Schedule

Company: Acme Mechanical Contractors **Project:** Mile-Hi Office Bldg **Job Number:** 7777 **Date:** 6/8/2022

Activity	2022						2023					Crew Sizes
	Jul	Aug	Sep	Oct	Nov	Dec	Jan	Feb	Mar	Apr	May	
Submittals	▓											
Purchasing		▓	▓									
U/G Plumbing				▓								
A/G Plumbing						▓	▓					
Fin. Plumbing								▓				
HVAC Piping						▓	▓	▓				
HVAC Ducting						▓	▓	▓	▓			
Equipment									▓			
Insulation									▓			
Temp. Controls										▓		
Start-up												

Construction Schedule

Company: Acme Mechanical Contractors **Project:** Mile-Hi Office Bldg **Job Number:** 7777 **Date:** 6/8/2022

Activity	2022						2023				Crew Sizes
	Jul	Aug	Sep	Oct	Nov	Dec	Jan	Feb	Mar	Apr	May	
Submittals												N/A
Purchasing												N/A
U/G Plumbing												1040 MH ÷ (173 x 2) = 3
A/G Plumbing												2075 MH ÷ (173 x 4) = 3
Fin. Plumbing												130 MH ÷ (173 x .5) = 1.5
HVAC Piping												2760 MH ÷ (173 x 4) = 4
HVAC Ducting												2790 MH ÷ (173 x 4) = 4
Equipment												690 MH ÷ (173 x 2) = 2
Insulation												Subcontract
Temp. Controls												Subcontract
Start-up												85 MH ÷ (173 x .5) = 1
Crew size/month			3	6	11	11	11	11.5	2	1		
Cumulative MH*			520	1159	3466	5373	7280	9143	9448	9573		MH Estimated = 9573
Actual MH used												
*Per estimate												
*Rounded												

Letter of Intent

On most large projects the contractor is required to submit drawings and technical data to the owner's representative for approval before work actually begins. Most of these submittals will be prepared by your suppliers and third-tier subs. Vendors and subs may be reluctant to prepare these documents without some assurance that they've been selected to do the work. That's the purpose of a letter of intent. It notifies a proposed vendor or subcontractor that you plan to contract with them when the owner approves the information submitted.

A letter of intent isn't a contract. It's an expressed intention to make a contract, which isn't a contract at all. But a vendor who gets your letter of intent and supplies all the information requested should get the contract. If you place the order with a different vendor, obviously, your letter of intent was worthless. That first vendor may be very reluctant to cooperate in the future.

A sample letter of intent follows this page.

ACME
Mechanical Contractors

7600 Oak Avenue - Smallville, U.S.A. 12345-6789
Voice 1-234-5678 / FAX 1-234-5680

December 20, 2021

Quikrap Insulation Company
320 Maple Boulevard
Smallville, U.S.A. 12345-6789

Attn: Mr. R. S. Quikrap

Dear Mr. Quikrap:

Acme Mechanical Contractors has been issued a subcontract to furnish and install the mechanical and plumbing systems for the proposed Mile-Hi Office Building to be located at 2600 Second Street in Smallville.

It is my intention to award your firm a subcontract to provide the thermal insulation for these systems in accordance with your proposal dated 12-13-2021 in the amount of $42,300.00.

Please prepare ten copies of your submittal data and forward them to me no later than January 15, 2022. Upon approval of your submittal, I will mail you the subcontract for your review and signature.

Thank you,

Bobby Thompson
Project Manager

CC: Project file

Submittal Data

The construction specifications may require that you submit manufacturer's technical data on certain equipment you plan to install. You may have to submit six to ten copies of this technical data for approval before buying the equipment or contracting with third-tier subcontractors.

First, find out how many submittal copies will be required. Request that number of copies plus at least two additional sets for your use. A subcontractor who provides only services (no materials) may have to submit a detailed written explanation of the work to be done.

The second step is to label each submittal with the paragraph number of the specification where that equipment, material or service is described. Type or write this number in the upper right corner of each submittal sheet. From this set of numbered submittals, prepare a submittal index, a list of submittals identified by title and paragraph number. This index should be arranged in specification paragraph order. A blank submittal index form follows this section.

I prefer to use a three-ring binder to hold submittals. That makes it easier to add and delete sections as needed. Be sure to include a cover sheet, usually on company letterhead, which identifies the project, job number and scope of work being submitted. A sample submittal cover sheet follows this section.

Purchase orders and subcontracts should not be written until an approved copy of the submittal brochure has been received.

ACME
Mechanical Contractors

7600 Oak Avenue - Smallville, U.S.A. 12345-6789
Voice 1-234-5678 / FAX 1-234-5680

Division 15

Heating, Ventilating, Air Conditioning and Plumbing Systems

Submittal Data for

Mile-Hi Office Building
2600 Second Street
Smallville, U.S.A.

Submittal Index

Specification Section	Specification Paragraph	Item Description	Proposed Vendor or Sub	As Specified	See Submittal Data Page

Billing Breakdown Worksheet

The format to be used for monthly progress billings is seldom specified in the contract documents. In any case, your bill has to include enough detail to satisfy the owner and the lender.

Before preparing your bill, ask the general contractor how much detail is required. Some general contractors will want detailed cost breakdowns showing costs for each system on each building floor. Others will require much less detail.

Find out if billings can include the cost of equipment and materials delivered and stored either on site or off site but not yet installed. If billings can include materials stored off site, do these materials have to be stored in a bonded warehouse? Do vendor's invoices have to accompany progress billings?

When you understand what's required for monthly progress billings, prepare a billing breakdown worksheet. A sample worksheet follows this section.

Column one, Activity Lists each work item in the order work will be performed. The first item listed will usually be mobilization. This includes the cost of the job site trailer, electrical and telephone hookups, office furniture and supplies and initial labor costs. Billings for mobilization costs are sometimes denied by the lender. Many contractors minimize these costs to increase the chance of approval. It's better to receive partial payment for these costs than none at all.

Column two, Cost ($) Lists the actual contractor's costs for each activity, including labor, material, equipment and sales taxes. Markup for overhead and profit are not included.

Column three, Factor This column shows the markup assigned to each activity. Activities scheduled to be completed first are usually assigned the highest markups. This is called front loading and is common in the construction industry. Most bills won't be paid for 30 to 60 days. When the bill is paid, the amount received probably won't include the percentage allowed for retention. Generally, retention is 10 percent and isn't released until the project is complete and accepted by the owner.

Front loading helps contractors carry the financial burden of work in progress. Few subcontractors have enough cash to pay all their bills when due and still wait months to collect for work that's been completed. Employees have to be paid in full and on time. Assigning a higher markup to work completed first accelerates the payment schedule and helps spread payments more evenly over the entire project. Front loading is routine with subcontractors and unpopular with owners and lenders. In practice, subcontractors have no choice but to place a higher price or higher markup on work completed first.

Note that figures in the Factor column are less than 1.0 for work done near the end of the project. That's because front loading doesn't change the contract price. It changes only when that money is received.

Column four, Sell ($) Values for this column are the product of the cost and factor columns. Use the figures in this column when submitting monthly progress invoices.

ACME
Mechanical Contractors

7600 Oak Avenue - Smallville, U.S.A. 12345-6789
Voice 1-234-5678 / FAX 1-234-5680

Billing Breakdown Worksheet
(CBC Bank Building - Job No. 7777)

Activity	Cost ($)	Factor	Sell ($)
Mobilization	2,000	1.50	3,000
Underground chilled and heated water piping	120,000	1.35	162,000
Underground plumbing & piping	87,000	1.35	117,450
Above ground chilled and heating water piping	115,000	1.20	138,000
Above ground plumbing & piping	87,000	1.20	104,400
HVAC ducting	96,000	1.20	115,200
Equipment	312,000	1.10	343,200
Duct and pipe insulation	119,000	1.00	119,000
Temperature controls	53,000	.90	47,700
Plumbing fixtures	43,000	.65	27,950
Water treatment	1,800	.55	990
Test and balance	13,000	.55	7,150
Validation	500	.50	250
	1,049,300		1,186,290

Estimated gross profit = $\dfrac{\$1{,}186{,}290 - \$1{,}049{,}300}{\$1{,}049{,}300} = 13\%$

Index

A

ABS, DWV pipe 150
   1/8 bend 150
   1/4 bend 150
   adapters 152
   bushings 152
   cleanouts 152
   closet bend 150
   closet flanges 150
   combinations 151
   couplings 153
   hanger assemblies 153
   P-traps 151
   reducers 153
   riser clamps 153
   solvent-weld joints 152-153
   tees 151
   wyes 151-152
ABS, DWV, test cap 152
A/C systems 286
Access doors, steel 121
Accessories
   iron filter 23
   water softener 23
Acid DWV systems 164
Actuator, damper 404
Adapters
   copper, DWV, soldered 148
   copper, press fit 88
   CPVC sprinkler pipe 184
   F.I.P., ABS 152
   PE-AL pipe 113
   PEX-AL pipe 113
   polypropylene pipe 166
   PVC, DWV 156
   PVC, Schedule 40 94
   PVC, Schedule 80 104
   PVC sewer, bell & spigot 163
   Schedule 5 steel, pressfit 236
   Schedule 10 steel, roll-grooved .. 275
   Schedule 40 steel, cut-grooved .. 281
   Schedule 40 steel, roll-grooved .. 268
   Type K copper, brazed 37
   Type K copper, soldered 47-48
   Type L copper, brazed 56
   Type L copper, soldered 64
   Type M copper, brazed 73
   Type M copper, soldered 81
Additional costs 8
Adjusting costs 7
Air admittance valve 328
Air balance software 338
Air balancing
   air handling units 402

centrifugal fans 403
diffusers 402
fan coil units 403
fume hoods 403
grilles 403
terminal boxes 402
Air compressor 406
   rental 420
Air conditioning
   budget estimates 435
   residential 286
Air conditioning condensate systems
   PVC, Schedule 40 91
   PVC, Schedule 80 101
   Type K copper, brazed 33
   Type K copper, soldered 43
   Type L copper, brazed 53
   Type L copper, soldered 61
   Type M copper, brazed 70
   Type M copper, soldered 78
Air conditioning units 288
Air cooled condenser 426
   condensing unit 212
Air grilles, return 333
Air handling equipment
   air conditioner 288
   exhaust fans 326
   housings 331
   ventilators 326
Air handling units 289
   accessories 290
   air balancing 402
   coil connection 297-298
   removal 424
Air mixing box, removal 423
Air separators, Rolaitrol type 201
Air vents 121
Alarm valves 171
Apartments, HVAC estimates 437
Area drains 168
   installation costs 435
Arresters, water hammer 133
As-built drawings 421
Assemblies
   air conditioning 286
   forced air heating 287
Auditoriums, HVAC estimates 437

B

Backfill costs, trenching 418
Backflow preventers
   double check 120
   reduced pressure 119

Backhoes, rental 420
Balancing valves PEX-AL pipe ... 114, 118
Ball valves
   copper, press fit 88
   PEX-AL pipe 114, 118
   pipe and plumbing specialty ... 126-127
   PVC, Schedule 40 97
   PVC, Schedule 80 107
   PVC, Solid body, EDPM 97, 107
   PVC, Solid body, threaded ... 97, 107
   PVC, Tru-union, threaded 97, 107
   PVC, Union type, Solvent
     weld 98, 107
   Schedule 40 steel, threaded 230
   Schedule 80 steel, threaded 252
   Type K copper, brazed 40
   Type K copper, soldered 50
   Type L copper, brazed 59
   Type L copper, soldered 67
   Type M copper, brazed 76
   Type M copper, soldered 84
Banks, HVAC estimates 437
Bar sinks 29
   estimating 435
Barber shops, HVAC estimates .. 437
Base wage 6
Baseboard fins 204
Bathroom
   fans 328-329
   fixtures 28
   heaters 328
Bathroom sink, disconnect 433
Bathtubs 28
   disconnect 433
   estimating 435
Beauty shops, HVAC estimates .. 437
Bell & spigot pipe, PVC 159
Benders, hydraulic, rental 420
Bends
   ABS, DWV pipe 150
   cast iron, no-hub 136-138
   class 110 DI, cement lined 411
   class 150 cast iron 412
   class 153 DI, cement lined 407
   class 153 DI, double cement
     lined 409
   class 2400 or 3000 asbestos
     cement 413
   copper, DWV, soldered 147
   polypropylene pipe 164
   PVC, DWV 159-160
   PVC sewer, bell & spigot 160
Bevel machines, rental 420
Billing breakdown worksheet ... 461-462

Biomass fired
  boilers 319-320
  central airspace heater 321
Black steel pipe 177
  assemblies 176
Blowers, centrifugal 324
Boiler
  blowdown 203
  burners 195
  stack 203
  trim 204
Boiler connections 190
  boiler pumps 205
Boilers
  biomass fired 319-320
  pulse type 322
  removal 425
  steam heating 189
Boilers, commercial
  accessories 204
  adjusting 198
  cast iron 185
  chemical feed pump 202
  combustion controls 197-198
  combustion train 192
  components 204
  deaerator/condenser 201
  electrical service 196
  feedwater pumps 197
  firebox 187
  firetube 186-187
  fuel train piping 193-194
  packaged, feedwater systems 202
  pumping unit 201
  refractory 197
  stacks 203
  water softening systems 198
  watertube 188
Boilers, gas fired
  cast iron 189
  steel 188
Bolt and gasket sets
  pipe and plumbing specialty 122
  polypropylene pipe 167
  PVC, Schedule 40 99
  PVC, Schedule 80 109
  Schedule 10 steel, roll-grooved 279
  Schedule 40 steel, cut-grooved 284
  Schedule 40 steel, roll-grooved 271
  Schedule 40 steel, threaded 232
  Schedule 40 steel, welded 222
  Schedule 80 steel, threaded 253
  Schedule 80 steel, welded 245
  Schedule 160 steel, full face 264
  Schedule 160 steel, ring face 264
  Type K copper, brazed 41
  Type K copper, soldered 52
  Type L copper, brazed 60
  Type L copper, soldered 68

Type M copper, brazed 77
Type M copper, soldered 85
Boom lifts, rental 420
Booster heaters 25
Bore holes, geothermal 317
Bowling alleys, HVAC estimates 437
Branch pipe and fittings, sprinkler 182
Brazed joint pipe
  Type K copper 33
  Type L copper 53
  Type M copper 70
Bucket steam trap 126
Budget estimating 435
Buildings, HVAC estimates 437
Burners, dual fuel 195
Bushings
  ABS 152
  PVC, DWV 156
  PVC, Schedule 40 94
  PVC, Schedule 80 104
  Type K copper, brazed 38
  Type K copper, soldered 48
  Type L copper, brazed 57
  Type L copper, soldered 65
  Type M copper, brazed 74
  Type M copper, soldered 82
Butterfly valves
  pipe and plumbing specialty 127
  PVC, Schedule 40 96
  PVC, Schedule 80 106
  Schedule 10 steel, roll-grooved 277
  Schedule 40 steel, cut-grooved 283
  Schedule 40 steel, roll-grooved 270
  Schedule 40 steel, threaded 230
  Schedule 40 steel, welded 219-220
  Schedule 80 steel, threaded 251-252
  Schedule 80 steel, welded 243
  Type K copper, brazed 39
  Type K copper, soldered 50
  Type K & L copper, roll grooved 90
  Type L copper, brazed 58
  Type L copper, soldered 66
  Type M copper, brazed 75
  Type M copper, soldered 83

C

Calcium silicate pipe insulation 399
California Code, 2010 199
Can washers, installation costs 435
Caps
  cast iron, no-hub 139
  cast iron, threaded 181
  CPVC sprinkler pipe 184
  PE-AL pipe 114, 117
  PEX-AL pipe 114, 117
  PVC sewer, bell & spigot 163
  PVC, Schedule 40 95
  PVC, Schedule 80 105

roll grooved, Victaulic 178
Schedule 10 steel, roll-grooved 276
Schedule 40 steel, cut-grooved 282
Schedule 40 steel, roll-grooved 269
Schedule 40 steel, threaded 226
Schedule 40 steel, welded 217
Schedule 80 steel, threaded 250
Schedule 80 steel, welded 241
Schedule 160 steel, welded 261
Type K copper, brazed 38
Type K copper, soldered 48
Type L copper, brazed 57
Type L copper, soldered 65
Type M copper, brazed 74
Type M copper, soldered 82
Carbon steel fittings 214-265
Carbon steel pipe
  Schedule 40 180
  Schedule 80 238
Cast iron class 150 412
Cast iron DWV pipe, hub & spigot 141
  1/16 bend 142
  1/8 bend 141
  1/4 bend 141
  bends 142
  closet flanges 142
  combinations 143
  gaskets 145
  hanger assemblies 145
  P-traps 142
  reducers 144
  riser clamps 145
  sanitary tees 142
  tees 142
  wyes 143-144
Cast iron DWV pipe, mechanical
  joint 135
Cast iron DWV pipe, no-hub 135
  1/8 bend 135-136
  1/4 bend 136
  caps 139
  closet bends 136
  closet flanges 136
  combinations 138
  couplings 139
  crosses 139
  hanger assemblies 140
  horizontal assembly 135
  P-traps 136
  reducers 139
  riser clamps 140
  tees 138-139
  wyes 136-138
Cast iron sprinkler pipe fittings
  cap 181
  couplings 181
  cross 181
  ells 180
  plugs 181

reducers 180
    reducing tee 180
Ceiling diffusers 334-335
Central air space heater,
    biomass fired 321
Central dehumidification 300
Centrifugal blowers 324
Centrifugal fans, air balancing 403
Centrifugal pumps, HVAC 206
Centrifugal water-cooled chiller 211
Ceramic heater 304
Chain hoists, rental 420
Change estimates 438-448
    change order log 443
    example 439
    summary 442, 446
    take-off 440, 444
    worksheet 441, 445
Check valves 171
Chemical feed pump, boiler 202
Chemical feed system 202
Chemical systems
    polypropylene, DWV 164
    PVC, Schedule 40 91
    PVC, Schedule 80 101
Chilled water systems
    Schedule 10 steel, roll-grooved 273
    Schedule 40 steel, cut-grooved 280
    Schedule 40 steel, roll-grooved 266
    Schedule 40 steel, threaded 223
    Schedule 80 steel, threaded 247
    Type K copper, brazed 33
    Type K copper, soldered 43
    Type L copper, brazed 53
    Type L copper, soldered 61
    Type M copper, brazed 70
    Type M copper, soldered 80
Chillers
    centrifugal 211
    drinking fountain 30
    reciprocating 211
    removal 426
    water cooled, connection 211
Chlorinated polyvinyl chloride pipe ... 183
Churches, HVAC estimates 437
Circuit balance valves 127
Circulating pumps
    all bronze 120
    iron body 121
Clarifications 8
Classrooms, HVAC estimates 437
Cleanouts
    ABS 152
    ABS/PVC 169
    copper, DWV, soldered 147
    end-of-line 169
    floor 169
    installation costs 435
    PVC, DWV 157
    wall 169

Closed loop heat pump 306
Close-out items 421
Closet bends, ABS, DWV pipe 150
Closet flanges
    ABS, DWV pipe 150
    cast iron, hub & spigot 142
    cast iron, no-hub 136
    copper, DWV, soldered 147
    PVC, DWV 155-156
Clothes dryers, exhaust 327
Cocktail lounges, HVAC estimates ... 437
Coil connection, air handling
    unit 297-298
Coil, duct mounted, removal 429
Coils, reheat, HVAC 209
Cold water connections 26
Collars, galvanized steel 343
Combinations
    ABS 151
    cast iron, hub & spigot 143
    cast iron, no-hub 138
    copper, DWV, soldered 147
    polypropylene pipe 165
Combustion controls, boiler 197-198
Combustion monitoring 205
Combustion train, boiler 192
Come-alongs, rental 420
Commercial boilers 186-189
    combustion trains 192
    components and accessories 204
    connections 190
Commercial fans and blowers 324
Commercial fixture rough-ins 31
    group 32
Commercial water heaters 19, 20
Compaction, trenching 418
Compactors, rental 420
Companion flanges
    150 pound, threaded 122
    300 pound, threaded 122
    PVC 122
    PVC, Schedule 40 99
    PVC, Schedule 80 109
    Schedule 40 steel, threaded 232
    Type K copper, brazed 41
    Type K copper, soldered 51
    Type L copper, brazed 60
    Type L copper, soldered 68
    Type M copper, brazed 77
    Type M copper, soldered 85
    welding type 121
Composite pipe 111
    compression fittings 116
    compression joint fittings 117
    compression joints 116
    crimped joint 111
    crimped joint fittings 111-112
Compressed air systems
    Type K copper, brazed 33
    Type L copper, brazed 53

Computer rooms, HVAC estimates ... 437
Condenser units 205
Condenser water systems
    Schedule 40 steel, cut-grooved 280
    Schedule 40 steel, roll-grooved 266
Condensing units, air cooled 212
Condominiums, HVAC estimates 437
Configurable controller 338
Connections
    air handling unit, HVAC 297-298
    continuous waste 26
    fire department 174
    flexible duct 343
    hot and cold water supply 26
    indirect waste 26
    kitchen equipment 26
    Siamese 174
    water heaters 21
Connectors, pipe 123
    flexible 124
Construction schedule 453-455
Contents 3
Continuous waste connections 26
Control modules, pollution 199
Control valves
    2 way 405-406
    3 way 405-406
    electric 405
    pipe and plumbing specialty ... 132-133
    pneumatic 406
    PVC, Schedule 40 99
    PVC, Schedule 80 109
    Schedule 10 steel, roll-grooved 278
    Schedule 40 steel, cut-grooved 284
    Schedule 40 steel, roll-grooved 271
    Schedule 40 steel, threaded 231
    Schedule 40 steel, welded 221
    Schedule 80 steel, threaded 253
    Schedule 80 steel, welded ... 244-245
    Schedule 160 steel, flanged ... 263-264
    Schedule 160 steel, threaded 263
    Type K copper, brazed 41
    Type K copper, soldered 51
    Type L copper, brazed 60
    Type L copper, soldered 68
    Type M copper, brazed 77
    Type M copper, soldered 85
Controllers 338
Controls
    boiler 203
    HVAC 404
Cooling systems, residential 286
Cooling towers
    connection assembly 213
    demolition 427
    forced draft 213
    galvanized steel 212
    induced draft 212
    removal 427
Cooling units, variable volume 337

Copper fittings, roll grooved 89
Copper pipe 89
    ball valve, press fit 88
    coupling, press fit 86
    ells, press fit 86
    female adapter, press fit 88
    male adapter, press fit 88
    tee, press fit 87
    tee, reducing, press fit 87
    union, press fit 87
Copper pipe, DWV, soldered 146
    1/8 bend 146
    1/4 bend 147
    adapters 148
    assembly with riser 146
    cleanouts 147
    closet flanges 147
    combinations 147
    couplings 148
    crosses 147
    hanger assemblies 149
    horizontal assemblies 146
    P-traps 147
    reducers 148
    riser clamps 149
    tees 147
    test caps 148
    test tees 148
    wyes 147
Copper pipe, Type K brazed 33-34
    adapters 37
    bolt and gasket sets 41
    bushings 38
    caps 38
    companion flanges 41
    couplings 38
    ells 34-35
    hanger assemblies 42
    pressure gauges 41
    reducers 37
    riser clamp 42
    strainers 40
    tees 35-36
    thermometers with wells 41
    unions 38
    valves 39-41
Copper pipe, Type K soldered 43-44
    adapters 47-48
    bolt and gasket sets 52
    bushings 48
    caps 48
    companion flanges 51
    couplings 49
    ells 44-45
    hanger assemblies 52
    pressure gauges 52
    reducers 47
    riser clamps 52
    strainers 51
    tees 45-46
    thermometers with wells 52
    unions 48
    valves 49-51
Copper pipe, Type K & L 89-90
    coupling 90
    ells 89
    flange adapter 90
    reducers 90
    roll grooved 89
    tees 89
    valves 90
Copper pipe, Type L brazed 53-54
    adapters 56
    bolt and gasket sets 60
    bushings 57
    caps 57
    companion flanges 60
    couplings 57
    ells 54-55
    pressure gauges 60
    reducers 56
    strainers 59
    tees 55
    thermometers with wells 60
    unions 57
    valves 58-60
Copper pipe, Type L soldered 61-62
    adapters 64
    bolt and gasket sets 68
    bushings 65
    caps 65
    companion flanges 68
    couplings 65
    ells 62-63
    hanger assemblies 69
    maximum working pressure 61
    pressure gauges 68
    reducers 64
    riser clamps 69
    strainers 67
    tees 63
    thermometers with wells 68
    unions 65
    valves 66-68
Copper pipe, Type M brazed 70-71
    adapters 73
    bolt and gasket sets 77
    bushings 74
    caps 74
    companion flanges 77
    couplings 74
    ells 71
    maximum working pressure 70
    pressure gauges 77
    reducers 73
    strainers 76
    tees 72
    thermometers with wells 77
    unions 74
    valves 75-77
Copper pipe, Type M soldered 78-79
    adapters 81
    bolt and gasket sets 85
    bushings 82
    caps 82
    companion flanges 85
    couplings 82
    ells 79-80
    maximum working pressure 78
    pressure gauges 85
    reducers 81
    strainers 84
    tees 80
    thermometers with wells 85
    unions 82
    valves 83-85
Copper piping, removal 430
Correction factors 6
Countertop sinks/lavatories 28
Couplings
    ABS 153
    cast iron, no-hub 139
    cast iron, threaded 181
    copper, DWV, soldered 148
    copper, press fit 86
    CPVC sprinkler pipe 183
    galvanized steel spiral duct 359
    PE-AL pipe 112-113, 117
    PEX-AL pipe 113, 117
    polypropylene pipe 166
    PVC, DWV 156
    PVC, Schedule 40 95
    PVC, Schedule 80 105
    PVC sewer, bell & spigot 160
    roll grooved, Victaulic 179
    Schedule 5 steel, pressfit 235-236
    Schedule 10 steel, roll-grooved 276
    Schedule 40 steel, cut-grooved 282
    Schedule 40 steel, roll-grooved 269
    Schedule 40 steel, threaded 226
    Schedule 80 steel, threaded 251
    Type K copper, brazed 38
    Type K copper, soldered 49
    Type K & L copper, roll grooved 90
    Type L copper, brazed 57
    Type L copper, soldered 65
    Type M copper, brazed 74
    Type M copper, soldered 82
CPVC sprinkler pipe 183
    adapters 184
    cap 184
    coupling 183
    elbows 183
    fittings 184
    flange 184
    head adapter 184
    reducing tees 183
    tees 183
Craft codes 7
Craft@hrs 5

Cranes, rental 420
Crew composition 7
Crimp rings
   PE-AL pipe 115, 118
   PEX-AL pipe 115, 118
Croll-Reynolds 199
Cross linked PEX-AL 111
Cross linked Polyethylene-
   Aluminum pipe 116
Crosses
   cast iron, no-hub 139
   cast iron, threaded 181
   copper, DWV, soldered 147
   Schedule 40 steel, threaded 225
   Schedule 80 steel, threaded 250

D

Daily rental, equipment 420
Dampers
   actuator 404
   correction factors 340
   fire 340
   fusible plug 341
   rectangular 339
   round 339
Deaerator/condenser, boiler 201
Deck drains, installation costs 435
Deep well jet pump 313
   dehumidification 300
Dehumidifiers 291-292
Demolition 422
Department stores, HVAC
   estimates 437
Dielectric unions 38, 122
Diffusers
   air balancing 402
   ceiling 334-335
   removal 423
Dishwashers 25
Disinfection unit 25
Disposals, garbage 25
Domestic water iron filter 23
Domestic water softeners 22-24
Doors 121
Double check detector valves 171
Downblast ventilation 329
Drain, waste, vent pipe
   cast iron, hub & spigot 141
   cast iron, no-hub 135
   copper 146
   polypropylene 164
   PVC 154
Drains 168
Drawings, as-built 421
Drilling wells 316
Drinking fountains 30
   disconnect 433
   refrigerated 30

   removal 432
Drinking water tank 415-416
Drops and tees, ductwork 391-393
Dry valves 171
Dryers, exhaust 327
Dual-fuel burners 195
Duct insulation
   calcium silicate 399
   fiberglass 401
   removal 434
Duct lining
   calcium silicate 399
   fiberglass 401
Duct markers 421
Ductwork
   correction factors 340
   demolition 422
   removal 422
Ductwork specialties
   collars 343
   connections 343
   dampers 339-340
   flexible connections 343
   turning vanes 342
Ductwork, fiberglass
   fabrication labor 395
   installation costs 396
   vinyl cover 396
Ductwork, galvanized steel
   per pound installed 346-347
   rectangular 380
   rectangular 20 gauge 378-380
   rectangular 22 gauge 373-377
   rectangular 24 gauge 371-373
   rectangular 26 gauge 369-370
   rectangular fittings 381-393
   round fittings 394
   spiral 356-357
   spiral fittings 358-368
DWV pipe
   ABS 150
   cast iron 135
   cast iron, hub & spigot 141
   copper 146
   polypropylene 164
   polypropylene heat-fused 164
   PVC 154

E

EDPM valves 97, 107
Elastomeric gaskets 159
Elastomeric pipe insulation 400
Elbows, ductwork
   galvanized steel spiral duct 358
   rectangular, galvanized steel 381-390
   round, galvanized steel 394
Elbows, pipe
   black steel 177-178

   CPVC sprinkler 183
Electric water heaters 19
Electrical service for boilers 196
Ells
   cast iron, threaded 180
   copper, press fit 86
   PE-AL pipe 111-112, 116-117
   PEX-AL pipe 111-112, 116-117
   PVC, DWV 154
   PVC, Schedule 40 92
   PVC, Schedule 80 102-103
   Schedule 5 steel, pressfit 234-235
   Schedule 10 steel, roll-grooved 274
   Schedule 40 steel, cut-grooved 280
   Schedule 40 steel, roll-grooved 267
   Schedule 40 steel, threaded 224
   Schedule 40 steel, welded 215
   Schedule 80 steel, threaded 248
   Schedule 80 steel, welded 238
   Schedule 160 steel,
      threaded 257-258
   Schedule 160 steel, welded ... 257-258
   Type K copper, brazed 34-35
   Type K copper, soldered 44-45
   Type K & L copper, roll grooved ... 89
   Type L copper, brazed 54-55
   Type L copper, soldered 62-63
   Type M copper, brazed 71
   Type M copper, soldered 79-80
Emissions reduction module 199
Emissions sensing 205
Energy recovery
   ventilators 291-292
   wheel 301
Engraved nameplates 421
Enthalpy energy and heat recovery ... 301
Equipment
   nameplates 421
   plumbing 19
   rental costs 420
Equipment costs 7
Estimate detail sheet 16
Estimates, budget 435
Estimating
   accuracy 8
   guidelines 13
Exclusions 8
Exhaust 327
   clothes dryer 327
   fans 325-329
   wall hood 327
Exhauster arrays 330
Exhausters 329
Expansion tanks 123

F

F.O.B. 9
Fabrication, fiberglass ductwork 395

Fan coil units
   air balancing 403
   HVAC equipment 208
Fan coils 427
Fan controls 327
Fans
   attic 326
   bathroom 326-328
   ceiling exhaust 326-327
   ceiling mounted 326
   centrifugal air foil 324
   centrifugal utility 324
   commercial 326
   exhaust 326, 328-329
   exhaust, roof 325
   humidistat 327
   kitchen 327
   roof 326
   room ventilation 326
   speed controller 327
   thru-wall 326
   timer 327
   tube-axial 325
   vane-axial 324
   ventilation 326-327
   wall exhaust 326
   wall mounted 326
   washroom 326-327
Feedwater pumps, boiler 197
   feedwater systems 205
Fiberglass
   blanket 401
   ductwork 396
   pipe insulation 397-398
      pressure tank 23
      rigid board 401
      tank 24, 415
Filter, iron 24
Fire dampers 340
Fire department connection 174
Fire extinguisher 175
Fire extinguishing systems 322
Fire hydrant 175
Fire protection
   CPVC sprinkler pipe 183
   fire hose cabinets 173
   plastic sprinkler pipe 183
   pumps 174
   Siamese connections 174
   sprinkler fittings 177-181
   sprinkler heads 172
   sprinkler pipe 176-177
   steel pipe nipples 182
   switches 172
   valves 171
Fire pumps 174
Firebox boilers 187
Fire-rated doors 121
Firetube boilers 186-187

Fittings
   ductwork 359
   roll grooved 89
Fittings, pipe
   copper, DWV, soldered 146
   expansion tank 123
   M.I., 150 pound 224
   malleable iron, Schedule 40 266
   polypropylene 164
   PVC, DWV 154
   PVC sewer, bell & spigot 159
   PVC, Schedule 40 92
   PVC, Schedule 80 101-102
   Schedule 10 steel, roll-grooved ... 274
   Schedule 40 steel, cut-grooved ... 280
   Schedule 40 steel, roll-grooved ... 267
   Schedule 40 steel, threaded 224
   Schedule 40 steel, welded 215
   Schedule 80 steel, threaded 248
   Type K copper, brazed 42
   Type L copper, brazed 54
   Type L copper, soldered 62
   Type M copper, brazed 71
   Type M copper, soldered 79
Fixtures
   bathroom 28
   disconnect 433
   estimating costs 435
   removal 432
Flange adapter, Type K & L
   copper, roll grooved 90
Flanges
   CPVC sprinkler pipe 184
   polypropylene pipe 167
   roll grooved, Victaulic 179
   Schedule 10 steel, roll-grooved ... 276
   Schedule 40 steel, cut-grooved ... 282
   Schedule 40 steel, roll-grooved ... 269
   Schedule 40 steel, threaded 232
   Schedule 40 steel, welded 218
   Schedule 80 steel, threaded 253
   Schedule 80 steel, welded 242
   Schedule 160 steel, slip on 262
   Schedule 160 steel, weld neck ... 262
Flanges, companion
   150 pound, threaded 122
   150 pound, welding type 121
   300 pound, threaded 122
   PVC 122
Flashing
   pipe 167
   roof 123, 153, 158
Flat panel water heater 323
Flexible connections, ductwork 343
Flexible fiberglass duct 396
Flexible pipe connectors 124
Floor drains 168
   estimating 435

Floor sinks 30, 168
   estimating 435
Flues, water heater 21
Foot valve 314
Forced air heating, residential 287
Forced-draft cooling tower 213
Forklifts, rental 420
Forms and letters 438
Fringe benefits 6
Front-end loaders, rental 420
Fuel train piping 193-194
Fume hoods, air balancing 403
Furnace removal 425
Furnaces, residential 299
   high efficiency 299
   wall 300
   with A/C 299
Fusible plug dampers 341

G

Galvanized steel collars 343
Galvanized steel cooling tower 212
Galvanized steel ductwork
   installation costs 345
   per pound installed 346-347
   rectangular 369-380
   rectangular 20 gauge 377-380
   rectangular 22 gauge 373-377
   rectangular 24 gauge 371-373
   rectangular 26 gauge 370
   rectangular fittings 381-393
   round 394
   round elbow 394
   round snap-lock 394
   spiral 356-357
   spiral coupling 359
   spiral crosses 366-368
   spiral elbows 358
   spiral tees 360-365
Galvanized steel pipe sleeves 153
Garbage disposals 25
Gas furnaces
   high efficiency 299
   residential 299
   wall 300
   with A/C 299
Gas heaters 303
Gas trim connections 26
Gas valves 133
Gas water heaters 19-20
   tankless 20
Gaskets
   cast iron, hub & spigot 145
   elastomeric 159
Gate valves
   pipe and plumbing specialty ... 129-130
   PVC, Schedule 40 96

    PVC, Schedule 80 106
    Schedule 10 steel, roll-grooved....276
    Schedule 40 steel, cut-grooved....283
    Schedule 40 steel, roll-grooved....269
    Schedule 40 steel, threaded... 229-230
    Schedule 40 steel, welded 219
    Schedule 80 steel, threaded.........251
    Schedule 80 steel, welded 242
    Schedule 160 steel, flanged... 262-263
    Type K copper, brazed 39
    Type K copper, soldered 49
    Type L copper, brazed 58
    Type L copper, soldered.................66
    Type M copper, brazed 75
    Type M copper, soldered................83
Geothermal
    bore holes.....................................317
    heat pump............................. 306-308
    wells...316
Globe valves
    pipe and plumbing specialty... 130-131
    PVC, Schedule 40 96
    PVC, Schedule 80 106
    Schedule 10 steel, roll-grooved....277
    Schedule 40 steel, cut-grooved....283
    Schedule 40 steel, roll-grooved....270
    Schedule 40 steel, threaded.........230
    Schedule 40 steel, welded 219
    Schedule 80 steel, threaded.........251
    Schedule 80 steel, welded ... 242-243
    Schedule 160 steel, flanged.........263
    Type K copper, brazed 39
    Type K copper, soldered 49
    Type L copper, brazed 58
    Type L copper, soldered.................66
    Type M copper, brazed 75
    Type M copper, soldered................83
Grease and oil interceptors 25
Green sand filter................................23
Greywater tank 417
Grilles
    air balancing 403
    removal...423
    return air 333
Ground source heat pump 306

H

Hair and lint interceptors 25
Hanger assemblies
    ABS 153
    cast iron, hub & spigot..................145
    cast iron, no-hub..........................140
    copper, DWV, soldered 149
    polypropylene pipe 167
    PVC, DWV....................................157
    PVC, Schedule 40 100
    PVC, Schedule 80 110

    Schedule 10 steel, roll-grooved....279
    Schedule 40 steel, cut-grooved....285
    Schedule 40 steel, roll-grooved....272
    Schedule 40 steel, threaded.........232
    Schedule 40 steel, welded 222
    Schedule 80 steel, threaded.........254
    Schedule 80 steel, welded 246
    Schedule 160 steel.......................265
    Type K copper, brazed 42
    Type K copper, soldered 52
    Type L copper, soldered.................69
Hangers, pipe 124
    PE-AL 115, 118
    PEX-AL................................115, 118
    steel band 124
Hard water softener..................... 22-24
Head adapter, CPVC........................184
Headers, PEX-AL pipe 114
Heads, sprinkler 172
Heat/cool thermostat 405
Heat exchangers
    demolition 429
    HVAC..207
    removal...429
Heat pumps 307-311
    accessories..................................312
    air to air 309
    demolition 428
    geothermal...................................306
    removal...428
    split system..................................309
    supplemental electric heating
      coil 312
    thermostats..................................312
Heat reclaimer 204
Heat recovery 203, 205, 301
Heat recovery systems
    continuous blowdown 202
    stack waste 199
Heat recovery ventilators.................292
Heat transfer equipment..................210
Heat-A-Lamp®.................................329
Heaters
    bathroom 328
    biomass fired 321
    ceramic 304
    commercial 304
    gas fired......................................304
    infrared 304
    infrared bulb................................328
    infrared tube 304
    residential furnaces 299
    resistance 329
    unit..303
Heat-fusioned joint pipe,
    polypropylene 164
Heating systems..............................287
    estimating 435
    residential 287

Help 5
High rise offices, HVAC estimates ...437
Holding tank 415-416
Hood, duct kit 327
Hooks, pipe 125
Hose bibbs 133
Hot and cold water connections 26
Hot water boilers
    biomass fired 319-320
    connections, HVAC 190
    gas fired................................ 185-187
    gas fired, cast iron.......................189
    gas fired, steel 188
    high pressure...............................188
    oil fired...187
Hot water reheat coils 209
Hot water softener.............................24
Hot water systems
    piping 214
    Schedule 40 steel, cut-grooved....280
    Schedule 40 steel, roll-grooved....266
    Type K copper, brazed 33
    Type K copper, soldered 43
    Type L copper, brazed 53
    Type L copper, soldered.................61
    Type M copper, brazed 70
    Type M copper, soldered................78
Hot water tank
    disconnect 433
    removal..432
Hourly labor costs................................6
How to use this book...........................5
HRV (heat recovery
    ventilators)............................. 291-292
Hub & spigot C.I. pipe, DWV...........141
Humidistat control............................327
HVAC
    boiler connections........................190
    controls 404
    demolition 422
HVAC balancing
    air..403
    wet...404
HVAC equipment
    air conditioning units....................288
    air handling equipment 331
    air handling units 289-290
    boilers 185-189
    centrifugal blowers.......................324
    connections, air handling
      unit..................................... 297-298
    fan coil units 208
    heat exchangers 207
    heat transfer equipment 210
    pumps..206
    reheat coils 209
    unit heaters.................................210
    variable-air volume units 337

HVAC systems
    Type K copper, brazed 33
    Type K copper, soldered 43
    Type L copper, brazed 53
    Type L copper, soldered................ 61
    Type M copper, brazed 70
    Type M copper, soldered............... 78
Hydrant, fire.................................... 175
Hydraulic benders, rental 420

I

Indicator post.................................. 175
Indirect waste connections 26
Indirect water heater...................... 203
Induced-draft cooling tower 212
Infrared
    bulb heater.................................. 328
    heater 304
    tube heater 304
Injector.. 314
Installation costs, ductwork 396
Instructing, operating personnel 421
Instructions for this book 5
Insulation, pipe
    calcium silicate 399
    elastomeric 400
    fiberglass 397-398
Insulation removal 434
Insurance..6
Interceptors...................................... 25
Iron filter 23-24
    accessories................................. 23
Iron removal 23
Irrigation systems
    PVC, Schedule 40 91
    PVC, Schedule 80 101

J

Jet pump... 313

K

Kitchen equipment 25
    connections 26
Kitchen fixtures
    tailpiece connections 26
    trap connections 26
Kitchen sinks 29
    disconnect 433

L

Labor costs..6
Laboratories, HVAC estimates 437
Laboratory DWV pipe systems........ 164

Laundry sinks 29
    disconnect 433
Lavatories 28
    estimating 435
Lead flashing 123, 153, 158
LEED certification.... 185-186, 187-189,
    .... 286, 305, 308, 312, 319-320, 322
Letter of intent 456-457
Libraries, HVAC estimates 437
Line voltage thermostat................... 405
Lined ductwork, installed......... 346-347
Low voltage thermostat 405

M

Makeup air units.............................. 336
Malleable iron fittings
    150 pound.................................. 214
    300 pound.................................. 237
    Schedule 10 steel pipe................ 274
    Schedule 40 steel, roll-grooved....266
Manganese filters
    green sand.................................. 23
    iron... 23
Manhours...5
Manifolds, PEX-AL pipe 114
Manufacturing plants, HVAC
    estimates 437
Markers, pipe and duct.................... 421
Markets, HVAC estimates 437
Material costs 7
Material pricing conditions..................9
Materials, Equipment, and
    Tool form 10
Maximum working pressures
    Type K copper, brazed 33
    Type K copper, soldered 43
    Type L copper, brazed 53
    Type L copper, soldered................ 61
    Type M copper, brazed 70
    Type M copper, soldered............... 78
Mechanical joint coupling 139
Mechanical tee, roll grooved,
    Victaulic..................................... 179
Medical buildings, HVAC estimates...437
MET... 10
    worksheet 14
Miscellaneous tools
    PE-AL pipe 115, 118
    PEX-AL pipe 115, 118
MJ coupling 139
Molded stone
    mop sinks 29
    shower basins 28
Monthly rental, equipment.............. 420
Mop sink, disconnect....................... 433
Motels, HVAC estimates 437
Museums, HVAC estimates 437

N

Nail clips
    PE-AL pipe 115, 118
    PEX-AL pipe 115, 118
Nameplates, equipment 421
Nipples
    Schedule 40 steel, threaded... 227-229
    steel pipe, fire protection 182
Non-taxable fringe benefits 6
NPT pump, in-line.............. 25, 120-121
Nursing homes, HVAC estimates....437

O

Office buildings, HVAC estimates ...437
Office trailers, rental 420
Oil fuel train piping 205
Old estimates 12
Open loop heat pump.............. 307-308
O-rings
    PE-AL pipe 115, 118
    PEX-AL pipe 115, 118
Overflow drains 168
    estimating 435
Overhead and profit............................7

P

Packaged boiler feedwater
    systems 202
PE-AL pipe 111-115
PEX-AL pipe........................... 116-118
    adapters..................................... 113
    brass fittings 111
    caps 114, 117
    compression brass fittings........... 116
    couplings 112-113, 117
    crimp rings 115, 118
    crimped brass fittings 111
    ells 111-112, 116-117
    hangers...............................115, 118
    miscellaneous tools 115, 118
    nail clips..............................115, 118
    O-rings................................115, 118
    tees 112, 117
    valves 114, 118
Pipe
    connector............................ 123-124
    flashing 123, 153, 158, 167
    hangers..................................... 124
    hooks.. 125
    markers..................................... 421
    sleeves 125, 153, 158, 167
    sleeves, cut-grooved 285
Pipe insulation
    calcium silicate 399
    elastomeric 400

fiberglass 397-398, 415
   removal 434
Pipe machines, rental 420
Pipe sizes
   Type K copper, brazed 33
   Type K copper, soldered 43
   Type L copper, brazed 53
   Type L copper, soldered.................61
   Type M copper, brazed 70
   Type M copper, soldered...............78
Piping
   air handling unit coil 297-298
   cast iron 141
   class 110 DI, cement lined 411
   class 150 cast iron.......................412
   class 153 DI, cement lined 407
   class 153 DI, double cement
      lined..409
   class 2400 or 3000 asbestos
      cement......................................413
   copper..146
   CPVC sprinkler..............................183
   polypropylene 164
   PVC 154
   PVC, DWV......................................154
   Schedule 10 steel,
      roll-grooved 273-274
   Schedule 40 steel, cut-grooved....280
   Schedule 40 steel, roll-grooved....266
   Schedule 40 steel, threaded.........223
   Schedule 40 steel, welded 214
   Schedule 80 steel, threaded.........248
   Schedule 80 steel, welded ... 237-238
Piping, removal..................................430
Piping specialties..............................134
Piping systems
   chilled water...................................214
   hot water..214
   recirculating water 246
Planter drains 168
Plastic
   sewage pit 417
   sprinkler pipe 183
   sump pit...417
   tank..416
Plug dampers, fusible.......................341
Plugs
   cast iron, threaded........................181
   polypropylene pipe 166
   PVC, Schedule 40 95
   PVC, Schedule 80 105
   Schedule 40 steel, threaded.........226
   Schedule 80 steel, threaded.........250
Plumbing
   budget estimates 435
   equipment..26
   fixture costs 435
   fixture rough-in......................... 31-32

fixtures 32
   specialties....................................134
Plumbing fixture
   disconnect 433
   removal 432
Pollution control modules 199
Pollution control stack retrofit ...319, 321
Polyethylene sewage pit...................417
Polyethylene sump pit 417
Polyethylene-aluminum pipe ... 111-116
Polypropylene DWV pipe 164
   adapters..166
   bends..164
   bolt and gasket sets 167
   combinations 165
   couplings 166
   fittings...164
   flanges..167
   hanger assemblies 167
   heat-fused joint pipe 164
   plugs...166
   P-traps..164
   reducers..167
   riser clamps 167
   tees 165
   wyes 165-166
Polyvinyl chloride pipe
   Schedule 40....................................91
   Schedule 80...................................101
Potable water storage tank 416
Potable water systems
   PVC, Schedule 40 91
   PVC, Schedule 80 101
   Type K copper, brazed 33
   Type K copper, soldered 43
   Type L copper, brazed 53
   Type L copper, soldered.................61
   Type M copper, brazed 70
   Type M copper, soldered...............78
Pressfit
   ball valve, copper 88
   coupling, copper 86
   ells, copper 86
   female adapter, copper 88
   fittings...234
   male adapter, copper 88
   tee, copper......................................87
   tee, reducing, copper......................87
   Type O o-rings..............................234
   union, copper..................................87
Pressure controller 338
Pressure fiberglass tank.....................24
Pressure gauges
   dial-type 125
   PVC, Schedule 40 100
   PVC, Schedule 80 109
   Schedule 10 steel, roll-grooved....279
   Schedule 40 steel, cut-grooved....285

   Schedule 40 steel, roll-grooved....271
   Schedule 40 steel, threaded.........232
   Schedule 40 steel, welded 222
   Schedule 80 steel, threaded.........254
   Schedule 80 steel, welded 245
   Type K copper, brazed 41
   Type K copper, soldered 52
   Type L copper, brazed 60
   Type L copper, soldered.................68
   Type M copper, brazed 77
   Type M copper, soldered...............85
Pressure pump 174
Pressure reducing valves 133
Pressure switches 172, 314
Pressure tank 23-24, 314
Pressure/temperature taps
   Schedule 10 steel, roll-grooved....279
   Schedule 40 steel, cut-grooved....285
   Schedule 40 steel, roll-grooved....272
   Schedule 40 steel, threaded.........232
   Schedule 40 steel, welded 222
   Schedule 80 steel, threaded.........254
   Schedule 80 steel, welded 245
   Schedule 160 steel........................264
Price updates 5
Pricing, HVAC systems 435
Process systems
   PVC, Schedule 40 91-92
   PVC, Schedule 80 101-102
Project summary 12
Project summary worksheet..............15
Proposal, preparing............................13
P-traps
   ABS 151
   cast iron, hub & spigot..................142
   cast iron, no-hub..........................136
   copper, DWV, soldered 147
   polypropylene pipe 164
   PVC, DWV......................................156
Pulse type boilers.............................322
Pumping unit for boilers 201
Pumps
   boiler...205
   centrifugal.....................................206
   heat...................................... 306-308
   in-line..25
   in-line circulating.................. 120-121
   removal 431
   submersible 313
   sump, installation costs 435
   well water......................................313
Purchase order........................ 451-452
PVC
   valves, EDPM........................97, 107
   valves, threaded 97, 107
   valves, Tru-union 97, 107
   valves, union type, solvent
      weld 98, 107

PVC, DWV pipe 154
   adapters 156
   bushings 156
   cleanouts 157
   closet flanges 155-156
   couplings 156
   ells 154
   fittings 154
   hanger assemblies 157
   P-traps 156
   reducers 157
   riser clamps 157
   solvent-weld joints 154
   tees 155
   wyes 155
PVC, Schedule 40 pipe 92
   adapters 94
   assembly 91
   bolt and gasket sets 99
   bushings 94
   caps 95
   companion flange 99
   control valves 99
   couplings 95
   ells 92
   hanger assemblies 100
   plugs 95
   pressure gauges 100
   pressure/temperature taps 100
   riser clamps 100
   solvent-weld joints 91
   strainers 98-99
   tees 93
   thermometers with wells 100
   unions 95
   valves 96-99
PVC, Schedule 80 pipe 101-102
   adapters 104
   assembly 101-102
   bolt and gasket sets 109
   bushings 104
   caps 105
   companion flanges 109
   couplings 105
   ells 102-103
   hanger assemblies 110
   plugs 105
   pressure gauges 109
   pressure/temperature taps 110
   riser clamps 110
   solvent-weld joints 101
   strainers 108
   tees 103
   thermometers with wells 109
   unions 105
   valves 106-109
PVC sewer pipe, bell & spigot 159
   1/16 bend 159
   1/8 bend 159-160

   1/4 bend 160
   adapters 163
   caps 163
   couplings 160
   gasket joints 159
   reducers 163
   tees 162
   test plugs 163
   wyes 160-162

Q
Quotation sheet 17

R
Rainwater systems, PVC 159
Reciprocating water-cooled chiller 211
Recirculating water systems 246
Record of telephone conversation 18
Recorder, digital 405
Recording equipment 205
Rectangular duct, galvanized
   steel 345, 380
Rectangular elbow, galvanized
   steel 390
Reducers
   ABS 153
   cast iron, hub & spigot 144
   cast iron, no-hub 139
   cast iron, threaded 180
   class 110 DI, cement lined 411
   class 150 cast iron 412
   class 153 DI, cement lined 408
   class 153 DI, double cement
     lined 410
   class 2400 or 3000 asbestos
     cement 414
   copper, DWV, soldered 148
   galvanized steel spiral duct 359
   polypropylene pipe 167
   PVC 163
   PVC, DWV 157
   roll grooved, Victaulic 178
   Schedule 10 steel, roll-grooved 275
   Schedule 40 steel, cut-grooved 282
   Schedule 40 steel, roll-grooved 268
   Schedule 40 steel, threaded 225
   Schedule 40 steel, welded 217
   Schedule 80 steel, threaded 249
   Schedule 80 steel, welded ... 240-241
   Schedule 160 steel, welded ... 260-261
   Type K copper, brazed 37
   Type K copper, soldered 47
   Type K & L copper, roll grooved 90
   Type L copper, brazed 56
   Type L copper, soldered 64
   Type M copper, brazed 73

   Type M copper, soldered 81
Reducing costs 9
Reducing ells, Schedule 5 steel,
   pressfit 235
Reducing tees
   cast iron 180
   CPVC sprinkler pipe 183
   roll grooved, Victaulic 178
   Schedule 5 steel, pressfit 236
   Schedule 10 steel, roll-grooved 275
   Schedule 40 steel, cut-grooved 281
   Schedule 40 steel, roll-grooved 268
   Schedule 40 steel, threaded 225
   Schedule 80 steel, threaded 249
Reducing valves, pressure 133
Refractory, boiler 197
Refrigeration systems
   Type K copper, brazed 33
   Type L copper, brazed 53
Registers
   return 334
   supply 333
Reheat coils 209
Reheat units, variable volume 337
Removal costs
   air cooled condensers 426
   air handling units 424
   air mixing box 423
   boilers 425
   chillers 426
   cooling towers 427
   copper piping 430
   diffusers 423
   duct insulation 434
   duct mounted coils 429
   ductwork 422-423
   fan coils 427
   furnaces 425
   grilles 423
   heat exchangers 429
   heat pumps 428
   hot water tank 432
   pipe insulation 434
   plastic piping 430
   plumbing fixtures 432
   pumps 431
   roof top unit 424
   steel piping 430
   unit heaters 428
   valves 431
Rental costs, equipment 420
Residences, HVAC estimates 437
Residential fixture rough-ins 32
Residential furnaces 299-300
Residential water heaters 19-20
Resistance heater 329
Retail shops, HVAC estimates 437
Retrofit pollution control stack ... 319-321
Return air grilles 333

Return registers 334
Riser clamps
   ABS 153
   cast iron, hub & spigot 145
   cast iron, no-hub 140
   copper, DWV, soldered 149
   pipe and plumbing specialty 125
   polypropylene pipe 167
   PVC, DWV 157
   PVC, Schedule 40 100
   PVC, Schedule 80 110
   Schedule 10 steel, roll-grooved 279
   Schedule 40 steel, cut-grooved 285
   Schedule 40 steel, roll-grooved 272
   Schedule 40 steel, threaded 233
   Schedule 40 steel, welded 222
   Schedule 80 steel, threaded 254
   Schedule 80 steel, welded 246
   Schedule 160 steel 265
   Type K copper, brazed 42
   Type K copper, soldered 52
   Type L copper, soldered 69
Rolairtrol type air separators 201
Roll-grooved fittings 179
Roll-grooved joint, Schedule 40
   carbon steel 266
Roof
   drains 168
   exhaust fan 325
   fans 326
   flashing 123, 153, 158
Roof exhauster 326
Roof top unit, removal 424
Rough-ins
   commercial fixture 31
   commercial group 32
   residential 32
Round galvanized steel ductwork ... 394
Roustabouts, rental 420
Run and branch, tees, galvanized
   steel 365

S

Saddle tee, roll grooved, Victaulic ... 179
Safety, trenching 418
Sandstone, trenching 418
Sanitary tees
   cast iron DWV pipe 142
   polypropylene pipe 165
Schedule, construction 453-455
Schedule 5 carbon steel pipe,
   pressfit 234
   adapters 236
   couplings 235-236
   ells 234-235
   reducing ells 235
   reducing tees 236
   tees 236

Schedule 10 carbon steel pipe,
   roll-grooved 273
   adapters 275
   bolt and gasket sets 279
   caps 276
   couplings 276
   ells 274
   flanges 276
   hanger assemblies 279
   horizontal assembly 273
   pressure gauges 279
   pressure/temperature taps 279
   reducers 275
   reducing tees 275
   riser clamps 279
   strainers 278
   tees 274
   thermometers with wells 279
   valves 276-278
   vertical assembly 273
Schedule 40
   threadolet 134
   weldolet 134
Schedule 40 carbon steel pipe 215
Schedule 40 carbon steel pipe,
   cut grooved 280
   adapters 281
   bolt and gasket sets 284
   caps 282
   couplings 282
   ells 280
   flanges 282
   hanger assemblies 285
   pipe sleeves 285
   pressure gauges 285
   pressure/temperature taps 285
   reducers 282
   reducing tees 281
   riser clamps 285
   strainers 284
   tees 281
   thermometers with wells 285
   valves 283-284
Schedule 40 carbon steel pipe,
   roll-grooved 266-267
   adapters 268
   bolt and gasket sets 271
   caps 269
   couplings 269
   ells 267
   flanges 269
   hanger assemblies 272
   horizontal assembly 266
   pressure gauges 271
   pressure/temperature taps 272
   reducers 268
   reducing tees 268
   riser clamps 272
   strainers 271

   tees 267
   thermometers with wells 271
   valves 269-271
   vertical assembly 266
Schedule 40 carbon steel pipe,
   threaded 224
   bolt and gasket sets 232
   caps 226
   companion flanges 232
   control valves 231-232
   couplings 226
   crosses 225
   ells 224
   fire protection 180
   hanger assemblies 232
   horizontal assembly 223
   nipples 227-229
   plugs 226
   pressure gauges 232
   pressure/temperature taps 232
   reducers 225
   reducing tees 225
   riser clamps 233
   strainers 231
   tees 224
   thermometers with wells 232
   unions 226
   valves 227, 229-231
   vertical assembly 223
Schedule 40 carbon steel pipe,
   welded 214
   bolt and gasket sets 222
   caps 217-218
   companion flanges 218
   ells 215
   pressure gauges 222
   pressure/hanger assemblies 222
   pressure/temperature tap 222
   reducers 217
   riser clamp 222
   strainers 220-221
   tees 216
   thermometers with wells 222
   threadolets 218
   valves 219-221
   vertical assembly 214
   weldolets 218
Schedule 40 polypropylene pipe 164
Schedule 40 PVC pipe 91-92
Schedule 80
   PVC pipe 101
   threadolet 134
   weldolet 134
Schedule 80 carbon steel pipe,
   threaded 247-248
   bolt and gasket sets 253
   caps 250
   couplings 251

crosses 250
ells 248
flanges 253
hanger assemblies 254
horizontal assembly 247
plugs 250
pressure gauges 254
pressure/temperature taps 254
reducers 249
reducing tees 249
riser clamps 254
strainers 252-253
tees 249
thermometers with wells 254
unions 250
valves 251-253
vertical assembly 247
Schedule 80 carbon steel pipe,
  welded 238
  bolt and gasket sets 245
  caps 241
  ells 238
  flanges 242
  hanger assemblies 246
  horizontal assembly 237
  pressure gauges 245
  pressure/temperature tap 245
  reducers 240-241
  riser clamps 246
  strainers 244
  tees 239-240
  thermometers with wells 245
  threadolets 242
  unions 241
  valves 242-245
  vertical assembly 237
  weldolets 241
Schedule 160 carbon steel pipe,
  plain end 256
Schedule 160 carbon steel pipe,
  threaded 255
  ells 257-258
  horizontal assembly 255
  tees 259
  unions 260
  vertical assembly 256
Schedule 160 carbon steel pipe,
  welded 255
  bolt and gasket sets 264
  caps 261
  ells 257-258
  flanges 262
  hanger assemblies 265
  horizontal assembly 255
  pressure gauges 264
  pressure/temperature taps 264
  reducers 260-261
  riser clamps 265
  tees 259
  thermometers with wells 264

threadolets 262
unions 260
valves 262-264
weldolets 261
Schedule 80 PVC pipe 102
Scissor-lifts, rental 420
Scotch marine firetube boilers ... 186-187
Self-sticking markers 421
Sensor
  CO_2 405
  HVAC controls 404
Septic tank 415
Service sinks 29
  estimating 435
Sewage lift tank 417
Sewage tank 415
Sewer pipe, PVC bell & spigot 159
Shale, trenching 418
Shallow well water pump 313
Sheet metal 346-347
Shoring, trench 418
Shower stall
  disconnect 433
  removal 432
Showers 28
  estimating 435
Siamese connection 174
Silent check valves
  pipe and plumbing specialty 129
  PVC, Schedule 40 98
  PVC, Schedule 80 108
  Schedule 10 steel, roll-grooved ... 278
  Schedule 40 steel, cut-grooved ... 284
  Schedule 40 steel, roll-grooved ... 270
  Schedule 40 steel, threaded 231
  Schedule 40 steel, welded 220
  Schedule 80 steel, threaded 252
  Schedule 80 steel, welded 244
  Type K copper, brazed 40
  Type K copper, soldered 50
  Type L copper, brazed 59
  Type L copper, soldered 67
  Type M copper, brazed 76
  Type M copper, soldered 84
Sinks 28-30
  acrylic 29
  bar 29
  cast iron 29
  countertop 28
  disconnect 433
  exam room 29
  kitchen 29
  laboratory 29
  laundry 29
  medical 29
  molded stone 29
  mop 29
  removal 432
  slop 30
  stainless steel 29

Skip loaders, rental 420
Sleeves
  galvanized steel pipe 153, 158
  polypropylene pipe 167
Slop sink, disconnect 433
Slope, trench 418
Softener, water 22
Software 338
Solar water heater 323
Solder, soft 33, 43
Soldered joint fittings
  Type K copper 43
  Type L copper 61
Soldered joint pipe 43, 78
  copper, DMV 146
  Type K copper 43
  Type L copper 61
  Type M copper 78
Solenoid valves 204
Solvent-weld joint pipe
  PVC, DWV 154
  PVC, Schedule 40 91
  PVC, Schedule 80 101
Specialties, piping and plumbing 134
Speed controller, fan 327
Spin-ins, plain 343
Spiral crosses, galvanized
  steel 366-368
Spiral duct, galvanized
  steel 356-357, 366-368
Spiral tees, galvanized steel ... 363, 365
Sprinkler fittings 177-181
Sprinkler heads 172
Sprinkler systems
  black steel pipe 176-177
  branch pipe and fittings 182
  heads 172
  per head costs 170
  square foot costs 170
  switches 172
  valves 171
Square-foot costs, HVAC 435
Stack waste, heat recovery 199
Stainless steel
  doors 121
  sinks 29
Standard form subcontract 448
Steam boiler connections, HVAC ... 190
Steam boilers
  biomass fired 320
  connections 190
  gas fired 185
Steam heating boilers 189
Steam systems, piping 238
Steam traps 125, 204
Steel collars 343
Steel doors 121
Steel ductwork,
  galvanized 356-357, 359
  fittings 358

Steel pipe
- black 176-177
- cooling systems 234
- heating systems 234
- nipples, threaded 182, 229
- pressfit system 234
- process applications 234

Steel pipe fittings, Schedule 40
- steel, roll-grooved 266

Steel pipe nipples, threaded 227-229

Steel pipe, Schedule 5 pressfit 234
- adapters 236
- couplings 235-236
- ells 234-235
- reducing ells 235
- reducing tees 236
- tees 236

Steel pipe, Schedule 10
- roll-grooved 273-274
- adapters 275
- bolt and gasket sets 279
- caps 276
- control valves 278
- couplings 276
- ells 274
- flanges 276
- hanger assemblies 279
- horizontal assembly 273
- pressure gauges 279
- pressure/temperature taps 279
- reducers 275
- reducing tees 275
- riser clamps 279
- strainers 278
- tees 274
- thermometers with wells 279
- valves 276-278
- vertical assembly 273

Steel pipe, Schedule 40
- cut-grooved 280
- adapters 281
- bolt and gasket sets 284
- caps 282
- couplings 282
- ells 280
- flanges 282
- hanger assemblies 285
- pipe sleeves 285
- pressure gauges 285
- pressure/temperature taps 285
- reducers 282
- reducing tees 281
- riser clamps 285
- strainers 284
- tees 281
- thermometers with wells 285
- valves 283-284

Steel pipe, Schedule 40
- roll-grooved 266
- adapters 268
- bolt and gasket sets 271
- caps 269
- couplings 269
- ells 267
- flanges 269
- hanger assemblies 272
- pressure gauges 271
- pressure/temperature taps 272
- reducers 268
- reducing tees 268
- riser clamps 272
- strainers 271
- tees 267
- thermometers with wells 271
- valves 269-271

Steel pipe, Schedule 40
- threaded 224
- ball valves 230
- bolt and gasket sets 232
- butterfly valves 230
- caps 226
- companion flanges 232
- control valves 231
- couplings 226
- crosses 225
- ells 224
- hanger assemblies 232
- horizontal assembly 223
- nipples 227-229
- plugs 226
- pressure gauges 232
- pressure/temperature taps 232
- reducers 225
- riser clamps 233
- silent check valves 231
- strainers 231
- swing check valves 231
- tees 224-225
- thermometers with wells 232
- unions 226
- valves 229-230, 232
- vertical assembly 223

Steel pipe, Schedule 40
- welded 214-215
- bolt and gasket sets 222
- caps 217
- control valves 221
- ells 215
- flanges 218
- hanger assemblies 222
- pressure gauges 222
- pressure/temperature tap 222
- reducers 217
- riser clamp 222
- strainers 220-221
- tees 216
- thermometers 222
- threadolets 218
- unions 218
- valves 219-221
- weldolets 218

Steel pipe, Schedule 80
- threaded 247-248
- ball valves 252
- bolt and gasket sets 253
- butterfly valves 251-252
- caps 250
- control valves 253
- couplings 251
- crosses 250
- ells 248
- flanges 253
- gate valves 251
- globe valves 251
- hanger assemblies 254
- plugs 250
- pressure gauges 254
- pressure/temperature taps 254
- reducers 249
- reducing tees 249
- riser clamps 254
- silent check valves 252
- strainers 252-253
- swing check valves 252
- tees 249
- thermometers with wells 254
- unions 250
- valves 253

Steel pipe, Schedule 80
- welded 237-238
- bolt and gasket sets 245
- butterfly valves 243
- caps 241
- control valves 244-245
- ells 238
- flanges 242
- gate valves 242
- globe valves 242-243
- hanger assemblies 246
- pressure gauges 245
- pressure/temperature tap 245
- reducers 240-241
- riser clamps 246
- silent check valves 244
- strainers 244
- swing check valves 243
- tees 239-240
- thermometers with wells 245
- threadolets 242
- unions 241
- valves 245
- weldolets 241

Steel pipe, Schedule 160 256
- bolt and gasket sets 264
- ells 258
- hanger assemblies 265
- pressure gauges 264

475

pressure/temperature taps 264
riser clamps 265
thermometers with wells..............264
Steel pipe, Schedule 160
   plain end.......................................256
Steel pipe, Schedule 160
   threaded 256-260
   ells 257-258
   horizontal assembly......................255
   tees..259
   unions 260
   vertical assembly.........................256
Steel pipe, Schedule 160 welded....256
   caps 261
   ells 257-258
   flanges..262
   horizontal assembly......................255
   reducers................................ 260-261
   tees..259
   threadolets...................................262
   unions 260
   valves 262-264
   vertical assembly.........................256
   weldolets......................................261
Steel piping, removal.......................430
Storage vans, rental 420
Stores, HVAC estimates..................437
Strainers
   pipe and plumbing specialty... 131-132
   PVC, Schedule 40 98-99
   PVC, Schedule 80 108
   Schedule 10 steel, roll-grooved....278
   Schedule 40 steel, cut-grooved....284
   Schedule 40 steel, roll-grooved....271
   Schedule 40 steel, threaded........231
   Schedule 40 steel, welded ... 220-221
   Schedule 80 steel,
      threaded 252-253
   Schedule 80 steel, welded 244
   Type K copper, brazed...................40
   Type K copper, soldered 51
   Type L copper, brazed 59
   Type L copper, soldered.................67
   Type M copper, brazed 76
   Type M copper, soldered...............84
Subcontract
   change order 450
   forms...447
Submersible pump 313
Submittal data 458-459
Submittal index................................460
Suction diffusers..............................120
Sump pit 417
Sump pumps, installation costs.......435
Supermarkets, HVAC estimates......437
Supervision expense 6
Supervision valves
   flanged...170
   grooved..171
Supply registers...............................333

Supports, wall bracket....................124
Surplus materials.............................10
Swimming pool heat recovery
   ventilators......................................291
Swing check valves
   pipe and plumbing specialty... 127-128
   PVC, Schedule 40 98
   PVC, Schedule 80 108
   Schedule 10 steel, roll-grooved....277
   Schedule 40 steel, cut-grooved....283
   Schedule 40 steel, roll-grooved....270
   Schedule 40 steel, threaded........231
   Schedule 40 steel, welded 220
   Schedule 80 steel, threaded........252
   Schedule 80 steel, welded 243
   Schedule 160 steel, flanged.........263
   Type K copper, brazed...................40
   Type K copper, soldered 50
   Type L copper, brazed 59
   Type L copper, soldered.................67
   Type M copper, brazed 76
   Type M copper, soldered...............84
Switches, sprinkler system.............172

T

Table of contents...............................3
Tables
   budget estimates 435
   trenching costs 417-419
Tailpiece connections.......................26
Tank tee 314
Tankless water heaters 20, 203
Tanks
   above ground...............................416
   buried 415-416
   deep burial.......................... 415-416
   drinking water 415-416
   expansion 123
   fiberglass......................................415
   greywater......................................417
   heat/cool......................................405
   holding 415-416
   line voltage 405
   low voltage...................................405
   plastic 416-417
   polyethylene 416
   septic 415-416
   sewage...415
   sewage lift....................................417
   sewer..416
   shallow burial..................... 415-416
   sump...417
   swimming pool.............................291
   water...416
Taxable fringe benefits.......................6
Taxes...6
Tee, reducing
   copper, press fit.............................87

Tees
   ABS 151-152
   cast iron, hub & spigot.................142
   cast iron, no-hub..................138-139
   cast iron, threaded.......................180
   class 110 DI, cement lined...........411
   class 150 cast iron.......................412
   class 153 DI, cement lined...........407
   class 153 DI, double cement
      lined...410
   class 2400 or 3000 asbestos
      cement......................................414
   copper, DWV, soldered 147
   copper, press fit.............................87
   CPVC sprinkler pipe 183
   PE-AL pipe 112, 117
   PEX-AL pipe 112, 117
   polypropylene pipe 165
   PVC, DWV....................................155
   PVC, Schedule 40 93
   PVC, Schedule 80 103
   PVC sewer, bell & spigot.............162
   roll grooved, Victaulic 178-179
   Schedule 5 steel, pressfit 236
   Schedule 10 steel, roll-grooved....274
   Schedule 40 steel, cut-grooved....281
   Schedule 40 steel, roll-grooved....267
   Schedule 40 steel, threaded........224
   Schedule 40 steel, welded 216
   Schedule 80 steel, threaded........249
   Schedule 80 steel, welded ... 239-240
   Schedule 160 steel, threaded......259
   Schedule 160 steel, welded 259
   spiral, galvanized steel 360-365
   Type K copper, brazed............. 35-36
   Type K copper, soldered 45-46
   Type K & L copper, roll grooved 89
   Type L copper, brazed 55
   Type L copper, soldered.................63
   Type M copper, brazed 72
   Type M copper, soldered...............80
   with run and branch, galvanized
      steel...365
Terminal box controller....................338
Terminal boxes, air balancing 402
Test caps
   ABS DWV.....................................152
   copper, DWV, soldered 148
Test plugs, PVC sewer, bell
   & spigot 163
Test tees, copper, DWV, soldered ...148
Theaters, HVAC estimates..............437
Thermometers with wells
   pipe and plumbing specialty.........126
   PVC, Schedule 40 100
   PVC, Schedule 80 109
   Schedule 10 steel, roll-grooved....279
   Schedule 40 steel, cut-grooved....285
   Schedule 40 steel, roll-grooved....271
   Schedule 40 steel, threaded........232

Schedule 40 steel, welded 222
Schedule 80 steel, threaded.........254
Schedule 80 steel, welded 245
Schedule 160 steel......................264
Type K copper, brazed.................41
Type K copper, soldered 52
Type L copper, brazed 60
Type L copper, soldered................68
Type M copper, brazed 77
Type M copper, soldered................85
Thermostats, heat pump 312
Threadolets
   pipe and plumbing specialty.........134
   Schedule 40 steel, welded 218
   Schedule 80 steel, welded 242
   Schedule 160 steel, welded 262
Tier IV.......................................199
Timer, fan 327
Tin solder.................. 33, 43, 61, 78
Toilet
   disconnect 433
   removal..432
Tools............................115, 118
Trailers, office, rental....................420
Transceiver..................................338
Trap primers, installation costs 435
Traps with bushing connections........26
Traps, steam 125
Treatment tank 417
Triple duty valves 120
Trucks, rental.............................420
Tub
   disconnect 433
   removal..432
Tub and shower combinations 28
Tube-axial fan..............................325
Turning vanes..............................342
Tutorial...5
Type I and II PVC
   pipe............................ 91-92, 101-102
Type K copper pipe, brazed 33-34
   adapters..37
   bolt and gasket sets 41
   bushings 38
   caps 38
   companion flanges 41
   couplings 38
   ells 34
   hanger assemblies 42
   pressure gauges............................41
   pressure/temperature taps 42
   reducers.......................................37
   riser clamps..................................42
   strainers.......................................40
   tees 35-36
   thermometers with wells................41
   unions 38
   valves 39-41
Type K copper pipe, soldered 43-44
   adapters................................ 47-48

bolt and gasket sets 52
bushings 48
caps 48
companion flanges 51
couplings 49
ells 44-45
hanger assemblies 52
pressure gauges............................52
pressure/temperature taps 52
reducers.......................................47
riser clamps..................................52
strainers.......................................51
tees 45-46
thermometers with wells................52
unions 48
valves 49-51
Type K & L copper pipe,
  roll grooved
   coupling 90
   ells 89
   flange adapter...............................90
   reducers.......................................90
   tees 89
   valves 90
Type L copper pipe, brazed 53-54
   adapters..56
   bolt and gasket sets 60
   bushings 57
   cap...57
   companion flanges 60
   couplings 57
   ells 54
   pressure gauges............................60
   reducers.......................................56
   strainers.......................................59
   tees 55
   thermometers with wells................60
   unions 57
   valves 58-60
Type L copper pipe, soldered...... 61-62
   adapters..64
   bolt and gasket sets 68
   bushings 65
   caps 65
   companion flanges 68
   couplings 65
   ells 62-63
   hanger assemblies 69
   pressure gauges............................68
   pressure/temperature taps 68
   reducers.......................................64
   riser clamps..................................69
   strainers.......................................67
   tees 63
   thermometers with wells................68
   unions 65
   valves 66-68
Type M copper pipe, brazed 70-71
   adapters..73
   bolt and gasket sets 77

bushings 74
caps 74
couplings 74
ells 71
pressure gauges............................77
pressure/temperature taps 77
reducers.......................................73
strainers.......................................76
tees 72
thermometers with wells................77
unions 74
valves 75-77
Type M copper pipe, soldered..... 78-79
   adapters..81
   bolt and gasket sets 85
   bushings 82
   caps 82
   companion flanges 85
   couplings 82
   ells 79-80
   pressure gauges............................85
   pressure/temperature taps 85
   reducers.......................................81
   strainers.......................................84
   tees 80
   thermometers with wells................85
   unions 82
   valves 83-85

U

U-bolts, galvanized..........................124
Ultra-violet
   disinfection....................................25
   water treatment.............................25
Underground piping, PVC 159
Unions
   copper, press fit............................87
   dielectric.....................................122
   PVC, Schedule 40 95
   PVC, Schedule 80 105
   Schedule 40 steel, threaded.........226
   Schedule 40 steel, welded 218
   Schedule 80 steel, threaded.........250
   Schedule 80 steel, welded 241
   Schedule 160 steel, threaded.......260
   Schedule 160 steel, welded 260
   Type K copper, brazed..................38
   Type K copper, soldered 48
   Type L copper, brazed 57
   Type L copper, soldered................65
   Type M copper, brazed 74
   Type M copper, soldered...............82
Unit heaters............................210, 303
   connections 210
   demolition 428
   gas fired.......................................210
   hot water.....................................301
   HVAC connections 210

removal 428
steam 302
Upblast ventilation 330
Updates 5
Urinals 27
   disconnect 433
   estimating 435
Using this book 5
Utility fan 324
UV disinfection unit 25

V

Vacuum breakers 120, 204
   atmospheric 120
   hose connection 120
Value engineering 9
Valves
   air admittance 328
   alarm 171
   check, flanged 171
   check, grooved 171
   control 221
   double check detector 171
   dry 171
   fire protection 171
   PE-AL pipe 114, 118
   PEX-AL pipe 114, 118
   pipe and plumbing
      specialty 126-128, 133
   PVC, Schedule 40 96-99
   PVC, Schedule 80 106-109
   PVC, threaded 97, 107
   PVC, Tru-union 97, 107
   PVC, union type, solvent
      weld 98, 107
   removal 431
   Schedule 40 steel,
      cut-grooved 283-284
   Schedule 40 steel, roll-grooved 271
   Schedule 40 steel, threaded 232
   Schedule 40, welded 221
   Schedule 80 steel, threaded 253
   Schedule 80 steel, welded 245
   Schedule 160 steel, flanged 263
   solenoid 204
   solvent weld 97, 107
   sprinkler system 171
   supervision, flanged 170
   supervision, grooved 171

tags 421
triple duty 120
Type K copper, brazed 40
Type K copper, soldered 50
Type L copper, brazed 60
Type L copper, soldered 66-68
Type M copper, brazed 75-77
Type M copper, soldered 83-85
Vane-axial fan 324
Vans, storage, rental 420
Variable-air volume
   cooling units 337
   reheat units 337
Vent systems 435
   cast iron, hub & spigot 141
   cast iron, no-hub 141
   copper 146
   PVC, DWV 154
Ventilation
   ductwork 346-347
   exhausters 329
   fans 326-327
Ventilator, heat recovery 301
Vents, air 121
Verantis 199
Victaulic roll-grooved fittings 177-179

W - X - Y - Z

Wall exhauster 326
Wall fan 326
Wash fountains, installation costs ... 435
Waste heat controls 203
Waste systems
   cast iron, hub & spigot 141
   cast iron, no-hub 141
   copper 146
   PVC 154
   PVC, DWV 154
Water closets 27
   disconnect 433
   estimating 435
Water coil piping 294
Water connections, hot and cold 26
Water cooled chiller connection 211
Water hammer arresters 133
Water heaters
   commercial 19-20
   connections 21

estimating 435
residential 19-20
solar 323
tankless 20
tankless indirect 203
Water meters
   by-pass and connection
      assembly 119
   compound type 119
   turbine type 119
Water motor gong 175
Water pump 313
Water softener 22-24
   accessories 23
Water softening systems, boiler 198
Water source heat pump 307-308
Water storage tank 416
Water wells, drilling 316
Watertube boilers 188
Weekly rental, equipment 420
Welding machines, rental 420
Weldolets
   pipe and plumbing specialty 134
   Schedule 40 steel, welded 218
   Schedule 80 steel, welded 241
   Schedule 160 steel, welded 261
Wells
   drilling 316
   geothermal 316
   pipe 314
   water pump 313
Well-to-well heat pump 307-308
Wheel, heat recovery 301
Wireless transceiver 338
Wyes
   ABS 151-152
   cast iron, hub & spigot 143-144
   cast iron, no-hub 136-138
   class 150 cast iron 412
   class 153 DI, cement lined 408
   class 153 DI, double cement
      lined 410
   class 2400 or 3000 asbestos
      cement 414
   copper, DWV, soldered 147
   polypropylene pipe 165-166
   PVC, DWV 155
   PVC sewer, bell & spigot 160-162

Practical References for Builders

Basic Plumbing with Illustrations, Revised

This completely revised edition brings this comprehensive manual fully up-to-date with all the latest plumbing codes. It is the journeyman's and apprentice's guide to installing plumbing, piping, and fixtures in residential and light commercial buildings: how to select the right materials, lay out the job and do professional-quality plumbing work, use essential tools and materials, make repairs, maintain plumbing systems, install fixtures, and add to existing systems. Includes extensive study questions at the end of each chapter, and a section with all the correct answers.
384 pages, 8$^1/_2$ x 11, $44.75
eBook (PDF) also available; $22.37 at www.craftsman-book.com

Construction Contract Writer

Relying on a "one-size-fits-all" boilerplate construction contract to fit your jobs can be dangerous — almost as dangerous as a handshake agreement. *Construction Contract Writer* lets you draft a contract in minutes that precisely fits your needs and the particular job, and meets both state and federal requirements. You just answer a series of questions — like an interview — to construct a legal contract for each project you take on. Anticipate where disputes could arise and settle them in the contract before they happen. Include the warranty protection you intend, the payment schedule, and create subcontracts from the prime contract by just clicking a box. Includes a feedback button to an attorney on the Craftsman staff to help should you get stumped — *No extra charge.*
$149.95. *Download the Construction Contract Writer at*
http://www.constructioncontractwriter.com

CD Estimator

If your computer has *Windows*™ and a CD-ROM drive, CD Estimator puts at your fingertips over 150,000 construction costs for new construction, remodeling, renovation & insurance repair, home improvement, framing & finish carpentry, electrical, concrete & masonry, painting, earthwork and heavy equipment and plumbing & HVAC. Quarterly cost updates are available at no charge on the Internet. You'll also have the *National Estimator* program — a stand-alone estimating program for *Windows*™ that *Remodeling* magazine called a "computer wiz," and *Job Cost Wizard*, a program that lets you export your estimates to QuickBooks Pro for actual job costing. A 60-minute interactive video teaches you how to use this CD-ROM to estimate construction costs. And to top it off, to help you create professional-looking estimates, the disk includes over 40 construction estimating and bidding forms in a format that's perfect for nearly any *Windows*™ word processing or spreadsheet program. **CD Estimator is $149.50**

Plumber's Handbook Revised

This edition shows what will and won't pass inspection in drainage, vent, and waste piping, septic tanks, water supply, graywater recycling systems, pools and spas, fire protection, and gas piping systems. All tables, standards, and specifications are completely up-to-date with recent plumbing code changes. Covers common layouts for residential work, how to size piping, select and hang fixtures, practical recommendations, and trade tips. It's the approved reference for the plumbing contractor's exam in many states. Includes an extensive set of multiple-choice questions after each chapter, with answers and explanations in the back of the book, along with a complete sample plumber's exam. **384 pages, 8$^1/_2$ x 11, $44.50**
eBook (PDF) also available; $22.25 at www.craftsman-book.com

Building Code Compliance for Contractors & Inspectors

Have you ever failed a construction inspection? Have you ever dealt with an inspector who has his own interpretation of the Code and forces you to comply with it? This new book explains what it takes to pass inspections under the 2009 *International Residential Code*. It includes a Code checklist — with explanations and the Code section number — for every trade, covering some of the most common reasons why inspectors reject residential work. The author uses his 30 years' experience as a building code official to provide you with little-known information on what code officials look for during inspections. **232 pages, 8$^1/_2$ x 11, $32.50**
eBook (PDF) also available; $16.25 at www.craftsman-book.com

Uniform Plumbing Code Quick-Card 2021

This 6-page, laminated, fold-out guide provides the basic numbers, flow rates and formulas the plumber needs based on the *2021 Uniform Plumbing Code (UPC)*. Like a Cliffs Notes to the Plumbing Code.
6 pages, 8$^1/_2$ x 11, $9.95

Plumbing & HVAC Manhour Estimates

Hundreds of tested and proven manhours for installing just about any plumbing and HVAC component you're likely to use in residential, commercial, and industrial work. You'll find manhours for installing piping systems, specialties, fixtures and accessories, ducting systems, and HVAC equipment. If you estimate the price of plumbing, you shouldn't be without the reliable, proven manhours in this unique book.
224 pages, 5$^1/_2$ x 8$^1/_2$, $28.25

Paper Contracting: The How-To of Construction Management Contracting

Risk, and the headaches that go with it, have always been a major part of any construction project — risk of loss, negative cash flow, construction claims, regulations, excessive changes, disputes, slow pay — sometimes you'll make money, and often you won't. But many contractors today are avoiding almost all of that risk by working under a construction management contract, where they are simply a paid consultant to the owner, running the job, but leaving him the risk. This manual is the how-to of construction management contracting. You'll learn how the process works, how to get started as a CM contractor, what the job entails, how to deal with the issues that come up, when to step back, and how to get the job completed on time and on budget. Includes a link to free downloads of CM contracts legal in each state.
272 pages, 8$^1/_2$ x 11, $55.50
eBook (PDF) also available; $27.75 at www.craftsman-book.com

Planning Drain, Waste & Vent Systems

How to design plumbing systems in residential, commercial, and industrial buildings. Covers designing systems that meet code requirements for homes, commercial buildings, private sewage disposal systems, and even mobile home parks. Includes relevant code sections and many illustrations to guide you through what the code requires in designing drainage, waste, and vent systems. **192 pages, 8$^1/_2$ x 11, $39.95**

Uniform Plumbing Code Tables/Equations Quick Card 2021

This 6-page, laminated guide provides the basic numbers, flow rates and formulas the plumber needs based from *2021 Uniform Plumbing Code (UPC)*. **6 pages, 8½ x 11, $9.95**

International Plumbing Code Quick-Card 2021 *IPC*

In this unique quick-reference guide, a single, 6-page laminated card, you get most of the new International Plumbing Code essentials that you need to know, based on the current 2021 *IPC*. **6 pages, 8 x 5½, $9.95**

Construction Forms for Contractors

This practical guide contains 78 practical forms, letters and checklists, guaranteed to help you streamline your office, organize your jobsites, gather and organize records and documents, keep a handle on your subs, reduce estimating errors, administer change orders and lien issues, monitor crew productivity, track your equipment use, and more. Includes accounting forms, change order forms, forms for customers, estimating forms, field work forms, HR forms, lien forms, office forms, bids and proposals, subcontracts, and more. All are also on the CD-ROM included, in Excel spreadsheets, as formatted Rich Text that you can fill out on your computer, and as PDFs. **360 pages, 8$^1/_2$ x 11, $48.50**
eBook (PDF) also available; $24.25 at www.craftsman-book.com

National Construction Estimator

Current building costs for residential, commercial, and industrial construction. Estimated prices for every common building material. Provides manhours, recommended crew, and gives the labor cost for installation. Includes a free download of an electronic version of the book with *National Estimator*, a stand-alone *Windows*™ estimating program. Additional information and *National Estimator* ShowMe tutorial video is available on our website under the "Support" dropdown tab.
672 pages, 8$^1/_2$ x 11, $97.50. Revised annually
eBook (PDF) also available; $48.75 at www.craftsman-book.com

National Appraisal Estimator

An Online Appraisal Estimating Service. Produce credible single-family residence appraisals – in as little as five minutes. A smart resource for appraisers using the cost approach. Reports consider all significant cost variables and both physical and functional depreciation. For more information, visit www.craftsman-book.com/national-appraisal-estimator-online-software

Plumber's Exam Preparation Guide eBook

Hundreds of questions and answers to help you pass the apprentice, journeyman, or master plumber's exam. Questions are in the style of the actual exam. Gives answers for both the Standard and Uniform plumbing codes. Includes tips on studying for the exam and the best way to prepare yourself for examination day. **$19.00**

Insurance Restoration Contracting: Startup to Success

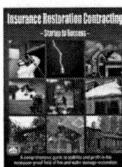

Insurance restoration — the repair of buildings damaged by water, fire, smoke, storms, vandalism and other disasters — is an exciting field of construction that provides lucrative work that's immune to economic downturns. And, with insurance companies funding the repairs, your payment is virtually guaranteed. But this type of work requires special knowledge and equipment, and that's what you'll learn about in this book. It covers fire repairs and smoke damage, water losses and specialized drying methods, mold remediation, content restoration, even damage to mobile and manufactured homes. You'll also find information on equipment needs, training classes, estimating books and software, and how restoration leads to lucrative remodeling jobs. It covers all you need to know to start and succeed as the restoration contractor that both homeowners and insurance companies call on first for the best jobs. **640 pages, 8½ x 11, $69.00**
eBook (PDF) also available; $34.50 at www.craftsman-book.com

Insurance Replacement Estimator

Insurance underwriters demand detailed, accurate valuation data. There's no better authority on replacement cost for single-family homes than the *Insurance Replacement Estimator*. In minutes you get an insurance-to-value report showing the cost of re-construction based on your specification. You can generate and save unlimited reports. For more details, visit www.craftsman-book.com/insurance-replacement-estimator-online-software

Craftsman eLibrary

Craftsman's eLibrary license gives you immediate access to 60+ PDF eBooks in our bookstore for 12 full months!
You pay only one low price. $129.99.
Visit www.craftsman-book.com **for more details.**

Home Building Mistakes & Fixes

This is an encyclopedia of practical fixes for real-world home building and repair problems. There's never an end to "surprises" when you're in the business of building and fixing homes, yet there's little published on how to deal with construction that went wrong - where out-of-square or non-standard or jerry-rigged turns what should be a simple job into a nightmare. This manual describes jaw-dropping building mistakes that actually occurred, from disastrous misunderstandings over property lines, through basement floors leveled with an out-of-level instrument, to a house collapse when a siding crew removed the old siding. You'll learn the pitfalls the painless way, and real-world working solutions for the problems every contractor finds in a home building or repair jobsite. Includes dozens of those "surprises" and the author's step-by-step, clearly illustrated tips, tricks and workarounds for dealing with them.
384 pages, 8½ x 11, $52.50
eBook (PDF) also available; $26.25 at www.craftsman-book.com

Craftsman Book Company
6058 Corte del Cedro
Carlsbad, CA 92011
☎ **Call me.**
1-800-829-8123
Fax (760) 438-0398

Name _____
e-mail address (for order tracking and special offers) _____
Company _____
Address _____
City/State/Zip _____   ○ This is a residence
Total enclosed_____(In California add 7.5% tax)
Free Media Mail shipping, within the US, when your check covers your order in full.

In A Hurry?
We accept phone orders charged to your
○ Visa, ○ MasterCard, ○ Discover or ○ American Express

Card#_____
Exp. date_____ CVV#_____ Initials_____

Tax Deductible: Treasury regulations make these references tax deductible when used in your work. Save the canceled check or charge card statement as your receipt.

Order online: www.craftsman-book.com
Free on the Internet! Download any of Craftsman's estimating databases for a 30-day free trial! www.craftsman-book.com/downloads

---

Writing contracts that comply with law in your state isn't easy. A contract that doesn't comply could leave you with no way to collect.
Construction Contract Writer has you covered. Download a trial today:
www.constructioncontractwriter.com

10-Day Money Back Guarantee | **Prices subject to change without notice**

- ○ 44.75 Basic Plumbing with Illustrations, Revised
- ○ 32.50 Building Code Compliance for Contractors & Inspectors
- ○ 149.50 CD Estimator
- ○ 48.50 Construction Forms for Contractors
- ○ 52.50 Home Building Mistakes & Fixes
- ○ 69.00 Insurance Restoration Contracting: Startup to Success
- ○ 9.95 Int'l Plumbing Code Quick-Card Based on 2021 *IPC*
- ○ 97.50 National Construction Estimator w/FREE *Natl Estimator* Download
- ○ 55.50 Paper Contracting: The How-To of Construction Management Contracting
- ○ 39.95 Planning Drain, Waste & Vent Systems
- ○ 44.50 Plumber's Handbook Revised
- ○ 28.25 Plumbing & HVAC Manhour Estimates
- ○ 9.95 Uniform Plumbing Code Quick-Card
- ○ 9.95 Uniform Plumbing Code Tables/Equations Quick-Card 2021
- ○ 98.25 National Plumbing & HVAC Estimator w/FREE *Natl Estimator* Download